# Plasticity for Structural Engineers

# TITLES IN THE SERIES

**Architectural Acoustics**
*M. David Egan*
ISBN 13: 978-1-932159-78-3, ISBN 10: 1-932159-78-9, 448 pages

**Earth Anchors**
*By Braja M. Das*
ISBN 13: 978-1-932159-72-1, ISBN 10: 1-932159-72-X, 242 pages

**Limit Analysis and Soil Plasticity**
*By Wai-Fah Chen*
ISBN 13: 978-1-932159-73-8, ISBN 10: 1-932159-73-8, 638 pages

**Plasticity in Reinforced Concrete**
*By Wai-Fah Chen*
ISBN 13: 978-1-932159-74-5, ISBN 10: 1-932159-74-6, 474 pages

**Plasticity for Structural Engineers**
*By Wai-Fah Chen & Da-Jian Han*
ISBN 13: 978-1-932159-75-2, ISBN 10: 1-932159-75-4, 606 pages

**Theoretical Foundation Engineering**
*By Braja M. Das*
ISBN 13: 978-1-932159-71-4, ISBN 10: 1-932159-71-1, 440 pages

**Theory of Beam-Columns, Volume 1: Space Behavior and Design**
*By Wai-Fah Chen & Toshio Atsuta*
ISBN 13: 978-1-932159-77-6, ISBN 10: 1-932159-77-0, 732 pages

**Theory of Beam-Columns, Volume 2: In-Plane Behavior and Design**
*By Wai-Fah Chen & Toshio Atsuta*
ISBN 13: 978-1-932159-76-2, ISBN 10: 1-932159-76-9, 513 pages

# Plasticity for Structural Engineers

by Wai-Fah Chen • Da-Jian Han

J.ROSS
PUBLISHING

ISBN-10: 1-932159-75-4
ISBN-13: 978-1-932159-75-2

Printed and bound in the U.S.A. Printed on acid-free paper
10 9 8 7 6 5 4 3 2 1

This J. Ross Publishing edition, first published in 2007, is an unabridged republication of the work originally published by Springer-Verlag, New York, in 1988.

**Library of Congress Cataloging-in-Publication Data**

Chen, Wai-Fah, 1936–
    Plasticity for structural engineers / By Wai-Fah Chen & Da-Jian Han.
        p. cm.
    Originally published: New York : Springer-Verlag, c1988.
    Includes index.
    ISBN-10: 1-932159-75-4 (pbk : alk. paper)
    ISBN-13: 978-1-932159-75-2 (pbk : alk. paper)
    1. Plasticity. 2. Structural design. I. Han, D. J. II. Title.
    TA418.14.C49 2007
    620.1'1233—dc22                                        2006101136

Phone: (954) 727-9333
Fax: (561) 892-0700
Web: www.jrosspub.com

# PREFACE

This book is intended primarily for structural engineers familiar with the processes of elastic and plastic analysis and design of framed structures in steel and reinforced concrete, but less familiar with the elastic and plastic behavior of structural elements under combined stresses. The more complex structural elements are necessary for the solution of more general structural problems under the general heading of "nonlinear analysis" by either the finite-element or finite-difference method. In this book, we have attempted to present the topic of structural plasticity in a manner that is simple, concise, and reasonably comprehensive, encompassing the classical theory of metal plasticity as well as the modern development of concrete plasticity. The scope of the book is indicated by the contents. It is divided into five parts. Part I examines, on the basis of simple test conditions, the elastic and plastic behaviors of metal and their possible generalizations under combined stresses. An understanding of stress and strain in three dimensions is essential for structural engineers to follow subsequent developments. To this end, index notation and the principles of stress and strain are developed briefly in the relevant parts of Chapters 1 to 3 of Part I.

Part II is concerned with the general developments of plastic stress-strain relations for perfectly plastic solids (Chapter 4) and for work-hardening plastic solids (Chapter 5). Part II ends with the detailed development of a constitutive equation that relates stress increments to total strain increments rather than to plastic strain increments, and that can be readily implemented for a finite-element or finite-difference code.

Part III deals with the application of the general theory of plasticity to metal. Constitutive formulations based on J2-theory together with procedures for their solution in a general nonlinear finite-element problem are discussed in some detail. The bounding surface theory recently developed for modeling the behavior of metal under cyclic loading is also presented. Part III closes with a brief discussion of the stress-strain relation for orthotropic materials.

Part IV deals with the application of the general theory of plasticity to reinforced concrete materials. It contains failure criteria of concrete materials and their constitutive modeling in the pre- and postfracture ranges. Computer implementation of these models together with model subroutines are included in a companion book entitled Theory, Problems, and CAE Softwares of Structural Plasticity by Chen and Zhang (1988). Part V on limk analysis is devoted to the general limit theorems and their application to metal and concrete structures and the interaction of these structures with ice and soil media. It covers various aspects of modern techniques of limh analysis, and the discussion is illustrated by many examples dealing with practical problems in structural engineering. The book can be used for courses of various lengths. The first six chapters can be reasonably covered in a three-hour one-semester course for the first-year graduate student who is learning about inelastic behavior of materials for the first time. In a course for graduate students who have already completed a course on plastic analysis of steel structures, the last two chapters on limit analysis can also be covered. The chapter on concrete plasticity may be skipped on a first reading. This part of the material is necessarily written at a slightly more advanced level, because it is directed toward the practicing engineer who is working on concrete structures in the general area of nonlinear analysis. The mathematics used here does not extend beyond the usual calculus so that the reader who has thoroughly studied the first six chapters has given himself most of the needed preparation. We have endeavored to give reasonably complete literature references to the topics covered in Part IV on concrete plasticity. The inclusion of a computer subroutine for a concrete model in the companion book cited previously is intended to encourage the reader to try out the proposed models.

Over the past years, Professor Chen has taught courses in plasticity and limit analysis at Lehigh Univershy and Purdue University and has also given a series of lectures at the Swiss Federal Institute of Technology, National Taiwan University, and University of Kassel. Early drafts of this book have been tested as classroom notes in these courses. The material on concrete plasticity was prepared more recently and was presented as guest lectures at the 1985 Workshop on Recent Developments in Solid Mechanics at Peking University, Beijing, China. Professor Chen wishes to thank Mr. Zhang Hung for preparing the Answers to Selected Problems as well as the Solution Manual during his course work on Structural Plastichy and later as a teaching and research assistant on this subject area in the School of Civil Engineering at Purdue University.

— W.F. CHEN & D J HAN
November, 1987

# CONTENTS

## PART II   PLASTIC STRESS-STRAIN RELATIONS

### Chapter 4  Stress-Strain Relations for Perfectly Plastic Materials .................179

### Chapter 5  Stress-Strain Relations for Work-Hardening Materials.................232

# Notation

## Stresses and Strains

| | |
|---|---|
| $\sigma_1, \sigma_2, \sigma_3$ | principal stresses, tensile stress positive |
| $\sigma_{ij}$ | stress tensor |
| $s_{ij}$ | stress deviator tensor |
| $\sigma$ | normal stress |
| $\tau$ | shear stress |
| $p = \frac{1}{3}I_1$ | hydrostatic pressure or spherical stress |
| $\sigma_{oct} = \frac{1}{3}I_1$ | octahedral normal stress |
| $\tau_{oct} = \sqrt{\frac{2}{3}J_2}$ | octahedral shear stress |
| $\sigma_m = \sigma_{oct}$ | mean normal stress |
| $\tau_m = \sqrt{\frac{2}{5}J_2}$ | mean shear stress |
| $s_1, s_2, s_3$ | principal stress deviators |
| $\epsilon_1, \epsilon_2, \epsilon_3$ | principal strains, tensile strain positive |
| $\epsilon_{ij}$ | strain tensor |
| $e_{ij}$ | strain deviator tensor |
| $\epsilon$ | normal strain |
| $\gamma$ | engineering shear strain |
| $\epsilon_v = I_1'$ | volumetric strain |
| $\epsilon_{oct} = \frac{1}{3}I_1'$ | octahedral normal strain |
| $\gamma_{oct} = 2\sqrt{\frac{2}{3}J_2'}$ | octahedral engineering shear strain |
| $e_1, e_2, e_3$ | principal strain deviators |

## Invariants

$$I_1 = \sigma_1 + \sigma_2 + \sigma_3 = \sigma_{ii} = \text{first invariant of stress tensor}$$

$$J_2 = \frac{1}{2}s_{ij}s_{ij}$$

$$= \frac{1}{6}[(\sigma_x - \sigma_y)^2 + (\sigma_y - \sigma_z)^2 + (\sigma_z - \sigma_x)^2] + \tau_{xy}^2 + \tau_{yz}^2 + \tau_{zx}^2$$

$$= \text{second invariant of stress deviator tensor}$$

$$J_3 = \tfrac{1}{3} s_{ij} s_{jk} s_{ki} = |s_{ij}| = \text{third invariant of stress deviator}$$

$$\cos 3\theta = \frac{3\sqrt{3}}{2} \frac{J_3}{J_2^{3/2}}, \text{ where } \theta \text{ is the angle of similarity defined in Fig. 2.9}$$

$$I_1' = \epsilon_1 + \epsilon_2 + \epsilon_3 = \text{first invariant of strain tensor}$$

$$\rho = \sqrt{2J_2} = \text{deviatoric length defined in Fig. 2.8}$$

$$\xi = \frac{1}{\sqrt{3}} I_1 = \text{hydrostatic length defined in Fig. 2.8}$$

$$J_2' = \tfrac{1}{2} e_{ij} e_{ij}$$

$$= \tfrac{1}{6}[(\epsilon_x - \epsilon_y)^2 + (\epsilon_y - \epsilon_z)^2 + (\epsilon_z - \epsilon_x)^2] + \epsilon_{xy}^2 + \epsilon_{yz}^2 + \epsilon_{zx}^2$$

$$= \text{second invariant of strain deviator tensor}$$

## Material Parameters

| | |
|---|---|
| $f_c'$ | uniaxial compressive cylinder strength ($f_c' > 0$) |
| $f_t'$ | uniaxial tensile strength ($f_c' = m f_t'$) |
| $f_{bc}'$ | equal biaxial compressive strength ($f_{bc}' > 0$) |
| $E$ | Young's modulus |
| $\nu$ | Poisson's ratio |

$K \qquad \dfrac{E}{3(1-2\nu)} = \text{bulk modulus}$

$G \qquad \dfrac{E}{2(1+\nu)} = \text{shear modulus}$

| | |
|---|---|
| $c, \phi$ | cohesion and friction angle in Mohr–Coulomb criterion |
| $\alpha, k$ | constants in Drucker–Prager criterion |
| $k$ | yield (failure) stress in pure shear |

## Miscellaneous

| | |
|---|---|
| $\{\ \}$ | vector |
| $[\ ]$ | matrix |
| $|\ |$ | determinant |
| $C_{ijkl}$ | material stiffness tensor |
| $D_{ijkl}$ | material compliance tensor |
| $f(\ )$ | failure criterion or yield function |
| $x, y, z$ or $x_1, x_2, x_3$ | Cartesian coordinates |

| | |
|---|---|
| $\delta_{ij}$ | Kronecker delta |
| $W(\epsilon_{ij})$ | strain energy density |
| $\Omega(\sigma_{ij})$ | complementary energy density |
| $l_{ij}$ | $\cos(x_i', x_j) =$ the cosine of the angle between the $x_i'$ and $x_j$ axes (see Section 1.5.3) |

# Part I: Fundamentals

# 1
# Introduction

## 1.1. Introduction

### 1.1.1. Role of Plasticity in Structural Engineering

The engineering design of large structures often involves a two-stage process: first, the internal force field acting on the structural material must be defined, and second, the response of the material to that force field must be determined. The first stage involves an analysis of the stresses acting within the structural elements; the second involves a knowledge of the properties of the structural material. The linear relationship between stress and strain in an idealized material forms the basis of the *mathematical theory of elasticity*, which has in turn been applied widely in practice to actual materials to estimate stress or strain in the structural elements under a specified working load condition. These stresses are restricted to be less than the specified *working* or *allowable* stress that is chosen as some fraction of the yield strength of the material. A safe design thus is evolved, not due to the adequacy of the structural analysis and the understanding of the properties of the material, but by reliance upon the experience of decades or centuries.

An actual structure is a very complex body with an extremely complicated state of stress. Many secondary stresses arise owing to fabrication, erection, and localization. The combination of unknown initial stress, secondary stresses, and stress concentration and redistribution due to discontinuities of the structure defy an idealized calculation based on the theory of elasticity. The *theory of plasticity* represents a necessary extension of the theory of elasticity and is concerned with the analysis of stresses and strains in the structure in the plastic as well as the elastic ranges. It furnishes more realistic estimates of load-carrying capacities of structures and provides a better understanding of the reaction of the structural elements to the forces induced in the material. An understanding of the role of the relevant mechanical variables that define the characteristic reaction of the material to the applied force is therefore essential to the engineer designing structures. These stress–strain relationships and their applications to structural engineering problems are developed and discussed in the following chapters. The more comprehensive this knowledge, the more exact will be the design and the more perfect will be the structure. This book attempts to achieve this goal for the case of structural analysis and design of metal and concrete structures.

## 1.1.2. Scope

Both the theory of elasticity and the theory of plasticity are phenomenological in nature. They are the formalization of experimental observations of the macroscopic behavior of a deformable solid and do not inquire deeply into the physical and chemical basis of this behavior.

A complete account of the theory and application of plasticity must deal with two equally important aspects: (1) the general technique used in the development of stress–strain relationships for inviscid elastic–plastic materials with work hardening as well as strain softening; and (2) the general numerical solution procedure for solving an elastic–plastic structural problem under the action of loads or displacements, each of which varies in a specified manner. Such an account will be given in the following chapters.

The first task of plasticity theory is to set up relationships between stress and strain under a complex stress state that can describe adequately the observed plastic deformation. This is a difficult task. However, deformational rules for metals that, in general, agree well with experimental evidence have been firmly established and successfully used in engineering applications. Moreover, in recent years, the methods of plasticity have also been extended and applied to study the deformational behavior of geological materials, such as rocks, soils, and concretes. The extension of plasticity theory to nonmetallic materials is probably the most active research subject in the field of mechanics of materials at present, and various material models have been developed.

The second task of the theory is to develop numerical techniques for implementing these stress–strain relationships in the analysis of structures. Because of the nonlinear nature of the plastic deformation rules, solutions of the basic equations of solid mechanics inevitably present considerable difficulties. However, in recent years, the rapid development of high-speed computers and modern techniques of finite-element analysis has provided the engineer with a powerful tool for the solution of virtually any nonlinear structural problem. It also has provoked newer developments and wider applications of the classical plasticity theory. Research activity in this field has increased tremendously during the last decade or so.

This book attempts to give a concise description of the basic concepts of the theory and its modern developments, as well as its computer implementations.

## 1.2. Historical Remarks

### 1.2.1. Pioneering Work

It is generally regarded that the origin of plasticity, as a branch of mechanics of continua, dates back to a series of papers from 1864 to 1872 by Tresca on the extrusion of metals, in which he proposed the first *yield condition*, which states that a metal yields plastically when the maximum shear stress attains a critical value. The actual formulation of the theory was done in 1870 by St. Venant, who introduced the basic *constitutive relations* for what today we would call rigid, perfectly plastic materials in plane stress. The salient feature of this formulation was the suggestion of a *flow rule* stating that the principal axes of the strain increment (or strain rate) coincide with the principal axes of stress. It remained for Levy later in 1870 to obtain the general equations in three dimensions. A generalization similar to the results of Levy was arrived at independently by von Mises in a landmark paper in 1913, accompanied by his well-known, pressure-insensitive yield criterion ($J_2$-theory, or octahedral shear stress yield condition).

In 1924, Prandtl extended the St. Venant–Levy–von Mises equations for the plane continuum problem to include the elastic component of strain, and Reuss in 1930 carried out their extension to three dimensions. In 1928, von Mises generalized his previous work for a rigid, perfectly plastic solid to include a general yield function and discussed the relation between the direction of plastic strain rate (increment) and the *regular or smooth* yield surface, thus introducing formally the concept of using the yield function as a *plastic potential* in the incremental stress–strain relations of flow theory. As is well known now, the von Mises yield function may be regarded as a plastic potential for the St. Venant–Levy–von Mises–Prandtl–Reuss stress-strain relations. The appropriate flow rule associated with the Tresca yield condition, which contains singular regimes (i.e., corners or discontinuities in derivatives with respect to stress), was discussed by Reuss in 1932 and 1933.

Since greater emphasis was placed on problems involving flow or perfect plasticity in the years before 1940, the development of incremental constitutive relationships for hardening materials proceeded more slowly. For example, in 1928, Prandtl attempted to formulate general relations for hardening behavior, and Melan, in 1938, generalized the foregoing concepts of perfect plasticity and gave incremental relations for hardening solids with smooth (regular) yield surface. Also, *uniqueness theorems* for elastic-plastic incremental problems were discussed by Melan in 1938 for both perfectly plastic and hardening materials based on some limiting assumptions.

## 1.2.2. Classical Theory

The nearly twenty years after 1940 saw the most intensive period of development of basic concepts and fundamental ingredients in what is now referred to as the *classical theory of metal plasticity*.

Independently of the work of Melan in 1938, Prager, in a significant paper published in 1949, arrived at a general framework (similar to that discussed by Melan in 1938) for the plastic constitutive relations for *hardening materials* with smooth (regular) yield surfaces. The yield function (also termed the *loading function*) and the *loading–unloading* conditions were precisely formulated. Such conditions as the *continuity* condition (near neutral loading), the *consistency* condition (for loading from plastic states), the *uniqueness* condition, and the condition of *irreversibility* of plastic deformation were formulated and discussed. Also, the interrelationship between the *convexity* of the (smooth) yield surface, the *normality* to the yield surface, and the uniqueness of the associated boundary-value problem was clearly recognized. In 1958, Prager further extended this general framework to include thermal effects (nonisothermal plastic deformation), by allowing the yield surface to change its shape with temperature.

A very significant concept of work hardening, termed the *material stability postulate*, was proposed by Drucker in 1951 and amplified in his further papers. With this concept, the plastic stress–strain relations together with many related fundamental aspects of the subject may be treated in a unified manner. We may note here that Drucker in 1959 also extended his postulate to include time-dependent phenomena such as creep and linear viscoelasticity. Based on this postulate, *uniqueness* of perfectly plastic and work-hardening solids has been proved, and various *variational theorems* have been formulated.

Postulates providing assumptions which play an equivalent role in the development of the framework of plasticity relations have been given by Hill in 1948, and extended by Bishop and Hill in 1951 in a study of polycrystalline aggregates, and by Ilyushin in 1961 through consideration of non-negative work in a cycle of straining (known as *Ilyushin's postulate* for material stability). However, it may be noted that the approach for developing the plasticity relations based entirely on Drucker's postulate seems so far to be the most plausible.

Precise formulations of the two fundamental *theorems of limit analysis* (the so-called upper-bound and lower-bound theorems) were given in two papers by Drucker, Greenberg, and Prager in 1951 and in 1952 for an elastic–perfectly (or ideally) plastic material, and by Hill (1951, 1952) from the point of view of rigid–ideally plastic materials. It appears, however, that the earliest reference to the theorems of limit analysis was probably due to Gvozdev in 1936; (a translation of his paper from Russian was provided by Haythornthwaite in 1960). The theorems are remarkably simple and in accord with intuition. Since then, the application of these theorems to the analysis of various classes of problems (e.g., beams and frames, plates and shells, metal-forming processes) has increased very rapidly (not only for metallic structures, but also for concrete and soil materials).

Further generalization of the plastic stress–strain relations for *singular yield surfaces* (i.e., in the presence of corners or discontinuities in the direction of the normal vector to the yield surface), as well as the uniqueness and variational theorems for such cases, is due to Koiter published in 1953. He introduced the device of using more than one yield (or loading) function in the stress–strain relationships, the plastic strain increment receiving a contribution from each active yield (loading) surface and falling within the fan of normals to the contributing surfaces. In this 1953 illuminating paper, Koiter has shown that the so-called slip theory of plasticity, introduced originally by Batdorf and Budiansky in 1949 and conceived as an alternative formulation to the classical flow theory, is a particular type of incremental (flow) theory with a singular yield condition composed of infinitely many *independently acting* regular (smooth or continuously differentiable), plane yield functions. The concept was further extended by Sanders in 1955, who also proposed a mechanism for the formulation of subsequent yield surfaces.

The introduction of the "corner" concept has ended a period of considerable controversy in which supposedly alternative frameworks for constitutive relations which admitted singular regimes in the yield surface were advanced. Further discussions on this subject may be found in the 1960 paper by Koiter and in the 1953 paper by Prager, among many others.

# 1.3. Plastic Behavior in Simple Tension and Compression

The simplest type of loading is represented by the uniaxial stress condition, e.g., the simple tension test, for which $\sigma_1 > 0$, $\sigma_2 = \sigma_3 = 0$, or the simple compression test, for which $\sigma_1 = \sigma_2 = 0$, $\sigma_3 < 0$. The well-known uniaxial stress–strain diagram, in which the axial principal stress $\sigma_1$ (or $\sigma_3$) is plotted against the axial strain $\epsilon_1$ (or $\epsilon_3$), affords a useful representation of the plastic as well as the elastic behavior.

## 1.3.1. Monotonic Loading

Figure 1.1a shows the typical curve for a simple tension specimen of mild steel. The initial elastic region generally appears as a straight line $OA$ where $A$ defines the *limit of proportionality*. On further straining, the relation between stress and strain is no longer linear but the material is still elastic,

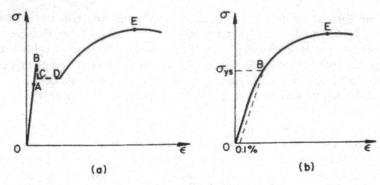

FIGURE 1.1. Stress–strain diagram for mild steel (a) and for some other metals (b).

and upon release of the load, the specimen reverts to its original length. The maximum stress point $B$ at which the load can be applied without causing any permanent deformation defines the *elastic limit*. Point $B$ is also called the *yield point*, for it marks the initiation of plastic or irreversible deformation. Usually, there is little difference between the proportional limit, $A$, and the elastic limit, $B$. Mild steel exhibits an upper yield point $B$ and a lower yield point $C$. Beyond point $C$, there is an extension at approximately constant load. The behavior in the flat region $CD$ is generally referred to as *plastic flow*. For most metals, however, neither a sharp yield point nor plastic flow is discernible, and a yield strength is generally defined by an *offset yield stress*, $\sigma_{ys}$, corresponding usually to a 0.1% definition of strain as shown in Fig. 1.1b. This offset yield stress is defined as the *initial yield stress*.

Above the yield point, the response of the material is both elastic and plastic. The slope of the curve decreases steadily, monotonically, and eventually failure of the specimen occurs at point $E$. A *ductile material* like mild steel is able to incur comparatively large strains without failure. On the other hand, cast iron is a *brittle material* since it fails after very little straining. It is generally considered when discussing the failure of metals that there are two types of failure modes: *cleavage type* such as exhibited by cast iron and *shear type* such as exhibited by mild steel. Failure characteristics of geological materials are much more complicated. They also depend on loading type: for example, concrete exhibits brittle behavior under tensile loadings, but under compressive loadings with confining pressure, it may exhibit a certain degree of ductility before failure.

## 1.3.2. Unloading and Reloading

Now, consider the test in which the specimen is first loaded monotonically to some value beyond the initial yield point and then completely unloaded.

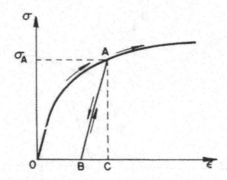

FIGURE 1.2. Loading, unloading, and reloading paths.

The behavior is as indicated in Fig. 1.2. When the stress is reduced, the strain decreases along an almost elastic unloading line $AB$ which is parallel to the initially linear region of the curve. When the load is again zero at the end of unloading, the strain is not zero; there remains a *residual strain* $OB$. The irrecoverable strain $OB$ is referred to as *plastic strain* while the recoverable strain $BC$ is the *elastic strain*. Now, if this specimen is reloaded, the stress–strain curve follows the reloading path $BA$, which is identical to the unloading path $AB$. The material is therefore elastic until the previous maximum stress at point $A$ is reached again. The stress $\sigma_A$ is regarded as the *subsequent yield stress*, beyond which further plastic deformation is induced and the stress–strain curve again follows the original one for monotonic loading.

For most materials, after the initial yield point has been reached, the stress–strain curve continues to rise although the slope becomes progressively less, until the slope falls to zero as failure occurs. Thus, the subsequent yield stress increases with further straining. This effect of the material being able to withstand a greater stress after plastic deformation is known as *strain hardening* or *work hardening*, in the sense that the material gets stronger the more it is strained or worked.

For some materials, such as concrete or rock in a simple compression test, there is a region beyond the failure or peak point in which the slope of the curve is negative. Such a behavior is called *strain softening*. This type of material gets weaker with a continuous straining beyond a certain limit or peak stress.

## 1.3.3. Reversed Loading

If we perform a simple compression test on a metal, we will obtain an almost identical stress–strain curve as in a simple tension test. However,

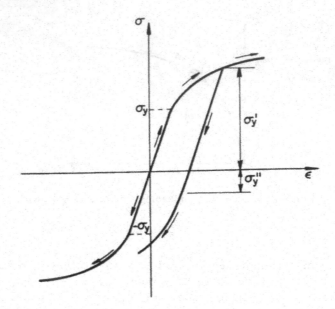

FIGURE 1.3. Bauschinger effect.

after a plastic prestraining in tension of a specimen, the stress–strain curve of this specimen in compression differs considerably from the curve which would be obtained on reloading of this specimen in tension, or on loading an undisturbed specimen in compression. As illustrated in Fig. 1.3, for the specimen with a preloading $\sigma_y'$ in tension, its corresponding compressive yielding occurs at a stress level $\sigma_y''$ which is less than the initial yield stress $\sigma_y$ and much less than the subsequent yield point $\sigma_y'$. This phenomenon is known as the *Bauschinger effect* and is usually present whenever there is a stress reversal.

It is evident from the previous discussion that there is no one-to-one correspondence between stress and strain in a plastically deformed solid. In other words, the strain is not a function of stress alone, but depends on the previous *loading history*. Thus, the material is *load path dependent*. This can be illustrated by the simple case of zero stress, when residual strains of different magnitudes can be established by varying the loading history with the stress starting and finishing at zero.

In this discussion so far, we have assumed that there is a single stress–strain curve for tension or compression, independent of the rate of straining. This assumption is referred to as *time independence*. It is reasonably true for structural metals at room temperature under a static loading condition. Rate effects are very important for materials under dynamic loading conditions. The case is not considered in this book.

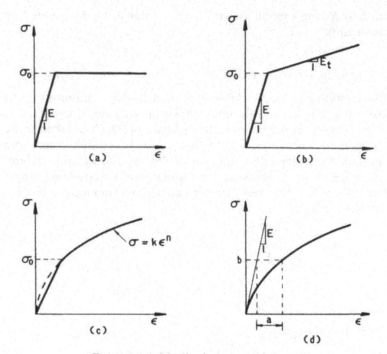

FIGURE 1.4. Idealized stress–strain curves.

## 1.4. Modeling of Uniaxial Behavior in Plasticity

### 1.4.1. Simplified Uniaxial Tensile Stress–Strain Curves

In order to obtain a solution to a deformation problem, it is necessary to idealize the stress–strain behavior of the material. The following idealized models deserve note.

#### 1.4.1.1. ELASTIC–PERFECTLY PLASTIC MODEL (FIG. 1.4a)

In some instances, it is permissible and convenient to neglect the effect of work hardening, assume that the plastic flow occurs as the stress has reached the yield stress $\sigma_0$. Thus, the uniaxial tension stress–strain relation may be expressed as

$$\epsilon = \frac{\sigma}{E} \qquad \text{for } \sigma < \sigma_0$$

$$\epsilon = \frac{\sigma}{E} + \lambda \qquad \text{for } \sigma = \sigma_0$$

(1.1)

where $E$ is Young's modulus, and $\lambda$ is a scalar to be determined and is greater than 0.

### 1.4.1.2. ELASTIC-LINEAR WORK-HARDENING MODEL (FIG. 1.4b)

In the elastic-linear work–hardening model, the continuous curve is approximated by two straight lines, thus replacing the smooth transition curve by a sharp breaking point, the ordinate of which is taken to be the elastic limit stress or the yield strength $\sigma_0$. The first straight-line branch of the diagram has a slope of Young's modulus, $E$. The second straight-line branch, representing in an idealized fashion the strain-hardening range, has a slope of $E_t < E$. The stress–strain relation for a monotonic loading in tension has the form

$$\epsilon = \frac{\sigma}{E} \qquad \text{for } \sigma \le \sigma_0$$

$$\epsilon = \frac{\sigma_0}{E} + \frac{1}{E_t}(\sigma - \sigma_0) \qquad \text{for } \sigma > \sigma_0$$

$$(1.2)$$

### 1.4.1.3. ELASTIC-EXPONENTIAL HARDENING MODEL (FIG. 1.4c)

Consider a power expression of the type

$$\sigma = E\epsilon \qquad \text{for } \sigma \le \sigma_0$$

$$\sigma = k\epsilon^n \qquad \text{for } \sigma > \sigma_0$$

$$(1.3)$$

where $k$ and $n$ are two characteristic constants of the material to be determined to best fit the experimentally obtained curve. If $\epsilon$ represents the total strain, the curve should pass through the point representing the yield stress and the corresponding elastic strain. The power expression (1.3) should be used only in the strain-hardening range (see Fig. 1.4c).

### 1.4.1.4. RAMBERG-OSGOOD MODEL (FIG. 1.4d)

The nonlinear stress–strain curve as shown in Fig. 1.4d has the following expression:

$$\epsilon = \frac{\sigma}{E} + a\left(\frac{\sigma}{b}\right)^n \qquad (1.4)$$

in which $a$, $b$, and $n$ are material constants. The initial slope of the curve takes the value of Young's modulus $E$ at $\sigma = 0$, and decreases monotonically with increasing loading. Since the model has three parameters, it allows for a better fit of real stress–strain curves.

## 1.4.2. Tangent Modulus $E_t$ and Plastic Modulus $E_p$

Because the elastic–plastic stress–strain response of a material is nonlinear in nature, an incremental procedure is generally adopted in solution of a deformation problem. We assume therefore that a *strain increment, $d\epsilon$*, consists of two parts: the *elastic strain increment, $d\epsilon^e$*, and the *plastic strain increment, $d\epsilon^p$* (see Fig. 1.5a), such that

$$d\epsilon = d\epsilon^e + d\epsilon^p \tag{1.5}$$

The *stress increment $d\sigma$* is related to the *strain increment $d\epsilon$* by

$$d\sigma = E_t\, d\epsilon \tag{1.6}$$

where $E_t$ is the *tangent modulus* which is changing during plastic deformation. In the case of uniaxial loading, $E_t$ is the current slope of the $\sigma$–$\epsilon$ curve (Fig. 1.5a). If we separate the plastic strain $\epsilon^p$ from the total strain $\epsilon$, then the plastic strain increment $d\epsilon^p$ and the stress increment $d\sigma$ are related by

$$d\sigma = E_p\, d\epsilon^p \tag{1.7}$$

where $E_p$ is referred to as the *plastic modulus*, which in the case of uniaxial loading is the slope of the $\sigma$–$\epsilon^p$ curve as shown in Fig. 1.5b. For the elastic strain increment $d\epsilon^e$, we have the usual relationship

$$d\sigma = E\, d\epsilon^e \tag{1.8}$$

where $E$ is the *elastic modulus*.

Substitution of $d\epsilon$ in Eq. (1.6), $d\epsilon^p$ in Eq. (1.7), and $d\epsilon^e$ in Eq. (1.8) into Eq. (1.5) leads to the relationship between the three moduli $E_t$, $E$, and $E_p$

$$\frac{1}{E_t} = \frac{1}{E} + \frac{1}{E_p} \tag{1.9}$$

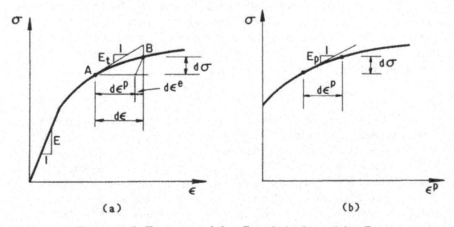

(a)                    (b)

FIGURE 1.5. Tangent modulus $E_t$ and plastic modulus $E_p$.

## 1.4.3. Hardening Rules

As described previously, the phenomenon whereby yield stress increases with further plastic straining is known as work hardening or strain hardening. To describe this behavior, we introduce a *hardening parameter* $\kappa$ to characterize various stages of hardening, and assume that the plastic modulus $E_p$ is a function of this hardening parameter $\kappa$ as

$$E_p = E_p(\kappa) \tag{1.10}$$

where $\kappa$ may be taken as the *plastic work* $W_p$

$$W_p = \int \sigma \, d\epsilon^P \tag{1.11}$$

or the *plastic strain* $\epsilon^P$ or, more realistically, the accumulative plastic strain

$$\epsilon_p = \int (d\epsilon^P \, d\epsilon^P)^{1/2}$$

which is the sum of the *effective plastic strain increments* defined by

$$|d\epsilon_p| = \sqrt{d\epsilon^P \, d\epsilon^P} \tag{1.12}$$

Since the uniaxial tensile $\sigma$-$\epsilon$ curve for a material is generally known from a simple test, the functional form of the plastic modulus $E_p$ in Eq. (1.10) can be determined from this test in terms of a given definition of the hardening parameter $\kappa$.

For a material element under a reversed loading condition, the subsequent yield stress is usually determined by one of the following three simple rules:

1. *Isotropic hardening rule*: The reversed compressive yield stress is assumed equal to the tensile yield stress. As illustrated in Fig. 1.6a, where $|\overline{B'C}| = |\overline{BC}|$, the reversed compressive yield stress $\sigma_{B'}$ is equal to the tensile yielding stress $\sigma_B$ before load reversal. Thus, the isotropic hardening rule neglects completely the Bauschinger effect, as it assumes that a raised yield point in tension carries over equally in compression. This hardening rule may be expressed mathematically in the form

$$|\sigma| = |\sigma(\kappa)| \tag{1.13}$$

where $\sigma(\kappa)$ is a function of the hardening parameter $\kappa$ and the parameter $\kappa$ must be defined such that it is always a non-negative scalar, such as the plastic work or accumulative plastic strain mentioned before.

2. *Kinematic hardening rule*: The elastic range is assumed to be unchanged during hardening. Thus, the kinematic hardening rule considers the Bauschinger effect to its full extent. Kinematic hardening for a linear hardening material is shown in Fig. 1.6b, where $|\overline{BB'}| = |\overline{AA'}|$. The center of the elastic region is moved along the straight line $aa'$. This hardening

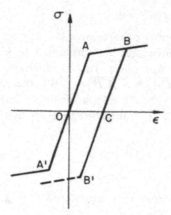

(a)  ISOTROPIC HARDENING

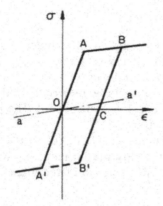

(b)  KINEMATIC HARDENING

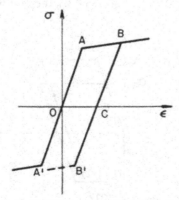

(c)  INDEPENDENT HARDENING

FIGURE 1.6. Hardening rules.

rule may be expressed mathematically in the form

$$|\sigma - c(\kappa)| = \sigma_0 \qquad (1.14)$$

where $c$ is a function of the hardening parameter $\kappa$.

3. *Independent hardening rule*: The material is assumed to be hardened independently in tension and in compression. This hardening rule is exemplified in Fig. 1.6c, where $\overline{BC} > \overline{OA}$, but $|\overline{CB'}| = |\overline{OA'}|$; the material has hardened only in tension, but it behaves like a virgin material under a reversed compressive loading condition. It may be expressed mathematically in the form

$$\sigma = \sigma_t(\kappa_t) \qquad \text{if } \sigma > 0$$

$$\sigma = \sigma_c(\kappa_c) \qquad \text{if } \sigma < 0$$

where $\kappa_t$ and $\kappa_c$ are hardening parameters accumulated during the tension and compression loading, respectively.

### 1.4.4. Examples

EXAMPLE 1.1. The behavior of a polycrystal metallic material composed of many monocrystals is analogous to a truss structure composed of many individual bars. Therefore, it is possible to use a simple truss model to simulate the elastic–plastic behavior of metallic materials. In this example, an overlay truss structure shown in Fig. 1.7 is considered. The Bauschinger effect will be simulated by the model.

In Fig. 1.7, two pairs of bar elements in parallel carry the load $P$. The vertical bars are made of elastic–perfectly plastic materials with different yield strength. Discuss the loading, unloading, and reloading characteristics of this structural model.

LOADING BEHAVIOR. With the load $P$ increasing from zero, the first two significant stages occur when bars 1 yield followed by the yielding of bars

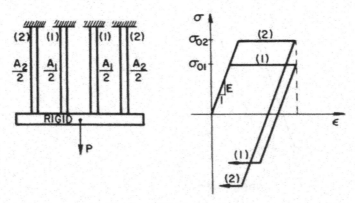

FIGURE 1.7. Overlay model.

2. Noting that both materials have the same elastic modulus, the load at first yield is found as

$$P_a = \sigma_{01} A_1 + \sigma_{01} A_2 \tag{1.15}$$

The equivalent stress can be expressed as

$$\sigma_a = \frac{\sigma_{01} A_1 + \sigma_{01} A_2}{A_1 + A_2} = \sigma_{01} \tag{1.16}$$

The corresponding strain is

$$\epsilon_a = \frac{\sigma_{01}}{E} \tag{1.17}$$

At the yielding of bars 2, the load, the stress, and the strain may be expressed as

$$P_b = \sigma_{01} A_1 + \sigma_{02} A_2 \tag{1.18}$$

$$\sigma_b = \frac{\sigma_{01} A_1 + \sigma_{02} A_2}{A_1 + A_2} \tag{1.19}$$

$$\epsilon_b = \frac{\sigma_{02}}{E} \tag{1.20}$$

UNLOADING BEHAVIOR. After this, further elongation of the bars does not result in any increase in load. Therefore, the next significant occurrence will be an unloading phase. During unloading, the modulus is the same as the initial modulus $E$. Therefore, load $P$ reaches zero when the strain has been reduced by an amount

$$\epsilon = \frac{\sigma_{01} A_1 + \sigma_{02} A_2}{E(A_1 + A_2)} \tag{1.21}$$

At this point, the stress in bars 1, $\sigma_1$, and that in bars 2, $\sigma_2$, are represented by

$$\sigma_1 = \sigma_{01} - E\epsilon = \frac{(\sigma_{01} - \sigma_{02}) A_2}{A_1 + A_2} \tag{1.22}$$

$$\sigma_2 = \sigma_{02} - E\epsilon = \frac{(\sigma_{02} - \sigma_{01}) A_1}{A_1 + A_2} \tag{1.23}$$

Because $\sigma_{02} > \sigma_{01}$, we have $\sigma_1 < 0$, $\sigma_2 > 0$, indicating that there exists a residual compression in the lower-yield-strength bars 1 and a residual tension in the higher-yield-strength bars 2 with the applied load being reduced to zero.

Since we assume that the bar itself is elastic–perfectly plastic, compressive yield will occur in bars 1 when the strain has been reduced by an amount

$$\epsilon = 2\frac{\sigma_{01}}{E} \qquad (1.24)$$

At this instance, the load in bars 1 is

$$P_1 = -\sigma_{01}A_1 \qquad (1.25)$$

The load in bars 2 is

$$P_2 = (\sigma_{02} - 2\sigma_{01})A_2 \qquad (1.26)$$

The equivalent stress is

$$\sigma_d = \frac{(\sigma_{02} - 2\sigma_{01})A_2 - \sigma_{01}A_1}{A_1 + A_2} \qquad (1.27)$$

Noting that the reversed yield load in Eq. (1.27) is less in magnitude than the initial yield load in Eq. (1.16), it follows therefore that bars 1 yield much earlier than they would have on initial loading, because they are already in compression when $P = 0$ as indicated in Eq. (1.22).

Compressive yielding occurs in bars 2 when the strain has been reduced by the amount

$$\epsilon = 2\frac{\sigma_{02}}{E} \qquad (1.28)$$

At this point, the load and the equivalent stress can be expressed as

$$P_e = -\sigma_{01}A_1 - \sigma_{02}A_2 \qquad (1.29)$$

$$\sigma_e = \frac{-\sigma_{01}A_1 - \sigma_{02}A_2}{A_1 + A_2} \qquad (1.30)$$

The $\sigma$–$\epsilon$ curve for this loading path is shown in Fig. 1.8 by 0–a–b–c–d–e. If more bar elements with different yield strengths are included in the structural model, a more smooth $\sigma$–$\epsilon$ curve will be obtained.

RELOADING BEHAVIOR. Consider now the case in which compressive loading is terminated at point $h$ before compressive plastic flow in bars 2 begins. Reloading of the structural system in tension will follow the linear response with the initial modulus $E$ but bars 1 will eventually yield again in tension at point $i$ and plastic flow begins at point $f$. The $\sigma$–$\epsilon$ curve for this loading cycle is shown in Fig. 1.8 by $f$–$g$–$h$–$i$.

This structural model with an assemblage of bars of different yield strengths may be considered, qualitatively, to represent a real specimen with slip planes of different strengths, and therefore explains why a real specimen generally exhibits the Bauschinger effect.

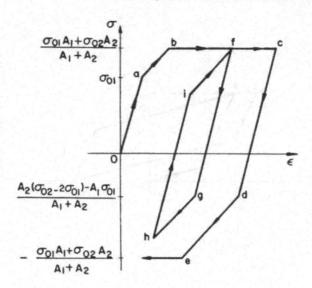

FIGURE 1.8. Loading and unloading characteristics of the overlay model.

EXAMPLE 1.2. The $\sigma$-$\epsilon$ response in simple tension for an elastic-linear hardening plastic material is approximated by the following expressions.

$$\sigma = \sigma_0 + m\epsilon^p \qquad \text{for } \sigma \geq \sigma_0$$

$$\epsilon^e = \frac{\sigma}{E}$$

where $\sigma_0 = 207$ MPa, $E = 207$ GPa, and $m = 25.9$ GPa. A material sample is first stretched to a total strain $\epsilon = 0.007$, is subsequently returned to its initial strain-free state ($\epsilon = 0$) by continued compressive stressing, and then is unloaded and reloaded in tension again to reach the same strain $\epsilon = 0.007$ (see Fig. 1.9). Sketch the stress-strain curve for the following hardening rules: (i) isotropic hardening, (ii) kinematic hardening, (iii) independent acting tensile and compressive hardening.

SOLUTION. According to the definition of the plastic modulus in Eq. (1.7), we have

$$E_p = \frac{d\sigma}{d\epsilon^p} = m = 25,900 \text{ MPa}$$

and the tangent modulus $E_t$ is found from Eq. (1.9) as

$$E_t = \frac{1}{\dfrac{1}{E} + \dfrac{1}{E_p}} = \frac{1}{\dfrac{1}{207,000} + \dfrac{1}{25,900}} = 23,020 \text{ MPa}$$

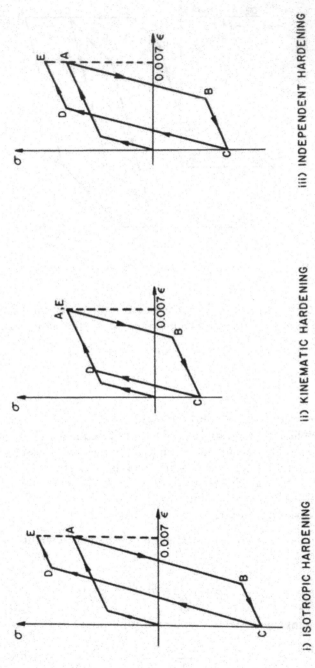

i) ISOTROPIC HARDENING            ii) KINEMATIC HARDENING            iii) INDEPENDENT HARDENING

FIGURE 1.9. Stress–strain curve for a strain cycle $\epsilon = 0 \to 0.007 \to 0 \to 0.007$.

which assumes the constant value 23,020 MPa for the linear hardening material.

With the material sample stretching, yielding occurs at the point with strain

$$\epsilon = \frac{\sigma_0}{E} = 0.001$$

Then, the sample is further stretched to point $A$ with strain $\epsilon = 0.007$, at which the stress $\sigma_A$ is found as

$$\sigma_A = \sigma_0 + \Delta\sigma$$

$$= \sigma_0 + E_t\Delta\epsilon$$

$$= 207 + 23,020(0.007 - 0.001) = 345 \text{ MPa}$$

The subsequent stresses are determined for the three hardening rules as follows:

(i) *Isotropic hardening case* [Fig. 1.9(i)]: During unloading and reversed loading in compression, the sample behaves elastically until it yields again in compression at point $B$. According to the isotropic hardening rule, we have

$$\sigma_B = -\sigma_A = -345 \text{ MPa}$$

$$\epsilon_B = \epsilon_A - 2\frac{\sigma_A}{E} = 0.007 - 2\left(\frac{345}{207,000}\right) = 0.00367$$

Now, the material sample is yielding until load reversal occurs at point $C$ as $\epsilon = 0$.

$$\sigma_C = \sigma_B + E_t\Delta\epsilon$$

$$= -345 + 23,020(0 - 0.00367) = -429 \text{ MPa}$$

Upon reversal of the straining, the material is elastic up to point $D$ at which

$$\sigma_D = 429 \text{ MPa}$$

$$\epsilon_D = \epsilon_C + \frac{2\sigma_D}{E} = 0 + 2\left(\frac{429}{207,000}\right) = 0.004145$$

As the strain $\epsilon$ reaches 0.007 at point $E$, the stress is

$$\sigma_E = \sigma_D + E_t\Delta\epsilon$$

$$= 429 + 23,020(0.007 - 0.004145)$$

$$= 495 \text{ MPa}$$

(ii) *Kinematic hardening case* [Fig. 1.9(ii)]: The yield stress at point $B$ is

$$\sigma_B = \sigma_A - 2\sigma_0 = 345 - 2(207) = -69 \text{ MPa}$$

$$\epsilon_B = \epsilon_A - 2\frac{\sigma_0}{E} = 0.007 - 2\left(\frac{207}{207,000}\right) = 0.005$$

At point $C$,

$$\sigma_C = \sigma_B + E_t \Delta\epsilon$$

$$= -69 + 23,020(0 - 0.005) = -184 \text{ MPa}$$

At point $D$, the sample yields again in tension at a stress

$$\sigma_D = \sigma_C + 2\sigma_0$$

$$= -184 + 2(207) = 230 \text{ MPa}$$

$$\epsilon_D = \epsilon_C + \frac{2\sigma_0}{E} = 0 + 2\left(\frac{207}{207,000}\right) = 0.002$$

At point $E$, the stress is

$$\sigma_E = \sigma_D + E_t \Delta\epsilon$$

$$= 230 + 23,020(0.007 - 0.002) = 345 \text{ MPa}$$

(iii) *Independent hardening case* [Fig. 1.9(iii)]: Because the material has not yielded in compression before, the yield stress at point $B$ is

$$\sigma_B = -\sigma_0 = -207 \text{ MPa}$$

$$\epsilon_B = \epsilon_A - \frac{\sigma_A}{E} - \frac{\sigma_0}{E}$$

$$= 0.007 - \frac{345}{207,000} - \frac{207}{207,000} = 0.00433$$

At point $C$,

$$\sigma_C = \sigma_B + E_t \Delta\epsilon$$

$$= -207 + 23,020(0 - 0.00433) = -307 \text{ MPa}$$

At point $D$, the material yields again in tension at a stress equal to $\sigma_A$, i.e.,

$$\sigma_D = \sigma_A = 345 \text{ MPa}$$

$$\epsilon_D = \epsilon_C - \frac{\sigma_C}{E} + \frac{\sigma_D}{E} = 0 - \frac{(-307)}{207,000} + \frac{345}{207,000}$$

$$= 0.00315$$

At point $E$, the stress is

$$\sigma_E = \sigma_D + E_t \Delta\epsilon$$

$$= 345 + 23,020(0.007 - 0.00315) = 434 \text{ MPa}$$

The stress–strain curves for each of the three cases are shown in Fig. 1.9.

EXAMPLE 1.3. Consider a nonlinear kinematic hardening material with a simple tension $\sigma$-$\epsilon$ curve obtained from a simple test in the form

$$\sigma = \sigma_0 + m(\epsilon^p)^n \qquad \text{for } \sigma \geq \sigma_0$$
$$\sigma = E\epsilon^e \qquad\qquad \text{for } \sigma < \sigma_0 \tag{1.31}$$

where $m = 500$ MPa, $n = 0.3$, $E = 70,000$ MPa, and $\sigma_0 = 200$ MPa. Equation (1.31) represents the elastoplastic behavior of the material under the simple tension stress path. This relationship cannot be used directly for a general loading path other than the simple tension stress path. To generalize it, we shall follow the discussion in Section 1.4.3 and introduce a non-negative hardening parameter $\kappa$ to represent the plastic deformation of the material under a general loading condition. The relationship between the hardening parameter $\kappa$ and the plastic modulus $E_p$, Eq. (1.10), is assumed to be of the same form as Eq. (1.31). This strategy extends and generalizes the elastoplastic behavior of a material observed in a simple loading test to the general loading case. This is a basic strategy used frequently in the theory of plasticity.

A material element is first loaded to point $A$ in tension at which $\sigma_A = 350$ MPa, and is then unloaded and continuously loaded in compression. State $B$ is the yield point in compression during reversed loading (Fig. 1.10).
(a) Find the expressions describing the $\sigma$-$\epsilon^p$ curve during plastically compressive loading starting from state $B$ for hardening parameter $\kappa$ defined as: case (i), $\kappa = \int (d\epsilon^p \, d\epsilon^p)^{1/2}$; case (ii), $\kappa = \epsilon^p$. Sketch the $\sigma$-$\epsilon^p$ curve.
(b) For $\kappa$ defined as case (iii), $\kappa = W_p = \int \sigma \, d\epsilon^p$, sketch the $\sigma$-$\epsilon^p$ curve using a simple incremental procedure.

SOLUTION. (a) In the initial plastic tension range, $d\epsilon^p > 0$, so we have

$$\kappa = \int_0^{\epsilon^p} (d\epsilon^p \, d\epsilon^p)^{1/2} = \int_0^{\epsilon^p} d\epsilon^p = \epsilon^p$$

which takes the same form as case (ii). From the given $\sigma$-$\epsilon$ relation (1.31) for simple tension, the plastic modulus $E_p$ can be expressed in terms of $\kappa$ as

$$E_p = \frac{d\sigma}{d\epsilon^p} = mn(\epsilon^p)^{n-1} = mn\kappa^{n-1} \tag{1.32}$$

In the compressive loading range starting from point $B$, the hardening parameter $\kappa$ for case (i) is the total accumulated plastic strain up to the current state $\epsilon^p$:

$$\kappa = \epsilon_A^p + (\epsilon_A^p - \epsilon^p) = 2\epsilon_A^p - \epsilon^p \tag{1.33a}$$

and for case (ii), it is $\epsilon^p$ itself:

$$\kappa = \epsilon^p \tag{1.33b}$$

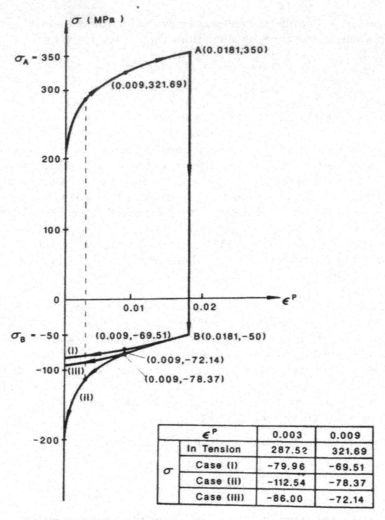

FIGURE 1.10. $\sigma$-$\epsilon^P$ relations for various definitions of $\kappa$.

The plastic modulus $E_p$ is given by

$$E_p = mn\kappa^{n-1} = mn(2\epsilon_A^P - \epsilon^P)^{n-1} \qquad \text{for case (i)} \qquad (1.34a)$$

$$E_p = mn(\epsilon^P)^{n-1} \qquad\qquad\qquad \text{for case (ii)} \qquad (1.34b)$$

Hence, the $\sigma$-$\epsilon^P$ curve in this range can be found by

$$\sigma = \sigma_B + \int_{\epsilon_B^P}^{\epsilon^P} E_p \, d\epsilon^P \qquad (1.35)$$

Substituting Eq. (1.34) for $E_p$ and noting that $\sigma_B = \sigma_A - 2\sigma_0$ and $\epsilon_B^p = \epsilon_A^p$, carrying out the integration leads to

$$\sigma = -m(2\epsilon_A^p - \epsilon^p)^n + m(\epsilon_A^p)^n + \sigma_A - 2\sigma_0 \qquad \text{for case (i)} \qquad (1.36a)$$

and

$$\sigma = m[(\epsilon^p)^n - (\epsilon_A^p)^n] + \sigma_A - 2\sigma_0 \qquad \text{for case (ii)} \qquad (1.36b)$$

At point $A$, the stress $\sigma_A$ is already given as $\sigma_A = 350$ MPa, so the strain can easily be determined by Eq. (1.31) as

$$\epsilon_A^p = 0.0181$$

Substituting the values of $m$, $n$, $\sigma_0$, $\sigma_A$, and $\epsilon_A^p$ into Eqs. (1.36a) and (1.36b) leads to

$$\sigma = -500(0.0362 - \epsilon^p)^{0.3} + 100.06 \qquad \text{for case (i)} \qquad (1.37a)$$

$$\sigma = 500(\epsilon^p)^{0.3} - 200.06 \qquad \text{for case (ii)} \qquad (1.37b)$$

The $\sigma - \epsilon^p$ curves for cases (i) and (ii) are sketched in Fig. 1.10.

(b) For case (iii), the hardening parameter $\kappa$ is defined by $W_p$, and according to assumption (1.10), the plastic modulus is now a function of $W_p$, i.e.,

$$E_p = E_p(W_p) \qquad (1.38)$$

From the given $\sigma - \epsilon$ relation (1.31), $W_p$ is given by

$$W_p = \int \sigma \, d\epsilon^p$$

$$= \int_0^{\epsilon^p} [\sigma_0 + m(\epsilon^p)^n] \, d\epsilon^p \qquad (1.39a)$$

$$= \sigma_0 \epsilon^p + \frac{m}{n+1}(\epsilon^p)^{n+1}$$

From Eq. (1.31) also we have

$$E_p = mn(\epsilon^p)^{n-1} \qquad (1.39b)$$

The relationship (1.38) is now established from Eqs. (1.39a) and (1.39b) by canceling out the variable $\epsilon_p$.

In the compressive loading range starting from point $B$, there is no explicit expression for $E_p = E(W_p)$. Thus, the $\sigma - \epsilon^p$ curve cannot be found by direct integration; we can only use the incremental procedure to find the $\sigma - \epsilon^p$ relation numerically. The steps in the calculation are the following:

1. Calculate $\sigma$, $\epsilon^p$, $W_p$, and $E_p$ at the starting point $B$.
2. Take the plastic strain increment $d\epsilon^p = -0.0015$ (first increment, $-0.0016$).

TABLE 1.1. Calculation of $\sigma$-$\epsilon^p$ relation by simple incremental procedure for case (iii) in Example 1.3.

| $\sigma$ | $\epsilon^p$ | $W_p$ | $E_p$ | $\dot{W}_p$ | $\dot{\sigma}$ |
|---|---|---|---|---|---|
| −50.00 | 0.0181 | 5.709 | 2487.23 | 0.0800 | −3.980 |
| −53.98 | 0.0165 | 5.789 | 2465.35 | 0.0810 | −3.698 |
| −57.68 | 0.0150 | 5.870 | 2443.93 | 0.0865 | −3.666 |
| −61.34 | 0.0135 | 5.957 | 2421.14 | 0.0920 | −3.632 |
| −64.98 | 0.0120 | 6.049 | 2397.99 | 0.0975 | −3.597 |
| −68.57 | 0.0105 | 6.146 | 2374.50 | 0.1029 | −3.562 |
| −72.13 | 0.0090 | 6.249 | 2349.89 | 0.1082 | −3.525 |
| −75.66 | 0.0075 | 6.357 | 2324.26 | 0.1135 | −3.486 |
| −79.15 | 0.0060 | 6.471 | 2298.50 | 0.1187 | −3.448 |
| −82.59 | 0.0045 | 6.590 | 2272.65 | 0.1239 | −3.409 |
| −86.00 | 0.0030 | 6.713 | 2245.99 | 0.1290 | −3.369 |
| −89.37 | 0.0015 | 6.842 | 2219.35 | 0.1341 | −3.329 |
| −92.70 | 0.0000 | 6.977 | 2192.04 | 0.1391 | −3.288 |

3. Calculate the stress increment $d\sigma = E_p\, d\epsilon^p$ and the plastic work increment $dW_p = \sigma\, d\epsilon_p$.
4. Update stress, plastic strain, and plastic work by adding up the corresponding increments: $\sigma + d\sigma$, $\epsilon^p + d\epsilon^p$, $W_p + dW_p$.
5. Update plastic modulus $E_p$ by solving Eq. (1.39a) for $\epsilon^p$, then substituting $\epsilon^p$ so obtained into Eq. (1.39b) to find $E_p$. Note here that $\epsilon^p$ is not the current plastic strain.
6. Go to step 2.

The results of this calculation are given in Table 1.1 and the corresponding $\sigma$-$\epsilon^p$ curve is shown in Fig. 1.10.

## 1.5. Index Notation

Because index notation allows a drastic reduction in the number of terms in an expression or equation and the simplification of the general formulation to a great extent, it is commonly used in the current literature when stress, strain, and constitutive equations are discussed. A basic knowledge of these notations is therefore essential in studying plasticity theory and constitutive modeling of materials. With such notations, the various stress-strain relationships can be expressed in a compact form, thereby allowing greater attention to be paid to physical principles rather than to the equations themselves.

## 1.5.1. Indicial Notation and Summation Convention

For the present, we restrict ourselves to right-handed Cartesian coordinate systems. In a three-dimensional space, a right-handed Cartesian coordinate system is pictured as a set of three mutually orthogonal axes denoted as the $x$-, $y$-, and $z$-axes. For future convenience, the axes are more conveniently designated as $x_1$-, $x_2$-, and $x_3$-axes, rather than the more familiar notation $x$, $y$, and $z$. The sketch shown in Fig. 1.11 is based on the right-handed notation, where the $x_2$- and $x_3$-axes lie in the plane of the paper and the $x_1$-axis is directed towards the reader.

### 1.5.1.1. INDICIAL NOTATION

In this coordinate system, a vector $\mathbf{V}$ is denoted as

$$\mathbf{V} = (v_1, v_2, v_3) = v_1 \mathbf{e}_1 + v_2 \mathbf{e}_2 + v_3 \mathbf{e}_3 \tag{1.40}$$

where $\mathbf{e}_1$, $\mathbf{e}_2$, and $\mathbf{e}_3$ are unit vectors as shown in Fig. 1.11, and $v_1$, $v_2$, and $v_3$ are three components of the vector. It is useful to abbreviate the latter by a single component with a generalized index. Thus, in the *indicial notation*, $v_i$ represents the components of vector $\mathbf{V}$. It is implicitly understood that the index $i$ ranges in value from 1 to 3 when $v_i$ is written for $\mathbf{V}$.

As an example, the statement $x_i = 0$ implies that each of the components $x_1$, $x_2$, $x_3$ of the vector $\mathbf{X}$ is zero, or $\mathbf{X}$ is a null vector. Similarly,

$$f(\mathbf{X}) = f(x_i) = f(x_j) = f(x_1, x_2, x_3) \tag{1.41}$$

The index may be freely chosen. Hence, $x_i$ and $x_j$ represent one and the same vector.

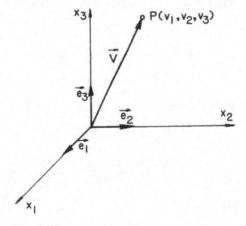

FIGURE 1.11. Right-handed Cartesian coordinate system.

### 1.5.1.2. SUMMATION CONVENTION

The *summation convention* is complementary to the indicial notation and allows for further brevity when dealing with summations. We adopt the following convention: whenever a subscript occurs twice in the same term, it is understood that the subscript is to be summed from 1 to 3. Consider, for example, the dot product of two vectors, **U** and **V**. It has the form

$$\mathbf{U} \cdot \mathbf{V} = u_1 v_1 + u_2 v_2 + u_3 v_3 = \sum_{i=1}^{3} u_i v_i \qquad (1.42)$$

The expression on the far right may be abbreviated as $u_i v_i$ since the summation always involves three components. The summation convention requires that the index $i$ be repeated, but eliminates the use of the summation symbol $\sum$. Again, however, the index itself may be freely chosen. Thus, $u_i v_i$ and $u_k v_k$ represent the same sum, $u_1 v_1 + u_2 v_2 + u_3 v_3$. Such repeated subscripts are often called *dummy* subscripts because of the fact that the particular letter used in the subscript is not important; thus, $u_i v_i = u_k v_k$.

In this context, it is necessary to point out that $u_i + v_i$ represents a vector sum, say, $w_i$, but not a scalar sum of any kind. Explicitly, the following equation is true:

$$(w_1, w_2, w_3) = (u_1 + v_1, u_2 + v_2, u_3 + v_3) \qquad (1.43)$$

but the following form is incorrect:

$$u_i + v_i = u_1 + v_1 + u_2 + v_2 + u_3 + v_3 \qquad (1.44)$$

Further, the index in a term of an equation or expression should occur only twice in this same term for the summation convention to be valid. An expression such as $u_i v_{ii}$ conveys no special sense.

The effectiveness of the convention is more apparent when it is applied to a set of three simultaneous equations. Consider the set

$$a_{11} x_1 + a_{12} x_2 + a_{13} x_3 = b_1$$
$$a_{21} x_1 + a_{22} x_2 + a_{23} x_3 = b_2 \qquad (1.45)$$
$$a_{31} x_1 + a_{32} x_2 + a_{33} x_3 = b_3$$

As a first stage in abbreviation, these may be written as

$$a_{1j} x_j = b_1$$
$$a_{2j} x_j = b_2 \qquad (1.46)$$
$$a_{3j} x_j = b_3$$

and, in the final stage, as

$$a_{ij} x_j = b_i \qquad (1.47)$$

In the first stage, the index $j$ assumes the values 1 to 3 and summation is understood on the left-hand side of the equations since the index is repeated. As mentioned before, the repeated index is referred to as a *dummy* index quite often because the choice of the letter for the index is immaterial. The three equations of the first stage may be represented as in the final stage by the use of the *free* index $i$. To be consistent, it is necessary to use the same index $i$ on both sides of the equation. The existence of one free index indicates that vectors are involved. Later, it will be seen that when two free indices appear, tensors are involved.

Based on the previous discussion, the simultaneous equations (1.45) may also be written as

$$a_{rs}x_s = b_r \tag{1.48}$$

As a review, equivalent vector and indicial (or component) forms are presented in Table 1.2 for study.

The divergence of vector $\mathbf{V}$ is $\nabla \cdot \mathbf{V}$ or the scalar sum

$$\nabla \cdot \mathbf{V} = \frac{\partial v_1}{\partial x_1} + \frac{\partial v_2}{\partial x_2} + \frac{\partial v_3}{\partial x_3} \tag{1.49}$$

In the summation convention,

$$\nabla \cdot \mathbf{V} = \frac{\partial v_i}{\partial x_i} \tag{1.50}$$

where $i$ is a dummy index.

The conventions regarding subscripts described above can now be summarized as a set of three rules:

Rule 1: If a subscript occurs precisely *once* in one term of an expression or equation, it is called a "*free index.*" This must occur precisely once in each term of the expression or equation.

Rule 2: If a subscript occurs precisely *twice* in one term of an expression or equation, it is called a "*dummy index.*" It is to be summed from 1 to 3. The dummy index may or may not occur precisely twice in any other term.

Rule 3: If a subscript occurs more than twice in one term of an expression or equation, it is a *mistake.*

TABLE 1.2. Equivalent vector and indicial notation.

| Vector | Components | Indicial notation |
|---|---|---|
| $\mathbf{V}$ | $(v_1, v_2, v_3)$ | $v_i$ |
| $\mathbf{U} + \mathbf{V}$ | $(u_1 + v_1, u_2 + v_2, u_3 + v_3)$ | $u_i + v_i$ |
| $\nabla\phi$ | $\left(\dfrac{\partial\phi}{\partial x_1}, \dfrac{\partial\phi}{\partial x_2}, \dfrac{\partial\phi}{\partial x_3}\right)$ | $\dfrac{\partial\phi}{\partial x_i}$ |

### 1.5.1.3. DIFFERENTIATION NOTATION

In subscript notation, we use a *comma* to indicate differentiation; thus, for example, the partial derivative form of Eq. (1.50) can be further simplified to the form $v_{i,i}$. The first subscript refers to the component of **V**, and the *comma* indicates the partial derivative with respect to the second subscript corresponding to the relevant coordinate axis. Thus

$$v_{i,i} = v_{1,1} + v_{2,2} + v_{3,3} \tag{1.51}$$

and the gradient of $\phi$ is conveniently written in the form $\phi_{,i}$, which indicates clearly the vector character of the gradient of $\phi$. The divergence of $\nabla\phi$ would be written as $\phi_{,ii} = \phi_{,11} + \phi_{,22} + \phi_{,33}$. It is a scalar, known as the Laplacian of $\phi$, and is often denoted by $\nabla^2\phi = \nabla \cdot \nabla\phi$.

## 1.5.2. The Symbol $\delta_{ij}$ (Kronecker Delta)

The *Kronecker delta* is a special matrix, denoted as $\delta_{ij}$:

$$\delta_{ij} = \begin{bmatrix} 1 & 0 & 0 \\ 0 & 1 & 0 \\ 0 & 0 & 1 \end{bmatrix} \tag{1.52}$$

Thus, the components of $\delta_{ij}$ are 1 if $i = j$, and 0 if $i \neq j$:

$$\delta_{11} = \delta_{22} = \delta_{33} = 1 \tag{1.53}$$

$$\delta_{12} = \delta_{21} = \delta_{13} = \delta_{31} = \delta_{23} = \delta_{32} = 0 \tag{1.54}$$

Further, the $\delta_{ij}$ matrix is symmetric since $\delta_{ij} = \delta_{ji}$. Note that, because of the summation convention,

$$\delta_{jj} = \delta_{11} + \delta_{22} + \delta_{33} = 3 \tag{1.55}$$

The Kronecker delta may be regarded as an operator and serves a useful function when so used. Consider, for example, the projection $\delta_{ij}v_j$. According to the summation convention, this yields an expansion of the vector

$$\delta_{i1}v_1 + \delta_{i2}v_2 + \delta_{i3}v_3 \quad \text{or} \quad v_i \tag{1.56}$$

This can be easily verified since on assigning values 1, 2, and 3 to $i$, the components obtained are $v_1$, $v_2$, and $v_3$. Hence,

$$\delta_{ij}v_j = v_i \tag{1.57}$$

This final answer may be visualized as the result of substituting $i$ for $j$ (or $j$ for $i$, if need be) in the quantity operated on (by the operator $\delta_{ij}$). It appears, therefore, that the application of $\delta_{ij}$ to $v_j$ has merely substituted $i$ for $j$ in $v_j$; the $\delta_{ij}$ symbol is therefore often called a *substitution* operator.

As another example, $\delta_{ij}\delta_{ji}$ represents a scalar sum in the summation convention. Using the concept of substitution operator,

$$\delta_{ij}\delta_{ji} = \delta_{ii} = \delta_{11} + \delta_{22} + \delta_{33} = 3 \tag{1.58}$$

Similarly,

$$\delta_{ij}a_{ji} = a_{ii} = a_{11} + a_{22} + a_{33} \tag{1.59}$$

Finally, noting that the dot product $e_i \cdot e_j$ is 1 if $i = j$ and 0 if $i \neq j$, then matching the components of $\delta_{ij}$, we can write

$$e_i \cdot e_j = \delta_{ij} \tag{1.60}$$

## 1.5.3. Transformation of Coordinates

### 1.5.3.1. DIRECTION COSINES

The values of the components of a vector $V$, designated by $v_1$, $v_2$, $v_3$ or simply $v_i$, are associated with the chosen set of coordinate axes. Often it is necessary to reorient the reference axes and evaluate the new values for the components of $V$ in the new coordinate system.

Let $x_i$ and $x_i'$ be two sets of right-handed Cartesian coordinate systems having the same origin. Vector $V$, then, has components $v_i$ and $v_i'$ in the two systems. Since the vector is the same, the components must be related through the cosines of the angles between the positive $x_i'$- and $x_i$-axes.

If $l_{ij}$ represents $\cos(x_i', x_j)$, that is, the cosines of the angles between $x_i'$ and $x_j$ axes for $i$ and $j$ ranging from 1 to 3, it can be shown as follows that $v_i' = l_{ij}v_j$. These cosines may be conveniently tabulated as in Table 1.3. It should be noted that the elements of $l_{ij}$ (a matrix) are not symmetrical, $l_{ij} \neq l_{ji}$. For example, $l_{12}$ is the cosine of the angle between $x_1'$ and $x_2$ and $l_{21}$ is that between $x_2'$ and $x_1$ (see Fig. 1.12). The angle is assumed to be measured from the primed system to the unprimed system.

### 1.5.3.2. RELATION BETWEEN THE $l_{ij}$

From the definition of $l_{ij}$, we know

$$l_{ij} = e_i' \cdot e_j \tag{1.61}$$

TABLE 1.3. Direction cosines ($l_{ij}$).

| | Axis | | |
|---|---|---|---|
| Axis | $x_1$ | $x_2$ | $x_3$ |
| $x_1'$ | $l_{11}$ | $l_{12}$ | $l_{13}$ |
| $x_2'$ | $l_{21}$ | $l_{22}$ | $l_{23}$ |
| $x_3'$ | $l_{31}$ | $l_{32}$ | $l_{33}$ |

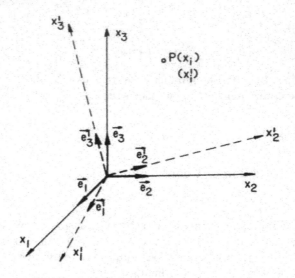

FIGURE 1.12. Transformation of coordinates.

The base (or unit) vector $e_i'$ may be expressed, in reference to the $x_i$-axes, as

$$e_i' = (e_i' \cdot e_1)e_1 + (e_i' \cdot e_2)e_2 + (e_i' \cdot e_3)e_3$$

$$= l_{i1}e_1 + l_{i2}e_2 + l_{i3}e_3 = l_{ij}e_j \tag{1.62}$$

Conversely,

$$e_i = l_{ji}e_j' \tag{1.63}$$

Hence,

$$e_i' \cdot e_j' = \delta_{ij} = l_{ir}e_r \cdot l_{jk}e_k = l_{ir}l_{jk}\delta_{rk} = l_{ir}l_{jr} \tag{1.64}$$

or

$$l_{ir}l_{jr} = \delta_{ij} \tag{1.65}$$

which implies the following six equations:

$$l_{11}^2 + l_{12}^2 + l_{13}^2 = 1$$

$$l_{21}^2 + l_{22}^2 + l_{23}^2 = 1$$

$$l_{31}^2 + l_{32}^2 + l_{33}^2 = 1$$

$$l_{11}l_{21} + l_{12}l_{22} + l_{13}l_{23} = 0 \tag{1.66}$$

$$l_{11}l_{31} + l_{12}l_{32} + l_{13}l_{33} = 0$$

$$l_{21}l_{31} + l_{22}l_{32} + l_{23}l_{33} = 0$$

Similarly,

$$\mathbf{e}_i \cdot \mathbf{e}_j = \delta_{ij} = l_{ri}\mathbf{e}'_r \cdot l_{kj}\mathbf{e}'_k = l_{ri}l_{kj}\delta_{rk} = l_{ri}l_{rj} \tag{1.67}$$

or

$$l_{ri}l_{rj} = \delta_{ij} \tag{1.68}$$

The arbitrary vector $\mathbf{V}$ can be expressed either in the form $v_i\mathbf{e}_i$ or $v'_i\mathbf{e}'_i$:

$$v'_i = \mathbf{V} \cdot \mathbf{e}'_i = v_j\mathbf{e}_j \cdot \mathbf{e}'_i = v_j\mathbf{e}_j \cdot l_{ik}\mathbf{e}_k = l_{ik}v_j\delta_{jk} = l_{ij}v_j \tag{1.69}$$

or

$$v'_i = l_{ij}v_j \tag{1.70}$$

Conversely,

$$v_i = \mathbf{V} \cdot \mathbf{e}_i = v'_j\mathbf{e}'_j \cdot l_{ri}\mathbf{e}'_r = l_{ri}v'_j\delta_{jr} = l_{ji}v'_j \tag{1.71}$$

or

$$v_i = l_{ji}v'_j \tag{1.72}$$

In a similar manner, if the point $P$ (Fig. 1.12) has coordinates $x_i$ in the unprimed system and $x'_i$ in the primed system, then

$$x'_i = l_{ij}x_j \quad \text{and} \quad x_i = l_{ji}x'_j \tag{1.73}$$

It follows that

$$l_{ij} = \frac{\partial x'_i}{\partial x_j} = \frac{\partial x_j}{\partial x'_i} \tag{1.74}$$

EXAMPLE 1.4. Table 1.4 gives the direction cosines $(l_{ij})$ for the $x_i$ and $x'_i$ coordinate systems. Show that the point $(0, 1, -1)$ in the $x_i$ system coincides with the point $(-\frac{29}{25}, \frac{4}{5}, -\frac{3}{25})$ in the $x'_i$ system.

TABLE 1.4. Direction cosines $(l_{ij})$ for Example 1.4.

| Axis | Axis | | |
|------|------|------|------|
|      | $x_1$ | $x_2$ | $x_3$ |
| $x'_1$ | $\frac{12}{25}$ | $-\frac{9}{25}$ | $\frac{4}{5}$ |
| $x'_2$ | $\frac{3}{5}$ | $\frac{4}{5}$ | $0$ |
| $x'_3$ | $-\frac{16}{25}$ | $\frac{12}{25}$ | $\frac{3}{5}$ |

SOLUTION. The relations between the coordinates of the point in the two coordinate systems $x_i$ and $x_i'$ are given by [Eq. (1.73)]

$$x_i' = l_{ij}x_j$$

where $x_j = (0, 1, -1)$. Hence, the $x_i'$ coordinates of the point can be calculated. For $i = 1$,

$$x_1' = l_{1j}x_j$$

Substituting from Table 1.4, we obtain

$$x_1' = l_{11}x_1 + l_{12}x_2 + l_{13}x_3$$
$$= (\tfrac{12}{25})(0) + (-\tfrac{9}{25})(1) + (\tfrac{4}{5})(-1)$$
$$= -\tfrac{29}{25}$$

Similarly,

$$x_2' = l_{2j}x_j = l_{21}x_1 + l_{22}x_2 + l_{23}x_3$$

and

$$x_3' = l_{3j}x_j = l_{31}x_1 + l_{32}x_2 + l_{33}x_3$$

which upon substituting from Table 1.4 give

$$x_2' = \tfrac{4}{5} \quad \text{and} \quad x_3' = -\tfrac{3}{25}$$

Therefore, the point $(0, 1, -1)$ in the $x_i$ system coincides with the point $(-\tfrac{29}{25}, \tfrac{4}{5}, -\tfrac{3}{25})$ in the $x_i'$ system.

## 1.5.4. Definition of Cartesian Tensors

In the preceding section, we proved that a vector at any point in a space is completely determined by a knowledge of its three components. If we know the components of a vector $v_i$ in the $x_i$ coordinate system, then the components of the same vector in the $x_i'$ coordinate system can be obtained by the transformation $v_i' = l_{ij}v_j$. This transformation equation holds for any vector, whether it is a physical quantity such as velocity or force, a geometric quantity such as a radius vector from the origin, or a less easily visualized quantity such as the gradient of a scalar. For example, if

$$G_i = \frac{\partial \phi}{\partial x_i} \tag{1.75}$$

then

$$G_i' = \frac{\partial \phi}{\partial x_i'} = \frac{\partial \phi}{\partial x_k} \frac{\partial x_k}{\partial x_i'} = l_{ik}G_k \tag{1.76}$$

The foregoing transformation rule, in which each new vector component in a new coordinate system is a linear combination of the old components, is very convenient and of considerable use. In the following, we adopt it as the definition of a vector, thus replacing the previous definition of a vector as a quantity possessing direction and magnitude. The basic reason for adopting this new definition of a vector is that it can be easily generalized to apply to more complicated physical quantities called *tensors* whereas the "magnitude and direction" definition cannot.

In the following, we first define a *tensor* of the first order to be a set of three quantities (called its components) possessing the property that if their values at a fixed point in any coordinate system $x_i$ are $v_i$, then their values at this point in any other coordinate system $x_i'$ are given by the relationship $v_i' = l_{ij}v_j$. An equivalent statement is, of course, $v_i = l_{ji}v_j'$. Since all vectors transform according to this law, vectors are tensors of the first order. A scalar, such as temperature, has the same value irrespective of the coordinate system used to specify it at a point, and hence a scalar is unaffected by transformations and is defined as a tensor of order zero. A *first-order tensor* (or a vector) is a set of $3^1 = 3$ components, and a *zero-order tensor* (or a scalar) is a set of $3^0 = 1$ component.

The definition is now extended to higher-order tensors similarly. A *second-order tensor* is defined as a set of $3^2 = 9$ components, such that if their values at a given point are $a_{ij}$ in a coordinate system $x_i$, their values $a_{ij}'$ at the same point in any other coordinate system $x_i'$ are given by

$$a_{ij}' = l_{im}l_{jn}a_{mn} \tag{1.77}$$

A second-order tensor may be interpreted to be defined completely by three vectors just as a vector is completely defined by three scalars. It will subsequently appear that the quantities expressing the state of stress at a point in a body form a second-order tensor. In other words, the state of stress at a point is completely defined by three *stress vectors*.

A *third-order tensor* is a set of $3^3 = 27$ components, such that if their values at a given point are $a_{ijk}$ in a coordinate system $x_i$, their values $a_{ijk}'$ in any other coordinate system $x_i'$ are given by

$$a_{ijk}' = l_{im}l_{jn}l_{kp}a_{mnp} \tag{1.78}$$

Tensors may be of any order; the general transformation equation is evident from the previous definitions. All such tensors are called *Cartesian tensors* because of the restriction to Cartesian coordinate systems.

As an example, suppose that the nine "components" of a second-order tensor are known:

$$a_{11} = 1, \qquad a_{12} = -1, \qquad a_{32} = 2, \qquad \text{all other } a_{ij} = 0$$

in the coordinate system $x_i$. Consider a new coordinate system $x'_i$, related to the $x_i$ system by the direction cosines $(l_{ij})$ table given below:

| New axis | Old axis | | |
|:---:|:---:|:---:|:---:|
| | $x_1$ | $x_2$ | $x_3$ |
| $x'_1$ | $\dfrac{1}{\sqrt{2}}$ | $\dfrac{1}{\sqrt{2}}$ | 0 |
| $x'_2$ | $\dfrac{-1}{\sqrt{2}}$ | $\dfrac{1}{\sqrt{2}}$ | 0 |
| $x'_3$ | 0 | 0 | 1 |

The new components $a'_{ij}$ in the $x'_i$ system are then given by

$$a'_{11} = l_{1k}l_{1r}a_{kr}$$
$$= l_{11}l_{11}a_{11} + l_{11}l_{12}a_{12} + l_{13}l_{12}a_{32} + 0$$
$$= \tfrac{1}{2}(1) + \tfrac{1}{2}(-1) + 0 = 0 \qquad (1.79)$$

Similarly, $a'_{12} = -1$, $a'_{32} = \sqrt{2}$, and so on.

## 1.5.5. Properties of Tensors

Operations on tensors parallel those on vectors.

### 1.5.5.1. EQUALITY

Two tensors $A$ and $B$ are defined to be equal when their respective components are equal. For example, the condition for equality of tensors $a_{ij}$ and $b_{ij}$ is that

$$a_{ij} = b_{ij} \qquad (1.80)$$

### 1.5.5.2. ADDITION

The sum or difference of two tensors of the same order is a tensor, also of the same order, which is defined by adding or subtracting the corresponding components of the two tensors. For example, if two second-order tensors $a_{ij}$ and $b_{ij}$ are added, the resulting nine quantities $c_{ij}$ also comprise a second-order tensor defined by

$$c_{ij} = a_{ij} + b_{ij} \qquad (1.81)$$

It is obvious that the sum or difference of two tensors of different order cannot be defined.

### 1.5.5.3. TENSOR EQUATIONS

As previously mentioned, a tensor equation that is true in one coordinate system is true in all systems, for if two tensors satisfy $a_{ij} = b_{ij}$ in the $x_i$

system, we can define $c_{ij} = a_{ij} - b_{ij}$ in all systems. Then, by the preceding reasoning that the difference of two tensors of the same order is a tensor, also of the same order, it follows that $c_{ij}$ is a tensor of second order. Now, $c_{ij}$ vanishes in the $x_i$ system, and hence in all systems. This can also be seen easily from the fact that $c'_{ij}$ in any system is a linear combination of the $c_{ij}$.

### 1.5.5.4. MULTIPLICATION

Multiplication of a tensor $a_{ij}$ by a scalar quantity $\alpha$ yields a tensor $b_{ij}$ of the same order:

$$b_{ij} = \alpha a_{ij} \tag{1.82}$$

Consider the two tensors $a_i$ of order one and $b_{ij}$ of order two. We may define a new set of quantities $c_{ijk}$ by a process called tensor multiplication:

$$c_{ijk} = a_i b_{jk} \tag{1.83}$$

It is, of course, understood that a similar rule of definition is to be used in other coordinate systems.

$$\begin{aligned}
c'_{ijk} &= a'_i b'_{jk} \\
&= (l_{im} a_m)(l_{jn} l_{ko} b_{no}) \\
&= l_{im} l_{jn} l_{ko} a_m b_{no} \\
&= l_{im} l_{jn} l_{ko} c_{mno}
\end{aligned} \tag{1.84}$$

It follows from Eq. (1.84) that $c_{ijk}$ is a third-order tensor. In general, tensor multiplication yields a new tensor whose order is the sum of the orders of the original tensors.

### 1.5.5.5. CONTRACTION

Consider the tensor $a_{ijk}$—a set of 27 quantities. If we give two indices the same letter, say, replacing the $j$ by a $k$, resulting in $a_{ikk}$, then only three quantities remain, each being the sum of three of the original components. It is easy to show that this set of three quantities is a first-order tensor. For the third-order tensor $a_{ijk}$, we have

$$a'_{ijk} = l_{ip} l_{jq} l_{kr} a_{pqr}$$

and therefore

$$\begin{aligned}
a'_{ikk} &= l_{ip}(l_{kq} l_{kr}) a_{pqr} \\
&= l_{ip} \delta_{qr} a_{pqr} \\
&= l_{ip} a_{prr}
\end{aligned} \tag{1.85}$$

which is the transformation rule for the first-order tensor; that is, $a_{ikk}$ is a first-order tensor.

TABLE 1.5. Examples of tensor properties.

| Tensor | Order | Remarks |
|---|---|---|
| $u_i + v_i$ | 1 | Addition |
| $cd$ | 0 | Multiplication |
| $cu_i$ | 1 | Multiplication |
| $u_i v_j$ | 2 | Multiplication |
| $u_i a_{jk}$ | 3 | Multiplication |
| $u_i v_i$ | 0 | Scalar (or dot) product |
| $u_i u_i$ | 0 | (Length)$^2$ |
| $a_{ii} = a_{11} + a_{22} + a_{33}$ | 0 | First invariant of $a_{ij}$ |
| $u_i a_{rk}$ | 1 | Contraction |
| $u_{i,j}$ | 2 | Differentiation |
| $u_{i,i} = u_{1,1} + u_{2,2} + u_{3,3}$ | 0 | Divergence, $\nabla \cdot \mathbf{U}$ |

### 1.5.5.6. EXAMPLES

Suppose that $c$ and $d$ are scalars, $u_i$ or $v_i$ are the three components of a vector, and $a_{ij}$ are the nine components of a second-order tensor. Then we have the results given in Table 1.5.

### 1.5.6. Isotropic Tensors

A tensor is isotropic if its components have the same value in all coordinate systems. A scalar (tensor of order zero) is a simple example. The tensor $\delta_{ij}$ is isotropic. For $\delta_{ij}$, the transformation rule yields

$$\delta'_{ij} = l_{ir} l_{js} \delta_{rs} = l_{ir} l_{jr} = \delta_{ij} \tag{1.86}$$

which is the definition for the second-order isotropic tensor.

It can be shown that any second-order isotropic tensor must be of the form of a constant times $\delta_{ij}$, and the most general fourth-order isotropic tensor has the form:

$$a_{ijkl} = \alpha \delta_{ij} \delta_{kl} + \beta \delta_{ik} \delta_{jl} + \gamma \delta_{il} \delta_{jk} \tag{1.87}$$

EXAMPLE 1.5. If $\phi$ is a scalar, show the following:

(a) $\phi_{,i}$ is a first-order tensor.
(b) $\phi_{,ij}$ is a second-order tensor.
(c) $\phi_{,kk}$ is a scalar (zero-order tensor).

SOLUTION. Since $\phi$ is a scalar,

$$\phi(\text{in } x_i \text{ system}) = \phi'(\text{in } x'_i \text{ system}) \tag{1.88}$$

(a) Define

$$G_i = \frac{\partial \phi}{\partial x_i} = \phi_{,i} \tag{1.89}$$

Thus,

$$G'_i = \frac{\partial \phi'}{\partial x'_i} = \frac{\partial \phi}{\partial x'_i} = \frac{\partial \phi}{\partial x_j} \frac{\partial x_j}{\partial x'_i} \tag{1.90}$$

From Eqs. (1.74) and (1.90),

$$G'_i = l_{ij} G_j$$

or

$$\phi'_{,i} = l_{ij} \phi_{,j} \quad (i \text{ is a free index}) \tag{1.91}$$

Hence $\phi_{,i}$ is a first-order tensor.

(b) Define

$$c_{ij} = \frac{\partial^2 \phi}{\partial x_i \partial x_j} = \phi_{,ij} \tag{1.92}$$

Thus,

$$c'_{ij} = \frac{\partial^2 \phi'}{\partial x'_i \partial x'_j} = \frac{\partial}{\partial x'_i}\left(\frac{\partial \phi}{\partial x_k} \frac{\partial x_k}{\partial x'_j}\right)$$

or

$$c'_{ij} = \frac{\partial}{\partial x'_i}\left(\frac{\partial \phi}{\partial x_k} l_{jk}\right)$$

$$= \frac{\partial}{\partial x_m}\left(\frac{\partial \phi}{\partial x_k} l_{jk}\right)\frac{\partial x_m}{\partial x'_i}$$

Hence,

$$c'_{ij} = \frac{\partial^2 \phi}{\partial x_m \partial x_k} l_{jk} l_{im} \tag{1.93}$$

or, using Eqs. (1.92) and (1.93), we get

$$\phi'_{,ij} = l_{im} l_{jk} \phi_{,mk} \tag{1.94}$$

That is, $\phi_{,ij}$ is a second-order tensor.

(c) Replacing subscript $j$ by $i$ in Eq. (1.94), we have

$$\phi'_{,ii} = l_{im} l_{ik} \phi_{,mk} \tag{1.95}$$

But from Eq. (1.65)

$$l_{im} l_{ik} = \delta_{mk} \tag{1.96}$$

Substituting Eq. (1.95) into Eq. (1.96), we get

$$\phi'_{,ii} = \delta_{mk} \phi_{,mk}$$

or

$$\phi'_{,ii} = \phi_{,mm}$$

Hence, $\phi_{,ii}$ is a scalar (zero-order tensor).

## 1.6. Summary

The theory of plasticity is concerned with the analysis of stresses and strains in the plastic range of ductile materials, especially metals. This chapter introduced the fundamental concepts of plasticity theory by discussing the uniaxial stress–strain behavior of metals. Important concepts, such as elastic deformation, plastic deformation, yielding, plastic flow, hardening, softening, and the Bauschinger effect for reversed loading have all been illustrated. The most significant feature of plastic deformation is its *irreversibility* and *load path dependence*. For a hardening material, a hardening parameter relating to plastic work or plastic strain was introduced to record the history of loading. The plastic modulus is then related to this hardening parameter. The general characteristics of materials discussed in this chapter in terms of the uniaxial stress–strain behavior of metals can be used in the later chapters to extract rather far-reaching information regarding the two-dimensional and three-dimensional stress–strain relations of materials in general and metals and concretes in particular.

The relationship between stress and strain in general loading cases characterizing the properties of a material is generally referred to in the open literature as the *constitutive relation*. The following chapters are concerned with the general techniques used in the necessary extension of stress–strain behavior in uniaxial conditions to the three-dimensional situation. Generally, six stress components and six strain components are involved in constitutive equations. To simplify the mathematical expressions, we shall use index notations. A concise introduction to tensor notations in preparation for the subsequent discussion has also been given in this chapter.

### References

Chen, W.F., and G.Y. Baladi, 1985. *Soil Plasticity: Theory and Implementation*, Elsevier, Amsterdam, The Netherlands.

Chen, W.F., 1982. *Plasticity in Reinforced Concrete*, McGraw-Hill, New York.

Chen, W.F., and A.F. Saleeb, 1982. *Constitutive Equations for Engineering Materials, Volume 1: Elasticity and Modeling*, Wiley-Interscience, New York.

### PROBLEMS

1.1. The $\sigma$–$\epsilon$ response in simple tension for a material is approximated by the following form of the Ramberg–Osgood formula

$$\epsilon = \epsilon^e + \epsilon^p = \frac{\sigma}{E} + \left(\frac{\sigma}{b}\right)^n$$

(a) Find the tangent modulus $E_t$ and the plastic modulus $E_p$ as functions of stress $\sigma$ and of plastic strain $\epsilon^p$.

(b) Find the plastic work $W_p$ as a function of stress $\sigma$ and of plastic strain $\epsilon^p$.

(c) Express the stress $\sigma$ and the plastic modulus $E_p$ in terms of the plastic work $W_p$.

(d) What is the initial yield stress?

(e) Assuming $n = 1$, sketch the $\sigma$-$\epsilon$ curve for loading followed by a complete unloading.

(f) Assuming $n = 5$, find the offset tensile stresses for the permanent offsets $\epsilon^P = 0.1\%$ and $\epsilon^r = 0.2\%$, respectively.

1.2. For the material of Problem 1.1, assume $n = 4$, $E = 73,000$ MPa, and $b = 800$ MPa. A material element is prestrained in tension up to a state with $\epsilon^r = 0.015$ and is subsequently unloaded and then reversely loaded until plastic flow in compression commences; further compressive yielding continues until $\epsilon^P = -0.015$. The material is assumed to follow: (i) the isotropic hardening rule; (ii) the independent hardening rule, both with the plastic modulus $E_p$ taken to depend on a single hardening modulus $\kappa$ defined as $\kappa = \int (d\epsilon^P \, d\epsilon^P)^{1/2}$.

(a) Find the stress at the initiation of compressive yielding.
(b) Sketch the $\sigma$-$\epsilon^P$ curve.

1.3. For the material of Problem 1.1, assume $n = 3$, $E = 69,000$ MPa, and $b = 690$ MPa. A material element is firstly strained in tension up to a State 1 with $W_p = 113.85$ kN $\cdot$ m/m$^3$ and is subsequently unloaded and reversely loaded until plastic flow in compression commences at State 2. Further, it is loaded with a stress increment $d\sigma = -2.07$ MPa up to State 3, and then with another stress increment $d\sigma = -2.07$ MPa up to State 4. After that, the element is unloaded and loaded in tension again until plastic flow occurs at State 5. The material is assumed to follow the isotropic hardening rule with the plastic modulus $E_p$ taken to depend on a hardening parameter $\kappa$ defined as $\kappa = W_p$.

(a) Find the tensile stress $\sigma_1$ and the plastic strain $\epsilon_1^P$ at State 1.
(b) Find the stress $\sigma$, strain $\epsilon$, plastic strain $\epsilon^P$, plastic work $W_p$, and plastic modulus $E_p$ at States 2, 3, and 4, respectively.
(c) Find the stress $\sigma_5$ and plastic modulus $E_p$ at State 5.

1.4. For the overlay material model of Example 1.1 (see Fig. 1.7), assume that the material parameters are selected as $A_1 = \frac{2}{3}$, $A_2 = \frac{1}{3}$, $\sigma_{01} = 138$ MPa, $\sigma_{02} = 345$ MPa, and $E = 69,000$ MPa. The strains at points $c$ and $f$ in Fig. 1.8 are taken to be $\epsilon_c = 0.013$ and $\epsilon_f = 0.011$, and State $h$ is assumed to correspond to a compressive stress in bars 2 of value $\sigma_{02}/2$.

(a) Find the residual stresses in bars 1 and 2 when $\sigma = 0$ along the unloading path $f$-$g$ and during reloading along path $h$-$i$?
(b) Determine the stress in bars 2, corresponding to States $g$ and $i$.
(c) What are the values of the stress in bars 2, $\sigma_2$, and the strain $\epsilon$ when stress in bars 1 is completely relieved (i.e., $\sigma_1 = 0$) during unloading along path $f$-$g$ and during reloading along path $h$-$i$.
(d) For the $\sigma$-$\epsilon$ paths in Fig. 1.8, plot bar stresses $\sigma_1$ vs. $\sigma_2$, showing the line of equivalent stress $\sigma = 0$.

1.5. An initially unstressed and unstrained element of the same linear strain-hardening material as in Example 1.2 is subjected to different loading histories which produce the stress paths given below. For each of the three hardening rules considered in Example 1.2, find the final strain state, $\epsilon$, and the corresponding $\epsilon^P$ attained at the end of each loading path. In the following, stress $\sigma$ is in MPa.

(i) $\sigma = 0 \rightarrow 414 \rightarrow -414 \rightarrow 0 \rightarrow 414$
(ii) $\sigma = 0 \rightarrow 621 \rightarrow 0$

For each case, show schematic representations of the stress–strain paths followed in the $\sigma-\epsilon$ and $\sigma-\epsilon^p$ spaces.

1.6. The $\sigma-\epsilon$ response in simple tension for an elastic–plastic material is approximated by the piecewise linear curve expressed as

| | | |
|---|---|---|
| Elastic | $\sigma = E\epsilon$ | $(\epsilon < \epsilon_0)$ |
| Elastic–plastic | $\sigma = \sigma_0 + E_{t_1}(\epsilon - \epsilon_0)$ | $(\epsilon_0 \leq \epsilon \leq \epsilon_1)$ |
| | $\sigma = \sigma_1 + E_{t_2}(\epsilon - \epsilon_1)$ | $(\epsilon_1 \leq \epsilon \leq \epsilon_2)$ |
| Perfectly plastic | $\sigma = \sigma_2$ | $(\epsilon > \epsilon_2)$ |

where the material constants are given as $\sigma_0 = 207$ MPa, $\epsilon_0 = 0.001$; $\sigma_1 = 414$ MPa, $\epsilon_1 = 0.005$; $\sigma_2 = 587$ MPa, $\epsilon_2 = 0.013$. An element of the material is prestrained in tension up to a State $A$ with $\epsilon_A = 0.015$ and is subsequently unloaded until plastic flow in compression commences at State $C$. Further compressive yielding continues until $\epsilon^p = 0$. The material is assumed to harden kinematically, with the plastic modulus taken to depend on a single hardening parameter $\kappa$ as defined below.

(a) Sketch the loading–unloading–reverse loading $\sigma-\epsilon^p$ curves for each of the following assumptions: (i) $\kappa = \int (d\epsilon^p \, d\epsilon^p)^{1/2}$; (ii) $\kappa = \epsilon^p$ (for $\epsilon^p \geq 0$); (iii) $\kappa = \int \sigma \, d\epsilon^p$; (iv) $\kappa = \int (\sigma - \alpha) \, d\epsilon^p$ where $\alpha$ (in stress units) is the center of the current elastic region.

(b) What are the values $\sigma$ and $\epsilon$ when $\epsilon^p = 0$ during reversed (compression) flow, for each of the four assumptions in (a).

1.7. A bar with two fixed ends is subjected to an axial force $P$ at the point with the left-end distance equal to $a$ and the right-end distance equal to $b$ and $a < b$, as shown in Fig. P1.7. The bar is made of an elastic-perfectly plastic material with yield stress $\sigma_y$. The axial force $P$ is first increased from zero until plastic flow occurs in the entire bar, and then is unloaded to zero, followed by a reloading in the reverse direction.

(a) Determine the elastic and plastic limit loads $P_e$ and $P_p$ during the loading.

(b) Determine the residual stress and plastic strain in the bar when the axial load $P$ is unloaded to zero.

(c) Determine the plastic limit load $P_p$ during the reversed loading.

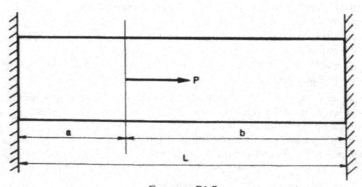

FIGURE P1.7.

(d) Sketch the $P$ vs. $u$ curve for the complete load-reversed load cycle for the case $b = 2a$, where $u$ is the axial displacement of the bar at the load point.

1.8. Using the table of direction cosines ($l_{ij}$) as given in Example 1.4 (Table 1.4), show that the following two planes coincide:

$$2x_1 - \tfrac{1}{3}x_2 + x_3 = 1 \qquad \text{in } (x_i) \text{ system}$$

$$\tfrac{47}{25}x_1' + \tfrac{14}{15}x_2' - \tfrac{21}{25}x_3' = 1 \qquad \text{in } (x_i') \text{ system}$$

1.9. If $B_i = A_i/\sqrt{(A_jA_j)}$, show that $B_i$ is a unit vector.

1.10. Given the relations

$$\sigma_{ij} = s_{ij} + \tfrac{1}{3}\sigma_{kk}\delta_{ij}$$

$$J_2 = \tfrac{1}{2}s_{ij}s_{ji}$$

where $\sigma_{ij}$ and $s_{ij}$ are symmetric second-order tensors, show that (a) $s_{ii} = 0$ and (b) $\partial J_2/\partial \sigma_{ij} = s_{ij}$.

1.11. Prove that there is no pair of vectors $A_i$ and $B_i$ such that $\delta_{ij} = A_iB_j$.

1.12. Show that an arbitrary second-order tensor $\sigma_{ij}$ may be written in the form

$$\sigma_{ij} = s_{ij} + \alpha\delta_{ij}$$

where $s_{ii} = 0$.

1.13. Consider a truss consisting of three bars with cross-sectional area $A$ and subjected to a vertical load $P$ at joint $D$ as shown in Fig. P1.13. The bars are made of an elastic-perfectly plastic material with elastic modulus $E$ and yield stress $\sigma_0$. As $P$ is continuously increased, bar 2 is first yielded at State $a$ ($P_a$, $\Delta_a$), and subsequently bars 1 and 3 are yielded at State $b$ ($P_b$, $\Delta_b$). Plastic flow occurs at a constant load $P_b$ until joint $D$ has an additional amount of displacement $\sigma_0 L/E$ (State $c$); then $P$ is completely unloaded (State $d$). Afterward, $P$ is increased in a reversed direction until all three bars are reversely yielded again (State $f$).

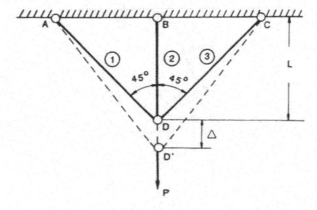

FIGURE P1.13.

(a) Find the residual stresses and residual strains at state $d$ for three truss bars.
(b) Find the reversed yield load.
(c) Plot the load vs. displacement ($P$ vs. $\Delta$) curve for the whole loading-unloading-reversed loading program.
(d) Referring to the parallel bar model of Example 1.1, what conclusion can be drawn from this truss model?

## Answers to Selected Problems

1.1. (d) 0; (f) $\sigma_{0.1\%} = 0.25b$, $\sigma_{0.2\%} = 0.289b$.

1.2. (a) (i) $\sigma_c = -280$ MPa; (ii) $\sigma_c = 0$.
(b) Equation of $\sigma$-$\epsilon^p$ curve on reversed loading is (i) $\sigma = -800(0.03 - \epsilon^p)^{1/4}$;
(ii) $\sigma = -800(0.03 - \epsilon^p)^{1/4} + 280$.

1.3. (a) $\sigma_1 = 84.07$ MPa, $\epsilon_1^p = 0.001806$.
(b) $\sigma_2 = -84.07$ MPa, $\epsilon_2 = 0.000588$, $\epsilon_2^p = \epsilon_1^p$, $(W_p)_2 = 113.85$ kN · m/m$^3$, $(E_p)_2 = 15,504$ MPa; $\sigma_3 = -86.15$ MPa, $\epsilon_3 = 0.000424$, $\epsilon_3^p = 0.001672$, $(W_p)_3 = 125.3$ kN · m/m$^3$, $(E_p)_3 = 14,793$ MPa; $\sigma_4 = -88.22$ MPa, $\epsilon_4 = 0.000254$, $\epsilon_4^p = 0.001532$, $(W_p)_4 = 137.5$ kN · m/m$^3$, $(E_p)_4 = 14,117$ MPa.
(c) $\sigma_5 = 88.22$ MPa, $(E_p)_5 = 14,117$ MPa.

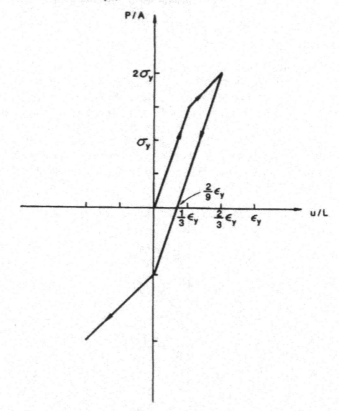

Figure S1.7.

1.4. (a) Path $c$-$d$ and Path $f$-$g$: $\sigma_1 = -69$ MPa, $\sigma_2 = 138$ MPa; Path $h$-$i$: $\sigma_1 = 11.5$ MPa, $\sigma_2 = -23$ MPa.
   (b) At Point $g$: $\sigma_2 = 69$ MPa; at Point $i$: $\sigma_2 = 103.5$ MPa.
   (c) For Path $f$-$g$: $\sigma_2 = 207$ MPa; for Path $h$-$i$: $\sigma_2 = -34.5$ MPa.

1.5. Isotropic hardening: (i) $\epsilon = 0 \rightarrow 0.01 \rightarrow 0.006 \rightarrow 0.008 \rightarrow 0.01$, $\epsilon^p = 0 \rightarrow 0.008 \rightarrow 0.008 \rightarrow 0.008 \rightarrow 0.008$; (ii) $\epsilon = 0 \rightarrow 0.019 \rightarrow 0.016$, $\epsilon^p = 0 \rightarrow 0.016 \rightarrow 0.016$.
   Kinematic hardening: (i) $\epsilon = 0 \rightarrow 0.01 \rightarrow -0.01 \rightarrow -0.008 \rightarrow 0.01$, $\epsilon^p = 0 \rightarrow 0.008 \rightarrow -0.008 \rightarrow -0.008 \rightarrow 0.008$; (ii) $\epsilon = 0 \rightarrow 0.019 \rightarrow 0.008$, $\epsilon^p = 0 \rightarrow 0.016 \rightarrow 0.008$.
   Independent hardening: (i) $\epsilon = 0 \rightarrow 0.01 \rightarrow -0.002 \rightarrow 0 \rightarrow 0.002$, $\epsilon^p = 0 \rightarrow 0.008 \rightarrow 0 \rightarrow 0 \rightarrow 0$; (ii) $\epsilon = 0 \rightarrow 0.019 \rightarrow 0.016$, $\epsilon^p = 0 \rightarrow 0.016 \rightarrow 0.016$.

1.6. (b) (i) $\sigma = 172.6$ MPa, $\epsilon = 0.0008$; (ii) $\sigma = -207.1$ MPa, $\epsilon = -0.001$; (iii) $\sigma = 43.3$ MPa, $\epsilon = 0.0002$; (iv) same as (i).

1.7. Figure S1.7.

1.13. Figure S1.13.

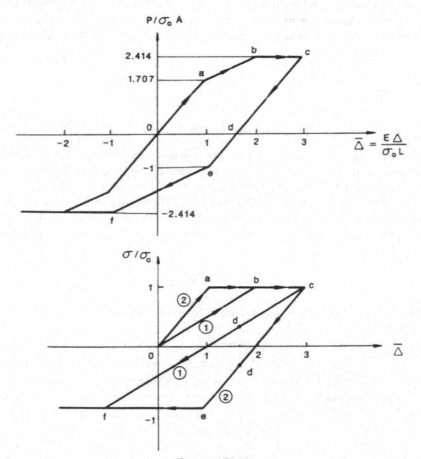

FIGURE S1.13.

# 2
# Yield and Failure Criteria

In Chapter 1, we have discussed the characteristics of the uniaxial behavior of a material, introducing some important concepts in plasticity theory for the uniaxial case. The main task of the subsequent chapters is to generalize these concepts to a combined state of stress. This chapter deals with the limits of elasticity and the limits of strength under all possible combinations of stresses. Before we proceed to this subject, an analysis of the state of combined stresses is first introduced to provide the necessary background for the subsequent study.

## 2.1. Stress

### 2.1.1. Stress at a Point and the Stress Tensor

As we know, stress is defined as the intensity of internal forces acting between particles of a body across imaginary internal surfaces. Consider a surface area $\Delta A$ passing through a point $P_0$ with a unit vector $\mathbf{n}$ normal to the area $\Delta A$ as shown in Fig. 2.1. Let $\mathbf{F}_n$ be the resultant force due to the action across the area $\Delta A$ of the material from one side onto the other side of the cut plane $\mathbf{n}$. Then the stress vector at point $P_0$ associated with the cut plane $\mathbf{n}$ is defined by

$$\lim_{\Delta A \to 0} \frac{\mathbf{F}_n}{\Delta A} = \mathbf{T} = \overset{n}{\mathbf{T}} \tag{2.1}$$

The state of stress at a point is defined as the totality of *all* stress vectors $\overset{n}{\mathbf{T}}$ at that point.

Since we can make an infinite number of cuts through a point, we have an infinite number of values of $\overset{n}{\mathbf{T}}$ which, in general, are different from each other. This infinite number of values of $\overset{n}{\mathbf{T}}$ characterizes the state of stress (or the stress state) at the point. Fortunately, as shown later, there is no need to know all the values of the stress vectors on the infinite numbers of planes containing the point. If the stress vectors $\overset{1}{\mathbf{T}}, \overset{2}{\mathbf{T}}$, and $\overset{3}{\mathbf{T}}$ on three

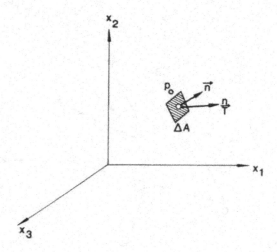

FIGURE 2.1. Stress vector $\overset{n}{\vec{T}}$ at point $P_0$ associated with cut **n**.

mutually perpendicular planes are known, as shown in Fig. 2.2, the stress vector on any plane containing this point can be found from equilibrium conditions at that point.

Figure 2.3 shows an element $OABC$ with the stress vectors $\overset{-1}{\vec{T}}, \overset{-2}{\vec{T}}, \overset{-3}{\vec{T}}$, and $\overset{n}{\vec{T}}$ acting on its faces $OBC$, $OAC$, $OAB$, and $ABC$, respectively. Stress vector $\overset{-1}{\vec{T}}(\overset{-2}{\vec{T}}, \overset{-3}{\vec{T}})$ represents the stress acting across the cut plane normal to unit vector $e_1(e_2, e_3)$ from the negative side onto the positive side.

The unit vector **n** can be written in the component form

$$\mathbf{n} = (n_1, n_2, n_3) \tag{2.2}$$

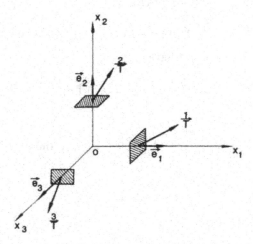

FIGURE 2.2. Stress vectors on three mutually perpendicular planes at a point.

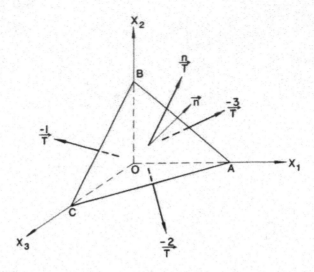

FIGURE 2.3. Stress vectors acting on arbitrary plane **n** and on the coordinate planes.

where the direction cosines $n_1$, $n_2$, and $n_3$ are given by

$$n_1 = \cos(\mathbf{e}_1, \mathbf{n})$$
$$n_2 = \cos(\mathbf{e}_2, \mathbf{n})$$ 
$$n_3 = \cos(\mathbf{e}_3, \mathbf{n})$$

(2.3)

Let $A$ be the area of $\triangle ABC$. Then the area of the face perpendicular to the $x_i$-axis, denoted by $A_i$, is given by

$$A_i = A \cos(\mathbf{e}_i, \mathbf{n}) = A n_i$$

(2.4)

From equilibrium of the body $OABC$ (Fig. 2.3) and using Eq. (2.4), we get

$$\overset{n}{\mathbf{T}}(A) + \overset{-1}{\mathbf{T}}(An_1) + \overset{-2}{\mathbf{T}}(An_2) + \overset{-3}{\mathbf{T}}(An_3) = 0$$

(2.5)

Dividing Eq. (2.5) by $A$, we have

$$\overset{n}{\mathbf{T}} = -\overset{-1}{\mathbf{T}} n_1 - \overset{-2}{\mathbf{T}} n_2 - \overset{-3}{\mathbf{T}} n_3$$

(2.6)

But

$$\overset{-i}{\mathbf{T}} = -\overset{i}{\mathbf{T}} \quad \text{for } i = 1, 2, \text{ and } 3$$

Therefore,

$$\overset{n}{\mathbf{T}} = \overset{1}{\mathbf{T}} n_1 + \overset{2}{\mathbf{T}} n_2 + \overset{3}{\mathbf{T}} n_3$$

(2.7)

or in the $xyz$ coordinate system,

$$\overset{n}{\mathbf{T}} = \overset{x}{\mathbf{T}} n_x + \overset{y}{\mathbf{T}} n_y + \overset{z}{\mathbf{T}} n_z$$

(2.8)

Equation (2.7) or (2.8) expresses the stress vector $\overset{n}{\mathbf{T}}$ at any point associated with the cut plane **n** in terms of the stress vectors on the planes perpendicular to the three coordinate axes, $x_1$, $x_2$, and $x_3$, at the same point. Therefore, it is clear that the three stress vectors $\overset{1}{\mathbf{T}}$, $\overset{2}{\mathbf{T}}$, and $\overset{3}{\mathbf{T}}$ define the state of stress at a point completely.

The stress vector, $\overset{n}{\mathbf{T}}$, of course, need not be perpendicular to the plane on which it acts. In practice, therefore, the stress vector $\overset{n}{\mathbf{T}}$ is decomposed into two components, one normal to the plane **n**, called the *normal stress*, and the other parallel to this plane, called the *shearing stress*.

The stress vectors associated with each of the three coordinate planes $x_1$, $x_2$, and $x_3$ are also decomposed into components in the direction of the three coordinate axes. For example, the stress vector $\overset{1}{\mathbf{T}}$ associated with the coordinate plane $x_1$, has three stress components: normal stress $\sigma_{11}$, and shearing stresses $\sigma_{12}$ and $\sigma_{13}$ in the direction of the three coordinate axes $x_1$, $x_2$, and $x_3$, respectively, as shown in Fig. 2.4. Thus, we have

$$\overset{1}{\mathbf{T}} = \sigma_{11}\mathbf{e}_1 + \sigma_{12}\mathbf{e}_2 + \sigma_{13}\mathbf{e}_3 \tag{2.9}$$

$$\overset{1}{\mathbf{T}} = \sigma_{1j}\mathbf{e}_j \tag{2.10}$$

Similarly, for the coordinate planes $x_2$ and $x_3$,

$$\overset{2}{\mathbf{T}} = \sigma_{2j}\mathbf{e}_j \tag{2.11}$$

$$\overset{3}{\mathbf{T}} = \sigma_{3j}\mathbf{e}_j \tag{2.12}$$

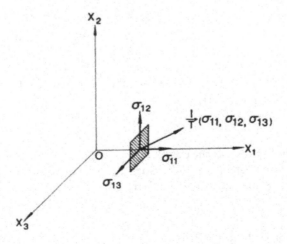

FIGURE 2.4. Components of a stress vector associated with the coordinate plane normal to $x_1$.

In general,

$$\overset{i}{T} = \sigma_{ij} e_j \tag{2.13}$$

where $\sigma_{ij}$ denotes the $j$-th component of the stress vector $\overset{i}{T}$ acting on an area element (at $P$) whose normal is in the direction of the positive $x_i$-axis (Fig. 2.4).

The nine quantities $\sigma_{ij}$ required to define the three stress vectors $\overset{1}{T}, \overset{2}{T}$, and $\overset{3}{T}$, are called the components of the *stress tensor*, which is given by

$$\sigma_{ij} = \begin{bmatrix} \overset{1}{T} \\ \overset{2}{T} \\ \overset{3}{T} \end{bmatrix} = \begin{bmatrix} \sigma_{11} & \sigma_{12} & \sigma_{13} \\ \sigma_{21} & \sigma_{22} & \sigma_{23} \\ \sigma_{31} & \sigma_{32} & \sigma_{33} \end{bmatrix} \tag{2.14}$$

where $\sigma_{11}$, $\sigma_{22}$, and $\sigma_{33}$ are normal components of stress and $\sigma_{12}$, $\sigma_{21}$, ... are shearing components of stress.

The components of the stress tensor can be written using von Karman's notation in the form

$$\sigma_{ij} = \begin{bmatrix} \sigma_x & \tau_{xy} & \tau_{xz} \\ \tau_{yx} & \sigma_y & \tau_{yz} \\ \tau_{zx} & \tau_{zy} & \sigma_z \end{bmatrix} \tag{2.15}$$

where $\sigma$ represents a normal component of stress, and $\tau$ represents a shearing component of stress. Also, the symbols $\sigma_{xx}, \sigma_{xy}, \ldots$ may be used instead of $\sigma_{ij}$ to designate the components of the stress tensor in Eqs. (2.14) and (2.15). Thus, the following forms are dual notations for the stress tensor $\sigma_{ij}$:

$$\sigma_{ij} = \begin{bmatrix} \sigma_{11} & \sigma_{12} & \sigma_{13} \\ \sigma_{21} & \sigma_{22} & \sigma_{23} \\ \sigma_{31} & \sigma_{32} & \sigma_{33} \end{bmatrix} = \begin{bmatrix} \sigma_{xx} & \sigma_{xy} & \sigma_{xz} \\ \sigma_{yx} & \sigma_{yy} & \sigma_{yz} \\ \sigma_{zx} & \sigma_{zy} & \sigma_{zz} \end{bmatrix} = \begin{bmatrix} \sigma_x & \tau_{xy} & \tau_{xz} \\ \tau_{yx} & \sigma_y & \tau_{yz} \\ \tau_{zx} & \tau_{zy} & \sigma_z \end{bmatrix} \tag{2.16}$$

Upon substitution of Eq. (2.13) into Eq. (2.7), the components of the stress vector $\overset{n}{T}$ can be written as

$$\overset{n}{T}_i = \sigma_{ji} n_j \tag{2.17}$$

From the consideration of the equilibrium of moments of a material element, it can be shown that the stress tensor, $\sigma_{ij}$, is symmetric, that is, $\sigma_{ij} = \sigma_{ji}$. Thus, Eq. (2.17) may be conveniently rewritten as

$$\overset{n}{T}_i = \sigma_{ij} n_j \qquad i = 1, 2, 3 \tag{2.18}$$

where $\sigma_{ij}$ is given by Eq. (2.16).

Equation (2.18) expresses the components of the stress vector acting on an arbitrary plane **n** at a given point in terms of the components of the stress tensor, $\sigma_{ij}$, at that point. It therefore follows that $T_i$ for any $n_i$ may be calculated from a knowledge of the nine basic quantities $\sigma_{ij}$.

In Eq. (2.18), $\overset{n}{T_i}$ and $n$, are vectors. From this equation, we can show that $\sigma_{ij}$ is a tensor of second order; that is, the stress components $\sigma_{ij}$ in the $x_i$ system and the components $\sigma'_{ij}$ in the $x'_i$ system are related by following equations:

$$\sigma'_{ij} = l_{im}l_{jn}\sigma_{mn} \tag{2.19}$$

and

$$\sigma_{ij} = l_{mi}l_{nj}\sigma'_{mn} \tag{2.20}$$

where the $l_{ij}$ are the direction cosines shown in Table 1.3.

## 2.1.2. Cauchy's Formulas for Stresses

Equations (2.7) and (2.18) derived in the preceding section are different forms of Cauchy's formulas for stresses. In practice, however, it is desirable to express directly the normal and shear stress components, $\sigma_n$ and $S_n$, respectively, of any stress vector $\overset{n}{T}$ acting on an arbitrary plane $n$ at a given point in terms of the components of the stress tensor $\sigma_{ij}$ at that point. The magnitude of the normal stress component is given by

$$\sigma_n = \overset{n}{T} \cdot n = \overset{n}{T_i} n_i \tag{2.21}$$

Substituting from Eq. (2.18) for $\overset{n}{T_i}$, Eq. (2.21) then becomes

$$\sigma_n = \sigma_{ij} n_i n_j \tag{2.22}$$

The magnitude of the shear stress component is given by

$$S_n^2 = (\overset{n}{T})^2 - \sigma_n^2 \tag{2.23}$$

where, from Eq. (2.18), $(\overset{n}{T})^2$ is obtained as

$$(\overset{n}{T})^2 = \overset{n}{T} \cdot \overset{n}{T} = \overset{n}{T_i}\overset{n}{T_i} = (\sigma_{ij}n_j)(\sigma_{ik}n_k) \tag{2.24}$$

or

$$(\overset{n}{T})^2 = \sigma_{ij}\sigma_{ik}n_jn_k \tag{2.25}$$

Equations (2.22) and (2.23), for the determination of the normal and shearing components of stress acting on an arbitrary plane $n$, are the most useful forms of Cauchy's formulas for stresses.

The vector $\sigma_n$ is in the direction of the normal vector $n$, and the vector $S_n$ lies in the plane formed by the two vectors $\overset{n}{T}$ and $n$.

EXAMPLE 2.1. The state of stress at a point is represented by the given stress tensor $\sigma_{ij}$:

$$\sigma_{ij} = \begin{bmatrix} -80 & 16 & 26 \\ 16 & 26 & -28 \\ 26 & -28 & -36 \end{bmatrix} \text{(units of stress)}$$

For a plane with unit normal $\mathbf{n} = (\frac{1}{4}, \frac{1}{2}, \sqrt{11}/4)$, calculate:

(a) The magnitude of the stress vector, $\overset{n}{\mathbf{T}}$, for plane $\mathbf{n}$.
(b) The normal and shear stress components, $\sigma_n$ and $S_n$, for plane $\mathbf{n}$.

SOLUTION. (a) The components $\overset{n}{T_i}$ of the stress vector $\overset{n}{\mathbf{T}}$ are calculated using Eq. (2.18), which gives

$$\overset{n}{T_1} = \sigma_{1j}n_j = \sigma_{11}n_1 + \sigma_{12}n_2 + \sigma_{13}n_3 = 9.56$$

Similarly,

$$\overset{n}{T_2} = \sigma_{2j}n_j = -6.22$$

$$\overset{n}{T_3} = \sigma_{3j}n_j = -37.35$$

Thus, the magnitude of the stress vector $\overset{n}{\mathbf{T}}$ is given by

$$|\overset{n}{\mathbf{T}}| = [(\overset{n}{T_1})^2 + (\overset{n}{T_2})^2 + (\overset{n}{T_3})^2]^{1/2} = 39.10$$

(b) Substituting into Eq. (2.22), we get

$$\sigma_n = \sigma_{ij}n_in_j = \sigma_{11}n_1^2 + \sigma_{22}n_2^2 + \sigma_{33}n_3^2 + 2(\sigma_{12}n_1n_2 + \sigma_{23}n_2n_3 + \sigma_{31}n_3n_1)$$

or

$$\sigma_n = -31.69$$

Thus, the magnitude of the shear stress component, $S_n$, is calculated using Eq. (2.23),

$$|S_n| = [(\overset{n}{\mathbf{T}})^2 - \sigma_n^2]^{1/2} = [(39.10)^2 - (-31.69)^2]^{1/2} = 22.90$$

## 2.1.3. Principal Stresses and Invariants of the Stress Tensor

Suppose that the direction $\mathbf{n}$ at a point in a body is so oriented that the resultant stress, stress vector $\overset{n}{\mathbf{T}}$, associated with direction is in the same direction as the unit normal $\mathbf{n}$; that is, $\overset{n}{\mathbf{T}} = \sigma_n$ and $S_n = 0$ (no shear stress). The plane $\mathbf{n}$ is then called a *principal plane* at the point, its normal direction $\mathbf{n}$ is called a *principal direction*, and the normal stress $\sigma_n$ is called a *principal stress*. At every point in a body, there exist at least three principal directions. From the definition, we have

$$\overset{n}{\mathbf{T}} = \sigma\mathbf{n} \tag{2.26}$$

or in component form

$$\overset{n}{T_i} = \sigma n_i \tag{2.27}$$

Substituting for $\overset{n}{T_i}$ from Eq. (2.18) leads to

$$\sigma_{ij} n_j = \sigma n_i \tag{2.28}$$

which implies the following three equations:

$$\sigma_{11} n_1 + \sigma_{12} n_2 + \sigma_{13} n_3 = \sigma n_1$$
$$\sigma_{21} n_1 + \sigma_{22} n_2 + \sigma_{23} n_3 = \sigma n_2 \tag{2.29}$$
$$\sigma_{31} n_1 + \sigma_{32} n_2 + \sigma_{33} n_3 = \sigma n_3$$

or in von Karman's notation

$$(\sigma_x - \sigma) n_x + \tau_{xy} n_y + \tau_{xz} n_z = 0$$
$$\tau_{yx} n_x + (\sigma_y - \sigma) n_y + \tau_{yz} n_z = 0 \tag{2.30}$$
$$\tau_{zx} n_x + \tau_{zy} n_y + (\sigma_z - \sigma) n_z = 0$$

These three linear simultaneous equations are homogeneous for $n_x$, $n_y$, and $n_z$. In order to have a nontrivial solution, the determinant of the coefficients must vanish:

$$\begin{vmatrix} \sigma_x - \sigma & \tau_{xy} & \tau_{xz} \\ \tau_{yx} & \sigma_y - \sigma & \tau_{yz} \\ \tau_{zx} & \tau_{zy} & \sigma_z - \sigma \end{vmatrix} = 0 \tag{2.31}$$

so that this requirement determines the value of $\sigma$. There are, in general, three roots, $\sigma_1$, $\sigma_2$, and $\sigma_3$. Since the basic equation was $\overset{n}{T_i} = \sigma n_i$, these three possible values of $\sigma$ are the three possible magnitudes of the normal stress corresponding to zero shear stress. In the abbreviated notation, Eqs. (2.30) and (2.31) have the forms

$$(\sigma_{ij} - \sigma \delta_{ij}) n_j = 0 \tag{2.32}$$

and

$$|\sigma_{ij} - \sigma \delta_{ij}| = 0 \tag{2.33}$$

Expanding Eq. (2.31) leads to the *characteristic equation*

$$\sigma^3 - I_1 \sigma^2 + I_2 \sigma - I_3 = 0 \tag{2.34}$$

where

$$I_1 = \text{sum of the diagonal terms of } \sigma_{ij}$$

or

$$I_1 = \sigma_{11} + \sigma_{22} + \sigma_{33} = \sigma_x + \sigma_y + \sigma_z \tag{2.35}$$

$$I_2 = \text{sum of the cofactors of diagonal terms of } \sigma_{ij}$$

or

$$I_2 = \begin{vmatrix} \sigma_{22} & \sigma_{23} \\ \sigma_{32} & \sigma_{33} \end{vmatrix} + \begin{vmatrix} \sigma_{11} & \sigma_{13} \\ \sigma_{31} & \sigma_{33} \end{vmatrix} + \begin{vmatrix} \sigma_{11} & \sigma_{12} \\ \sigma_{21} & \sigma_{22} \end{vmatrix}$$

$$= \begin{vmatrix} \sigma_y & \tau_{yz} \\ \tau_{zy} & \sigma_z \end{vmatrix} + \begin{vmatrix} \sigma_x & \tau_{xz} \\ \tau_{zx} & \sigma_z \end{vmatrix} + \begin{vmatrix} \sigma_x & \tau_{xy} \\ \tau_{yx} & \sigma_y \end{vmatrix} \tag{2.36}$$

$I_3 = $ determinant of $\sigma_{ij}$

or

$$I_3 = \begin{vmatrix} \sigma_{11} & \sigma_{12} & \sigma_{13} \\ \sigma_{21} & \sigma_{22} & \sigma_{23} \\ \sigma_{31} & \sigma_{32} & \sigma_{33} \end{vmatrix} = \begin{vmatrix} \sigma_x & \tau_{xy} & \tau_{xz} \\ \tau_{yx} & \sigma_y & \tau_{yz} \\ \tau_{zx} & \tau_{zy} & \sigma_z \end{vmatrix} \tag{2.37}$$

From the property of the roots of a cubic equation, it can be shown that [refer to Eq. (2.34)],

$$I_1 = \sigma_1 + \sigma_2 + \sigma_3$$
$$I_2 = \sigma_1\sigma_2 + \sigma_2\sigma_3 + \sigma_3\sigma_1 \tag{2.38}$$
$$I_3 = \sigma_1\sigma_2\sigma_3$$

where $\sigma_1$, $\sigma_2$, and $\sigma_3$ are the roots of Eq. (2.34).

The cubic Eq. (2.34) must therefore be the same whether we derive it from $x$, $y$, $z$ coordinates or from the principal directions 1, 2, 3. Hence quantities $I_1$, $I_2$, and $I_3$ are the *invariants of the stress tensor*; that is, their values would be the same regardless of rotation of the coordinate axes.

Substituting $\sigma_1$, $\sigma_2$, and $\sigma_3$ in turn into Eq. (2.32), and also employing the identity

$$n_1^2 + n_2^2 + n_3^2 = 1 \tag{2.39}$$

we can determine the components $(n_1, n_2, n_3)$ of the unit normal $n_i$ corresponding to each value of $\sigma$ (principal stress),

$$\mathbf{n}^{(1)} = (n_1^{(1)}, n_2^{(1)}, n_3^{(1)}) \qquad \text{for } \sigma = \sigma_1$$
$$\mathbf{n}^{(2)} = (n_1^{(2)}, n_2^{(2)}, n_3^{(2)}) \qquad \text{for } \sigma = \sigma_2 \tag{2.40}$$
$$\mathbf{n}^{(3)} = (n_1^{(3)}, n_2^{(3)}, n_3^{(3)}) \qquad \text{for } \sigma = \sigma_3$$

These three directions are called *principal directions* at the point.

The need for Eq. (2.39) arises from the fact that when $\sigma$ in Eq. (2.32) is set equal to, say, $\sigma_1$, Eq. (2.33) implies from linear-algebra theory that at most two of the three equations (2.32) can be independent. It can be shown that if the three $\sigma$-roots are all different, *exactly* two of the three equations are independent. The special case in which two or more $\sigma$-roots coincide may subsequently be treated as a limiting case. In the meantime, we need only the resulting fact that whether two or only one of Eqs. (2.32) is independent, at least one solution $n_i^{(1)}$ satisfying Eqs. (2.32) and also Eq. (2.39) exists. Similarly, an $n_i^{(2)}$ corresponding to $\sigma_2$ and an $n_i^{(3)}$ corresponding to $\sigma_3$ may be found.

## 2.1.4. Principal Shear Stresses and Maximum Shear Stress

In describing the stress state at a point, let us take the principal axes 1, 2, and 3 as the reference axes instead of the general $x_1$, $x_2$, $x_3$ coordinate system. Note that on these coordinate planes 1, 2, and 3, all shear stresses are zero (Fig. 2.5). So, the magnitude of the stress acting on an arbitrary plane $n$ at this point as given by Eq. (2.25) turns out to be

$$(\overset{n}{T})^2 = \sigma_1^2 n_1^2 + \sigma_2^2 n_2^2 + \sigma_3^2 n_3^2 \tag{2.41}$$

Equation (2.22) gives the normal stress component as

$$\sigma_n = \sigma_1 n_1^2 + \sigma_2 n_2^2 + \sigma_3 n_3^2 \tag{2.42}$$

From Eq. (2.23), the magnitude of the shear stress component is expressed as

$$S_n^2 = (\overset{n}{T})^2 - \sigma_n^2 = (\sigma_1^2 n_1^2 + \sigma_2^2 n_2^2 + \sigma_3^2 n_3^2) - (\sigma_1 n_1^2 + \sigma_2 n_2^2 + \sigma_3 n_3^2)^2 \tag{2.43}$$

The condition for $n$ is given by

$$n_1^2 + n_2^2 + n_3^2 = 1 \tag{2.44}$$

From Eqs. (2.42) and (2.44), and eliminating $n_3$, we obtain $\sigma_n$ as a function of $n_1$ and $n_2$. For stationary values of $\sigma_n$, letting $\partial \sigma_n / \partial n_1 = 0$ and $\partial \sigma_n / \partial n_2 = 0$, we can show that $\sigma_n = \sigma_3$ is a stationary value. Similarly, we can also prove $\sigma_1$ and $\sigma_2$ are stationary values of the normal stress $\sigma_n$.

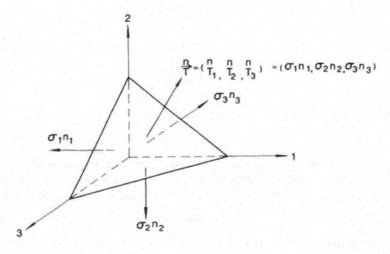

FIGURE 2.5. Components of stress on any plane referred to principal stress axes 1, 2, and 3.

Now, let us examine the stationary values of shear stress $S_n$. From Eqs. (2.43) and (2.44), and eliminating $n_3$, we obtain

$$S_n^2 = (\sigma_1^2 - \sigma_3^2)n_1^2 + (\sigma_2^2 - \sigma_3^2)n_2^2 + \sigma_3^2 - [(\sigma_1 - \sigma_3)n_1^2 + (\sigma_2 - \sigma_3)n_2^2 + \sigma_3]^2$$

(2.45)

Hence, for stationary values of $S_n$, we have

$$\frac{1}{2}\frac{\partial S_n^2}{\partial n_1} = (\sigma_1 - \sigma_3)n_1\{(\sigma_1 - \sigma_3) - 2[(\sigma_1 - \sigma_3)n_1^2 + (\sigma_2 - \sigma_3)n_2^2]\} = 0 \qquad (2.45a)$$

and

$$\frac{1}{2}\frac{\partial S_n^2}{\partial n_2} = (\sigma_2 - \sigma_3)n_2\{(\sigma_2 - \sigma_3) - 2[(\sigma_1 - \sigma_3)n_1^2 + (\sigma_2 - \sigma_3)n_2^2]\} = 0 \qquad (2.45b)$$

Assuming that $\sigma_1$, $\sigma_2$, and $\sigma_3$ are distinct and $\sigma_1 > \sigma_2 > \sigma_3$, we obtain the conditions that can satisfy Eqs. (2.45a,b), and Eq. (2.44) in the following:

(i)                         $$n_1 = n_2 = 0, \; n_3 = \pm 1 \qquad (2.46)$$

Equation (2.43) gives $S_n = 0$, a minimum value, and this shear stress component $S_n$ acts on the principal plane with the normal in the direction of axis 3.

(ii)            $$n_1 = 0, \; n_2 = \pm\frac{1}{\sqrt{2}}, \text{ and } n_3 = \pm\frac{1}{\sqrt{2}} \qquad (2.47)$$

These values define two planes passing through the principal axis of $\sigma_1$ at an angle of 45° to the principal axes of $\sigma_2$ and $\sigma_3$. The stationary value of $S_n$ in this case is given by

$$S_n^2 = \tfrac{1}{4}(\sigma_2 - \sigma_3)^2 \qquad (2.48)$$

or

$$|S_n| = \tfrac{1}{2}|\sigma_2 - \sigma_3| \qquad (2.49)$$

(iii)           $$n_2 = 0, \; n_1 = \pm\frac{1}{\sqrt{2}}, \text{ and } n_3 = \pm\frac{1}{\sqrt{2}} \qquad (2.50)$$

The stationary value of $S_n$ in this case is

$$|S_n| = \tfrac{1}{2}|\sigma_1 - \sigma_3| \qquad (2.51)$$

These values of $n_1$, $n_2$, and $n_3$ define two planes passing through the principal axis of $\sigma_2$ at an angle of 45° to the principal axes of $\sigma_1$ and $\sigma_3$. Similarly, we can determine another stationary value of shear stress $S_n$ given by

$$|S_n| = \tfrac{1}{2}|\sigma_1 - \sigma_2| \qquad (2.52)$$

This shear stress acts on planes passing through the principal axis of $\sigma_3$ at an angle of 45° to the principal axes of $\sigma_1$ and $\sigma_2$ ($n_1 = \pm 1/\sqrt{2}$, $n_2 = \pm 1/\sqrt{2}$, $n_3 = 0$).

Stationary values $\frac{1}{2}|\sigma_1-\sigma_2|$, $\frac{1}{2}|\sigma_2-\sigma_3|$, $\frac{1}{2}|\sigma_1-\sigma_3|$ are called *principal shear stresses* since they occur on planes which bisect the angle between principal planes. It should be noted that these principal shear planes are not pure shear planes; the normal stresses on the principal shear planes can be calculated using Eq. (2.41) and the corresponding values of $n_1$, $n_2$, and $n_3$. The largest value of the principal shear stresses, called the *maximum shearing stress*, $\tau_{max}$, is equal to $\frac{1}{2}|\sigma_1-\sigma_3|$ for $\sigma_1 > \sigma_2 > \sigma_3$.

## 2.1.5. Stress Deviation Tensor and Its Invariants

It is convenient in material modeling to decompose the stress tensor into two parts, one called the *spherical* or the *hydrostatic stress tensor* and the other called the *stress deviator tensor*. The hydrostatic stress tensor is the tensor whose elements are $p\delta_{ij}$, where $p$ is the mean stress and is given by

$$p = \tfrac{1}{3}\sigma_{kk} = \tfrac{1}{3}(\sigma_x + \sigma_y + \sigma_z) = \tfrac{1}{3}I_1 \tag{2.53}$$

From Eq. (2.53), it is apparent that $p$ is the same for all possible orientations of the axes; hence, it is called the *spherical* or the *hydrostatic stress*. The *stress deviator tensor* $s_{ij}$ is defined by subtracting the spherical state of stress from the actual state of stress. Therefore, we have

$$\sigma_{ij} = s_{ij} + p\delta_{ij} \tag{2.54}$$

$$s_{ij} = \sigma_{ij} - p\delta_{ij} \tag{2.55}$$

Equation (2.55) gives the required definition of the stress deviator tensor $s_{ij}$. The components of this tensor are given by

$$s_{ij} = \begin{bmatrix} s_{11} & s_{12} & s_{13} \\ s_{21} & s_{22} & s_{23} \\ s_{31} & s_{32} & s_{33} \end{bmatrix} = \begin{bmatrix} (\sigma_{11}-p) & \sigma_{12} & \sigma_{13} \\ \sigma_{21} & (\sigma_{22}-p) & \sigma_{23} \\ \sigma_{31} & \sigma_{32} & (\sigma_{33}-p) \end{bmatrix} \tag{2.56}$$

or, using von Karman's notation,

$$s_{ij} = \begin{bmatrix} s_x & s_{xy} & s_{xz} \\ s_{yx} & s_y & s_{yz} \\ s_{zx} & s_{zy} & s_z \end{bmatrix} = \begin{bmatrix} (\sigma_x-p) & \tau_{xy} & \tau_{xz} \\ \tau_{yx} & (\sigma_y-p) & \tau_{yz} \\ \tau_{zx} & \tau_{zy} & (\sigma_z-p) \end{bmatrix} \tag{2.57}$$

Note that $\delta_{ij}=0$ and $s_{ij}=\sigma_{ij}$ for $i \neq j$ in Eq. (2.55).

It is apparent that subtracting a constant normal stress in all directions does not change the principal directions. The principal directions are therefore the same for the stress deviator tensor as for the original stress tensor. In terms of the principal stresses, the stress deviator tensor $s_{ij}$ is

$$s_{ij} = \begin{bmatrix} \sigma_1-p & 0 & 0 \\ 0 & \sigma_2-p & 0 \\ 0 & 0 & \sigma_3-p \end{bmatrix} \tag{2.58}$$

or

$$s_{ij} = \begin{bmatrix} \dfrac{2\sigma_1 - \sigma_2 - \sigma_3}{3} & 0 & 0 \\[2ex] 0 & \dfrac{2\sigma_2 - \sigma_3 - \sigma_1}{3} & 0 \\[2ex] 0 & 0 & \dfrac{2\sigma_3 - \sigma_1 - \sigma_2}{3} \end{bmatrix} \qquad (2.59)$$

To obtain the invariants of the stress deviator tensor $s_{ij}$, a similar derivation is followed as was used to derive Eq. (2.34). Thus, we can write

$$|s_{ij} - s\delta_{ij}| = 0 \qquad (2.60)$$

or

$$s^3 - J_1 s^2 - J_2 s - J_3 = 0 \qquad (2.61)$$

where $J_1$, $J_2$, and $J_3$ are the invariants of the stress deviator tensor. Using Eq. (2.54) and definitions similar to those given in Eqs. (2.35) to (2.37), the invariants $J_1$, $J_2$, and $J_3$ may be expressed in different forms in terms of the components of $s_{ij}$ or its principal values, $s_1$, $s_2$, and $s_3$, or alternatively, in terms of the components of the stress tensor $\sigma_{ij}$ or its principal values, $\sigma_1$, $\sigma_2$, and $\sigma_3$. Thus, we have

$$J_1 = s_{ii} = s_{11} + s_{22} + s_{33} = s_1 + s_2 + s_3 = 0 \qquad (2.62)$$

$$\begin{aligned} J_2 &= \tfrac{1}{2} s_{ij} s_{ji} = \tfrac{1}{2}(s_{11}^2 + s_{22}^2 + s_{33}^2 + s_{12} s_{21} + s_{21} s_{12} + \cdots) = \tfrac{1}{2}(s_1^2 + s_2^2 + s_3^2) \\ &= \tfrac{1}{2}(s_{11}^2 + s_{22}^2 + s_{33}^2 + 2\sigma_{12}^2 + 2\sigma_{23}^2 + 2\sigma_{31}^2) \\ &= -s_{11} s_{22} - s_{22} s_{33} - s_{33} s_{11} + \sigma_{12}^2 + \sigma_{23}^2 + \sigma_{31}^2 = -(s_1 s_2 + s_2 s_3 + s_3 s_1) \\ &= \tfrac{1}{6}[(s_{11} - s_{22})^2 + (s_{22} - s_{33})^2 + (s_{33} - s_{11})^2] + \sigma_{12}^2 + \sigma_{23}^2 + \sigma_{31}^2 \\ &= \tfrac{1}{6}[(\sigma_x - \sigma_y)^2 + (\sigma_y - \sigma_z)^2 + (\sigma_z - \sigma_x)^2] + \tau_{xy}^2 + \tau_{yz}^2 + \tau_{zx}^2 \\ &= \tfrac{1}{6}[(\sigma_1 - \sigma_2)^2 + (\sigma_2 - \sigma_3)^2 + (\sigma_3 - \sigma_1)^2] \qquad (2.63) \end{aligned}$$

$$J_3 = \tfrac{1}{3} s_{ij} s_{jk} s_{ki} = \begin{vmatrix} s_x & \tau_{xy} & \tau_{xz} \\ \tau_{yx} & s_y & \tau_{yz} \\ \tau_{zx} & \tau_{zy} & s_z \end{vmatrix} = \tfrac{1}{3}(s_1^3 + s_2^3 + s_3^3) = s_1 s_2 s_3 \qquad (2.64)$$

It can be shown that the invariants $J_1$, $J_2$, and $J_3$ are related to the invariants $I_1$, $I_2$, and $I_3$ of the stress tensor $\sigma_{ij}$ through the following relations:

$$J_1 = 0$$

$$J_2 = \tfrac{1}{3}(I_1^2 - 3I_2) \qquad (2.65)$$

$$J_3 = \tfrac{1}{27}(2I_1^3 - 9I_1 I_2 + 27 I_3)$$

One advantage of using the stress deviator tensor is now apparent. The first invariant of this tensor, $J_1$, is always zero. This can also be seen by taking the sum of the diagonal elements in Eq. (2.56) or (2.58).

It can be shown that the necessary and sufficient condition for a state of stress $\sigma_{ij}$ to be a pure shear state is that $\sigma_{ii} = 0$, or its first invariant $I_1 = 0$ (see Problem 2.11). Therefore, the stress deviator tensor $s_{ij}$ is a state of pure shear.

### 2.1.6. Octahedral Stresses

An *octahedral* (*stress*) *plane* is a plane whose normal makes equal angles with each of the principal axes of stress. Thus, the planes with normal $\mathbf{n} = (n_1, n_2, n_3) = |1/\sqrt{3}|(1, 1, 1)$ in the principal coordinate system are called *octahedral planes*. Note that we can have eight octahedral planes, as shown in Fig. 2.6, with $OA = OB = OC = OA' = OB' = OC'$. Referring to the principal stress axes, 1, 2, and 3, the stress tensor $\sigma_{ij}$ is written as

$$\sigma_{ij} = \begin{bmatrix} \sigma_1 & 0 & 0 \\ 0 & \sigma_2 & 0 \\ 0 & 0 & \sigma_3 \end{bmatrix} \tag{2.66}$$

The normal component of a stress vector at point $O$ associated with any direction, $\mathbf{n}$, can be obtained by Cauchy's formula in Eq. (2.22),

$$\sigma_n = \sigma_{ij} n_i n_j$$

or

$$\sigma_n = \sigma_1 n_1 n_1 + \sigma_2 n_2 n_2 + \sigma_3 n_3 n_3 \tag{2.67}$$

Therefore, the normal stress on a face of the octahedron will be

$$\sigma_{oct} = \sigma_1 n_1^2 + \sigma_2 n_2^2 + \sigma_3 n_3^2 = \tfrac{1}{3}(\sigma_1 + \sigma_2 + \sigma_3) = \tfrac{1}{3}I_1 \tag{2.68}$$

Note that the magnitude of $\sigma_{oct}$ on all the eight faces is the same and that the quantity $\sigma_{oct}$ is the mean normal stress (or hydrostatic stress).

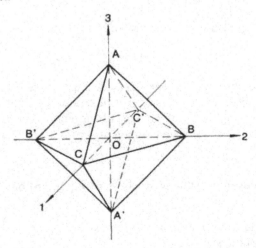

FIGURE 2.6. Octahedral planes in principal coordinate system.

The shear stress on a face of the octahedron, $\tau_{oct}$, can be obtained from the formula in Eq. (2.23):

$$\tau_{oct}^2 = (\overset{n}{T}_{oct})^2 - \sigma_{oct}^2 \tag{2.69}$$

Using Eq. (2.24) to calculate $(\overset{n}{T}_{oct})^2$, we get

$$(\overset{n}{T}_{oct})^2 = \sigma_1^2 n_1^2 + \sigma_2^2 n_2^2 + \sigma_3^2 n_3^2 = \tfrac{1}{3}(\sigma_1^2 + \sigma_2^2 + \sigma_3^2) \tag{2.70}$$

Therefore,

$$\tau_{oct}^2 = \frac{1}{3}(\sigma_1^2 + \sigma_2^2 + \sigma_3^2) - \frac{1}{3^2}(\sigma_1 + \sigma_2 + \sigma_3)^2$$

$$= \frac{1}{9}[(\sigma_1 - \sigma_2)^2 + (\sigma_2 - \sigma_3)^2 + (\sigma_3 - \sigma_1)^2] \tag{2.71}$$

Recalling the results we have obtained for the principal shear stresses, $\tau_{oct}$ can be expressed as

$$\tau_{oct}^2 = \tfrac{4}{9}(\tau_{12}^2 + \tau_{23}^2 + \tau_{31}^2) \tag{2.72}$$

where $\tau_{12}$, $\tau_{23}$, and $\tau_{31}$ are the principal shear stresses. Therefore,

$$\tau_{oct} = \tfrac{2}{3}(\tau_{12}^2 + \tau_{23}^2 + \tau_{31}^2)^{1/2} = \sqrt{\tfrac{2}{3}J_2} \tag{2.73}$$

where $J_2$ is an invariant of the stress deviator tensor. In terms of the invariants of the stress tensor, the octahedral shear stress can be written as [see Eq. (2.65)]

$$\tau_{oct} = \frac{\sqrt{2}}{3}(I_1^2 - 3I_2)^{1/2} \tag{2.74}$$

and in terms of general nonprincipal stresses, it becomes [see Eq. (2.63)]

$$\tau_{oct} = \tfrac{1}{3}[(\sigma_x - \sigma_y)^2 + (\sigma_y - \sigma_z)^2 + (\sigma_z - \sigma_x)^2 + 6(\tau_{xy}^2 + \tau_{yz}^2 + \tau_{zx}^2)]^{1/2} \tag{2.75}$$

which gives the *octahedral shear stress* at a point in terms of the stress components referred to an arbitrary set of coordinate axes, $x$, $y$, and $z$.

Note that the magnitude of $\tau_{oct}$ on all the eight faces is the same and that the quantity $\tau_{oct}$ is somewhat an average principal shear stress as given by Eq. (2.73).

EXAMPLE 2.2. Given the state of stress $\sigma_{ij}$ in Eq. (2.76) below, calculate the following:

(a) The octahedral normal and shear stresses.
(b) The hydrostatic stress.
(c) The stress deviator tensor, $s_{ij}$.

$$\sigma_{ij} = \begin{bmatrix} 1 & 2 & 4 \\ 2 & 2 & 1 \\ 4 & 1 & 3 \end{bmatrix} \text{(units of stress)} \qquad (2.76)$$

SOLUTION. (a) The first invariant $I_1$ is calculated from Eq. (2.35):

$$I_1 = \sigma_{ii} = 1 + 2 + 3 = 6$$

Therefore, from Eq. (2.68), we have

$$\sigma_{oct} = \tfrac{1}{3}I_1 = 2$$

Using Eq. (2.75) for $\tau_{oct}$, we get

$$\tau_{oct} = \tfrac{1}{3}[(1-2)^2 + (2-3)^2 + (3-1)^2 + 6(4+1+16)]^{1/2} = 3.83$$

(b) The hydrostatic (mean) stress is given by Eq. (2.53):

$$p = \tfrac{1}{3}(6) = 2$$

(c) The stress deviator tensor $s_{ij}$ is obtained from Eq. (2.55):

$$s_{ij} = \sigma_{ij} - p\delta_{ij}$$

$$s_{ij} = \begin{bmatrix} 1 & 2 & 4 \\ 2 & 2 & 1 \\ 4 & 1 & 3 \end{bmatrix} - 2\begin{bmatrix} 1 & 0 & 0 \\ 0 & 1 & 0 \\ 0 & 0 & 1 \end{bmatrix}$$

$$= \begin{bmatrix} -1 & 2 & 4 \\ 2 & 0 & 1 \\ 4 & 1 & 1 \end{bmatrix}$$

Since $s_{ij}$ is a pure shear state, as a check, the condition $s_{ii} = 0$ is found to be satisfied.

EXAMPLE 2.3. The states of stress $\sigma_{ij}^{(1)}$ and $\sigma_{ij}^{(2)}$ at two different points in a body are given by Eqs. (2.77) and (2.78) below. Determine which state is more critical to yielding if the following criteria of yielding are used:

(a) Octahedral normal stress, $\sigma_{oct}$.
(b) Octahedral shear stress, $\tau_{oct}$.
(c) Maximum shear stress, $\tau_{max}$.

$$\sigma_{ij}^{(1)} = \begin{bmatrix} 10 & 0 & 3 \\ 0 & 3 & 0 \\ 3 & 0 & 2 \end{bmatrix} \text{(units of stress)} \qquad (2.77)$$

$$\sigma_{ij}^{(2)} = \begin{bmatrix} 3 & 0 & 0 \\ 0 & -7 & 0 \\ 0 & 0 & -5 \end{bmatrix} \text{(units of stress)} \qquad (2.78)$$

SOLUTION. (a) From Eq. (2.68), $\sigma_{oct}$ can be calculated in both cases. Thus we have

$$\sigma_{oct}^{(1)} = \tfrac{1}{3}(10+3+2) = 5$$
$$\sigma_{oct}^{(2)} = \tfrac{1}{3}(3-7-5) = -3$$

Therefore, based on $\sigma_{oct}$, yielding occurs first at the first point.

(b) Using Eq. (2.75), we have

$$\tau_{oct}^{(1)} = \tfrac{1}{3}[49+1+64+6(9)]^{1/2} = 4.32$$
$$\tau_{oct}^{(2)} = \tfrac{1}{3}[100+4+64]^{1/2} = 4.32$$

Thus, based on $\tau_{oct}$, yielding will occur at both points at the same time.

(c) Following the procedure given in Section 2.1.3, we can find the principal stresses of the first state of stress. The results are

$$\sigma_1^{(1)} = 11, \qquad \sigma_2^{(1)} = 3, \qquad \sigma_3^{(1)} = 1$$

Equation (2.78) represents a principal state of stress in which

$$\sigma_1^{(2)} = 3, \qquad \sigma_2^{(2)} = -5, \qquad \sigma_3^{(2)} = -7$$

Thus, the maximum shear stresses are given by Eq. (2.51):

$$\tau_{max}^{(1)} = \left| \frac{(11)-(1)}{2} \right| = 5$$

$$\tau_{max}^{(2)} = \left| \frac{(3)-(-7)}{2} \right| = 5$$

Again, based on $\tau_{max}$, yielding will occur at both points at the same time.

## 2.1.7. Physical Interpretations of Stress Invariants $I_1$ and $J_2$

There are several interpretations of stress invariants $I_1$ and $J_2$, one of which has been shown by Eqs. (2.68) and (2.73). Namely, $I_1/3$ is the octahedral normal stress $\sigma_{oct}$, while $\sqrt{2/3}J_2$ is the octahedral shear stress. Other interpretations are presented in the following sections.

### 2.1.7.1. ELASTIC STRAIN ENERGY

The total elastic strain energy $W$ per unit volume of a linear elastic material can be divided into two parts, associated respectively with the change in volume, $W_1$, and with the change in shape, $W_2$:

$$W = W_1 + W_2 \tag{2.79}$$

where

$$W_1 = \text{dilatational energy} = \frac{1-2v}{6E} I_1^2 \tag{2.80}$$

$$W_2 = \text{distortional energy} = \frac{1+v}{E} J_2 \tag{2.81}$$

and where $E$ and $\nu$ are the modulus of elasticity and Poisson's ratio, respectively. The invariants $I_1$ and $J_2$ are seen to be directly proportional to the energy of dilatation and the energy of distortion, respectively.

### 2.1.7.2. MEAN STRESSES

Consider an infinitesimal spherical element of volume. At any point on the surface of this sphere, the stress vector on the tangent plane has a shear stress component $\tau_s$ and a normal stress component $\sigma_s$. The mean value of the normal stress $\sigma_s$ over the spherical surface can be defined by

$$\sigma_m = \lim_{s \to 0} \left( \frac{1}{S} \int_S \sigma_s \, dS \right) \tag{2.82}$$

where $S$ denotes the surface of the sphere. Evaluation of this expression gives

$$\sigma_m = \tfrac{1}{3}(\sigma_1 + \sigma_2 + \sigma_3) = \tfrac{1}{3} I_1 \tag{2.83}$$

For the shear stress $\tau_s$ on the surface of the sphere, the mean value of $\tau_s$ can be based upon stresses existing on all possible planes of orientation through the point by carrying out the averaging process over the spherical surface. Since the sign of shear stress has no significance with respect to the physical mechanism of failure, it is expedient to take the average in the sense of the root mean. Thus,

$$\tau_m = \lim_{s \to 0} \left( \frac{1}{S} \int_S \tau_s^2 \, dS \right)^{1/2} \tag{2.84}$$

Carrying out the indicated operations leads to

$$\tau_m = \frac{1}{\sqrt{15}} [(\sigma_1 - \sigma_2)^2 + (\sigma_2 - \sigma_3)^2 + (\sigma_3 - \sigma_1)^2]^{1/2} \tag{2.85}$$

or, in terms of the invariant $J_2$,

$$\tau_m = \sqrt{\tfrac{2}{5} J_2} \tag{2.86}$$

### 2.1.7.3. ROOT MEAN OF THE PRINCIPAL SHEAR STRESS

Equations (2.49), (2.51), and (2.52) give the principal shear stresses, whose root mean is

$$\sqrt{\frac{1}{3} \left[ \left( \frac{\sigma_1 - \sigma_2}{2} \right)^2 + \left( \frac{\sigma_2 - \sigma_3}{2} \right)^2 + \left( \frac{\sigma_1 - \sigma_3}{2} \right)^2 \right]} = \frac{\sqrt{2J_2}}{2} \tag{2.87}$$

## 2.1.8. Mohr's Circles for Three-Dimensional Stress Systems

Mohr's diagram is a useful graphical representation of the stress state at a point. In this graphical representation, the state of stress at a point is represented by the Mohr circle diagram, in which the abscissa, $\sigma_n$, and ordinate, $S_n$, of each point give the normal and shear stress components, respectively, acting on a particular cut plane with a fixed normal direction.

In the general three-dimensional case, for a given state of stress at a point, the values of the principal stresses $\sigma_1$, $\sigma_2$, and $\sigma_3$ must first be calculated from Eq. (2.34), and the corresponding principal axes are calculated from Eqs. (2.40). Once the values of $\sigma_1$, $\sigma_2$, and $\sigma_3$ are known, a Mohr circle diagram can be constructed as shown in Fig. 2.7, for the case $\sigma_1 > \sigma_2 > \sigma_3$. In this figure, the centers of the three Mohr circles $C_1$, $C_2$, and $C_3$ have the coordinates $[\frac{1}{2}(\sigma_2 + \sigma_3), 0]$, $[\frac{1}{2}(\sigma_1 + \sigma_3), 0]$, and $[\frac{1}{2}(\sigma_1 + \sigma_2), 0]$, respectively. The three radii $R_1$, $R_2$, and $R_3$ are equal to $\frac{1}{2}(\sigma_2 - \sigma_3)$, $\frac{1}{2}(\sigma_1 - \sigma_3)$, and $\frac{1}{2}(\sigma_1 - \sigma_2)$, respectively. Associated with the cut plane **n** at the considered point with respect to the principal coordinate system, the corresponding normal and shear stresses can be plotted as a point in the $\sigma_n$-$S_n$ stress space. Let us consider the positive values of $S_n$, that is, in the upper half of the $\sigma_n$-$S_n$ stress space.

Assuming that the components of the unit normal **n** are $n_1$, $n_2$, and $n_3$ in the direction of the principal axes 1, 2, and 3, respectively, and that $\sigma_1 > \sigma_2 > \sigma_3$, Eqs. (2.42) and (2.43) give

$$\sigma_n^2 + S_n^2 = (\overset{n}{T})^2 = \sigma_1^2 n_1^2 + \sigma_2^2 n_2^2 + \sigma_3^2 n_3^2 \tag{2.88}$$

$$\sigma_n = \sigma_1 n_1^2 + \sigma_2 n_2^2 + \sigma_3 n_3^2 \tag{2.89}$$

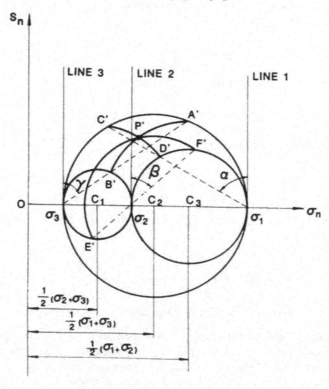

FIGURE 2.7. Mohr's circles in three-dimensional case ($\sigma_1 > \sigma_2 > \sigma_3$).

For the unit vector $\mathbf{n}$, we have

$$n_1^2 + n_2^2 + n_3^2 = 1 \qquad (2.90)$$

Solving Eqs. (2.88) to (2.90) for $n_1^2$, $n_2^2$, and $n_3^2$ leads to

$$n_1^2 = \frac{S_n^2 + (\sigma_n - \sigma_2)(\sigma_n - \sigma_3)}{(\sigma_1 - \sigma_2)(\sigma_1 - \sigma_3)} \qquad (2.91)$$

$$n_2^2 = \frac{S_n^2 + (\sigma_n - \sigma_3)(\sigma_n - \sigma_1)}{(\sigma_2 - \sigma_3)(\sigma_2 - \sigma_1)} \qquad (2.92)$$

$$n_3^2 = \frac{S_n^2 + (\sigma_n - \sigma_1)(\sigma_n - \sigma_2)}{(\sigma_3 - \sigma_1)(\sigma_3 - \sigma_2)} \qquad (2.93)$$

Since $\sigma_1 > \sigma_2 > \sigma_3$, and the left-hand sides of Eqs. (2.91) to (2.93) are non-negative, it follows that

$$S_n^2 + (\sigma_n - \sigma_2)(\sigma_n - \sigma_3) \geq 0 \qquad (2.94)$$

$$S_n^2 + (\sigma_n - \sigma_3)(\sigma_n - \sigma_1) \leq 0 \qquad (2.95)$$

$$S_n^2 + (\sigma_n - \sigma_1)(\sigma_n - \sigma_2) \geq 0 \qquad (2.96)$$

which may be rewritten as

$$S_n^2 + [\sigma_n - \tfrac{1}{2}(\sigma_2 + \sigma_3)]^2 \geq \tfrac{1}{4}(\sigma_2 - \sigma_3)^2 \qquad (2.97)$$

$$S_n^2 + [\sigma_n - \tfrac{1}{2}(\sigma_1 + \sigma_3)]^2 \leq \tfrac{1}{4}(\sigma_1 - \sigma_3)^2 \qquad (2.98)$$

$$S_n^2 + [\sigma_n - \tfrac{1}{2}(\sigma_1 + \sigma_2)]^2 \geq \tfrac{1}{4}(\sigma_1 - \sigma_2)^2 \qquad (2.99)$$

Relations (2.97) to (2.99) show that the admissible values of $\sigma_n$ and $S_n$ lie inside, or on the boundaries of, the region bounded by the circles $C_1$, $C_2$, and $C_3$, as shown in Fig. 2.7.

For any fixed value of $n_1$, eliminating $n_2$ and $n_3$ from Eqs. (2.88) to (2.90) gives

$$[\sigma_n - \tfrac{1}{2}(\sigma_2 + \sigma_3)]^2 + S_n^2 = \tfrac{1}{4}(\sigma_2 - \sigma_3)^2 + n_1^2(\sigma_1 - \sigma_2)(\sigma_1 - \sigma_3) \qquad (2.100)$$

Therefore, for a given value of $n_1$, the point $(\sigma_n, S_n)$ corresponding to this particular value of $n_1$ lies on the arc $C'D'$ as shown in Fig. 2.7. To construct this arc, we draw line 1 parallel to the '$S_n$-axis passing through the point $(\sigma_1, 0)$ and measure an angle $\alpha = \cos^{-1} n_1$ from that line. This line making an angle $\alpha$ with line 1 intersects circles $C_2$ and $C_3$ in points $C'$ and $D'$, respectively. Using $[\tfrac{1}{2}(\sigma_2 + \sigma_3), 0]$ as the center, we draw the arc $C'D'$.

Similarly, for a fixed value of $n_2$, eliminating $n_1$ and $n_3$ from Eqs. (2.88) to (2.90) gives

$$[\sigma_n - \tfrac{1}{2}(\sigma_1 + \sigma_3)]^2 + S_n^2 = \tfrac{1}{4}(\sigma_1 - \sigma_3)^2 + n_2^2(\sigma_2 - \sigma_1)(\sigma_2 - \sigma_3) \qquad (2.101)$$

Thus, $(\sigma_n, S_n)$ corresponding to this particular value of $n_2$ lies on the arc $E'F'$ in Fig. 2.7. This arc $E'F'$ is drawn from the center $[\tfrac{1}{2}(\sigma_1 + \sigma_3), 0]$ between the points of intersection, $E'$ and $F'$, of circles $C_1$ and $C_3$, respectively, with the line making an angle $\beta = \cos^{-1} n_2$ with line 2.

Finally, for a fixed value of $n_3$, the relation between the values of $(\sigma_n, S_n)$ for this particular value of $n_3$ is given by

$$[\sigma_n - \tfrac{1}{2}(\sigma_1 + \sigma_2)]^2 + S_n^2 = \tfrac{1}{4}(\sigma_1 - \sigma_2)^2 + n_3^2(\sigma_3 - \sigma_1)(\sigma_3 - \sigma_2) \quad (2.102)$$

and in this case, the point $(\sigma_n, S_n)$ lies on arc $A'B'$ in Fig. 2.7.

For a given point $P$ with known values of $n_1$, $n_2$, and $n_3$, one can find $(\sigma_n, S_n)$ corresponding to these values graphically. Since only two values of $n_1$, $n_2$, and $n_3$ are independent, we can use any two values, for example, $n_1$ and $n_3$, to determine the values $(\sigma_n, S_n)$ corresponding to these values. For a fixed value of $n_1$, we construct the arc $C'D'$. Similarly, for a fixed value of $n_3$, we construct the arc $A'B'$, as shown in Fig. 2.7. The point of intersection, $P'$, of the two arcs gives the required values $\sigma_n$ and $S_n$ corresponding to the given values $n_1$, $n_2$, and $n_3$. The third value, $n_2$, is used to check the procedure since the third arc $E'F'$ must pass through the same point $P'$.

## 2.1.9. Haigh–Westergaard Stress Space

This geometric representation of the stress state at a point is very useful in the study of plasticity theory and failure criteria. Since the stress tensor $\sigma_{ij}$ has six independent components, it is, of course, possible to consider these components as positional coordinates in a six-dimensional space. However, this is too difficult to deal with. The simplest alternative is to take the three principal stresses $\sigma_1$, $\sigma_2$, $\sigma_3$ as coordinates and represent the stress state at a point as a point in this three-dimensional stress space. This space is called the *Haigh–Westergaard stress space*. In this principal stress space, every point having coordinates $\sigma_1$, $\sigma_2$, and $\sigma_3$ represents a possible stress state. Any two stress states at a point $P$ which differ in the orientation of their principal axes, but not in the principal stress values, would then be represented by the same point in the three-dimensional stress space. This implies that this type of stress space representation is focused primarily on the geometry of stress and not on the orientation of the stress state with respect to the material body.

Consider the straight line $ON$ passing through the origin and making the same angle with each of the coordinate axes, as shown in Fig. 2.8. Then, for every point on this line, the state of stress is one for which $\sigma_1 = \sigma_2 = \sigma_3$. Thus, every point on this line corresponds to a hydrostatic or spherical state of stress, while the deviatoric stresses, $s_1 = (2\sigma_1 - \sigma_2 - \sigma_3)/3$, etc., are equal to zero. This line is therefore termed the *hydrostatic axis*. Furthermore, any plane perpendicular to $ON$ is called the *deviatoric plane*. Such a plane has the form

$$\sigma_1 + \sigma_2 + \sigma_3 = \sqrt{3}\xi \quad (2.103)$$

where $\xi$ is the distance from the origin to the plane measured along the

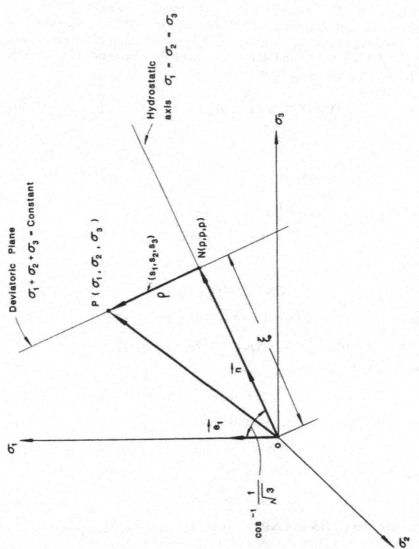

FIGURE 2.8. Haigh–Westergaard stress space.

normal $ON$. The particular deviatoric plane passing through the origin $O$,

$$\sigma_1 + \sigma_2 + \sigma_3 = 0 \tag{2.104}$$

is called the $\pi$-plane.

Consider an arbitrary state of stress at a given point with stress components $\sigma_1$, $\sigma_2$, and $\sigma_3$. This state of stress is represented by point $P(\sigma_1, \sigma_2, \sigma_3)$ in the principal stress space in Fig. 2.8. The stress vector $OP$ can be decomposed into two components, the component $ON$ in the direction of the unit vector $n = (1/\sqrt{3}, 1/\sqrt{3}, 1/\sqrt{3})$ and the component $NP$ perpendicular to $ON$ (parallel to the $\pi$-plane). Thus,

$$|ON| = OP \cdot n = (\sigma_1, \sigma_2, \sigma_3) \cdot \left(\frac{1}{\sqrt{3}}, \frac{1}{\sqrt{3}}, \frac{1}{\sqrt{3}}\right) \tag{2.105}$$

or

$$|ON| = \frac{1}{\sqrt{3}}(\sigma_1 + \sigma_2 + \sigma_3) = \frac{I_1}{\sqrt{3}} = \sqrt{3}p \tag{2.106}$$

The components of vector $NP$ are given by

$$NP = OP - ON \tag{2.107}$$

But

$$ON = |ON|n = (p, p, p) \tag{2.108}$$

Therefore, substituting from Eq. (2.108) into Eq. (2.107),

$$NP = (\sigma_1, \sigma_2, \sigma_3) - (p, p, p) = [(\sigma_1 - p), (\sigma_2 - p), (\sigma_3 - p)] \tag{2.109}$$

which, using Eq. (2.55), reduces to

$$NP = (s_1, s_2, s_3) \tag{2.110}$$

Hence, the length $\rho$ of vector $NP$ is given by

$$\rho = |NP| = (s_1^2 + s_2^2 + s_3^2)^{1/2} = \sqrt{2J_2} \tag{2.111}$$

or, by Eq. (2.73),

$$\rho = |NP| = \sqrt{3}\tau_{oct} \tag{2.112}$$

Thus, the vectors $ON$ and $NP$ represent the hydrostatic components ($p\delta_{ij}$) and the deviatoric stress components ($s_{ij}$), respectively, of the state of stress ($\sigma_{ij}$) represented by point $P$ in Fig. 2.8.

Now consider the projections of vector $NP$ and the coordinate axes $\sigma_i$ on a deviatoric plane as shown in Fig. 2.9. In this figure, the axes $\sigma_1'$, $\sigma_2'$, and $\sigma_3'$ are the projections of the axes $\sigma_1$, $\sigma_2$, and $\sigma_3$ on the deviatoric plane, and $NP$ is the projection of vector $NP$ on the same plane. Since the

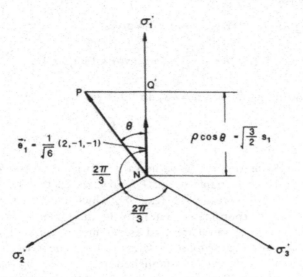

FIGURE 2.9. State of stress at a point projected on a deviatoric plane.

unit vector $e_1'$ in the direction of the $\sigma_1'$-axis has components $(1/\sqrt{6}) \times (2, -1, -1)$ with respect to the axes $\sigma_1$, $\sigma_2$, and $\sigma_3$, then the projection of vector **NP** in the direction of the unit vector $e_1'$, denoted by $NQ'$, is given by

$$NQ' = \rho \cos \theta = \mathbf{NP} \cdot e_1' = (s_1, s_2, s_3) \cdot \frac{1}{\sqrt{6}} (2, -1, -1)$$

or

$$\rho \cos \theta = \frac{1}{\sqrt{6}} (2s_1 - s_2 - s_3) \tag{2.113}$$

Substituting for $s_2 + s_3 = -s_1$, we have

$$\rho \cos \theta = \sqrt{\tfrac{3}{2}} s_1 \tag{2.114}$$

Substituting for $\rho$ from Eq. (2.111) into Eq. (2.114) results in

$$\cos \theta = \frac{\sqrt{3}}{2} \frac{s_1}{\sqrt{J_2}} \tag{2.115}$$

Using the trigonometric identity $\cos 3\theta = 4 \cos^3 \theta - 3 \cos \theta$ and substituting for $\cos \theta$ from Eq. (2.115) leads to

$$\cos 3\theta = 4 \left( \frac{\sqrt{3}}{2} \frac{s_1}{\sqrt{J_2}} \right)^3 - 3 \left( \frac{\sqrt{3}}{2} \frac{s_1}{\sqrt{J_2}} \right)$$

or

$$\cos 3\theta = \frac{3\sqrt{3}}{2J_2^{3/2}} (s_1^3 - s_1 J_2) \tag{2.116}$$

Substituting for $J_2 = -(s_1s_2 + s_2s_3 + s_3s_1)$ gives

$$\cos 3\theta = \frac{3\sqrt{3}}{2J_2^{3/2}} [s_1^3 + s_1^2(s_2 + s_3) + s_1s_2s_3] \qquad (2.117)$$

Finally, substituting for $s_2 + s_3 = -s_1$ and $J_3 = s_1s_2s_3$, we get

$$\cos 3\theta = \frac{3\sqrt{3}}{2} \frac{J_3}{J_2^{3/2}} \qquad (2.118)$$

Equation (2.118) shows that the value of $\cos 3\theta$ is an invariant related to the deviatoric stress invariants $J_2$ and $J_3$. Now, we see that a state of stress $(\sigma_1, \sigma_2, \sigma_3)$ can be expressed by $(\xi, \rho, \theta)$, which are referred to as the Haigh–Westergaard coordinates. Later, in the discussion of the yield and failure conditions, $\xi$, $\rho$, and $\theta$ are used as parameters required to represent the yield and failure functions in stress space. The relations between $(\sigma_1, \sigma_2, \sigma_3)$ and $(\xi, \rho, \theta)$ can be established in the following manner.
From Eq. (2.115), we know

$$s_1 = \frac{2}{\sqrt{3}} \sqrt{J_2} \cos \theta \qquad (2.119)$$

In a similar manner, the deviatoric stress components $s_2$ and $s_3$ can also be obtained in terms of the angle $\theta$. From Fig. 2.9, these components are given by

$$s_2 = \frac{2}{\sqrt{3}} \sqrt{J_2} \cos \left( \frac{2\pi}{3} - \theta \right) \qquad (2.120)$$

$$s_3 = \frac{2}{\sqrt{3}} \sqrt{J_2} \cos \left( \frac{2\pi}{3} + \theta \right) \qquad (2.121)$$

These relations are satisfied only if the angle lies in the range (for $\sigma_1 \geq \sigma_2 \geq \sigma_3$)

$$0 \leq \theta \leq \frac{\pi}{3} \qquad (2.122)$$

In view of Eqs. (2.58), (2.103), (2.111), and (2.119)–(2.121), the three principal stresses of $\sigma_{ij}$ are therefore given by

$$\begin{Bmatrix} \sigma_1 \\ \sigma_2 \\ \sigma_3 \end{Bmatrix} = \begin{Bmatrix} p \\ p \\ p \end{Bmatrix} + \frac{2}{\sqrt{3}} \sqrt{J_2} \begin{Bmatrix} \cos \theta \\ \cos(\theta - 2\pi/3) \\ \cos(\theta + 2\pi/3) \end{Bmatrix}$$

$$= \frac{1}{\sqrt{3}} \begin{Bmatrix} \xi \\ \xi \\ \xi \end{Bmatrix} + \sqrt{\frac{2}{3}} \rho \begin{Bmatrix} \cos \theta \\ \cos(\theta - 2\pi/3) \\ \cos(\theta + 2\pi/3) \end{Bmatrix} \qquad (2.123)$$

## 2.1.10. Equation of Equilibrium

For any volume $V$ of a material body and having $S$ as the surface area of $V$, as shown in Fig. 2.10, we have the following condition of equilibrium:

$$\int_S \overset{n}{T}_i \, dS + \int_V F_i \, dV = 0 \tag{2.124}$$

Substituting $\overset{n}{T}_i$ from Eq. (2.18), Eq. (2.124) may be written as

$$\int_S \sigma_{ij} n_j \, dS + \int_V F_i \, dV = 0 \tag{2.125}$$

Using the divergence theorem

$$\int_S u_i n_i \, dS = \int_V u_{i,i} \, dV \tag{2.126}$$

Eq. (2.125) can be expressed as

$$\int_V (\sigma_{ij,j} + F_i) \, dV = 0 \tag{2.127}$$

For an arbitrary volume,

$$\sigma_{ij,j} + F_i = 0 \tag{2.128}$$

Equation (2.128) may be written in the $(x, y, z)$ notation as

$$\frac{\partial \sigma_x}{\partial x} + \frac{\partial \tau_{xy}}{\partial y} + \frac{\partial \tau_{xz}}{\partial z} + F_x = 0$$

$$\frac{\partial \tau_{yx}}{\partial x} + \frac{\partial \sigma_y}{\partial y} + \frac{\partial \tau_{yz}}{\partial z} + F_y = 0 \tag{2.129}$$

$$\frac{\partial \tau_{zx}}{\partial x} + \frac{\partial \tau_{zy}}{\partial y} + \frac{\partial \sigma_z}{\partial z} + F_z = 0$$

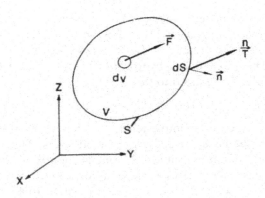

FIGURE 2.10. Equilibrium of a material body.

## 2.2. Yield Criteria Independent of Hydrostatic Pressure

### 2.2.1. General Considerations

The yield criterion defines the elastic limits of a material under combined states of stress. As we know, the elastic limit in a simple tension test is the yield stress $\sigma_0$, while in a simple shear test, it is the yield stress $\tau_0$. In general, the elastic limit or yield stress is a function of the state of stress, $\sigma_{ij}$. Hence, the yield condition can generally be expressed as

$$f(\sigma_{ij}, k_1, k_2, \ldots) = 0 \qquad (2.130)$$

where $k_1, k_2, \ldots$ are material constants, which, like $\sigma_0$ and $\tau_0$, are to be determined experimentally.

For isotropic materials, the orientation of the principal stresses is immaterial, and the values of the three principal stresses suffice to describe the state of stress uniquely. A yield criterion therefore consists in a relation of the form

$$f(\sigma_1, \sigma_2, \sigma_3, k_1, k_2, \ldots) = 0 \qquad (2.131)$$

We have shown that the three principal stresses $\sigma_1$, $\sigma_2$, and $\sigma_3$ can be expressed in terms of the combinations of the three stress invariants $I_1, J_2$, and $J_3$, where $I_1$ is the first invariant of the stress tensor $\sigma_{ij}$ and $J_2$ and $J_3$ are the second and third invariants of the deviatoric tensor $s_{ij}$. Thus, one can replace Eq. (2.131) by

$$f(I_1, J_2, J_3, k_1, k_2, \ldots) = 0 \qquad (2.132)$$

Furthermore, these three particular principal invariants are directly related to Haigh–Westergaard coordinates $\xi, \rho, \theta$ in stress space [see Eq. (2.123)]. Therefore, Eq. (2.132) can also be rewritten as

$$f(\xi, \rho, \theta, k_1, k_2, \ldots) = 0 \qquad (2.133)$$

Yield criteria of materials should be determined experimentally. An important experimental fact for metals, shown by Bridgman and others [see Hill (1950)], is that the influence of hydrostatic pressure on yielding is not appreciable. The absence of a hydrostatic pressure effect means that the yield function can be reduced to the form

$$f(J_2, J_3, k_1, k_2, \ldots) = 0 \qquad (2.134)$$

A stress–strain curve in simple tension does not, in itself, provide any information on the behavior under combined stress. The combined stress tests, analogous to simple tension, are termed proportional or radial loading tests. In these tests, all stresses are increased proportionately. In a biaxial state of stress, for example, $\sigma_1$ and $\sigma_2$ are increased so as to keep the ratio $\sigma_1/\sigma_2$ constant. It seems that we would need to perform a number of tests in order to construct a yield locus. However, we will show that one point on the yield locus may give rise to twelve points (Fig. 2.11) if the material (1) is isotropic, (2) is hydrostatic pressure independent, and (3) has equal yield stresses in tension and compression.

Now suppose that a material yields in a state of stress, $(3\sigma, \sigma, 0)$. Point $A_1(3\sigma, \sigma, 0)$ in Fig. 2.11 then lies on the yield locus on the $\sigma_1$-$\sigma_2$ plane. If the material is isotropic, there is no reason why we should not relabel the axes in an alternative way. We thus conclude that point $A_2(\sigma, 3\sigma, 0)$ also lies on the yield locus. Further, if the material has the same response to tension and compression, points $A_3(-3\sigma, -\sigma, 0)$ and $A_4(-\sigma, -3\sigma, 0)$ will also lie on the yield locus. Now considering $A_1$ and $A_2$ or $A_3$ and $A_4$, we see that they are mirror images about a line $aa'$ bisecting the $\sigma_1$ and $\sigma_2$ axes. Similarly, $A_1$ and $A_4$ or $A_2$ and $A_3$ are symmetric about another line $bb'$ perpendicular to line $aa'$. Hence, there are two symmetric axes for the yield locus.

Moreover, if hydrostatic pressure has no effect on yielding, we can add a hydrostatic state of stress, $(h, h, h)$ say, to a yield stress state to generate another yield point. For example, if a hydrostatic pressure $(-3\sigma, -3\sigma, -3\sigma)$ is added to the yield stress point $(3\sigma, \sigma, 0)$, then the stress state $(0, -2\sigma, -3\sigma)$ is another yield point. Now, we alter its coordinates such that a yield point $B_1(-2\sigma, -3\sigma, 0)$ is obtained on the $\sigma_1$-$\sigma_2$ plane. Similarly, one can get another new yield point $C_1(2\sigma, -\sigma, 0)$ on the $\sigma_1$-$\sigma_2$ plane by adding $(-\sigma, -\sigma, -\sigma)$ to $(3\sigma, \sigma, 0)$ and altering the coordinates correspondingly. Finally, because of symmetry, points $B_1$ and $C_1$, like point $A_1$, can generate four points $B_1, B_2, B_3, B_4$ and $C_1, C_2, C_3, C_4$, respectively, lying on the

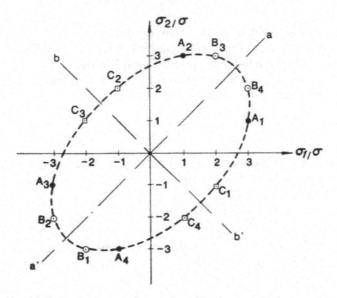

FIGURE 2.11. Yield locus on $\sigma_1$-$\sigma_2$ plane ($\sigma_3 = 0$) generated from one test point, $A_1$.

yield locus. Now, we have generated a total of twelve yield points on the $\sigma_1-\sigma_2$ plane from one test point. Connecting these points with a smooth curve, we construct a yield locus as shown in Fig. 2.11. Noting that this locus is generated from only one radial test point, it can be considered an approximation of the yield function of a biaxial state of stress for a material with isotropy, with the same response to tension and compression, and with no hydrostatic pressure effect on yielding.

We have discussed so far the general form and some characteristics of a yield function. The very useful yield criteria of Tresca and von Mises for metals will be studied in the following sections.

## 2.2.2. The Tresca Yield Criterion

Historically, the first yield criterion for a combined state of stress for metals was that proposed in 1864 by Tresca, who suggested that yielding would occur when the maximum shearing stress at a point reaches a critical value $k$. Stating this in terms of principal stresses (see Section 2.1.4), one-half of the greatest absolute value of the differences between the principal stresses taken in pairs must be equal to $k$ at yield, namely,

$$\text{Max}(\tfrac{1}{2}|\sigma_1-\sigma_2|, \tfrac{1}{2}|\sigma_2-\sigma_3|, \tfrac{1}{2}|\sigma_3-\sigma_1|) = k \qquad (2.135)$$

where the material constant $k$ may be determined from the simple tension test. Then

$$k = \frac{\sigma_0}{2} \qquad (2.136)$$

in which $\sigma_0$ is the yield stress in simple tension.

There are six different expressions in various regions of the $\sigma_1-\sigma_2$ plane, depending upon the relative magnitudes and the signs of $\sigma_1$ and $\sigma_2$ (see Fig. 2.12). In the first quadrant, between the $\sigma_1$-axis and the bisector of the two axes, the order of the stresses requires that

$$\tau_{max} = \frac{\sigma_1}{2}$$

Hence, the yield criterion becomes $\sigma_1 = \sigma_0$ and gives the line $AB$. In the same quadrant, between the bisector and the $\sigma_2$-axis, we have

$$\tau_{max} = \frac{\sigma_2}{2}$$

and the yield criterion $\sigma_2 = \sigma_0$ is represented by the line $BC$. In the second quadrant, we have

$$\tau_{max} = \frac{\sigma_2 - \sigma_1}{2}$$

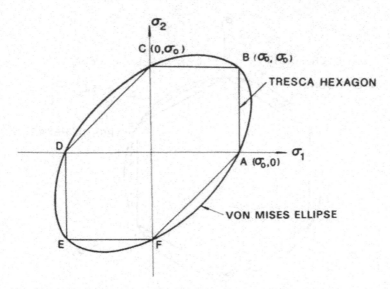

FIGURE 2.12. Yield criteria matched in tension in the coordinate plane $\sigma_3 = 0$.

Thus, the yield criterion becomes $\sigma_2 - \sigma_1 = \sigma_0$, and line $CD$ is obtained. By proceeding similarly for the third and fourth quadrants, it can be found that the yield locus for plane stress is a hexagon $ABCDEF$ as shown in Fig. 2.12.

To represent the yield surface in the principal stress space, Eq. (2.123) is used here for the principal stresses. Assuming the ordering of stresses to be $\sigma_1 > \sigma_2 > \sigma_3$, we can rewrite Eq. (2.135) in the form

$$\frac{1}{2}(\sigma_1 - \sigma_3) = \frac{1}{\sqrt{3}}\sqrt{J_2}\left[\cos\theta - \cos\left(\theta + \frac{2}{3}\pi\right)\right] = k \qquad (0 \le \theta \le 60°) \qquad (2.137)$$

Expanding this equation and noting Eq. (2.136), we obtain the Tresca criterion in terms of stress invariants,

$$f(J_2, \theta) = 2\sqrt{J_2}\sin(\theta + \tfrac{1}{3}\pi) - \sigma_0 = 0 \qquad (0 \le \theta \le 60°) \qquad (2.138)$$

or identically in terms of the variables $\xi$, $\rho$, $\theta$,

$$f(\rho, \theta) = \sqrt{2}\rho\sin(\theta + \tfrac{1}{3}\pi) - \sigma_0 = 0 \qquad (2.139)$$

Since the hydrostatic pressure has no effect on the yield surface, Eq. (2.138) or Eq. (2.139) must be independent of hydrostatic pressure $I_1$ or $\xi$, representing a cylindrical surface whose generator is parallel to the hydrostatic axis. On the deviatoric plane, Eq. (2.138) or Eq. (2.139) is a straight line passing through point $A$ (with $\theta = 0$, and $\rho = \sqrt{2/3}\sigma_0$) and point $B$ (with $\theta = 60°$ and the same $\rho$ as point $A$), as shown in Fig. 2.13. This is one sector of the yield locus on the deviatoric plane. Each of the five other possible orderings of the magnitudes of the principal stresses gives similar lines in the appropriate sectors of the yield locus on the deviatoric plane, and a regular hexagon

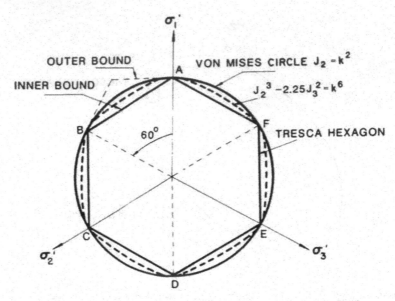

FIGURE 2.13. Yield criteria matched in tension in a deviatoric plane.

*ABCDEF* is thus obtained. Now we can see that the yield surface is a regular hexagonal prism in principal stress space, as shown in Fig. 2.14. The yield locus for a biaxial state of stress shown in Fig. 2.12 is the intersection of the cylinder with the coordinate plane $\sigma_3 = 0$.

Isotropy means that there is no need to draw the yield surface in a general stress space ($\sigma_{ij}$). Nevertheless, some intersections of particular planes with

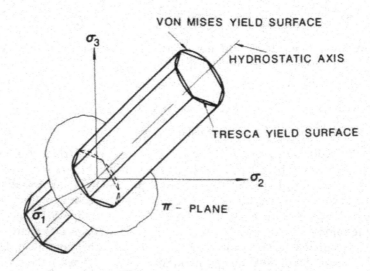

FIGURE 2.14. Yield surfaces in principal stress space.

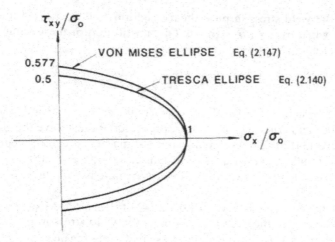

FIGURE 2.15. Intersection of the $\sigma_x$-$\tau_{xy}$ plane with the yield surface.

the surface in general stress space are of interest, e.g., the intersection with the $\sigma_x$-$\tau_{xy}$ plane. In simpler language, the latter intersection is the yield locus for combined normal stress and shear (Fig. 2.15), which is an ellipse

$$\sigma_x^2 + 4\tau_{xy}^2 = \sigma_0^2 \tag{2.140}$$

It is of interest to note that the invariant form of Eq. (2.137) can also be expressed explicitly in terms of the invariants $J_2$ and $J_3$ as

$$f(J_2, J_3) = 4J_2^3 - 27J_3^2 - 36k^2J_2^2 + 96k^4J_2 - 64k^6 = 0 \tag{2.141}$$

## 2.2.3. The von Mises Yield Criterion

Although the maximum shearing stress criterion is simple, it does not reflect any influence of the intermediate principal stress. The octahedral shearing stress or the strain energy of distortion is a convenient alternative choice to the maximum shearing stress as the key variable for causing yielding of materials which are pressure independent. The von Mises yield criterion, dating from 1913, is based on this alternative. It states that yielding begins when the octahedral shearing stress reaches a critical value $k$. From Eq. (2.73), it must have the form

$$\tau_{oct} = \sqrt{\tfrac{2}{3}J_2} = \sqrt{\tfrac{2}{3}}k \tag{2.142}$$

which reduces to the simple form

$$f(J_2) = J_2 - k^2 = 0 \tag{2.143}$$

or, written in terms of principal stresses,

$$(\sigma_1 - \sigma_2)^2 + (\sigma_2 - \sigma_3)^2 + (\sigma_3 - \sigma_1)^2 = 6k^2 \tag{2.144}$$

where $k$ is the yield stress in pure shear. Yielding will occur in a uniaxial tension test when $\sigma_1 = \sigma_0$, $\sigma_2 = \sigma_3 = 0$. On substitution of these values into Eq. (2.144), one finds

$$k = \frac{\sigma_0}{\sqrt{3}} \tag{2.145}$$

We know from our earlier discussions that for pressure-independent materials, the yield criterion for an isotropic material must have the general form of Eq. (2.134). It follows that the simplest mathematical form compatible with this requirement is Eq. (2.143). This equation represents a circular cylinder whose intersection with the deviatoric plane is a circle with radius $\rho = \sqrt{2}k$.

Note that the constant $k$, both in Eq. (2.143) for the von Mises criterion and in Eq. (2.135) for the Tresca criterion, is the yield stress in pure shear. However, relations between the yield stress in simple tension, $\sigma_0$, and the parameter $k$ defined by Eq. (2.136) of the Tresca criterion and by Eq. (2.145) of the von Mises criterion are different. If the two criteria are made to agree for a simple tension yield stress $\sigma_0$, the ratio of the yield stress in shear, $k$, between the von Mises and Tresca criteria is $2/\sqrt{3} = 1.15$, and graphically, the von Mises circle circumscribes the Tresca hexagon as shown in Fig. 2.13. However, if the two criteria are made to agree for the case of pure shear (same $k$ value), the circle will inscribe the hexagon.

The von Mises criterion for a biaxial state of stress is represented by the intersection of the circular cylinder with the coordinate plane $\sigma_3 = 0$, i.e.,

$$\sigma_1^2 + \sigma_2^2 - \sigma_1 \sigma_2 = \sigma_0^2 \tag{2.146}$$

which is an ellipse shown in Fig. 2.12. The intersection of the von Mises surface in general stress space with the $\sigma_x$-$\tau_{xy}$ plane is also an ellipse, given by

$$\sigma_x^2 + 3\tau_{xy}^2 = \sigma_0^2 \tag{2.147}$$

as shown in Fig. 2.15.

EXAMPLE 2.4. A thin-walled steel cylindrical vessel with diameter $D = 50.8$ cm and wall thickness $t = 6.35$ mm is subjected to an interior pressure $p$ as shown in Fig. 2.16a. The ends of the tube are closed. The yield stress of the steel is $\sigma_0 = 225$ MPa. According to (a) the Tresca criterion and (b) the von Mises criterion, find the pressure $p_y$ under which the vessel begins to yield.

SOLUTION. The state of stress for an element at the wall of the thin-walled pressure vessel is considered biaxial as shown in Fig. 2.16b, in which the circumferential stress $\sigma_c$ and the axial stress $\sigma_a$ are given by

$$\sigma_c = \frac{pD}{2t}, \qquad \sigma_a = \frac{pD}{4t} \tag{2.148}$$

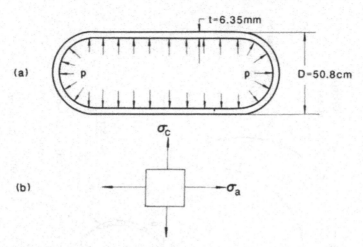

FIGURE 2.16. (a) A thin-walled cylindrical pressure vessel. (b) An element at the wall considered to be in a state of biaxial stress.

although the interior pressure acting on the wall causes a local compressive stress equal to this pressure $p$. Actually, a triaxial state of stress exists on the inside of the vessel. However, for a thin-walled cylindrical vessel, $D/t \gg 1$, this latter stress, $\sigma_r = p$, is much smaller than $\sigma_a$ and $\sigma_c$ and thus is ignored.

(a) Tresca criterion: Clearly, the ordering of the principal stresses is

$$\sigma_1 = \sigma_c, \ \sigma_2 = \sigma_a, \ \sigma_3 = \sigma_r = 0 \qquad (2.149)$$

Hence, the Tresca yield condition is represented as

$$\tau_{max} = \frac{\sigma_1 - \sigma_3}{2} = \frac{\sigma_c}{2} = \frac{\sigma_0}{2} \qquad (2.150)$$

Substituting Eq. (2.148) into Eq. (2.150), solve for the yield pressure $p_y$ as

$$p_y = \frac{2 t \sigma_0}{D} = \frac{(2)(0.00635)(225)}{0.508} = 5.625 \text{ MPa}$$

(b) von Mises criterion: For a state of biaxial stress, the von Mises criterion of Eq. (2.146) is applied, i.e.,

$$\sigma_a^2 + \sigma_c^2 - \sigma_a \sigma_c = \sigma_0^2 \qquad (2.151)$$

Substitution of Eq. (2.148) into Eq. (2.151) yields

$$\frac{p^2 D^2}{16 t^2} + \frac{p^2 D^2}{4 t^2} - \frac{p^2 D^2}{8 t^2} = \sigma_0^2$$

The pressure satisfying the above condition is obtained as

$$p_y = \frac{4}{\sqrt{3}} \frac{t}{D} \sigma_0 = \frac{4}{\sqrt{3}} \frac{(0.00635)(225)}{0.508} = 6.495 \text{ MPa}$$

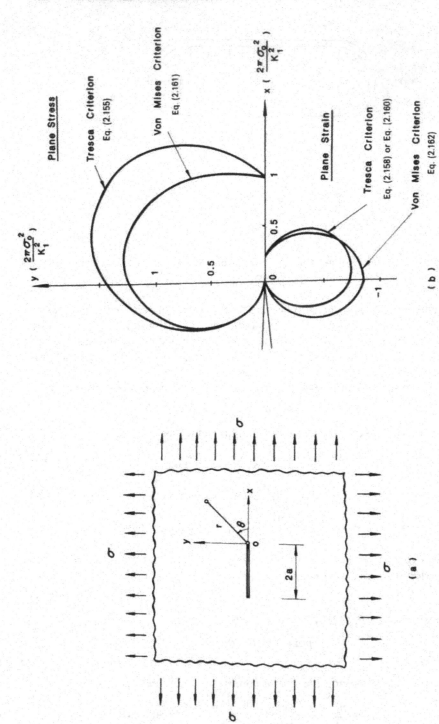

FIGURE 2.17. (a) A plane with a line crack stretched at infinity. (b) Elastic–plastic boundary near the crack tip (considered symmetrical about the x-axis).

EXAMPLE 2.5. A plane with a line crack of length $2a$ is subjected to a biaxial stress at infinity as shown in Fig. 2.17a. If the origin of the coordinate system is located at the crack tip, the stress field near the crack tip can be expressed by

$$\sigma_x = \frac{K_1}{\sqrt{2\pi r}} \cos\frac{\theta}{2}\left[1 - \sin\frac{\theta}{2}\sin\frac{3\theta}{2}\right]$$

$$\sigma_y = \frac{K_1}{\sqrt{2\pi r}} \cos\frac{\theta}{2}\left[1 + \sin\frac{\theta}{2}\sin\frac{3\theta}{2}\right] \tag{2.152}$$

$$\tau_{xy} = \frac{K_1}{\sqrt{2\pi r}} \sin\frac{\theta}{2}\cos\frac{3\theta}{2}\cos\frac{\theta}{2}$$

where $K_1$ is the *stress intensity factor*. Determine the plastic zone boundary based on (a) the Tresca criterion and (b) the von Mises criterion.

SOLUTION. (a) Tresca criterion: First we must find the principal stresses from Mohr's circle as below:

$$\sigma_{1,2} = \frac{1}{2}(\sigma_x + \sigma_y) \pm \sqrt{\left(\frac{\sigma_x - \sigma_y}{2}\right)^2 + \tau_{xy}^2}$$

From substitution, we obtain

$$\sigma_1 = \frac{K_1}{\sqrt{2\pi r}} \cos\frac{\theta}{2}\left(1 + \sin\frac{\theta}{2}\right)$$

$$\sigma_2 = \frac{K_1}{\sqrt{2\pi r}} \cos\frac{\theta}{2}\left(1 - \sin\frac{\theta}{2}\right) \tag{2.153}$$

(i) Plane stress case: From $\sigma_3 = 0$, we have

$$\sigma_1 > \sigma_2 > \sigma_3 = 0 \qquad \text{for } 0 \le \theta \le \pi$$

and the governing yield criterion is

$$\sigma_1 = \frac{K_1}{\sqrt{2\pi r}} \cos\frac{\theta}{2}\left(1 + \sin\frac{\theta}{2}\right) = \sigma_0 \tag{2.154}$$

The plastic zone boundary is obtained as

$$r = \frac{K_1^2}{2\pi\sigma_0^2} \cos^2\frac{\theta}{2}\left(1 + \sin\frac{\theta}{2}\right)^2 \tag{2.155}$$

(ii) Plane strain case:

$$\sigma_3 = \nu(\sigma_x + \sigma_y) = \nu(\sigma_1 + \sigma_2) \tag{2.156}$$

For $\nu < 0.5$, $\sigma_1$ is always the largest of the principal stresses. However, for $\sigma_2$ and $\sigma_3$, there are two possibilities, i.e.,

$$\sigma_1 > \sigma_2 > \sigma_3$$

or

$$\sigma_1 > \sigma_3 > \sigma_2$$

depending on the value of Poisson's ratio $\nu$.

If $\sigma_1 > \sigma_2 > \sigma_3$, the condition

$$\sigma_1(1-\nu) - \nu\sigma_2 = \sigma_0 \tag{2.157}$$

governs yielding. Substituting Eq. (2.153) into Eq. (2.157) and rearranging leads to the following expression for the boundary of the plastic zone:

$$r_1 = \frac{K_1^2}{2\pi\sigma_0^2} \cos^2 \frac{\theta}{2} \left[(1-2\nu) + \sin\frac{\theta}{2}\right]^2 \tag{2.158}$$

On the other hand, if $\sigma_1 > \sigma_3 > \sigma_2$, the yield condition becomes

$$\sigma_1 - \sigma_2 = \sigma_0 \tag{2.159}$$

and the plastic zone boundary is obtained as

$$r_2 = \frac{K_1^2}{2\pi\sigma_0^2} \sin^2 \theta \tag{2.160}$$

In summary, the real plastic zone boundary $r$ is determined by taking the larger value of $r_1$ and $r_2$ given by Eq. (2.158) or Eq. (2.160) respectively, depending on the value of Poisson's ratio $\nu$.

(b) von Mises criterion: We have already determined the three principal stresses $\sigma_1$, $\sigma_2$, and $\sigma_3$. The procedure to obtain the plastic zone boundary is quite straightforward.

(i) Plane stress case: Substituting Eq. (2.153) and $\sigma_3 = 0$ into the von Mises yield condition given by Eqs. (2.144) and (2.145) and rearranging leads to

$$r = \frac{K_1^2}{2\pi\sigma_0^2} \cos^2 \frac{\theta}{2} \left(1 + 3 \sin^2 \frac{\theta}{2}\right) \tag{2.161}$$

(ii) Plane strain case: Using Eq. (2.156) for $\sigma_3$ and following the same procedure, we get

$$r = \frac{K_1^2}{2\pi\sigma_0^2} \cos^2 \frac{\theta}{2} \left[(1-2\nu)^2 + 3 \sin^2 \frac{\theta}{2}\right] \tag{2.162}$$

The plastic zone boundaries given by Eqs. (2.155), (2.158), (2.160), (2.161) and (2.162) are sketched for $\nu = 0.25$ in terms of the dimensionless ratio $r(2\pi\sigma_0^2 / K_1^2)$ in Fig. 2.17b.

## 2.2.4. Comments on the Tresca and von Mises Criteria

For an isotropic material whose yielding is independent of hydrostatic pressure, the yield criterion must be a cylinder with generator parallel to the hydrostatic axis. Hence, the complete shape of the yield surface is determined by the cross section with the deviatoric plane. Further, if the yield stresses in tension and in compression are equal, such a cross section must have the sixfold symmetry shown in Fig. 2.13. It follows that a typical section can be determined experimentally by exploring only one of the typical 30° sectors. On the basis of energy considerations, it can be shown that for a wide class of materials, the yield surface must be convex (see Chapters 3 and 4). If we accept the fact that the yield surface is convex, it must lie between the two hexagons shown in Fig. 2.13. The inner Tresca hexagon is obviously a lower bound on the yield curve, and the von Mises cylinder gives a somewhat average value between the outer and inner bounds.

In short, on the basis of the four assumptions of (1) isotropy, (2) hydrostatic pressure independence, (3) equal yield stresses in tension and compression, and (4) convexity, the general shape of the yield surface can be well defined, and the von Mises cylinder cannot deviate much from the actual yield surface $f(J_2, J_3) = 0$.

As a matter of fact, there have been many experimental results showing that the yield points fall between the Tresca hexagon and the von Mises circle and closer to the latter. Osgood in 1947, among many others, performed radial loading tests on thin-wall aluminum tubes, and the results were correlated to an equivalent shearing stress defined by

$$\tau_{eq} = \sqrt[6]{\tfrac{2}{3}(J_2^3 - 2.25 J_3^2)} = \tau_{oct}\sqrt[6]{1 - 2.25 J_3^2/J_2^3} \qquad (2.163)$$

A plot of $\tau_{eq}$ against the octahedral shear strain $\gamma_{oct}$, as shown in Fig. 2.18, shows that the equivalent shearing stress $\tau_{eq}$ is a good parameter for

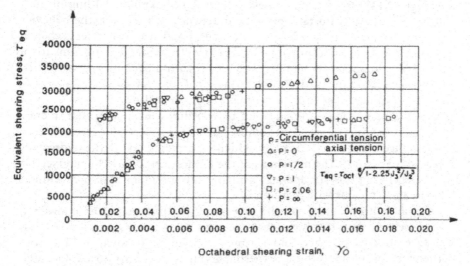

FIGURE 2.18. Osgood's test on thin-walled tubes under tension and interior pressure.

characterizing yielding as well as hardening of the material. The yield criterion is therefore expressed as

$$J_2^3 - 2.25 J_3^2 = k^6 \tag{2.164}$$

where $k$ is the yield stress in pure shear, which is related to the yield stress $\sigma_0$ in simple tension by

$$k = \sqrt[6]{\tfrac{2}{81}}\,\sigma_0 = 0.54 \sigma_0 \tag{2.165}$$

Comparing to $k = 0.5\sigma_0$ of Tresca and $k = 0.577\sigma_0$ of von Mises, the yield stress in pure shear falls between the values predicted by Tresca and von Mises. The yield curve of Eq. (2.165) as plotted in Fig. 2.13 does lie between the Tresca hexagon and the von Mises circle and passes through most of the experimental points.

## 2.3. Failure Criterion for Pressure-Dependent Materials

### 2.3.1. Characteristics of the Failure Surface of an Isotropic Material

Failure of a material is usually defined in terms of its load-carrying capacity. However, for perfectly plastic materials, yielding itself implies failure, so the yield stress is also the limit of strength.

As in the case of the yield criteria, a general form of the failure criteria can be given by Eq. (2.130) for anisotropic materials and by Eqs. (2.131) through (2.133) for isotropic ones. As we already know, yielding of most ductile metals is hydrostatic pressure independent. However, the behavior of many nonmetallic materials, such as soils, rocks, and concrete, is characterized by its hydrostatic pressure dependence. Therefore, the stress invariant $I_1$ or $\xi$ should not be omitted from Eq. (2.132) and Eq. (2.133), respectively.

The general shape of a failure surface, $f(I_1, J_2, J_3) = 0$ or $f(\xi, \rho, \theta) = 0$, in a three-dimensional stress space can be described by its cross-sectional shapes in the deviatoric planes and its meridians in the meridian planes. The cross sections of the failure surface are the intersection curves between this surface and a deviatoric plane which is perpendicular to the hydrostatic axis with $\xi = \text{const}$. The meridians of the failure surface are the intersection curves between this surface and a plane (the meridian plane) containing the hydrostatic axis with $\theta = \text{const}$.

For an isotropic material, the labels 1, 2, 3 attached to the coordinate axes are arbitrary; it follows that the cross-sectional shape of the failure surface must have a threefold symmetry of the type shown in Fig. 2.19b. Therefore, when performing experiments, it is necessary to explore only the sector $\theta = 0°$ to $60°$, the other sectors being known by symmetry.

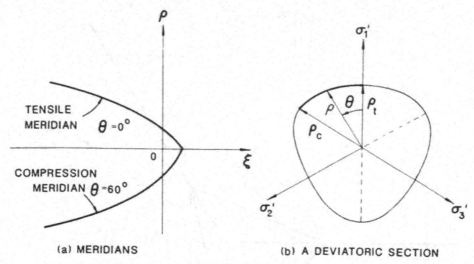

(a) MERIDIANS                    (b) A DEVIATORIC SECTION

FIGURE 2.19. General shape of the failure surface for an isotropic material.

The typical sector shown in Fig. 2.19b by a heavy line corresponds to the regular ordering of the principal stresses, $\sigma_1 \geq \sigma_2 \geq \sigma_3$. Within this ordering, there are two extreme cases:

$$\sigma_1 = \sigma_2 > \sigma_3 \tag{2.166}$$

and

$$\sigma_1 > \sigma_2 = \sigma_3 \tag{2.167}$$

corresponding to $\theta_1 = 60°$ and $\theta_2 = 0°$, respectively. To show this, we substitute Eqs. (2.166) and (2.167) into Eq. (2.115) and get

$$\cos \theta_1 = \frac{\sqrt{3}}{2} \frac{s_1}{\sqrt{J_2}} = \frac{2\sigma_1 - \sigma_1 - \sigma_3}{2\sqrt{3}\sqrt{\frac{2}{6}(\sigma_1 - \sigma_3)^2}} = \frac{1}{2}$$

and

$$\cos \theta_2 = \frac{2\sigma_1 - \sigma_3 - \sigma_3}{2\sqrt{3}\sqrt{\frac{2}{6}(\sigma_1 - \sigma_3)^2}} = 1$$

respectively. The meridian corresponding to $\theta_1 = 60°$ is called the *compression meridian* in that Eq. (2.166) represents a stress state corresponding to a hydrostatic stress state with a compressive stress superimposed in one direction. The meridian determined by $\theta = 0°$, corresponding to Eq. (2.167), represents a hydrostatic stress state with a tensile stress superimposed in one direction and is therefore called the *tensile meridian*.

Furthermore, the meridian determined by $\theta = 30°$ is sometimes called the *shear meridian*. It also follows from the definition of cos $\theta$ in Eq. (2.115) that this equation is fulfilled for $\theta = 30°$, when the stresses are $\sigma_1, (\sigma_1 + \sigma_3)/2,$

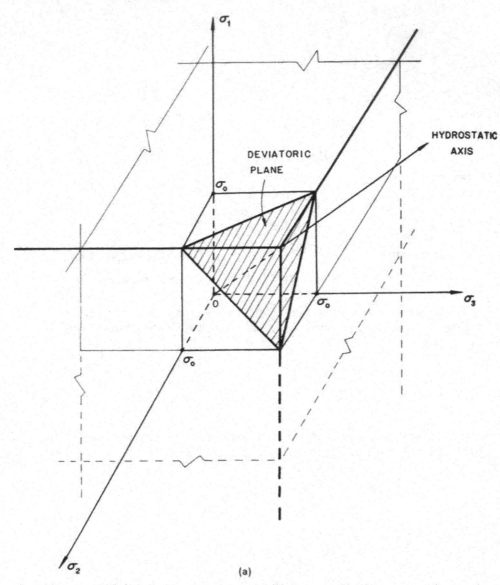

FIGURE 2.20. (a) Rankine maximum-principal-stress criterion; cross sections of Rankine criterion: (b) meridian plane ($\theta = 0°$); (c) $\pi$-plane. Figure continues on next page.

and $\sigma_3$, which is a pure shear state $\frac{1}{2}(\sigma_1 - \sigma_3, 0, \sigma_3 - \sigma_1)$ with a hydrostatic stress state $\frac{1}{2}(\sigma_1 + \sigma_3)$ superimposed.

Based on the above considerations, a general shape of the failure surface for an isotropic material may be illustrated in Haigh–Westergaard stress space as shown in Fig. 2.19a. We shall consider this in more detail in the following discussion of some simple failure criteria.

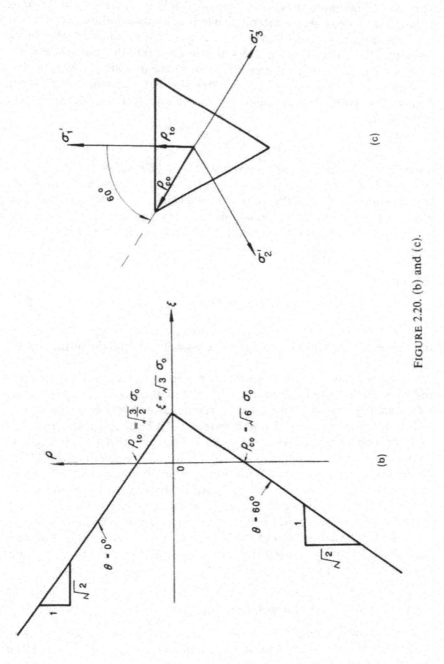

FIGURE 2.20. (b) and (c).

## 2.3.2. The Maximum-Tensile-Stress Criterion (Rankine)

The maximum-tensile-stress criterion of Rankine, dating from 1876, is generally accepted today to determine whether a tensile failure has occurred for a brittle material. According to this criterion, brittle failure takes place when the maximum principal stress at a point inside the material reaches a value equal to the tensile strength $\sigma_0$ as found in a simple tension test, regardless of the normal or shearing stresses that occur on other planes through this point. The equations for the failure surface defined by this criterion are

$$\sigma_1 = \sigma_0, \qquad \sigma_2 = \sigma_0, \qquad \sigma_3 = \sigma_0 \qquad (2.168)$$

which result in three planes perpendicular to the $\sigma_1$, $\sigma_2$, and $\sigma_3$ axes, respectively as shown in Fig. 2.20a. This surface will be referred to as the tension-failure surface or the simple *tension cutoff*. When the variables $\xi$, $\rho$, $\theta$ or $I_1$, $J_2$, $\theta$ are used, the failure surface can be fully described by the following equations within the range $0 \leq \theta \leq 60°$ using Eq. (2.123).

$$f(I_1, J_2, \theta) = 2\sqrt{3J_2} \cos \theta + I_1 - 3\sigma_0 = 0 \qquad (2.169)$$

or identically

$$f(\xi, \rho, \theta) = \sqrt{2}\rho \cos \theta + \xi - \sqrt{3}\sigma_0 = 0 \qquad (2.170)$$

Figures (2.20b and c) show the cross-sectional shape on the $\pi$-plane ($\xi = 0$) and the tensile ($\theta = 0°$) and compressive ($\theta = 60°$) meridians of the failure surface.

As we know, some of the nonmetallic materials, such as concrete, rocks, and soils, have a good compressive strength. Under compression loading with confining pressure, this kind of material may even exhibit some ductile and shear failure behavior. Under tension loads, however, a brittle failure behavior with a very low tensile strength is generally observed. Hence, the Rankine criterion is sometimes combined with the Tresca or the von Mises criterion to approximate the failure behavior of such materials. The combined criteria are referred to as the Tresca or the von Mises criterion with a tension cutoff, and their graphical representations consist of two surfaces, corresponding to a combined behavior of shear failure in compression and tensile failure in tension. An example of such failure surfaces is shown in Fig. 2.21, in which the compressive strength is assumed three times as large as the tensile strength.

## 2.3.3. The Mohr–Coulomb Criterion

Mohr's criterion, dating from 1900, may be considered as a generalized version of the Tresca criterion. Both criteria are based on the assumption that the maximum shear stress is the only decisive measure of impending failure. However, while the Tresca criterion assumes that the critical value

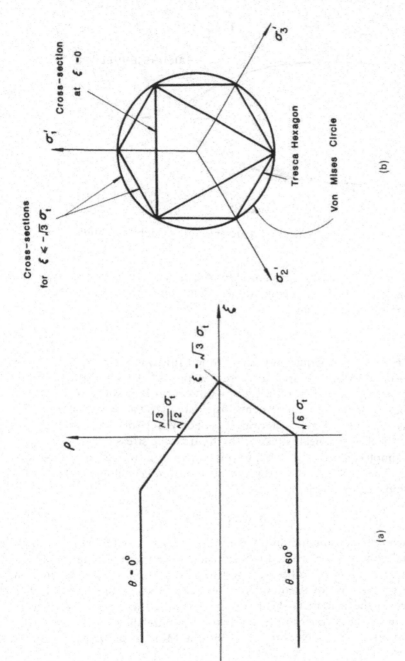

(b)

(a)

FIGURE 2.21. Tresca and von Mises criteria with tension cutoff: (a) meridian section ($\theta = 0°$); (b) cross sections.

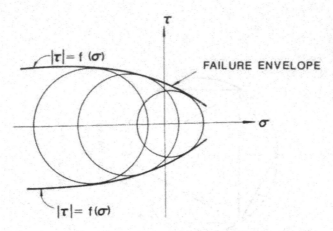

FIGURE 2.22. Graphical representation of Mohr's criterion.

of the shear stress is a constant, Mohr's failure criterion considers the limiting shear stress $\tau$ in a plane to be a function of the normal stress $\sigma$ in the same plane at a point, i.e.,

$$|\tau| = f(\sigma) \qquad (2.171)$$

where $f(\sigma)$ is an experimentally determined function.

In terms of Mohr's graphical representation of the state of stress, Eq. (2.171) means that failure of material will occur if the radius of the largest principal circle is tangent to the envelope curve $f(\sigma)$ as shown in Fig. 2.22. In constrast to the Tresca criterion, it is seen that Mohr's criterion allows for the effect of the mean stress or the hydrostatic stress.

The simplest form of the Mohr envelope $f(\sigma)$ is a straight line, illustrated in Fig. 2.23. The equation for the straight-line envelope is known as Coulomb's equation, dating from 1773,

$$|\tau| = c - \sigma \tan \phi \qquad (2.172)$$

in which $c$ is the cohesion and $\phi$ is the angle of internal friction; both are material constants determined by experiment. The failure criterion associated with Eq. (2.172) will be referred to as the Mohr–Coulomb criterion. In the special case of frictionless materials, for which $\phi = 0$, Eq. (2.172) reduces to the maximum-shear-stress criterion of Tresca, $\tau = c$, and the cohesion becomes equal to the yield stress in pure shear $c = k$.

From Eq. (2.172) and for $\sigma_1 \geq \sigma_2 \geq \sigma_3$, the Mohr–Coulomb criterion can be written as

$$\frac{1}{2}(\sigma_1 - \sigma_3) \cos \phi = c - \left[ \frac{1}{2}(\sigma_1 + \sigma_3) + \frac{\sigma_1 - \sigma_3}{2} \sin \phi \right] \tan \phi \qquad (2.173)$$

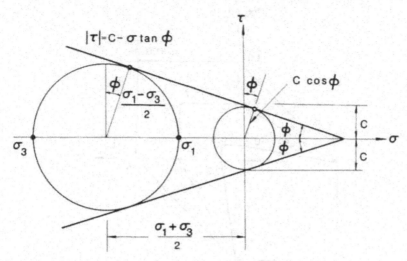

FIGURE 2.23. Mohr–Coulomb criterion: with straight line as failure envelope.

or rearranging

$$\sigma_1 \frac{1+\sin\phi}{2c\cos\phi} - \sigma_3 \frac{1-\sin\phi}{2c\cos\phi} = 1 \qquad (2.174)$$

If we define

$$f_c' = \frac{2c\cos\phi}{1-\sin\phi} \qquad (2.175)$$

and

$$f_t' = \frac{2c\cos\phi}{1+\sin\phi} \qquad (2.176)$$

Eq. (2.174) is further reduced to

$$\frac{\sigma_1}{f_t'} - \frac{\sigma_3}{f_c'} = 1 \qquad \text{for } \sigma_1 \geq \sigma_2 \geq \sigma_3 \qquad (2.177)$$

It is clear from Eq. (2.177) that $f_t'$ is the strength in simple tension while $f_c'$ is the strength in simple compression.

It is sometimes convenient to introduce a parameter $m$, where

$$m = \frac{f_c'}{f_t'} = \frac{1+\sin\phi}{1-\sin\phi} \qquad (2.178)$$

Then Eq. (2.177) can be written in the slope–intercept form

$$m\sigma_1 - \sigma_3 = f_c' \qquad \text{for } \sigma_1 \geq \sigma_2 \geq \sigma_3 \qquad (2.179)$$

Similarly to what we have done for the Tresca criterion, $\sigma_1 - \sigma_3 = \sigma_0$, the failure locus for the Mohr–Coulomb criterion in the $\sigma_1 - \sigma_2$ plane can be sketched based on Eq. (2.179) for several values of $m$. The failure loci are irregular hexagons as shown in Fig. 2.24.

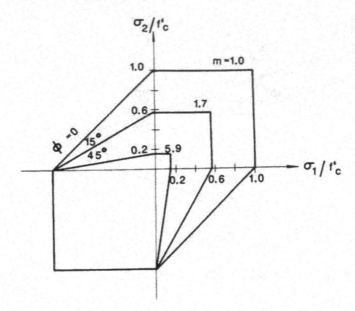

FIGURE 2.24. Mohr–Coulomb criterion in the coordinate plane $\sigma_3 = 0$.

To demonstrate the shape of the three-dimensional failure surface of the Mohr–Coulomb criterion, we again use Eq. (2.123) and rewrite Eq. (2.174) in the following form:

$$f(I_1, J_2, \theta) = \frac{1}{3} I_1 \sin \phi + \sqrt{J_2} \sin \left( \theta + \frac{\pi}{3} \right)$$

$$+ \frac{\sqrt{J_2}}{\sqrt{3}} \cos \left( \theta + \frac{\pi}{3} \right) \sin \phi - c \cos \phi = 0 \qquad (2.180)$$

or identically in terms of variables $\xi, \rho, \theta$:

$$f(\xi, \rho, \theta) = \sqrt{2}\xi \sin \phi + \sqrt{3}\rho \sin \left( \theta + \frac{\pi}{3} \right)$$

$$+ \rho \cos \left( \theta + \frac{\pi}{3} \right) \sin \phi - \sqrt{6}c \cos \phi = 0 \qquad (2.181)$$

with $0 \le \theta \le \pi/3$.

In principal stress space, this gives an irregular hexagonal pyramid. Its meridians are straight lines (Fig. 2.25a), and its cross section in the $\pi$-plane is an irregular hexagon (Fig. 2.25b). Only two characteristic lengths are required to draw this hexagon: the lengths $\rho_{t0}$ and $\rho_{c0}$, which can be obtained directly from Eq. (2.181) with $\xi = 0$, $\theta = 0°$, $\rho = \rho_{t0}$ and $\xi = 0$, $\theta = 60°$, $\rho = \rho_{c0}$. Using Eqs. (2.175) and (2.176), we have the following alternative forms for

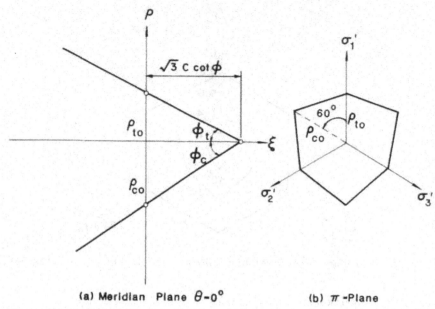

(a) Meridian Plane $\theta$-0°          (b) $\pi$ -Plane

FIGURE 2.25. Graphical representation of Mohr–Coulomb criterion in principal stress space.

$\rho_{to}$ and $\rho_{co}$ on the $\pi$-plane:

$$\rho_{to} = \frac{2\sqrt{6}c \cos \phi}{3+\sin \phi} = \frac{\sqrt{6}f_c'(1-\sin \phi)}{3+\sin \phi} \tag{2.182}$$

$$\rho_{co} = \frac{2\sqrt{6}c \cos \phi}{3-\sin \phi} = \frac{\sqrt{6}f_c'(1-\sin \phi)}{3-\sin \phi} \tag{2.183}$$

and the ratio of these lengths is given by

$$\frac{\rho_{to}}{\rho_{co}} = \frac{3-\sin \phi}{3+\sin \phi} \tag{2.184}$$

A family of Mohr–Coulomb cross sections in the $\pi$-plane for several values of $\phi$ is shown in Fig. 2.26, where the stresses have been normalized with respect to the compressive strength $f_c'$. Obviously, the hexagons shown in Fig. 2.24 are the intersections of the pyramid with the coordinate plane $\sigma_3 = 0$. When $f_c' = f_t'$ (or equivalently, when $\phi = 0$ or $m = 1$), the hexagon becomes identical with Tresca's hexagon, as it should.

To obtain a better approximation when tensile stresses occur, it is sometimes necessary to combine the Mohr–Coulomb criterion with a maximum-tensile-strength cutoff. It should be noted that this combined criterion is a three-parameter criterion. We need two stress states to determine the values of $c$ and $\phi$ and one stress state to determine the maximum tensile stress.

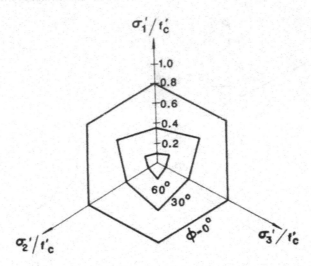

FIGURE 2.26. Failure curves for Mohr–Coulomb criterion in the deviatoric planes.

### 2.3.4. The Drucker–Prager Criterion

As we have seen, the Mohr–Coulomb failure criterion can be considered a generalized Tresca criterion accounting for the hydrostatic pressure effect. The Drucker–Prager criterion, formulated in 1952, is a simple modification of the von Mises criterion, where the influence of a hydrostatic stress component on failure is introduced by inclusion of an additional term in the von Mises expression to give

$$f(I_1, J_2) = \alpha I_1 + \sqrt{J_2} - k = 0 \qquad (2.185)$$

Using variables $\xi$ and $\rho$ leads to

$$f(\xi, \rho) = \sqrt{6}\alpha\xi + \rho - \sqrt{2}k = 0 \qquad (2.186)$$

where $\alpha$ and $k$ are material constants. When $\alpha$ is zero, Eq. (2.186) reduces to the von Mises criterion.

The failure surface of Eq. (2.186) in principal stress space is clearly a right-circular cone. Its meridian and cross section on the $\pi$-plane are shown in Fig. 2.27.

The Mohr–Coulomb hexagonal failure surface is mathematically convenient only in problems where it is obvious which one of the six sides is to be used. If this information is not known in advance, the corners of the hexagon can cause considerable difficulty and give rise to complications in obtaining a numerical solution. The Drucker–Prager criterion, as a smooth approximation to the Mohr–Coulomb criterion, can be made to match the latter by adjusting the size of the cone. For example, if the Drucker–Prager circle is made to agree with the outer apices of the Mohr–Coulomb hexagon, i.e., the two surfaces are made to coincide along the compression meridian

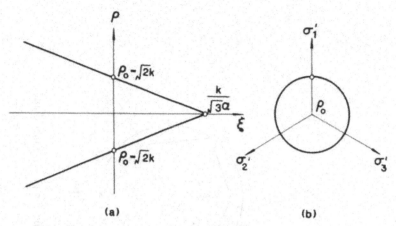

(a)                                                    (b)

FIGURE 2.27. Drucker–Prager criterion: (a) meridian plane, $\theta = 0°$; (b) $\pi$-plane.

$p_c$, where $\theta = 60°$, then the constants $\alpha$ and $k$ in Eq. (2.185) are related to the constants $c$ and $\phi$ in Eq. (2.174) by

$$\alpha = \frac{2 \sin \phi}{\sqrt{3}(3 - \sin \phi)}, \qquad k = \frac{6c \cos \phi}{\sqrt{3}(3 - \sin \phi)} \qquad (2.187)$$

The cone corresponding to the constants in Eq. (2.187) circumscribes the hexagonal pyramid and represents an outer bound on the Mohr-Coulomb failure surface (Fig. 2.28). On the other hand, the inner cone passes through the tension meridian $p_t$, where $\theta = 0$, and will have the constants

$$\alpha = \frac{2 \sin \phi}{\sqrt{3}(3 + \sin \phi)}, \qquad k = \frac{6c \cos \phi}{\sqrt{3}(3 + \sin \phi)} \qquad (2.188)$$

However, the approximation given by either the inner or the outer cone to the Mohr-Coulomb failure surface can be poor for certain stress states. Other approximations made to match another meridian, say, the shear meridian, may be better.

The Drucker-Prager criterion for a biaxial stress state is represented by the intersection of the circular cone with the coordinate plane of $\sigma_3 = 0$. Substituting $\sigma_3 = 0$ into Eq. (2.185) leads to

$$\alpha(\sigma_1 + \sigma_2) + \sqrt{\tfrac{1}{3}(\sigma_1^2 - \sigma_1\sigma_2 + \sigma_2^2)} = k \qquad (2.189)$$

or rearranging

$$(1 - 3\alpha^2)(\sigma_1^2 + \sigma_2^2) - (1 + 6\alpha^2)\sigma_1\sigma_2 + 6k\alpha(\sigma_1 + \sigma_2) - 3k^2 = 0 \qquad (2.190)$$

which is an off-center ellipse as shown in Fig. 2.29.

EXAMPLE 2.6. A material has a tensile strength $f_t'$ equal to one-tenth of its compressive strength $f_c'$. Consider a material element subjected to a combination of normal stress $\sigma$ and shear stress $\tau$. On the basis of (a) the

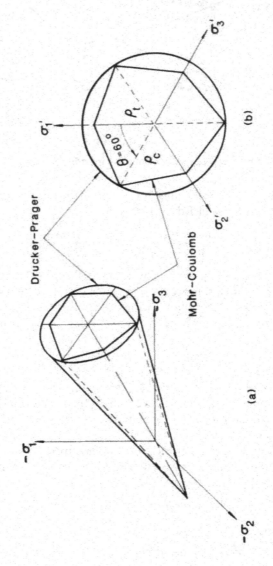

FIGURE 2.28. Drucker–Prager and Mohr–Coulomb criteria matched along the compressive meridian: (a) in principal stress space; (b) in the deviatoric plane.

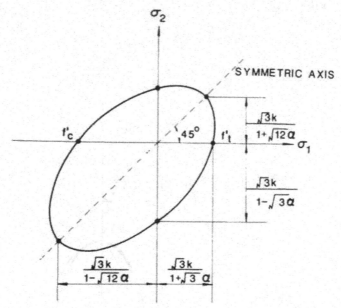

FIGURE 2.29. Drucker–Prager criterion in the coordinate plane $\sigma_3 = 0$.

Mohr–Coulomb criterion and (b) the Drucker–Prager criterion, sketch the interaction curves which govern the failure of the element.

SOLUTION. (a) Mohr–Coulomb criterion: To use the failure condition of Eq. (2.177), we must find the principal stresses from Mohr's circle as

$$\frac{\sigma}{2} + \sqrt{\left(\frac{\sigma}{2}\right)^2 + \tau^2} = \sigma_1 > 0$$

$$\frac{\sigma}{2} - \sqrt{\left(\frac{\sigma}{2}\right)^2 + \tau^2} = \sigma_3 < 0 \tag{2.191}$$

and the stress in the direction perpendicular to the $\sigma_1$-$\sigma_3$ plane is zero,

$$\sigma_2 = 0$$

Substituting Eq. (2.191) into Eq. (2.177) yields

$$\frac{\sigma + \sqrt{\sigma^2 + 4\tau^2}}{2f_t'} - \frac{\sigma - \sqrt{\sigma^2 + 4\tau^2}}{2f_c'} = 1 \tag{2.192}$$

Noting that $f_t' = \frac{1}{10}f_c'$ and rearranging, one gets

$$\left[\frac{\sigma + \frac{9}{20}f_c'}{\frac{11}{20}f_c'}\right]^2 + \left(\frac{\tau}{f_c'/\sqrt{40}}\right)^2 = 1 \tag{2.193}$$

which is an ellipse as shown in Fig. 2.30.

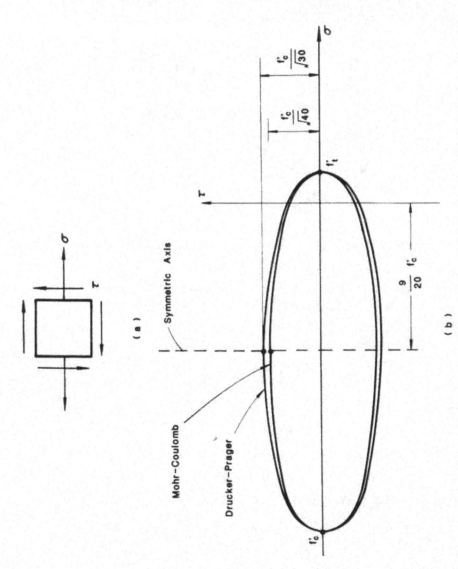

FIGURE 2.30. (a) An element subjected to normal stress $\sigma$ and shear stress $\tau$. (b) Failure curves based on Mohr–Coulomb and Drucker–Prager criteria.

(b) Drucker–Prager criterion: The material constants $\alpha$ and $k$ can be determined from the given tensile failure stress $f'_t$ and compression failure strength $f'_c$. Substituting stress states $(\sigma_1 = f'_t, \sigma_2 = \sigma_3 = 0)$ and $(\sigma_1 = \sigma_2 = 0, \sigma_3 = -f'_c)$ into the failure condition of Eq. (2.185), one gets

$$\alpha f'_t + \frac{1}{\sqrt{3}} f'_t - k = 0$$

$$-\alpha f'_c + \frac{1}{\sqrt{3}} f'_c - k = 0$$

(2.194)

Noting that $f'_c = 10 f'_t$ and solving Eq. (2.194) for $k$ and $\alpha$ leads to

$$k = \frac{2}{11\sqrt{3}} f'_c, \qquad \alpha = \frac{9}{11\sqrt{3}}$$

(2.195)

For the stress state $(\sigma, \tau)$, $I_1 = \sigma$, $J_2 = \frac{1}{3}\sigma^2 + \tau^2$, Eq. (2.185) becomes

$$\alpha\sigma + \sqrt{\tfrac{1}{3}\sigma^2 + \tau^2} - k = 0$$

(2.196)

Substituting Eq. (2.195) into Eq. (2.196) and rearranging, we obtain the failure condition for the given stress state as

$$\left[\frac{\sigma + \frac{9}{20} f'_c}{\frac{11}{20} f'_c}\right]^2 + \left[\frac{\tau}{f'_c/\sqrt{30}}\right]^2 = 1$$

(2.197)

which is also an ellipse as shown in Fig. 2.30.

# 2.4. Anisotropic Failure/Yield Criteria

Although most materials can be treated as isotropic approximately, strictly speaking, all materials are anisotropic to some extent; that is, the material properties are not the same in every direction. The general form of the failure/yield criteria for anisotropic materials has been expressed by Eq. (2.130). However, the definite form of the function $f(\sigma_{ij}, k_1, k_2, \ldots)$ depends very much on the characteristics of the material.

## 2.4.1. A Yield Criterion for Orthotropic Materials

An orthotropic material has three mutually orthogonal planes of symmetry at every point. The intersection of these planes are known as the principal axes of anisotropy. The yield criterion proposed by Hill (1950), when referred to these axes, has the form

$$f(\sigma_{ij}) = a_1(\sigma_y - \sigma_z)^2 + a_2(\sigma_z - \sigma_x)^2 + a_3(\sigma_x - \sigma_y)^2$$
$$+ a_4\tau_{yz}^2 + a_5\tau_{zx}^2 + a_6\tau_{xy}^2 - 1 = 0$$

(2.198)

where $a_1, a_2, \ldots, a_6$ are material parameters. Equation (2.198) is a quadratic expression of the stresses, representing some kind of energy that governs yielding of the orthotropic materials. The Hill criterion is therefore considered an extended form of the distortion-energy criterion of von Mises. The omission of the linear terms and the appearance of only differences between normal stress components in the yield criterion implies the assumptions that the material responses are equal in tension and compression and that a hydrostatic stress does not influence yielding.

The material parameters may be determined from three simple tension tests in the directions of the principal axes of anisotropy and three simple shear tests along the planes of symmetry. Denote the tensile strengths as $X$, $Y$, and $Z$, corresponding to the $x$-, $y$-, and $z$-axes, and the shear strengths as $S_{23}$, $S_{31}$, and $S_{12}$, corresponding to the three coordinate planes. Substituting these six states of stress into Eq. (2.198) and solving for the parameters, we obtain

$$2a_1 = \frac{1}{Y^2} + \frac{1}{Z^2} - \frac{1}{X^2}$$

$$2a_2 = \frac{1}{Z^2} + \frac{1}{X^2} - \frac{1}{Y^2}$$

$$2a_3 = \frac{1}{X^2} + \frac{1}{Y^2} - \frac{1}{Z^2}$$

$$\text{(2.199)}$$

$$a_4 = \frac{1}{S_{23}^2}$$

$$a_5 = \frac{1}{S_{31}^2}$$

$$a_6 = \frac{1}{S_{12}^2}$$

If the material is transversely isotropic (rotational symmetry about the $z$-axis), Eq. (2.198) must remain invariant for arbitrary $x$-, $y$-axes of reference. It follows that the parameters must satisfy the relations:

$$a_1 = a_2, \qquad a_4 = a_5, \qquad a_6 = 2(a_1 + 2a_3) \qquad \text{(2.200)}$$

For a complete isotropy,

$$6a_1 = 6a_2 = 6a_3 = a_4 = a_5 = a_6 \qquad \text{(2.201)}$$

and Eq. (2.198) reduces to the von Mises criterion.

## 2.4.2. A Criterion for Ice Crushing Failure

Ice is columnar-grained in structure. It may be treated as an orthotropic material. However, the strength of ice is sensitive to hydrostatic pressure. Its tensile strength is much lower than its compressive strength. The Hill criterion of Eq. (2.198) cannot model such behavior, and therefore, is not applicable to ice. A yield function including linear terms of normal stresses and having the following form has been proposed:

$$f(\sigma_{ij}) = a_1(\sigma_y - \sigma_z)^2 + a_2(\sigma_z - \sigma_x)^2 + a_3(\sigma_x - \sigma_y)^2 + a_4\tau_{yz}^2$$
$$+ a_5\tau_{zx}^2 + a_6\tau_{xy}^2 + a_7\sigma_x + a_8\sigma_y + a_9\sigma_z - 1 = 0 \qquad (2.202)$$

This function, being a special case of the $n$-type yield functions presented by Pariseau (1968), can describe materials with differing tensile and compressive strengths and predicts a nonlinear (parabolic) increase in strength with confining pressure. If the material is completely anisotropic, nine independent strength measurements would be required to determine the coefficients of Eq. (2.202). Any isotropy, such as transverse isotropy, will reduce the number of required tests. Obviously, this is the case for an ice sheet. Its strength within the horizontal plane is isotropic (see Fig. 2.31). This implies that the coefficients in Eq. (2.202) are not independent, but are subjected to the restrictions

$$a_1 = a_2, \qquad a_4 = a_5, \qquad a_7 = a_8, \qquad a_6 = 2(a_1 + 2a_3) \qquad (2.203)$$

which is similar to Eq. (2.200). Thus, Eq. (2.202) reduces to

$$f(\sigma_{ij}) = a_1[(\sigma_y - \sigma_z)^2 + (\sigma_z - \sigma_x)^2] + a_3(\sigma_x - \sigma_y)^2 + a_4(\tau_{yz}^2 + \tau_{zx}^2)$$
$$+ 2(a_1 + 2a_3)\tau_{xy}^2 + a_7(\sigma_x + \sigma_y) + a_9\sigma_z - 1 = 0 \qquad (2.204)$$

This criterion is employed by Ralston (1977) for ice crushing failure analysis. The values of the coefficients $a_1$, $a_3$, $a_7$, and $a_9$ can be determined from

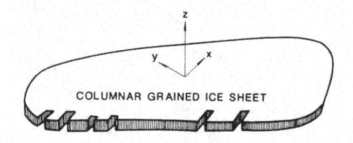

UNCONFINED STRENGTH TESTS

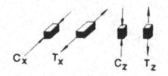

FIGURE 2.31. An example of transversely isotropic material with different tensile and compressive strengths.

compressive and tensile strength measurements as follows:

$$a_1 = \frac{1}{2C_z T_z}, \qquad a_3 = \frac{1}{T_x C_x} - \frac{1}{2C_z T_z}$$

$$a_7 = \frac{1}{T_x} - \frac{1}{C_x}, \qquad a_9 = \frac{1}{T_z} - \frac{1}{C_z} \tag{2.205}$$

where $T_x$, $C_x$, $T_z$, $C_z$ are the absolute values of the horizontal and vertical tensile and compressive strengths, respectively (see Fig. 2.31). The value of $a_4$ could be determined from either a shear test or a compression test on a sample inclined away from the vertical direction.

The ice strength data used in Ralston's work are

$$T_x = 1.01 \text{ MPa}, \qquad T_z = 1.21 \text{ MPa}$$

$$C_x = 7.11 \text{ MPa}, \qquad C_z = 13.5 \text{ MPa} \tag{2.206}$$

As can be seen, the compressive strengths are 7 to 11 times as large as the tensile strengths. With the uniaxial strengths given, the coefficients are calculated from Eq. (2.205) as

$$a_1 = 3.06 \times 10^{-2} \text{ MPa}^{-2}, \qquad a_3 = 10.9 \times 10^{-2} \text{ MPa}^{-2}$$

$$a_7 = 84.9 \times 10^{-2} \text{ MPa}^{-1}, \qquad a_9 = 75.2 \times 10^{-2} \text{ MPa}^{-1} \tag{2.207}$$

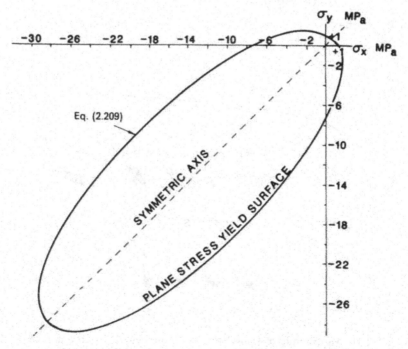

FIGURE 2.32. Yield curve under plane stress condition.

In contrast, if the strength in tension is equal to that in compression, i.e., $T_x = C_x$ and $T_z = C_z$, the coefficients $a_7$ and $a_9$ vanish, so the linear terms disappear. In this case, Eq. (2.202) reduces to Eq. (2.198) for an orthotropic material without hydrostatic stress effect. Now, we assume a plane stress condition, i.e.,

$$\sigma_z = \tau_{yz} = \tau_{xz} = 0$$

With this assumption, the yield function (2.204) further reduces to

$$a_1(\sigma_x^2 + \sigma_y^2) + a_3(\sigma_x - \sigma_y)^2 + 2(a_1 + 2a_3)\tau_{xy}^2 + a_7(\sigma_x + \sigma_y) = 1 \qquad (2.208)$$

If $x$ and $y$ are the principal stress directions, then $\tau_{xy}$ vanishes and we arrive at

$$a_1(\sigma_x^2 + \sigma_y^2) + a_3(\sigma_x - \sigma_y)^2 + a_7(\sigma_x + \sigma_y) = 1 \qquad (2.209)$$

The yield surface given by Eq. (2.209) with the coefficients given in Eq. (2.207) is plotted in Fig. 2.32. This yield surface is a long narrow ellipse symmetric about the line $\sigma_x = \sigma_y$.

## 2.5. Summary

This chapter deals with the yield/failure criteria of materials under combined stress conditions. The stress analysis of Section 2.1 at the beginning of this chapter gives the necessary background for the later discussion. As we have already seen, a better understanding of the yield/failure criteria can be obtained if their geometric shapes can be described intuitively in the principal stress space.

There are four types of criteria introduced in this chapter: isotropic and independent of hydrostatic pressure; isotropic and dependent on hydrostatic pressure; orthotropic and independent of hydrostatic pressure; and orthotropic and dependent on hydrostatic pressure. Characteristics of these criteria may be summarized as follows:

*Isotropic without hydrostatic pressure effect.* This type of yield function, $f(J_2, J_3, k_1, k_2, \ldots) = 0$, independent of the variable $I_1$ or $\xi$, suggests that the shearing stress is the only decisive factor which governs yielding. Its geometric shape is a cylinder with meridians parallel to the hydrostatic axis. The cross sections with the deviatoric planes have sixfold symmetry. The Tresca and von Mises criteria are the most useful yield functions of this type.

*Isotropic with hydrostatic pressure effect.* The yield/failure functions of this type depend on both the hydrostatic stress and the shearing stresses, having the form $f(I_1, J_2, J_3, k_1, k_2, \ldots) = 0$. The corresponding yield/failure surface may have either curved or straight meridians, which are no longer parallel to the hydrostatic axis. Its deviatoric sections generally have threefold symmetry. The Mohr–Coulomb and Drucker–Prager criteria, including a linear term in $I_1$ and $\xi$ in their expressions and having conical shapes in the principal stress space, are the simplest functions of this type.

*Orthotropic with and without hydrostatic pressure effect.* It should be noted that for anisotropic materials, the yield function, $f(\sigma_{ij}, k_1, k_2, \ldots) = 0$, is established in a certain reference coordinate system which is fixed with respect to the orientation of the material body. We cannot change the reference coordinates without changing the form of the function. Thus, it is impossible to illustrate the yield surface in the three-dimensional principal stress space, as we have done for isotropic materials. The yield functions of Eq. (2.198) and Eq. (2.202), suitable only for materials with orthotropic anisotropy, are referenced from the three principal axes of anisotropy of the material. Equation (2.202) is different from Eq. (2.198) in that the linear terms of normal stresses are included in the yield function, reflecting the influence of hydrostatic pressure.

## References

Chen, W.F., 1982. *Plasticity in Reinforced Concrete*, McGraw-Hill, New York.

Chen, W.F., and A.F. Saleeb, 1982. *Constitutive Equations for Engineering Materials, Volume 1: Elasticity and Modeling*, Wiley-Interscience, New York.

Drucker, D.C., "The Relation of Experiments to Mathematical Theories of Plasticity," *Journal of Applied Mechanics* **16:** 349–357 (1949).

Hill, R., 1950. *The Mathematical Theory of Plasticity*, Oxford University Press, New York.

Pariseau, W.G., 1968. "Plasticity Theory for Anisotropic Rocks and Solids," *Proceedings of the Tenth Symposium of Rock Mechanics*, Chapter 10, University of Texas, Austin.

Ralston, T.D., 1977. "Yield and Plastic Deformation in Ice Crushing Failure" (Preprint), ICSI/AIDJEX Symposium on Sea Ice-Processes and Models, Seattle, Washington.

PROBLEMS

2.1. The stress tensor $\sigma_{ij}$ at a point is given by

$$\sigma_{ij} = \begin{bmatrix} 0 & 0 & 0 \\ 0 & 300 & 100\sqrt{3} \\ 0 & 100\sqrt{3} & 100 \end{bmatrix} \text{ (stress unit)}$$

Find:

(a) The magnitude of the normal and shear stresses on an area element whose unit normal vector is given by $n = (\frac{1}{2}, \frac{1}{2}, 1/\sqrt{2})$.

(b) The magnitude of the principal stresses.

(c) The orientation of the principal axes of stresses.

(d) The octahedral stresses.

(e) The maximum shear stress.

2.2. For the given stress tensor $\sigma_{ij}$:

$$\sigma_{ij} = \begin{bmatrix} \dfrac{3}{2} & -\dfrac{1}{2\sqrt{2}} & -\dfrac{1}{2\sqrt{2}} \\[2mm] -\dfrac{1}{2\sqrt{2}} & \dfrac{11}{4} & -\dfrac{5}{4} \\[2mm] -\dfrac{1}{2\sqrt{2}} & -\dfrac{5}{4} & \dfrac{11}{4} \end{bmatrix} \text{ (stress unit)}$$

(a) find the principal stresses and their associated principal directions;

(b) find the deviatoric stress tensor, $s_{ij}$, and the principal deviatoric stresses $s_1$, $s_2$, and $s_3$;

(c) determine the deviatoric stress invariants $J_1$, $J_2$, and $J_3$.

2.3. (a) If $s_1 > s_2 > s_3$, can $s_3 = 0$? Explain.

(b) Can $J_2$ be negative? Explain.

(c) Can $J_3$ be positive? Explain.

2.4. Show that the ratio $\tau_{oct}/\tau_{max}$ is bounded by

$$\frac{\sqrt{6}}{3} \leq \frac{\tau_{oct}}{\tau_{max}} \leq \frac{2\sqrt{2}}{3}$$

2.5. Prove the following relations:

(a) $J_2 = \frac{1}{3}I_1^2 - I_2$.

(b) $J_3 = I_3 - \frac{1}{3}I_1 I_2 + \frac{2}{27}I_1^3$.

(c) $\tau_{oct} = \frac{\sqrt{2}}{3}(I_1^2 - 3I_2)^{1/2}$

(d) $J_2 = -(s_1 s_2 + s_2 s_3 + s_3 s_1)$.

2.6. Given Cauchy's formula for stress at a point as

$$T_i = \sigma_{ij} n_j$$

where $T_i$ is the stress vector acting on a plane-area element with normal vector $n_i$, show that the stress components $\sigma_{ij}$ form a second-order tensor using the definition of tensors.

2.7. Show that subtracting a hydrostatic stress from a given state of stress does not change the principal directions.

2.8. Prove that the shear stress component $S_n$ acting on any plane passing through a given point is unchanged by the addition of hydrostatic tension or compression to the original state of stress.

2.9. The stress state at a point is given by

$$\sigma_{ij} = \begin{bmatrix} 30 & 45 & 60 \\ 45 & 20 & 50 \\ 60 & 50 & 10 \end{bmatrix} \text{ MPa}$$

(a) Determine the stress invariants $I_1$, $J_2$, $J_3$, and $\theta$.

(b) Based on the expressions for principal stresses in terms of the stress invariants $I_1$, $J_2$, and $\theta$, find the magnitudes of the principal stresses $\sigma_1$, $\sigma_2$, and $\sigma_3$.

2.10. (a) Show that

$$\frac{\partial J_3}{\partial \sigma_{ij}} = s_{ik}s_{kj} - \frac{2}{3}J_2\delta_{ij}$$

(b) Find $\partial\theta/\partial\sigma_{ij}$.

2.11. Show that if $I_1 = \sigma_x + \sigma_y + \sigma_z = 0$, the stress state $\sigma_{ij}$ is a pure shear state. Note that the state of stress is termed a pure shear state if there exists some coordinate system $0x'y'z'$ such that $\sigma'_x = \sigma'_y = \sigma'_z = 0$.

2.12. If a stress state is obtained by superposing two other stress states, show that:

(a) The maximum principal stress is not greater than the sum of the individual maximum principal stresses.
(b) The maximum shear stress is not greater than the sum of the individual maximum shear stresses.
(c) The resultant hydrostatic pressure component is simply the algebraic summation of the hydrostatic components of the two individual states, but the resultant shear component is the vectorial summation of the shear components of the two individual states.

2.13. Considering Eq. (2.61) for the principal deviatoric stresses,

$$s^3 - J_2s - J_3 = 0$$

and making the substitution $s = r \sin \psi$ leads to

$$\sin^3 \psi - \frac{J_2}{r^2} \sin \psi - \frac{J_3}{r^3} = 0$$

(a) Considering the similarity of the latter equation to the trigonometric identity,

$$\sin^3 \psi - \tfrac{3}{4} \sin \psi + \tfrac{1}{4} \sin 3\psi = 0$$

show that $r$ and $\psi$ are invariants related to $J_2$ and $J_3$ through

$$r = \frac{2}{\sqrt{3}}\sqrt{J_2} \quad \text{and} \quad \sin 3\psi = -\frac{3\sqrt{3}}{2}\frac{J_3}{J_2^{3/2}}$$

(b) Using the results obtained in (a) with Eqs. (2.111) and (2.118), prove that:
(i) $r = \sqrt{2/3}\rho$; (ii) $\psi = [\theta - (\pi/6)]$ and $\psi$ varies in the range $-\pi/6 \leq \psi \leq \pi/6$ for $0 \leq \theta \leq \pi/3$.
(c) For an arbitrary state of stress defined by the principal stresses $\sigma_1 \geq \sigma_2 \geq \sigma_3$ and considering the projection on the $\pi$-plane (as shown in Fig. 2.9), show the corresponding values of $\theta$ and $\psi$ for the following cases: (i) $\sigma_2 = \sigma_3$; (ii) $\sigma_2 = \sigma_1$; (iii) $\sigma_2 = \tfrac{1}{2}(\sigma_1 + \sigma_3)$. (Note: In many cases, the invariants $\psi$ or $\theta$ are conveniently used to replace $J_3$ in writing the general expressions for yield or failure functions.)

2.14. A metal yields when the maximum shear stress, $\tau_{max}$, reaches the value of 125 MPa. A material element of this metal is subjected to a biaxial state of stress:

$$\sigma_1 = \sigma, \qquad \sigma_2 = \alpha\sigma, \qquad \sigma_3 = 0$$

where $\alpha$ is a constant. For what values of $(\sigma, \alpha)$ will yielding occur?

2.15. A metal yields at a state of plane stress with

$$\sigma_x = 80 \text{ MPa}, \qquad \sigma_y = 40 \text{ MPa}, \qquad \tau_{xy} = 80 \text{ MPa}$$

Assume isotropy, independence of hydrostatic pressure, and equality of properties in tension and compression.

(a) Find all other biaxial states of stress at yield in $(\sigma_1, \sigma_2)$ space.
(b) Plot the results in part (a) in $(\sigma_1, \sigma_2)$ space and estimate the yield stress
  (i) in axial tension and (ii) in simple shear, and give limits of possible error of your estimates, based on convexity.
(c) Determine the yield stresses in (b), based on (i) the von Mises criterion; (ii) the Tresca criterion; (iii) $F(J_2, J_3) = J_2^3 - 2.25J_3^2 - k^6 = 0$.

2.16. A long steel circular tube of 25.4-cm diameter and 0.32-cm wall thickness is subjected to an interior pressure of 4.83 MPa. The ends of the tube are closed. The yield stress of the steel is 227 MPa. Find the additional axial tensile load $P$ which is needed to cause yielding of the tube, based on (i) the von Mises criterion; (ii) the Tresca criterion; (iii) $f(J_2, J_3) = J_2^3 - 2.25J_3^2 - k^6 = 0$.

2.17. The stress tensor at a point under the working load condition is given by

$$\sigma_{ij} = \begin{bmatrix} 25 & 50 & 0 \\ 50 & 100 & 0 \\ 0 & 0 & -50 \end{bmatrix} \text{ MPa}$$

The yield stress of the material is 250 MPa. Based on (a) the Tresca criterion and (b) the von Mises criterion, calculate the factor of safety of the stress state against failure (i) if all stresses are increased proportionally to reach the yield surface and (ii) if only the normal stress $\sigma_x$ is increased to a critical failure value at the yield surface.

2.18. A material fails under confined compression loading with stresses $\sigma_1 = \sigma_2 = -\frac{1}{3}f_c'$, $\sigma_3 = -2.0f_c'$, where $f_c'$ is its uniaxial compressive strength.

(a) Determine the constants $c$ and $\phi$ in terms of $f_c'$ for the Mohr–Coulomb criterion.
(b) Determine the constants $k$ and $\alpha$ for the Drucker–Prager criterion.
(c) Illustrate the Mohr–Coulomb and Drucker–Prager failure surfaces by plotting their cross sections in the $\pi$-plane and the meridian section in the meridian plane of $\theta = 0°$.
(d) Find the largest discrepancy in a deviatoric plane between the Mohr–Coulomb and Drucker–Prager criteria.
(e) Find the tensile strength $f_t'$ predicted by these two criteria.

2.19. Assume that the material described in Problem 2.18 has a tensile strength $f_t' = 0.1f_c'$. (a) Plot the failure surfaces defined by (i) the Mohr–Coulomb criterion with tension cutoff; (ii) the Drucker–Prager criterion with tension cutoff, in the meridian plane of $\theta = 0°$ and in the deviatoric planes, respectively. (b) Plot the intersections of the failure surfaces with the $\sigma_1$-$\sigma_2$ plane $(\sigma_3 = 0)$ and with the $\sigma_x$-$\tau_{xy}$ plane.

2.20. Find the constants $\alpha$ and $k$ of the Drucker-Prager criterion in terms of the cohesion, $c$, and the angle of internal friction, $\phi$, of the Mohr-Coulomb criterion for the following cases:

(a) to match the shear meridian;
(b) to match the simple compression and the simple tension points;
(c) to match the equal biaxial compression and simple tension points.

2.21. Prove Eq. (2.200) for transversely isotropic materials.

2.22. Because of the similarity of the elastic-plastic behavior of a truss structure to that of a polycrystal metal, we may simulate the elastic-plastic behavior of a metal by an analysis of a simple truss structure. A three-component truss structure is subjected to a pair of forces, $F_H$ and $F_V$, at Joint $D$ as shown in Fig. P2.22a. The three bars have the same cross-sectional area $A$, Young's modulus $E$, and yielding stress $\sigma_y$, and $A$ is considered large enough such that no buckling will occur. Define a two-dimensional force space $F_H$-$F_V$. A point with coordinate $(F_H/N_y, F_V/N_y)$, where $N_y = A\sigma_y$, in the force space represents a pair of forces $F_H$ and $F_V$ acting on the structure. Define the elastic limit state in the force space corresponding to the force state at which one of the three bars starts to yield, and the plastic limit state at which the structure reaches its plastic collapse state.

(a) Write explicitly the three sets of general equations required for a solution of this problem.
(b) Sketch the initial elastic limit locus in the force space (a simulation of the initial yield surface in the two-dimensional stress space).
(c) Show that the closed curve of Fig. 2.22b is the locus of the plastic collapse limit envelope (a simulation of the failure surface in the two-dimensional stress space).
(d) Calculate the stress history of the three bars for the load-unload cycle: $(F_H/N_y, F_V/N_y) = (0, 0) \rightarrow (0, 2) \rightarrow (0, 0)$.
(e) Sketch the subsequent elastic limit locus of the structure after it completes the load-unload cycle (kinematic hardening).
(f) What are the relationships and the differences between the elastic limit locus and the plastic collapse limit locus?

2.23. As discussed in Section 2.3.3, when the ratio of the two characteristic lengths, $\rho_{t0}/\rho_{c0}$ [Eq. (2.184)], of the Mohr-Coulomb failure surface on the $\pi$-plane becomes unity, the Mohr-Coulomb surface will be reduced to the Tresca surface. What surface will the Mohr-Coulomb surface be reduced to when $\rho_{t0}/\rho_{c0} = \frac{1}{2}$? Discuss the characteristics of the failure surface and plot the locus of the surface on a deviatoric plane and also show its tensile and compressive meridians on a meridian plane.

2.24. A long thin-walled steel cylindrical vessel with diameter $D$ and wall thickness $t$ is subjected to an internal pressure $p_1$ and an external pressure $p_2$, as shown in Fig. P2.24. Suppose that the external pressure $p_2$ does not contribute to the axial stress of the tube, and let $p_2 = rp_1$, $r \geq 0$. The tube begins to yield under $p_1 = p_0$, and $p_2 = 0$.

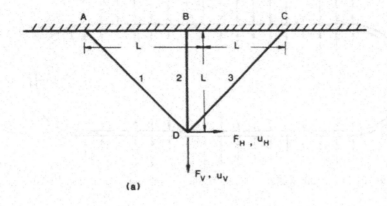

(a)

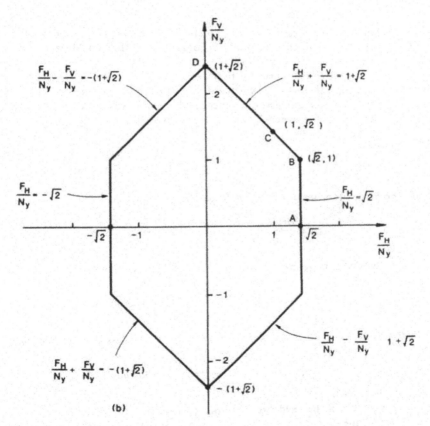

(b)

FIGURE P2.22. (a) Three-bar truss; (b) limit state locus of plastic collapse.

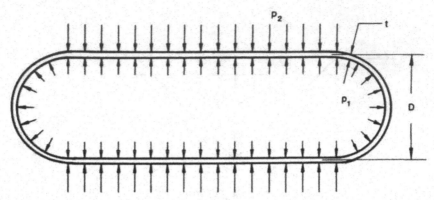

FIGURE P2.24. A thin-walled cylindrical vessel subjected to internal and external pressure.

(a) Find the limit pressure, $p_1 = p_y$, at which the tube begins to yield, in terms of $p_0$ and $r$ for the case $r > 0$, according to (i) the von Mises criterion and (ii) the Tresca criterion.
(b) Sketch the $p_y$ vs. $r$ curve for the two criteria.
(c) Find the value of $r$ at which the limit pressures predicted by the two criteria have the largest difference, and give the reason. Assume that the two criteria are matched at the pure shear yield point.

2.25. Show that

$$s_{ik}s_{kj}s_{il}s_{lj} = 2J_2^2$$

2.26. The lode parameter of stress is defined by

$$\mu_\sigma = \frac{2\sigma_2 - \sigma_1 - \sigma_3}{\sigma_1 - \sigma_3} = \frac{\tau_{12} - \tau_{23}}{\tau_{13}}, \qquad \sigma_1 \geq \sigma_2 \geq \sigma_3$$

(a) Show that the parameter $\mu_\sigma$ is related to the lode angle $\theta$ or $\psi$ as

$$\tan \psi = \tan\left(\theta - \frac{\pi}{6}\right) = \frac{\mu_\sigma}{\sqrt{3}}, \qquad |\theta| \leq 60°$$

(b) Show that the von Mises yield criterion can be expressed in terms of $\mu_\sigma$ as

$$\frac{2\tau_{max}}{\sigma_0} = \frac{\sigma_1 - \sigma_3}{\sigma_0} = \frac{2}{\sqrt{3 + \mu_\sigma^2}}, \qquad -1 \leq \mu_\sigma \leq 1$$

where $\sigma_0$ is the yield stress in uniaxial tension.
(c) Plot the $2\tau_{max}/\sigma_0$ vs. $\mu_\sigma$ curve and discuss the discrepancy between the von Mises and Tresca criteria.

2.27. Assume that the uniaxial compressive strength is $f_c'$ and the uniaxial tensile strength is $f_t' = f_c'/10$ for a concrete material. Predict the stress state at failure of a concrete cube specimen in the triaxial compression test shown in Fig. P2.27 for the following loading paths. Use the Mohr–Coulomb criterion.

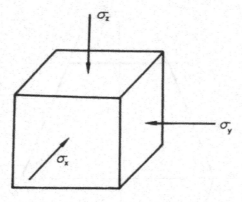

FIGURE P2.27. A concrete cube subjected to a triaxial compression stress state.

(a) Proportional loading, $\sigma_x = \sigma_y = -p$, $\sigma_z = -Ap$, where $A$ is a constant greater than 1. Express the stress state at failure in terms of $f'_c$ and $A$. What is the minimum value of $A$, $A_{min}$, such that if $A \le A_{min}$, the specimen will never fail?

(b) Increase $\sigma_x = \sigma_y$ first until the stress state $\sigma_x = \sigma_y = -Rf'_c$, $\sigma_z = 0$, where $R$ is a constant greater than zero, is reached; then, increase $\sigma_z$. Express the stress state at failure in terms of $f'_c$ and $R$. What is the maximum value of $R$, $R_{max}$, such that if $R \ge R_{max}$, the specimen will fail before reaching the state $\sigma_x = \sigma_y = -Rf'_c$.

2.28. The $\sigma$-$\epsilon$ response in a simple tension test for an elastic-linear hardening plastic material is approximated by the following expression

$$\sigma = \begin{cases} \sigma_0 + m\epsilon^p, & \text{for } \sigma \ge \sigma_0 \\ E\epsilon, & \text{for } \sigma < \sigma_0 \end{cases}$$

where $\sigma_0$ is the initial tensile yield stress. A material sample is first stretched to a state at which the accumulated plastic strain $\epsilon^p = \epsilon_0^p$ and is subsequently unloaded and reversely loaded to a compression state at which the accumulated plastic strain $\epsilon^p = 0$. According to (i) the isotropic hardening rule, (ii) the kinematic hardening rule, and (iii) the independent hardening rule, determine the current tensile and compression yield strengths of the material at this stress state.

## ANSWERS TO SELECTED PROBLEMS

2.1. (a) $\sigma_n = 247.5$, $S = 194.3$.

(b) $\sigma_1 = 400$, $\sigma_2 = \sigma_3 = 0$.

(c) $n^{(1)} = (0, \pm 0.866, \pm 0.5)$, $n^{(2)} = (0, \mp 0.5, \pm 0.866)$, $n^{(3)} = (\pm 1, 0, 0)$. Note that the principal directions 2 and 3 are not unique, and any two mutually perpendicular axes normal to axis 1 can be chosen.

(d) $\sigma_{oct} = 133.33$, $\tau_{oct} = 188.56$.

(e) $\tau_{max} = 200$.

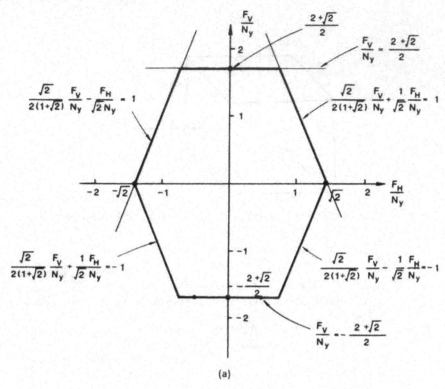

FIGURE S2.22. (a).

2.2. (a) $\sigma_1 = 4$, $\sigma_2 = 2$, $\sigma_3 = 1$, $\mathbf{n}^{(1)} = (0, \pm 1/\sqrt{2}, \mp 1/\sqrt{2})$, $\mathbf{n}^{(2)} = (\pm 1/\sqrt{2}, \mp \frac{1}{2}, \mp \frac{1}{2})$, $\mathbf{n}^{(3)} = (\pm 1/\sqrt{2}, \pm \frac{1}{2}, \pm \frac{1}{2})$.

(b) $s_1 = \frac{5}{3}$, $s_2 = -\frac{1}{3}$, $s_3 = -\frac{4}{3}$.

(c) $J_1 = 0$, $J_2 = 2.333$, $J_3 = 0.741$.

2.4. Hint: $\tau_{oct} = \sqrt{\frac{2}{3}J_2}$; $\tau_{max} = (\sigma_1 - \sigma_3)/2 = \sqrt{J_2/3}[\cos\theta - \cos(\theta + \frac{2}{3}\pi)]$, for $0° \le \theta \le 60°$.

2.9. (a) $I_1 = 60$, $J_2 = 8225$, $J_3 = 265{,}250$, $\theta = 7°30'$.

(b) $\begin{Bmatrix} \sigma_1 \\ \sigma_2 \\ \sigma_3 \end{Bmatrix} = \frac{1}{3} \begin{Bmatrix} I_1 \\ I_1 \\ I_1 \end{Bmatrix} + \frac{2}{\sqrt{3}}\sqrt{J_2} \begin{Bmatrix} \cos\theta \\ \cos(\theta - 2\pi/3) \\ \cos(\theta + 2\pi/3) \end{Bmatrix} = \begin{Bmatrix} 123.83 \\ -20.08 \\ -43.75 \end{Bmatrix}$ MPa

2.10. (a) $\dfrac{\partial J_3}{\partial \sigma_{ij}} = \dfrac{\partial}{\partial \sigma_{ij}}(\frac{1}{3}s_{mn}s_{nk}s_{km}) = \dfrac{\partial s_{mn}}{\partial \sigma_{ij}}s_{nk}s_{km}$

$= \dfrac{\partial}{\partial \sigma_{ij}}(\sigma_{mn} - \frac{1}{3}\sigma_{ll}\delta_{mn})s_{nk}s_{km}$

$= (\delta_{mi}\delta_{nj} - \frac{1}{3}\delta_{ij}\delta_{nm})s_{nk}s_{km} = s_{jk}s_{ki} - \frac{1}{3}s_{mk}s_{km}\delta_{ij}$

$= s_{ik}s_{kj} - \frac{2}{3}J_2\delta_{ij}$

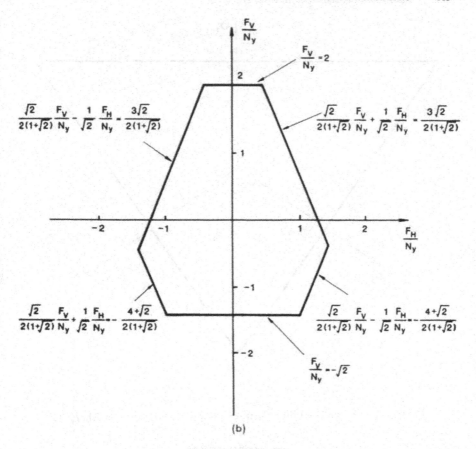

FIGURE S2.22. (b)

2.13. (c) Case (i): $\sigma_2 = \sigma_3$; $\theta = 0°$, $\psi = -30°$. Case (ii): $\sigma_2 = \sigma_1$; $\theta = 60°$, $\psi = 30°$. Case (iii): $\sigma_2 = \frac{1}{2}(\sigma_1 + \sigma_3)$; $\theta = 30°$, $\psi = 0°$.

2.14.

$$|\sigma| = \begin{cases} 2\tau_{max} & \text{for } 0 \le \alpha \le 1 \\ \dfrac{2}{\alpha}\,\tau_{max} & \text{for } \alpha > 1 \\ \dfrac{2}{1-\alpha}\,\tau_{max} & \text{for } \alpha < 0 \end{cases}$$

2.15. (a) $(142.5, -22.5)$,    $(-22.5, 142.5)$,    $(-142.5, 22.5)$,    $(22.5, -142.5)$, $(-165, -142.5)$, $(-142.5, -165)$, $(165, 142.5)$, $(142.5, 165)$, $(165, 22.5)$, $(22.5, 165)$, $(-165, -22.5)$, $(-22.5, -165)$.

(b) (i) $153.8\text{ MPa} \le \sigma_y \le 165\text{ MPa}$; (ii) $82.5\text{ MPa} \le \tau_y \le 102.5\text{ MPa}$.

(c) (i) $\sigma_y = 155\text{ MPa}$, $\tau_y = 89.4\text{ MPa}$; (ii) $\sigma_y = 165\text{ MPa}$, $\tau_y = 82.5\text{ MPa}$; (iii) $\sigma_y = 156.6\text{ MPa}$, $\tau_y = 84.5\text{ MPa}$.

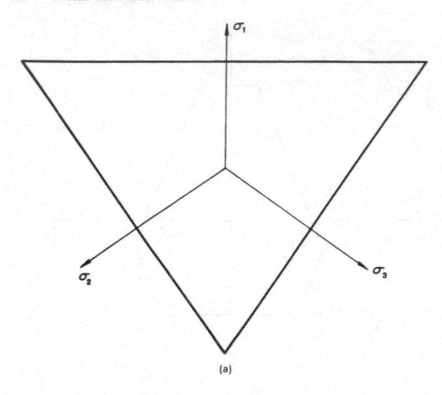

FIGURE S2.23. (a) The failure surface on a deviatoric plane with $I_1 < 0$.

2.16. (i) $P = 395$ kN; (ii) $P = 335$ kN; (iii) $P = 375$ kN.

2.17. (a) (i) S.F. = 1.43; (ii) S.F. = 7.0.
      (b) (i) S.F. = 1.60; (ii) S.F. = 8.81.

2.18. (a) $c = 0.2887 f'_c$, $\phi = 30°$.
      (b) $k = 0.3465 f'_c$, $\alpha = 0.2309$.
      (c) On $\pi$-plane, for Mohr-Coulomb: $\rho_{t0} = 0.3499 f'_c$, $\rho_{c0} = 0.4899 f'_c$; for Drucker-Prager: $\rho = 0.49 f'_c$.
          On the $\theta = 0°$ meridian plane, the coordinate of the apex: for Mohr-Coulomb, $\xi_0 = \sqrt{3} c \cot \phi = 0.8661 f'_c$; for Drucker-Prager, $\xi_0 = k/(\sqrt{3}\alpha) = 0.8664 f'_c$.
      (e) For Mohr-Coulomb, $f'_t = \frac{1}{3} f'_c$; for Drucker-Prager, $f'_t = 0.4287 f'_c$.

2.20. (a) $\alpha = \frac{1}{3} \sin \phi$, $k = c \cos \phi$.
      (b) $\alpha = (1/\sqrt{3}) \sin \phi$, $k = (2/\sqrt{3}) c \cos \phi$.
      (c) $\alpha = 2 \sin \phi / [\sqrt{3}(3 + \sin \phi)]$, $k = 6c \cos \phi / [\sqrt{3}(3 + \sin \phi)]$.

2.22. (b) Figure S2.22a; (e) Figure S2.22b.

2.23. Figures S2.23a and b.

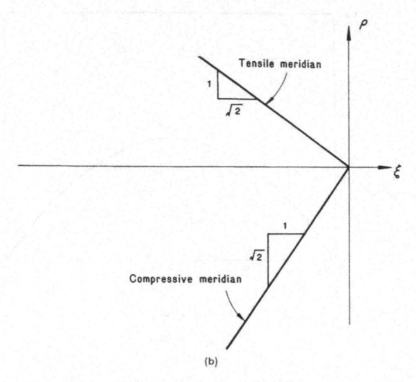

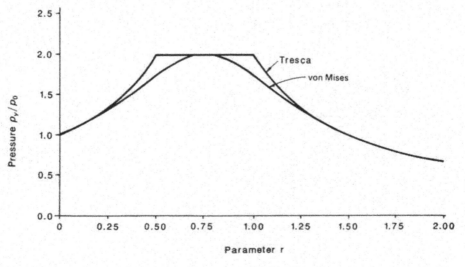

FIGURE S2.23. (b) The tensile and compressive meridians of the failure surface on a meridian plane.

FIGURE S2.24. $p_y$ vs. $r$ curves.

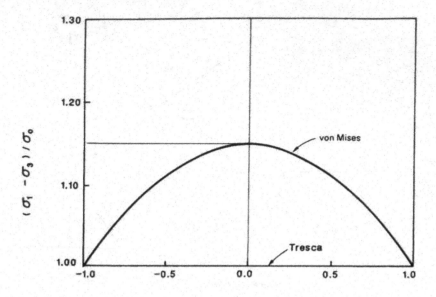

Lode Parameter $\mu_\sigma$

FIGURE S2.26. (c)

2.24. (a) (i) $p_1 = p_y = \sqrt{\dfrac{3}{4r^2 - 6r + 3}}\, p_0$

(ii) $p_y = \begin{cases} \dfrac{p_0}{1-r}, & 0 \le r \le \frac{1}{2} \\[2mm] 2p_0, & \frac{1}{2} \le r \le 1 \\[2mm] \dfrac{p_0}{1-\frac{1}{2}}, & 1 \le r \end{cases}$

(b) Figure S2.24.

(c) $r = \frac{1}{2}$ and $r = 1$.

2.26. (c) Figure S2.26c.

# 3
# Elastic Stress–Strain Relations

## 3.1. Strain

### 3.1.1. State of Strain at a Point

In the analysis of stress, a state of stress at a point can be found by making an infinite number of cuts through the point from which, for each cut, the associated *stress vector* is known. Similarly, the state of strain at a point is defined as the *totality* of *all* the changes in length of lines (fibers) of the material which pass through the point and also the *totality* of *all* the changes in angle between pairs of lines radiating from this point.

However, it will be shown later that the change in length of *any* line of the material which passes through the point and the change in angle between *any* two lines radiating from this point can be calculated once the changes in length and angle for the three lines parallel to a set of mutually perpendicular coordinate axes through this point are known.

Figure 3.1 shows an infinitesimal line element $OP$ at point $O$ in a body in its unstrained original position with length equal to unity. After deformation, the element is displaced to the new position $O'P'$, as shown in the figure. Notice that for a very small length of the line element, and for smooth variation of the deformations in the neighborhood of point $O$, the displaced element $O'P'$ remains straight. The *relative displacement vector* of point $P$ with respect to point $O$ is denoted by $\overset{n}{\delta}'$, where the vector $\mathbf{O'P''}$ is equal and parallel to the vector $\mathbf{OP}$ and the superscript $n$ indicates the direction of the fiber element $OP$ before deformation. Considering unit length fibers in the directions of the coordinate axes, $x_1$, $x_2$, and $x_3$, the corresponding relative displacement vectors for these lines are denoted by $\overset{1}{\delta}'$, $\overset{2}{\delta}'$, and $\overset{3}{\delta}'$, respectively. Alternatively, we can also use the dual notation $\overset{x}{\delta}'$, $\overset{y}{\delta}'$, and $\overset{z}{\delta}'$, respectively. Both notations are used interchangeably in this chapter.

In order to find the relation between the relative displacement vector $\overset{n}{\delta}'$ for any fiber with direction $\mathbf{n}$ and the relative displacement vectors $\overset{1}{\delta}'$, $\overset{2}{\delta}'$, and $\overset{3}{\delta}'$ for the three coordinate axes, a two-dimensional picture is considered. This is because a two-dimensional picture is easier to visualize, and the

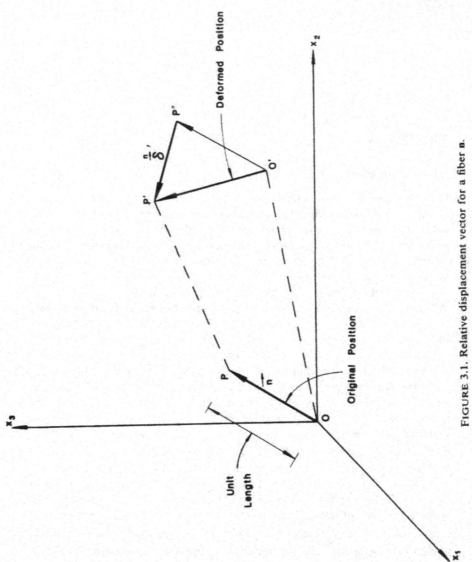

FIGURE 3.1. Relative displacement vector for a fiber **n**.

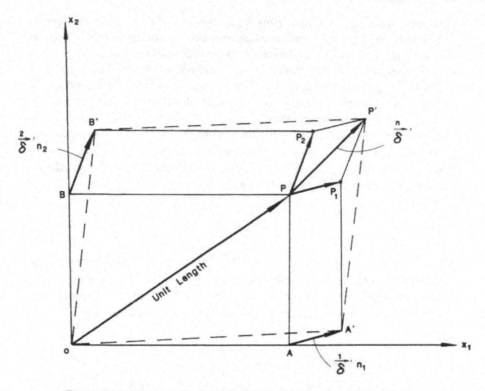

FIGURE 3.2. Relative displacement vectors in two-dimensional space.

extension to three dimensions is straightforward. In Fig. 3.2, the infinitesimal fiber **n** with unit length at point $O$ is shown in the $x_1$-$x_2$ plane. The projections of this fiber length on the axes $x_1$ and $x_2$ are $n_1$ and $n_2$, respectively, where $n_1$ and $n_2$ are the direction cosines of vector **n**. Thus, from Fig. 3.2, and because of the homogeneous state of deformations throughout the small region in the immediate vicinity of point $O$, we have

$$\mathbf{PP'} = \mathbf{PP_1} + \mathbf{PP_2}$$

or

$$\overset{n}{\boldsymbol{\delta}}{}' = \overset{1}{\boldsymbol{\delta}}{}' n_1 + \overset{2}{\boldsymbol{\delta}}{}' n_2 \tag{3.1}$$

For the general three-dimensional case, we have

$$\overset{n}{\boldsymbol{\delta}}{}' = \overset{1}{\boldsymbol{\delta}}{}' n_1 + \overset{2}{\boldsymbol{\delta}}{}' n_2 + \overset{3}{\boldsymbol{\delta}}{}' n_3 \tag{3.2}$$

Equation (3.2) is analogous to Eq. (2.7) for stresses, which expresses the stress vector, $\overset{n}{\mathbf{T}}$, at a given point acting on any plane **n** in terms of the stress vectors at that point on three particular planes perpendicular to the three coordinate axes. Unlike the stresses, however, the state of deformation or strain at a point $O$ cannot be completely defined simply by knowing the three relative displacement vectors $\overset{1}{\delta}{}'$, $\overset{2}{\delta}{}'$, and $\overset{3}{\delta}{}'$. We still have to separate the rigid-body displacements (translations and/or rotations of the body as a whole), if any, from these relative displacement vectors since the rigid-body displacements are of no interest in the analysis of strain. The separation procedure is given in the following for the case of infinitesimal deformations.

The relative displacement vectors associated with the three fibers in the direction of the coordinate axes $x_1$, $x_2$, and $x_3$ can be decomposed into components in the direction of the three coordinate axes. For example, the relative displacement vector $\overset{1}{\delta}{}'$, associated with the $x_1$ direction, has the three components $\epsilon'_{11}$, $\epsilon'_{12}$, and $\epsilon'_{13}$ in the direction of the three coordinate axes $x_1, x_2$, and $x_3$, respectively. Thus, Eq. (3.2) can be written in the component form

$$\overset{n}{\delta}{}'_i = \epsilon'_{ji} n_j \tag{3.3}$$

where the nine scalar quantities $\epsilon'_{ij}$ needed to define the three relative displacement vectors $\overset{1}{\delta}{}'$, $\overset{2}{\delta}{}'$, and $\overset{3}{\delta}{}'$ constitute a tensor. This tensor, called the *relative displacement tensor*, defines completely the relative displacement vector $\overset{n}{\delta}{}'$ of fiber $\mathbf{n}$. Using dual notations, this tensor is written as

$$\epsilon'_{ij} = \begin{bmatrix} \epsilon'_{11} & \epsilon'_{12} & \epsilon'_{13} \\ \epsilon'_{21} & \epsilon'_{22} & \epsilon'_{23} \\ \epsilon'_{31} & \epsilon'_{32} & \epsilon'_{33} \end{bmatrix} = \begin{bmatrix} \epsilon'_x & \epsilon'_{xy} & \epsilon'_{xz} \\ \epsilon'_{yx} & \epsilon'_y & \epsilon'_{yz} \\ \epsilon'_{zx} & \epsilon'_{zy} & \epsilon'_z \end{bmatrix} \tag{3.4}$$

In general, as can be seen from Eq. (3.4), the relative displacement tensor $\epsilon'_{ij}$ is *not* symmetric.

A rigid-body motion, as mentioned earlier, is characterized by the fact that the length of any line element joining any two points remains unchanged. In the following, the conditions on the coefficients $\epsilon'_{ij}$ that satisfy this requirement for rigid-body motion are derived. Consider the line element $\mathbf{OP} =$ unit vector $\mathbf{n}$, as shown in Fig. 3.1, and assume that after pure rigid-body motion, the element assumes the new position $O'P'$ as shown. Then,

$$|\mathbf{n}|^2 = |\mathbf{n} + \overset{n}{\delta}{}'|^2 = |\mathbf{n}|^2 + 2|\mathbf{n}||\overset{n}{\delta}{}'|$$

or

$$n_i n_i = (n_i + \overset{n}{\delta}{}'_i)(n_i + \overset{n}{\delta}{}'_i) = n_i n_i + 2 n_i \overset{n}{\delta}{}'_i$$

where the higher-order terms in $\overset{n}{\delta}{}'$ are neglected since only infinitesimal deformations are considered. Substituting for $\overset{n}{\delta}{}'$ from Eq. (3.3), we get

$$\mathbf{n} \cdot \overset{n}{\delta}{}' = n_i \overset{n}{\delta}{}'_i = n_i (\epsilon'_{ji} n_j) = 0$$

or, when written out in full,

$$\epsilon'_{ji}n_i n_j = \epsilon'_{11}n_1^2 + \epsilon'_{22}n_2^2$$

$$+ \epsilon'_{33}n_3^2 + (\epsilon'_{12} + \epsilon'_{21})n_1 n_2 + (\epsilon'_{23} + \epsilon'_{32})n_2 n_3 + (\epsilon'_{31} + \epsilon'_{13})n_3 n_1 = 0 \tag{3.5}$$

Since Eq. (3.5) must be true for all values of $n_1$, $n_2$, and $n_3$, the necessary and sufficient condition for the tensor $\epsilon'_{ij}$ to represent a rigid-body rotation is given by

$$\epsilon'_{11} = \epsilon'_{22} = \epsilon'_{33} = \epsilon'_{12} + \epsilon'_{21} = \epsilon'_{23} + \epsilon'_{32} = \epsilon'_{31} + \epsilon'_{13} = 0$$

or

$$\epsilon'_{ij} = -\epsilon'_{ji} \tag{3.6}$$

That is, for rigid-body rotation, the relative displacement tensor $\epsilon'_{ij}$ of Eq. (3.4) is *skew-symmetric.*

Now, every second-order tensor can be decomposed into the sum of a symmetric tensor and a skew-symmetric tensor. It follows, therefore, that if we decompose the tensor $\epsilon'_{ij}$ into symmetric and skew-symmetric parts, the skew-symmetric part represents rigid-body rotation, whereas the symmetric part represents *pure deformation.* Thus, we can write

$$\epsilon'_{ij} = \tfrac{1}{2}(\epsilon'_{ij} + \epsilon'_{ji}) + \tfrac{1}{2}(\epsilon'_{ij} - \epsilon'_{ji}) \tag{3.7}$$

or

$$\epsilon'_{ij} = \epsilon_{ij} + \omega_{ij} \tag{3.8}$$

where

$$\epsilon_{ij} = \tfrac{1}{2}(\epsilon'_{ij} + \epsilon'_{ji}) \tag{3.9}$$

$$\omega_{ij} = \tfrac{1}{2}(\epsilon'_{ij} - \epsilon'_{ji}) \tag{3.10}$$

Expanding both $\epsilon_{ij}$ and $\omega_{ij}$, we get

$$\epsilon_{ij} = \begin{bmatrix} \epsilon'_{11} & \tfrac{1}{2}(\epsilon'_{12} + \epsilon'_{21}) & \tfrac{1}{2}(\epsilon'_{13} + \epsilon'_{31}) \\ \tfrac{1}{2}(\epsilon'_{12} + \epsilon'_{21}) & \epsilon'_{22} & \tfrac{1}{2}(\epsilon'_{23} + \epsilon'_{32}) \\ \tfrac{1}{2}(\epsilon'_{31} + \epsilon'_{13}) & \tfrac{1}{2}(\epsilon'_{23} + \epsilon'_{32}) & \epsilon'_{33} \end{bmatrix} \tag{3.11}$$

$$\omega_{ij} = \begin{bmatrix} 0 & \tfrac{1}{2}(\epsilon'_{12} - \epsilon'_{21}) & \tfrac{1}{2}(\epsilon'_{13} - \epsilon'_{31}) \\ \tfrac{1}{2}(\epsilon'_{21} - \epsilon'_{12}) & 0 & \tfrac{1}{2}(\epsilon'_{23} - \epsilon'_{32}) \\ \tfrac{1}{2}(\epsilon'_{31} - \epsilon'_{13}) & \tfrac{1}{2}(\epsilon'_{32} - \epsilon'_{23}) & 0 \end{bmatrix} \tag{3.12}$$

The symmetric tensor $\epsilon_{ij}$ is called the *strain tensor* and the skew-symmetric tensor $\omega_{ij}$ is known as the *rotation tensor.* Now, if we substitute for $\epsilon'_{ij}$ from Eq. (3.8) into Eq. (3.3), we obtain

$$\overset{n}{\delta'_i} = \epsilon_{ji}n_j + \omega_{ji}n_j \tag{3.13}$$

The second part of Eq. (3.13) represents the rigid-body rotation whereas the first part represents the pure deformations.

The relative displacement vector corresponding to pure deformation is called the *strain vector.* The strain vector is denoted by $\overset{n}{\delta}$ and is given by

$$\overset{n}{\delta_i} = \epsilon_{ji}n_j = \epsilon_{ij}n_j \tag{3.14}$$

The relative displacement vector corresponding to rigid-body rotation is called the *rotation vector*. This vector is denoted by $\overset{n}{\Omega}$ and is given by

$$\overset{n}{\Omega}_i = \omega_{ji} n_j = -\omega_{ij} n_j \tag{3.15}$$

For pure deformation, Eq. (3.2) becomes

$$\overset{n}{\delta} = \overset{1}{\delta} n_1 + \overset{2}{\delta} n_2 + \overset{3}{\delta} n_3 \tag{3.16}$$

which gives the strain vector for any fiber with direction **n** in terms of the strain vectors of the three mutually perpendicular fibers in the direction of the coordinate axes $x_1$, $x_2$, and $x_3$, respectively. Thus, the three strain vectors $\overset{1}{\delta}$, $\overset{2}{\delta}$, and $\overset{3}{\delta}$ *characterize the state of strain at a point completely.*

These results for separation of rigid-body displacements from strain displacements can be easily visualized by considering a two-dimensional picture as shown in Fig. 3.3. In this figure, the original and final positions of two line fibers in the direction of the axes $x_1$ and $x_2$ are shown in $x_1$–$x_2$ plane. It can be seen that the final position of the fibers is obtained from the original position by superposing two separate processes of deformation. The first is due to pure deformation and the second represents the rigid-body rotation.

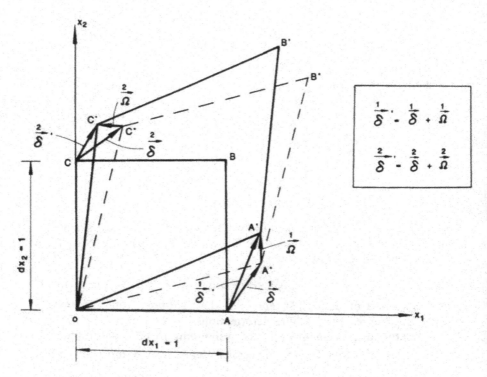

FIGURE 3.3. Strain and rotation vectors in two-dimensional case.

The formulas in Eqs. (3.14) and (3.16) are completely analogous to the corresponding formulas for stresses. Therefore, general results similar to those for stresses can be established.

### 3.1.2. Cauchy's Formulas for Strains

Here, as for stresses, the strain vector $\overset{n}{\delta}$ for any fiber $\mathbf{n}$ can be decomposed into two components, one in the direction of the fiber $\mathbf{n}$, called the *normal strain*, and the other lying in the plane normal to the fiber, called the *shear strain*. For instance, the strain vector $\overset{1}{\delta}$ has the following three strain components: normal strain $\epsilon_{11}$ and shear strains $\epsilon_{12}$ and $\epsilon_{13}$ in the direction of the three axes $x_1$, $x_2$, and $x_3$, respectively.

Consider any line fiber $\mathbf{n}$ at a point $P$ with strain vector $\overset{n}{\delta}$ having normal strain component $\epsilon_n$ and shear strain component $\epsilon_{ns}$, as shown in Fig. 3.4. Vector $\mathbf{n}$ has components $(n_1, n_2, n_3)$. The magnitude of the normal strain component $\epsilon_n$ is given by

$$\epsilon_n = \overset{n}{\delta} \cdot \mathbf{n} = \overset{n}{\delta_i} n_i \tag{3.17}$$

Substituting for $\overset{n}{\delta_i}$ from Eq. (3.14) and using the symmetry of tensor $\epsilon_{ij}$, we get

$$\epsilon_n = \epsilon_{ij} n_i n_j \tag{3.18}$$

Similarly, if the unit vector $\mathbf{s}$ normal to direction $\mathbf{n}$ has components $(s_1, s_2, s_3)$, the magnitude of the shear strain component $\epsilon_{ns}$ is given by

$$\epsilon_{ns} = \overset{n}{\delta} \cdot \mathbf{s} = \overset{n}{\delta_i} s_i$$

which upon substitution for $\overset{n}{\delta_i}$ from Eq. (3.14) becomes

$$\epsilon_{ns} = \epsilon_{ij} n_j s_i \tag{3.19}$$

Equations (3.18) and (3.19) are the required *Cauchy's formulas* for the determination of the normal and shear components of strain for an arbitrary fiber $\mathbf{n}$. Each of Eqs. (3.14), (3.16), (3.18), and (3.19) represents a particular form of Cauchy's formulas, but, in practice, for a given state of strain at a point, Eqs. (3.18) and (3.19) are the most useful forms for obtaining normal and shear strain components directly.

EXAMPLE 3.1. The state of strain at a point is defined by the strain tensor

$$\epsilon_{ij} = \begin{bmatrix} 0.00200 & 0.00183 & -0.00025 \\ 0.00183 & 0.00100 & -0.00125 \\ -0.00025 & -0.00125 & -0.00150 \end{bmatrix}$$

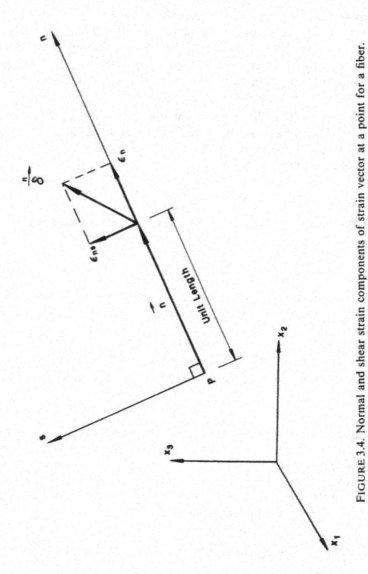

FIGURE 3.4. Normal and shear strain components of strain vector at a point for a fiber.

For a fiber with direction $n = (0, -1/\sqrt{5}, -2/\sqrt{5})$, calculate:

(a) The normal strain $\epsilon_n$ for the fiber.
(b) The magnitude of the strain vector $\overset{n}{\delta}$.
(c) The magnitude of the shear strain $\epsilon_{ns}$, where unit vector s has the components $(-1, 0, 0)$.

SOLUTION. (a) Substituting for the components of n and $\epsilon_{ij}$ in Eq. (3.18), we get

$$\epsilon_n = \frac{0.00100}{5} - \frac{0.00150(4)}{5} - \frac{0.00250(2)}{5} = -0.00200$$

(b) From Eq. (3.14), the components $\overset{n}{\delta}_i$ are calculated:

$$\overset{n}{\delta}_1 = (0.00183)\left(-\frac{1}{\sqrt{5}}\right) + (-0.00025)\left(-\frac{2}{\sqrt{5}}\right) = -0.00059$$

$$\overset{n}{\delta}_2 = (0.0010)\left(-\frac{1}{\sqrt{5}}\right) + (-0.00125)\left(-\frac{2}{\sqrt{5}}\right) = 0.00067$$

$$\overset{n}{\delta}_3 = (-0.00125)\left(-\frac{1}{\sqrt{5}}\right) + (-0.00150)\left(-\frac{2}{\sqrt{5}}\right) = 0.00190$$

Thus,

$$|\overset{n}{\delta}| = [(-0.00059)^2 + (0.00067)^2 + (0.00190)^2]^{1/2} = 0.00210$$

or $|\overset{n}{\delta}|$ may be calculated directly from the relation

$$|\overset{n}{\delta}|^2 = \overset{n}{\delta}_i\overset{n}{\delta}_i = \epsilon_{ij}\epsilon_{ik}n_jn_k = 4.41 \times 10^{-6}$$

(c) The magnitude of shear strain $\epsilon_{ns}$ is calculated using Eq. (3.19):

$$\epsilon_{ns} = \epsilon_{11}n_1s_1 + \epsilon_{22}n_2s_2 + \epsilon_{33}n_3s_3$$
$$+ \epsilon_{12}(n_2s_1 + n_1s_2) + \epsilon_{23}(n_3s_2 + n_2s_3) + \epsilon_{31}(n_3s_1 + n_1s_3)$$
$$= (0.00183)\left(-\frac{1}{\sqrt{5}}\right)(-1) + (-0.00025)\left(-\frac{2}{\sqrt{5}}\right)(-1) = 0.00060$$

Note that the shear strain $\epsilon_{ns}$ so calculated represents the projection of the resultant shear strain $\theta$ in the s direction. The resultant shear strain $\theta$ at this point for the fiber n must be calculated by the relation

$$\theta^2 = |\overset{n}{\delta}|^2 - \epsilon_n^2 = 0.41 \times 10^{-6}, \qquad \text{or } |\theta| = 0.00064$$

### 3.1.3. Engineering Shear Strains and Tensorial Shear Strains

The values of the shear strain components $\epsilon_{12}$, $\epsilon_{13}$, and $\epsilon_{23}$ (or $\epsilon_{xy}$, $\epsilon_{xz}$, and $\epsilon_{yz}$) are called the *tensorial shear strains*. In many applications, an alternative engineering definition for shear strains is commonly used. The *engineering shear strain*, $\gamma$, is defined as the total angle change between two fibers which before deformation were perpendicular to each other. Thus, from Fig. 3.3 for pure deformation, we have

$$\text{total angle change} = \epsilon_{12} + \epsilon_{21} = 2\epsilon_{12}$$

or

$$\gamma_{xy} = \gamma_{yx} = 2\epsilon_{12} = 2\epsilon_{xy} \tag{3.20}$$

and similarly for $\gamma_{xz} = \gamma_{zx}$ and $\gamma_{yz} = \gamma_{zy}$.

Thus, the shear strain tensor $\epsilon_{ij}$ is written in different notations as

$$\epsilon_{ij} = \begin{bmatrix} \epsilon_{11} & \epsilon_{12} & \epsilon_{13} \\ \epsilon_{21} & \epsilon_{22} & \epsilon_{23} \\ \epsilon_{31} & \epsilon_{32} & \epsilon_{33} \end{bmatrix} \equiv \begin{bmatrix} \epsilon_{xx} & \epsilon_{xy} & \epsilon_{xz} \\ \epsilon_{yx} & \epsilon_{yy} & \epsilon_{yz} \\ \epsilon_{zx} & \epsilon_{zy} & \epsilon_{zz} \end{bmatrix} \equiv \begin{bmatrix} \epsilon_x & \gamma_{xy}/2 & \gamma_{xz}/2 \\ \gamma_{yx}/2 & \epsilon_y & \gamma_{yz}/2 \\ \gamma_{zx}/2 & \gamma_{zy}/2 & \epsilon_z \end{bmatrix} \tag{3.21}$$

The different notations in Eq. (3.21) are used interchangeably in what follows when it seems more convenient in some specific applications.

### 3.1.4. Principal Strains

In the analysis of stress, it has been demonstrated that there exist at least three mutually orthogonal planes which have no shear stress acting on them, that is, the principal planes and their associated principal directions, known as principal axes. In the analysis of strain at a point, such *principal axes* also exist.

**Definition.** The principal direction or axis is the direction of any fiber **n** for which the strain vector $\overset{n}{\delta}$ is in the direction of the fiber **n**. For such a direction, the shear strains would be zero. This implies that fibers in these directions that are mutually perpendicular to each other before the motion remain perpendicular to each other after motion. The corresponding strains are called *principal strains*; thus, for the principal direction, we have

$$\overset{n}{\delta} = \epsilon \mathbf{n} \tag{3.22}$$

or

$$\overset{n}{\delta_i} = \epsilon n_i \tag{3.23}$$

Substituting for $\overset{n}{\delta_i}$ from Eq. (3.14) leads to

$$\epsilon_{ij} n_j = \epsilon n_i$$

or, writing $n_i$ as $\delta_{ij}n_j$ ($\delta_{ij}$ is a substitution operator), we get

$$\epsilon_{ij}n_j = \epsilon\delta_{ij}n_j \tag{3.24}$$

or

$$(\epsilon_{ij} - \epsilon\delta_{ij})n_j = 0 \tag{3.25}$$

For a non-trivial solution, we have

$$|\epsilon_{ij} - \epsilon\delta_{ij}| = 0 \tag{3.26}$$

which is exactly the same as the corresponding equation for principal stresses, with stresses replaced by strains [see Eq. (2.33)]. Therefore, all the remarks and derivations made for the stress tensor apply here just as well. Equation (3.26) can be written as

$$\begin{vmatrix} \epsilon_x - \epsilon & \epsilon_{xy} & \epsilon_{xz} \\ \epsilon_{yx} & \epsilon_y - \epsilon & \epsilon_{yz} \\ \epsilon_{zx} & \epsilon_{zy} & \epsilon_z - \epsilon \end{vmatrix} = 0 \tag{3.27}$$

which has three real roots, $\epsilon_1$, $\epsilon_2$, and $\epsilon_3$, corresponding to the three principal strains.

The characteristic equation given in Eq. (3.27) may be rewritten in the form

$$\epsilon^3 - I_1'\epsilon^2 + I_2'\epsilon - I_3' = 0 \tag{3.28}$$

where the strain invariants $I_1'$, $I_2'$, and $I_3'$ are given by

$$I_1' = \epsilon_{11} + \epsilon_{22} + \epsilon_{33} = \epsilon_x + \epsilon_y + \epsilon_z = \epsilon_{ii} \tag{3.29}$$

$$I_2' = \begin{vmatrix} \epsilon_{22} & \epsilon_{23} \\ \epsilon_{32} & \epsilon_{33} \end{vmatrix} + \begin{vmatrix} \epsilon_{11} & \epsilon_{13} \\ \epsilon_{31} & \epsilon_{33} \end{vmatrix} + \begin{vmatrix} \epsilon_{11} & \epsilon_{12} \\ \epsilon_{21} & \epsilon_{22} \end{vmatrix}$$

$$= \begin{vmatrix} \epsilon_y & \epsilon_{yz} \\ \epsilon_{zy} & \epsilon_z \end{vmatrix} + \begin{vmatrix} \epsilon_x & \epsilon_{xz} \\ \epsilon_{zx} & \epsilon_z \end{vmatrix} + \begin{vmatrix} \epsilon_x & \epsilon_{xy} \\ \epsilon_{yx} & \epsilon_y \end{vmatrix}$$

$$= \text{sum of principal two-rowed minors of } \epsilon_{ij} \tag{3.30}$$

$$I_3' = \begin{vmatrix} \epsilon_{11} & \epsilon_{12} & \epsilon_{13} \\ \epsilon_{21} & \epsilon_{22} & \epsilon_{23} \\ \epsilon_{31} & \epsilon_{32} & \epsilon_{33} \end{vmatrix} = \begin{vmatrix} \epsilon_x & \epsilon_{xy} & \epsilon_{xz} \\ \epsilon_{yx} & \epsilon_y & \epsilon_{yz} \\ \epsilon_{zx} & \epsilon_{zy} & \epsilon_z \end{vmatrix}$$

$$= \text{determinant of } \epsilon_{ij} \tag{3.31}$$

or, in terms of the principal strains $\epsilon_1$, $\epsilon_2$, and $\epsilon_3$,

$$I_1' = \epsilon_1 + \epsilon_2 + \epsilon_3$$
$$I_2' = \epsilon_1\epsilon_2 + \epsilon_2\epsilon_3 + \epsilon_3\epsilon_1 \tag{3.32}$$
$$I_3' = \epsilon_1\epsilon_2\epsilon_3$$

By substituting the principal strains $\epsilon_1$, $\epsilon_2$, and $\epsilon_3$, as evaluated by solving the cubic Eq. (3.28), into Eq. (3.25), the principal directions $\mathbf{n}^{(1)}$, $\mathbf{n}^{(2)}$, and $\mathbf{n}^{(3)}$ can be obtained, just as was done for the case of the stress tensor. Following the same procedures as in the analysis of stress, it can be shown that the principal strains assume the stationary values.

### 3.1.5. Principal Shear Strains

**Definitions.** The shear strains for some fibers at a point that assume stationary values are called the *principal shear strains*. To find the directions of such fibers, a similar procedure as for stresses is followed. Consider a line fiber at point $P$ with direction $\mathbf{n}$ and strain vector $\overset{n}{\delta}$ referred to principal strain axes. The normal strain component for this fiber is $\epsilon_n$ and the magnitude of the resultant shear strain component is denoted by $\theta_n$ (tensorial shear strain). Thus,

$$\theta_n^2 = \overset{n}{\delta}_i\overset{n}{\delta}_i - \epsilon_n^2$$

Substituting for $\overset{n}{\delta}_i$ and $\epsilon_n$ from Eqs. (3.14) and (3.18), respectively, in terms of the components $\epsilon_{ij}$ referred to the principal strain axes, we get

$$\theta_n^2 = (\epsilon_{ji}\epsilon_{ki}n_j n_k) - (\epsilon_1 n_1^2 + \epsilon_2 n_2^2 + \epsilon_3 n_3^2)^2$$

or

$$\theta_n^2 = (\epsilon_1^2 n_1^2 + \epsilon_2^2 n_2^2 + \epsilon_3^2 n_3^2) - (\epsilon_1 n_1^2 + \epsilon_2 n_2^2 + \epsilon_3 n_3^2)^2 \tag{3.33}$$

Comparing Eq. (3.33) with Eq. (2.43), it is seen that they are of identical form, with $S_n$ replaced by $\theta_n$ and the principal stresses by the principal strains. The principal shear strains and the corresponding directions can therefore be obtained in exactly the same manner as for stresses. Thus, designating the tensorial principal shear strains by $\theta_1$, $\theta_2$, and $\theta_3$, we can write

$$\theta_1 = \tfrac{1}{2}|\epsilon_2 - \epsilon_3|$$
$$\theta_2 = \tfrac{1}{2}|\epsilon_1 - \epsilon_3| \tag{3.34}$$
$$\theta_3 = \tfrac{1}{2}|\epsilon_1 - \epsilon_2|$$

and the engineering principal shear strains $\gamma_1$, $\gamma_2$, and $\gamma_3$ are given by

$$\gamma_1 = |\epsilon_2 - \epsilon_3|$$
$$\gamma_2 = |\epsilon_1 - \epsilon_3| \tag{3.35}$$
$$\gamma_3 = |\epsilon_1 - \epsilon_2|$$

The *maximum shear strain* is the largest value of the principal shear strains. Hence, for $\epsilon_1 > \epsilon_2 > \epsilon_3$, the maximum shear strain is given by

$$\gamma_{max} = 2\theta_{max} = |\epsilon_1 - \epsilon_3| \tag{3.36}$$

## 3.1.6. Octahedral Strains

The octahedral normal and shear strains of the octahedral fiber, that is, of a material fiber which before deformation is equally inclined with respect to the three principal strain axes 1, 2, and 3, are denoted by $\epsilon_{oct}$ and $\gamma_{oct}$, respectively. For an octahedral fiber, the unit vector $n$ has the components $(1/\sqrt{3}, 1/\sqrt{3}, 1/\sqrt{3})$. Thus, from Eq. (3.18), the octahedral normal strain $\epsilon_{oct}$ is given by

$$\epsilon_{oct} = \frac{1}{3}(\epsilon_1 + \epsilon_2 + \epsilon_3) = \frac{I_1'}{3} \qquad (3.37)$$

which represents the mean of the three principal strains.

The *octahedral shear strain*, $\gamma_{oct}$, with the engineering definition of shear strain, can be obtained from Eq. (3.33), with $\gamma_{oct} = 2\theta_{oct}$. Thus,

$$\gamma_{oct} = \tfrac{2}{3}[(\epsilon_1 - \epsilon_2)^2 + (\epsilon_2 - \epsilon_3)^2 + (\epsilon_3 - \epsilon_1)^2]^{1/2} \qquad (3.38)$$

In terms of strain invariants, the octahedral shear strain can be written as

$$\gamma_{oct} = \frac{2\sqrt{2}}{3}(I_1'^2 - 3I_2')^{1/2} \qquad (3.39)$$

and in terms of general nonprincipal strains, it becomes

$$\gamma_{oct} = \tfrac{2}{3}[(\epsilon_x - \epsilon_y)^2 + (\epsilon_y - \epsilon_z)^2 + (\epsilon_z - \epsilon_x)^2 + 6(\epsilon_{xy}^2 + \epsilon_{yz}^2 + \epsilon_{zx}^2)]^{1/2} \quad (3.40)$$

which expresses the octahedral shear strain in terms of the strain components referred to an arbitrary set of axes $x$, $y$, and $z$.

EXAMPLE 3.2. The strain tensor $\epsilon_{ij}$ at a point is given by

$$\epsilon_{ij} = \begin{bmatrix} -0.00100 & 0 & 0 \\ 0 & -0.00100 & 0.000785 \\ 0 & 0.000785 & 0.00200 \end{bmatrix} \qquad (3.41)$$

Calculate:

(a) The principal strains $\epsilon_1$, $\epsilon_2$, and $\epsilon_3$.
(b) The maximum shear strain $\gamma_{max}$.
(c) The octahedral strains.

SOLUTION. (a) Calculate the strain invariants $I_1'$, $I_2'$, and $I_3'$ from Eqs. (3.29) to (3.31):

$$I_1' = (-0.00100) + (-0.00100) + (0.00200) = 0$$

$$I_2' = (-0.00100)(0.00200) - (0.000785)^2$$

$$+ (-0.00100)(0.00200) + (-0.00100)^2 = -3.62 \times 10^{-6}$$

$$I_3' = \begin{vmatrix} -0.00100 & 0 & 0 \\ 0 & -0.00100 & 0.000785 \\ 0 & 0.000785 & 0.00200 \end{vmatrix} = 2.62 \times 10^{-9}$$

Thus, the characteristic equation becomes

$$\epsilon^3 - 3.62 \times 10^{-6}\epsilon - 2.62 \times 10^{-9} = 0$$

or

$$(\epsilon + 10^{-3})(\epsilon^2 - 1 \times 10^{-3}\epsilon + 2.62 \times 10^{-6}) = 0$$

and the three principal strains are obtained as:

$$\epsilon_1 = 0.00219, \quad \epsilon_2 = -0.00100, \quad \text{and} \quad \epsilon_3 = -0.00119$$

CHECK. Substitute the values of $\epsilon_1$, $\epsilon_2$, and $\epsilon_3$ into Eqs. (3.32) to check the obtained results:

$$I_1' = \epsilon_1 + \epsilon_2 + \epsilon_3 = 0$$

$$I_2' = \epsilon_1\epsilon_2 + \epsilon_2\epsilon_3 + \epsilon_3\epsilon_1$$

$$= (0.00219)(-0.00100) + (-0.00100)(-0.00119) + (-0.00119)(0.00219)$$

$$= -3.62 \times 10^{-6}$$

$$I_3' = \epsilon_1\epsilon_2\epsilon_3 = (0.00219)(-0.00100)(-0.00119) = 2.62 \times 10^{-9}$$

(b) The maximum shear strain $\gamma_{max}$ is calculated from Eq. (3.36):

$$\gamma_{max} = |0.00219 + 0.00119| = 0.00338$$

(c) Since the strain invariants $I_1'$ and $I_2'$ were calculated previously, the octahedral strains $\epsilon_{oct}$ and $\gamma_{oct}$ are obtained directly from Eqs. (3.37) and (3.39), respectively:

$$\epsilon_{oct} = \frac{I_1'}{3} = 0$$

$$\gamma_{oct} = \frac{2\sqrt{2}}{3}[0 - 3(-3.62 \times 10^{-6})]^{1/2} = 0.00311$$

### 3.1.7. Strain Deviator Tensor

As in the case of the stress tensor, the strain tensor $\epsilon_{ij}$ can be decomposed into two parts, a spherical part associated with a change in volume, and a deviatoric part associated with a change in shape (distortion). That is,

$$\epsilon_{ij} = e_{ij} + \tfrac{1}{3}\epsilon_{kk}\delta_{ij} \qquad (3.42)$$

where $e_{ij}$ is defined here as the *deviatoric strain tensor* and $\tfrac{1}{3}\epsilon_{kk} = (\epsilon_x + \epsilon_y + \epsilon_z)/3$ is the *mean* or the *hydrostatic strain*. Thus, the deviatoric strain tensor $e_{ij}$ becomes

$$e_{ij} = \begin{bmatrix} e_x & e_{xy} & e_{xz} \\ e_{yx} & e_y & e_{yz} \\ e_{zx} & e_{zy} & e_z \end{bmatrix} = \begin{bmatrix} (2\epsilon_x - \epsilon_y - \epsilon_z)/3 & \epsilon_{xy} & \epsilon_{xz} \\ \epsilon_{yx} & (2\epsilon_y - \epsilon_z - \epsilon_x)/3 & \epsilon_{yz} \\ \epsilon_{zx} & \epsilon_{zy} & (2\epsilon_z - \epsilon_x - \epsilon_y)/3 \end{bmatrix}$$

(3.43)

or, in terms of the principal strains,

$$e_{ij} = \begin{bmatrix} (2\epsilon_1 - \epsilon_2 - \epsilon_3)/3 & 0 & 0 \\ 0 & (2\epsilon_2 - \epsilon_3 - \epsilon_1)/3 & 0 \\ 0 & 0 & (2\epsilon_3 - \epsilon_1 - \epsilon_2)/3 \end{bmatrix}$$

(3.44)

Note that the condition for a state of pure shear strain is the same as that for a state of pure shear stress; that is, the necessary and sufficient condition for pure shear deformation is that $\epsilon_{kk} = 0$. It follows that $e_{ij}$ is a *pure shear state* and $e_{ij}$ and $\epsilon_{ij}$ have the same principal axes.

If we consider a unit cube with edges directed along the principal axes of strain, 1, 2, and 3, then after deformations, since there is no shear strain for the principal axes, these three axes remain mutually orthogonal after motion; the cube becomes a rectangular parallelepiped with edges $(1+\epsilon_1)$, $(1+\epsilon_2)$, and $(1+\epsilon_3)$. The relative change in volume, $\epsilon_v$, is given by

$$\epsilon_v = \frac{\Delta V}{V} = (1+\epsilon_1)(1+\epsilon_2)(1+\epsilon_3) - 1$$

(3.45)

For small strains,

$$\epsilon_v = \frac{\Delta V}{V} = \epsilon_1 + \epsilon_2 + \epsilon_3 = I_1' = \epsilon_x + \epsilon_y + \epsilon_z$$

(3.46)

Therefore, the spherical part of a strain tensor is proportional to the volume change $\epsilon_v = \epsilon_{kk}$. The relative change in volume (or the volume change per unit volume), $\epsilon_v$, given in Eq. (3.46) is called the *dilatation* or simply the *volume change*.

The invariants of the strain deviator tensor $e_{ij}$ are analogous to those obtained for the stress deviator tensor $s_{ij}$. These deviatoric strain invariants appear in the cubic equation of the determinantal equation $|e_{ij} - e\delta_{ij}| = 0$.

$$e^3 - J_1'e^2 - J_2'e - J_3' = 0$$

(3.47)

where

$$J_1' = e_{ii} = e_x + e_y + e_z = e_1 + e_2 + e_3 = 0$$

(3.48)

$$\begin{aligned} J_2' &= \tfrac{1}{2} e_{ij} e_{ij} = -(e_1 e_2 + e_2 e_3 + e_3 e_1) \\ &= \tfrac{1}{6}[(\epsilon_x - \epsilon_y)^2 + (\epsilon_y - \epsilon_z)^2 + (\epsilon_z - \epsilon_x)^2] + \epsilon_{xy}^2 + \epsilon_{yz}^2 + \epsilon_{zx}^2 \\ &= \tfrac{1}{6}[(\epsilon_1 - \epsilon_2)^2 + (\epsilon_2 - \epsilon_3)^2 + (\epsilon_3 - \epsilon_1)^2] \\ &= \tfrac{1}{2}(e_x^2 + e_y^2 + e_z^2 + 2e_{xy}^2 + 2e_{yz}^2 + 2e_{zx}^2) \end{aligned}$$

(3.49)

$$J_3' = \tfrac{1}{3} e_{ij} e_{jk} e_{ki} = \begin{vmatrix} e_x & e_{xy} & e_{xz} \\ e_{yx} & e_y & e_{yz} \\ e_{zx} & e_{zy} & e_z \end{vmatrix} = \tfrac{1}{3}(e_1^3 + e_2^3 + e_3^3) = e_1 e_2 e_3 \qquad (3.50)$$

in which $e_1$, $e_2$, and $e_3$ are the principal values of the deviatoric strain tensor. Also it can be shown that the invariants $J_1'$, $J_2'$, and $J_3'$ are related to the strain invariants, $I_1'$, $I_2'$, and $I_3'$, through the following relations:

$$J_1' = 0$$

$$J_2' = \tfrac{1}{3}(I_1'^2 - 3I_2') \qquad (3.51)$$

$$J_3' = \tfrac{1}{27}(2I_1'^3 - 9I_1'I_2' + 27I_3')$$

Finally, it can be seen that the octahedral shear strain $\gamma_{oct}$ is related to the second invariant of the deviatoric strain tensor, $J_2'$, as was the case for stresses:

$$\gamma_{oct} = 2\sqrt{\tfrac{2}{3} J_2'} \qquad (3.52)$$

## 3.1.8. Strain–Displacement Relationships

Let the coordinates of a material particle $P$ in a body in the initial (undeformed) position be denoted by $x_i(x_1, x_2, x_3)$ referred to the fixed axes $x_1$, $x_2$, and $x_3$, as shown in Fig. 3.5. The cordinates of the same particle after deformation are denoted by $\xi_i(\xi_1, \xi_2, \xi_3)$ with respect to axes $x_1$, $x_2$, and $x_3$. Figure 3.5 shows two neighboring points $P$ and $Q$ with coordinates $x_i$ and $x_i + dx_i$, respectively, before deformation, and the length of element $PQ$ is denoted by $ds_0$. After deformation, the two points are deformed to points $P'$ and $Q'$ with coordinates $\xi_i$ and $\xi_i + d\xi_i$, respectively, and the length of element $P'Q'$ becomes $ds$. The displacement vector of $P$ is denoted by $u_i$, as shown. Thus, we have

$$ds_0^2 = dx_i\, x_i \qquad (3.53)$$

$$ds^2 = d\xi_i\, d\xi_i \qquad (3.54)$$

and

$$\xi_i = x_i + u_i \qquad (3.55)$$

Then

$$d\xi_i = dx_i + u_{i,j}\, dx_j \qquad (3.56)$$

or

$$d\xi_i = (\delta_{ij} + u_{i,j})\, dx_j \qquad (3.57)$$

where $u_{i,j} = \partial u_i / \partial x_j$.

It is evident that the equality of $ds^2$ and $ds_0^2$ is the necessary and sufficient condition for rigid-body motion; hence, the difference $ds^2 - ds_0^2$ can be taken

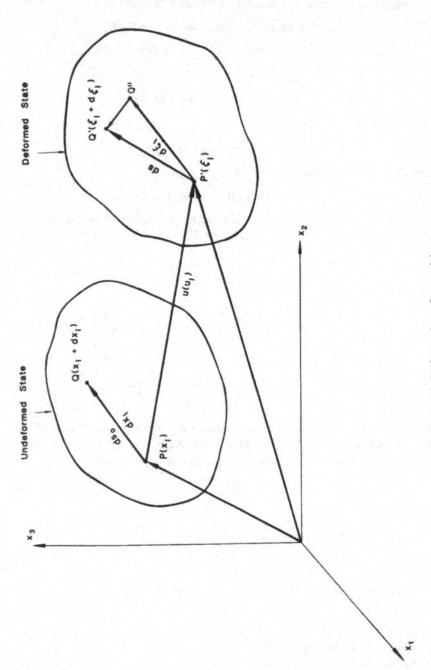

FIGURE 3.5. Deformation of a particle.

as a measure of the strains. From Eqs. (3.53), (3.54), and (3.57), we obtain

$$ds^2 - ds_0^2 = d\xi_i \, d\xi_i - dx_i \, dx_i = d\xi_r \, d\xi_r - dx_r \, dx_r$$
$$= (\delta_{ri} + u_{r,i}) \, dx_i (\delta_{rj} + u_{r,j}) \, dx_j - dx_r \, dx_r$$

or

$$ds^2 - ds_0^2 = (u_{i,j} + u_{j,i} + u_{r,i}u_{r,j}) \, dx_i \, dx_j$$

which can be written as

$$ds^2 - ds_0^2 = 2\epsilon_{ij} \, dx_i dx_j \tag{3.58}$$

where the tensor $\epsilon_{ij}$ is defined as

$$\epsilon_{ij} = \tfrac{1}{2}(u_{i,j} + u_{j,i} + u_{r,i}u_{r,j}) \tag{3.59}$$

If the displacements $u_i$ and their derivatives are small, then the second-order terms in Eq. (3.59) can be neglected. The strain tensor is obtained as

$$\epsilon_{ij} = \tfrac{1}{2}(u_{i,j} + u_{j,i}) \tag{3.60}$$

or in matrix notation

$$[\epsilon_{ij}] = \begin{bmatrix} \dfrac{\partial u}{\partial x} & \dfrac{1}{2}\left(\dfrac{\partial u}{\partial y}+\dfrac{\partial v}{\partial x}\right) & \dfrac{1}{2}\left(\dfrac{\partial u}{\partial z}+\dfrac{\partial w}{\partial x}\right) \\[2ex] \dfrac{1}{2}\left(\dfrac{\partial u}{\partial y}+\dfrac{\partial v}{\partial x}\right) & \dfrac{\partial v}{\partial y} & \dfrac{1}{2}\left(\dfrac{\partial v}{\partial z}+\dfrac{\partial w}{\partial y}\right) \\[2ex] \dfrac{1}{2}\left(\dfrac{\partial u}{\partial z}+\dfrac{\partial w}{\partial x}\right) & \dfrac{1}{2}\left(\dfrac{\partial v}{\partial z}+\dfrac{\partial w}{\partial y}\right) & \dfrac{\partial w}{\partial z} \end{bmatrix} \tag{3.61}$$

Note that Eq. (3.56) can be used to obtain an expression for the relative displacement tensor, $\epsilon'_{ij}$, in terms of displacement derivatives. Since from the definition of the relative displacement vector $\overset{n}{\delta'_i}$, we have

$$\overset{n}{\delta'_i} = \frac{d\xi_i - dx_i}{ds_0}$$

Then, from Eq. (3.56) we obtain the result

$$\overset{n}{\delta'_i} = \frac{u_{i,j} \, dx_j}{ds_0} = u_{i,j}n_j$$

Comparing this equation to Eq. (3.3), we conclude that

$$\epsilon'_{ji} = u_{i,j} \tag{3.62}$$

Using Eq. (3.9) or (3.11), we obtain the same expressions for the strain tensor $\epsilon_{ij}$ as given by Eq. (3.60) or (3.61).

Similarly, substituting Eq. (3.62) into Eq. (3.10) or (3.12) yields the expression for the rotation tensor

$$\omega_{ij} = \tfrac{1}{2}(u_{j,i} - u_{i,j}) \tag{3.63}$$

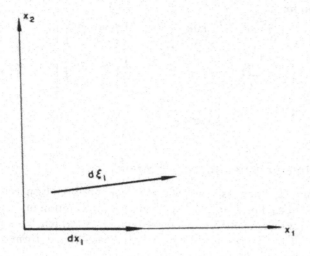

( a ) Strain  Component  $\epsilon_{11}$

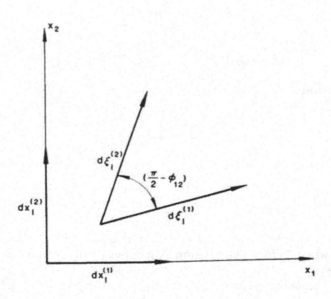

( b ) Strain  Component  $\epsilon_{12}$

FIGURE 3.6. Physical interpretation of strain components.

or in matrix notation,

$$
\omega_{ij} = \begin{bmatrix} 0 & -\dfrac{1}{2}\left(\dfrac{\partial u}{\partial y}-\dfrac{\partial v}{\partial x}\right) & -\dfrac{1}{2}\left(\dfrac{\partial u}{\partial z}-\dfrac{\partial w}{\partial x}\right) \\[2ex] \dfrac{1}{2}\left(\dfrac{\partial u}{\partial y}-\dfrac{\partial v}{\partial x}\right) & 0 & -\dfrac{1}{2}\left(\dfrac{\partial v}{\partial z}-\dfrac{\partial w}{\partial y}\right) \\[2ex] \dfrac{1}{2}\left(\dfrac{\partial u}{\partial z}-\dfrac{\partial w}{\partial x}\right) & \dfrac{1}{2}\left(\dfrac{\partial v}{\partial z}-\dfrac{\partial w}{\partial y}\right) & 0 \end{bmatrix} \tag{3.64}
$$

### 3.1.9. Physical Interpretation for Small Strains

For small deformation, there is a simple physical interpretation for the components of strain $\epsilon_{ij}$. For example, the normal strain component $\epsilon_{11}$ can be interpreted as follows. Consider the line element $dx_i$ which lies along the $x_1$-axis before deformation as shown in Fig. 3.6a. After deformation, the element is denoted by $d\xi_i$. Thus, we have

$$
dx_i = (ds_0, 0, 0)
$$

Substituting for $dx_i$ in Eq. (3.58), we get

$$
ds^2 - ds_0^2 = 2\epsilon_{ij}\, dx_i\, dx_j = 2\epsilon_{11}\, dx_1\, dx_1 = 2\epsilon_{11}\, ds_0^2
$$

or

$$
(ds + ds_0)(ds - ds_0) = 2\epsilon_{11}\, ds_0^2
$$

which can be written as

$$
\frac{ds - ds_0}{ds_0} = \left(\frac{2ds_0}{ds + ds_0}\right)\epsilon_{11}
$$

But for small deformation, $ds$ is nearly equal to $ds_0$; therefore,

$$
\epsilon_{11} = \frac{ds - ds_0}{ds_0} \tag{3.65}
$$

Thus, $\epsilon_{11}$ respesents the extension or change in length per unit length of a line element which before deformation is parallel to the $x_1$-axis. Strain components $\epsilon_{22}$ and $\epsilon_{33}$ have similar interpretations.

The shear component $\epsilon_{12}$ can also be interpreted by considering the two line elements $dx_i^{(1)}$ and $dx_i^{(2)}$ originally parallel to the $x_1$ and $x_2$ axes, respectively, as shown in Fig. 3.6b. Denote the total decrease in the right angle between the two lines after deformation by $\phi_{12}$. Thus, we have

$$
\cos\left(\frac{\pi}{2} - \phi_{12}\right) = \frac{d\xi_i^{(1)}\, d\xi_i^{(2)}}{|d\xi_i^{(1)}||d\xi_i^{(2)}|}
$$

Using Eq. (3.56), we get

$$\cos\left(\frac{\pi}{2}-\phi_{12}\right)=\frac{(dx_i^{(1)}+u_{i,k}\,dx_k^{(1)})(dx_i^{(2)}+u_{i,t}\,dx_t^{(2)})}{|d\xi_i^{(1)}||d\xi_i^{(2)}|}$$

$$=\frac{dx_i^{(1)}\,dx_i^{(2)}+(u_{i,k}+u_{k,i}+u_{r,i}u_{r,k})\,dx_i^{(1)}\,dx_k^{(2)}}{|d\xi_i^{(1)}||d\xi_i^{(2)}|}$$

But $dx_i^{(1)}\,dx_i^{(2)}=0$ (two orthogonal vectors); then

$$\cos\left(\frac{\pi}{2}-\phi_{12}\right)=\frac{2\epsilon_{ik}\,dx_i^{(1)}\,dx_k^{(2)}}{|d\xi_i^{(1)}||d\xi_i^{(2)}|}=\frac{2\epsilon_{12}}{(1+\epsilon_{11})(1+\epsilon_{22})}$$

which for small deformation reduces to

$$\cos\left(\frac{\pi}{2}-\phi_{12}\right)=2\epsilon_{12} \tag{3.66}$$

But $\cos(\pi/2-\phi_{12})\cong\phi_{12}$; hence, $\epsilon_{12}$ represents one-half the decrease in the right angle between two line elements that before deformation are parallel to the $x_1$- and $x_2$-axis. Similar derivations can be made for the components $\epsilon_{13}$ and $\epsilon_{23}$. Consequently, the off-diagonal terms in Eq. (3.61) represent shear deformation. Physical visualization of the meaning of the partial derivatives in Eq. (3.61) leads to the same interpretation in terms of fractional elongations and angle changes previously obtained more generally.

## 3.1.10. Equations of Strain Compatibility

In the analysis of stress, it has been pointed out that we must establish the equilibrium equations to ensure that the body is always in an equilibrium state. In the analysis of strain, however, there must be some conditions to be imposed on the strain components so that the deformed body remains continuous. This can be illustrated by considering, for example, Eq. (3.60); namely,

$$u_{i,j}+u_{j,i}=2\epsilon_{ij} \tag{3.67}$$

For given displacements, $u_i$, the strain components, $\epsilon_{ij}$, can be determined from Eq. (3.67). On the other hand, for prescribed strain components $\epsilon_{ij}$, Eq. (3.67) represents a system of partial differential equations for the determination of the displacement components $u_i$. Since there are six equations for three unknown functions $u_i$, we cannot expect in general that the system of Eqs. (3.67) will have a solution if the strain components $\epsilon_{ij}$ are arbitrarily chosen. Therefore, in order to have single-valued continuous displacement function $u_i$, some restrictions must be imposed on the strain components $\epsilon_{ij}$. Such restrictions are called compatibility conditions. It can be shown that the compatibility equations for a simply connected region may be written in the form

$$\epsilon_{ij,kl}+\epsilon_{kl,ij}-\epsilon_{ik,jl}-\epsilon_{jl,ik}=0 \tag{3.68}$$

or, expanding these expressions, we get

$$\frac{\partial^2 \epsilon_x}{\partial y^2} + \frac{\partial^2 \epsilon_y}{\partial x^2} = 2 \frac{\partial^2 \epsilon_{xy}}{\partial x \, \partial y}$$

$$\frac{\partial^2 \epsilon_y}{\partial z^2} + \frac{\partial^2 \epsilon_z}{\partial y^2} = 2 \frac{\partial^2 \epsilon_{yz}}{\partial y \, \partial z}$$

$$\frac{\partial^2 \epsilon_z}{\partial x^2} + \frac{\partial^2 \epsilon_x}{\partial z^2} = 2 \frac{\partial^2 \epsilon_{zx}}{\partial z \, \partial x}$$

$$\frac{\partial}{\partial x} \left( -\frac{\partial \epsilon_{yz}}{\partial x} + \frac{\partial \epsilon_{zx}}{\partial y} + \frac{\partial \epsilon_{xy}}{\partial z} \right) = \frac{\partial^2 \epsilon_x}{\partial y \, \partial z}$$

$$\frac{\partial}{\partial y} \left( -\frac{\partial \epsilon_{zx}}{\partial y} + \frac{\partial \epsilon_{xy}}{\partial z} + \frac{\partial \epsilon_{yz}}{\partial x} \right) = \frac{\partial^2 \epsilon_y}{\partial z \, \partial x}$$

$$\frac{\partial}{\partial z} \left( -\frac{\partial \epsilon_{xy}}{\partial z} + \frac{\partial \epsilon_{yz}}{\partial x} + \frac{\partial \epsilon_{zx}}{\partial y} \right) = \frac{\partial^2 \epsilon_z}{\partial x \, \partial y}$$

(3.69)

These six compatibility equations are the necessary and sufficient conditions required to ensure that the strain components give single-valued continuous displacements for a simply connected region.

## 3.2. Linear Elastic Isotropic Stress–Strain Relation— Hooke's Law

### 3.2.1. Introduction

Figure 3.7a shows an elastic material body of volume $V$ and surface area $A$. The part of the surface area where surface tractions $T_i$ are prescribed is denoted by $A_T$ and that where surface displacements $\bar{u}_i$ are prescribed is denoted by $A_u$. When the body forces $F_i$ and the surface forces $T_i$ and surface displacement $\bar{u}_i$ act upon the body, the resulting stresses $\sigma_{ij}$ must satisfy the equilibrium equations

$$\sigma_{ij,j} + F_i = 0 \qquad \text{in } V \tag{3.70a}$$

$$\sigma_{ij} n_j = T_i \qquad \text{at } A_T \tag{3.70b}$$

and the resulting strains, $\epsilon_{ij}$, and displacements, $u_i$, must satisfy the geometry conditions

$$\epsilon_{ij} = \tfrac{1}{2}(u_{i,j} + u_{j,i}) \qquad \text{in } V \tag{3.71a}$$

$$u_i = \bar{u}_i \qquad \text{at } A_u \tag{3.71b}$$

For a material element in volume $V$, the three equations of equilibrium, Eq. (3.70a), and the six equations of compatibility between the strains and

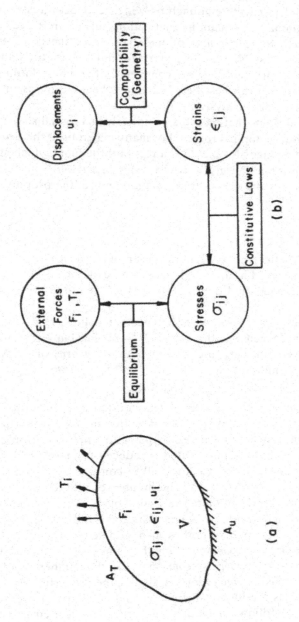

FIGURE 3.7. Establishment of a solid mechanics problem. (a) Variables; (b) interrelationships of variables.

displacements, Eq. (3.71a), represent a total of nine equations involving fifteen unknowns (six stresses, six strains, and three displacements). The insufficiency in the total number of available equations is made up for by a set of six material-dependent relationships which connect stresses with strains. These additional six equations or relationships are referred to as the *constitutive equations or relations* of materials. Once the constitutive relation for a material is established, the general formulation for the solution of a solid mechanics problem can be completed. The interrelations of variables ($\sigma_{ij}$, $\epsilon_{ij}$, and $u_i$) can best be illustrated schematically as in Fig. 3.7b for the case of a static analysis.

An *elastic* material is one that recovers completely its original shape and size upon the removal of applied forces. For many materials at the *working* load level, the elastic range also includes a region throughout which stress and strain have a *linear* relationship, as we have shown previously in Chapter 1. This linear portion of the stress-strain relation ends at the proportional *limit*, and its general form is given by

$$\sigma_{ij} = C_{ijkl}\epsilon_{kl} \tag{3.72}$$

where $C_{ijkl}$ is the *material elastic constant* tensor. It may also be remarked here that Eq. (3.72) is the simplest generalization of the linear dependence of stress on strain observed in the familiar Hooke's experiment in a simple tension test, and consequently Eq. (3.72) is often referred to as the *generalized Hooke's law*.

Since both $\sigma_{ij}$ and $\epsilon_{kl}$ are second-order tensors, it follows that $C_{ijkl}$ is a fourth-order tensor. In general, there are $(3)^4 = 81$ constants for such a tensor $C_{ijkl}$. However, since $\sigma_{ij}$ and $\epsilon_{kl}$ are both *symmetric*, one has the following symmetry conditions:

$$C_{ijkl} = C_{jikl} = C_{ijlk} = C_{jilk} \tag{3.73}$$

Hence, the maximum number of independent constant is reduced to 36.

For a *Green elastic* material, it is shown later that the four subscripts of the elastic constants can be considered as pairs $C_{(ij)(kl)}$. As a result, the number of independent constants needed is reduced from 36 to 21. That is, if we know these 21 constants, we know all 81 constants. If, in addition, we have a plane of elastic symmetry, the number of elastic constants is reduced further from 21 to 13. If there is a second plane of elastic symmetry orthogonal to the first, the number of elastic constants is reduced still further. The second plane of symmetry implies also symmetry about the third orthogonal plane (*orthotropic symmetry*) and the number of elastic constants is reduced to 9. For a *transversely isotropic* material, the number is reduced to 5. Further, if we specify *cubic symmetry*, that is, the properties along the x-, y-, and z-directions are identical, then we cannot distinguish between directions x, y, and z. It follows that it takes only three independent constants to describe the elastic behavior of such a material. Finally, if we have a solid whose elastic properties are *not* a function of direction at all, then we need only *two* independent elastic constants to describe its behavior.

## 3.2.2. Isotropic Linear Elastic Stress–Strain Relations

For an *isotropic* material, the elastic constants in Eq. (3.72) must be the same for all directions. Thus, tensor $C_{ijkl}$ must be an isotropic fourth-order tensor. It can be shown that the most general form for the isotropic tensor $C_{ijkl}$ is given by (Section 1.5.6)

$$C_{ijkl} = \lambda \delta_{ij}\delta_{kl} + \mu(\delta_{ik}\delta_{jl} + \delta_{il}\delta_{jk}) + \alpha(\delta_{ik}\delta_{jl} - \delta_{il}\delta_{jk}) \tag{3.74}$$

where $\lambda$, $\mu$, and $\alpha$ are scalar constants. Now, since $C_{ijkl}$ must satisfy the symmetry conditions in Eqs. (3.73), we have $\alpha = 0$ in Eq. (3.74). Thus, Eq. (3.74) must take the form

$$C_{ijkl} = \lambda \delta_{ij}\delta_{kl} + \mu(\delta_{ik}\delta_{jl} + \delta_{il}\delta_{jk}) \tag{3.75}$$

From Eqs. (3.72) and (3.75), we get

$$\sigma_{ij} = \lambda \delta_{ij}\delta_{kl}\epsilon_{kl} + \mu(\delta_{ik}\delta_{jl} + \delta_{il}\delta_{jk})\epsilon_{kl}$$

or

$$\sigma_{ij} = \lambda \epsilon_{kk}\delta_{ij} + 2\mu\epsilon_{ij} \tag{3.76}$$

Hence, for an isotropic linear elastic material, there are only *two* independent material constants, $\lambda$ and $\mu$, which are called *Lamé's constants*.

Conversely, strains $\epsilon_{ij}$ can be expressed in terms of stresses in the constitutive relation of Eq. (3.76). For Eq. (3.76), one has

$$\sigma_{kk} = (3\lambda + 2\mu)\epsilon_{kk}$$

or

$$\epsilon_{kk} = \frac{\sigma_{kk}}{3\lambda + 2\mu} \tag{3.77}$$

Substituting this value of $\epsilon_{kk}$ into Eq. (3.76) and solving for $\epsilon_{ij}$, we get

$$\epsilon_{ij} = \frac{-\lambda \delta_{ij}}{2\mu(3\lambda + 2\mu)}\sigma_{kk} + \frac{1}{2\mu}\sigma_{ij} \tag{3.78}$$

Equations (3.76) and (3.78) are the general forms of the constitutive relation for an isotropic linear elastic material. An important consequence of these equations is that for an isotropic material, the *principal directions* of the stress and strain tensors *coincide*.

## 3.2.3. Generalized Isotropic Hooke's Law Based on Experimental Evidence

Consider a simple tension test as shown in Fig. 3.8a. The only nonzero stress component, $\sigma_x = \sigma$, causes axial strain $\epsilon_x$ according to

$$\epsilon_x = \frac{\sigma_x}{E} \tag{3.79}$$

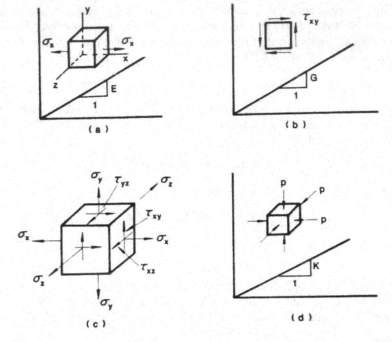

FIGURE 3.8. Stress-strain relations for isotropic linear elastic materials. (a) Simple tension test; (b) simple shear test; (c) general multiaxial state; (d) hydrostatic compression test.

and transverse strains according to

$$\epsilon_y = \epsilon_z = -\nu\epsilon_x = -\frac{\nu\sigma_x}{E} \tag{3.80}$$

where $E$ is Young's modulus and $\nu$ Poisson's ratio. It should be noted that the normal stress $\sigma_x$ produces no shear strain. On the other hand, in a pure shear test (Fig. 3.8b), the shear stress $\tau_{xy}$ produces no normal strain but only the shear strain $\gamma_{xy}$ as

$$\gamma_{xy} = \frac{\tau_{xy}}{G} \tag{3.81}$$

where $G$ is the *shear modulus* of elasticity.

Consider now a three-dimensional isotropic rectangular element, subjected to a triaxial state of stress (Fig. 3.8c). According to the small-deformation assumption, the principle of superposition applies under multiaxial stressing. Then the simultaneous action of $\sigma_x$, $\sigma_y$, and $\sigma_z$ and $\tau_{yz}$, $\tau_{zx}$,

$\tau_{xy}$ leads to the following strains:

$$\epsilon_x = \frac{1}{E}[\sigma_x - \nu(\sigma_y + \sigma_z)]$$

$$\epsilon_y = \frac{1}{E}[\sigma_y - \nu(\sigma_z + \sigma_x)]$$

$$\epsilon_z = \frac{1}{E}[\sigma_z - \nu(\sigma_x + \sigma_y)] \qquad (3.82)$$

$$\gamma_{yz} = \frac{1}{G}\tau_{yz}$$

$$\gamma_{zx} = \frac{1}{G}\tau_{zx}$$

$$\gamma_{xy} = \frac{1}{G}\tau_{xy}$$

Equation (3.82) is referred to as the generalized Hooke's law for isotropic materials. The elastic constants $E$, $\nu$, and $G$ can be demonstrated to be related by

$$G = \frac{E}{2(1+\nu)} \qquad (3.83)$$

To show this, refer again to the element subjected to pure shear (Fig. 3.8b). A pure shear stress can be expressed in terms of the principal stresses $\tau_{xy}$ and $-\tau_{xy}$ acting on planes making an angle of 45° with the shear planes. On the other hand, the corresponding normal strain in the principal direction can also be obtained in terms of shear strain $\gamma_{xy}$. Then, applying Hooke's law, Eq. (3.83) follows. The details are left out here for the reader to work out.

Using the index notation, we can rewrite Eq. (3.82) in a concise form

$$\epsilon_{ij} = \frac{1+\nu}{E}\sigma_{ij} - \frac{\nu}{E}\sigma_{kk}\delta_{ij} \qquad (3.84)$$

Proceeding in a similar manner as in Section 3.2.2, we can solve Eq. (3.84) for stresses $\sigma_{ij}$ and obtain

$$\sigma_{ij} = \frac{E}{(1+\nu)}\epsilon_{ij} + \frac{\nu E}{(1+\nu)(1-2\nu)}\epsilon_{kk}\delta_{ij} \qquad (3.85)$$

Comparing the constitutive relations (3.76) and (3.85), Lamé's constants $\mu$ and $\lambda$ can be expressed in terms of $E$ and $\nu$ as

$$\mu = G = \frac{E}{2(1+\nu)} \qquad (3.86a)$$

and

$$\lambda = \frac{\nu E}{(1+\nu)(1-2\nu)} \tag{3.86b}$$

or, conversely, Young's modulus $E$ and Poisson's ratio $\nu$ can be expressed in terms of $\mu$ and $\lambda$ as

$$E = \frac{\mu(3\lambda + 2\mu)}{\lambda + \mu} \tag{3.87a}$$

$$\nu = \frac{\lambda}{2(\lambda + \mu)} \tag{3.87b}$$

By using Eqs. (3.86a) and (3.86b) in Eq. (3.75), the tensor of elastic moduli, $C_{ijkl}$, can be expressed in terms of $E$ and $\nu$:

$$C_{ijkl} = \frac{E}{2(1+\nu)} \left[ \frac{2\nu}{(1-2\nu)} \delta_{ij}\delta_{kl} + \delta_{ik}\delta_{jl} + \delta_{il}\delta_{jk} \right] \tag{3.88}$$

Note that Eq. (3.84) may be written in a compact form:

$$\epsilon_{ij} = D_{ijkl}\sigma_{kl} \tag{3.89a}$$

and

$$D_{ijkl} = \frac{(1+\nu)}{2E} \left[ -\frac{2\nu}{1+\nu} \delta_{ij}\delta_{kl} + \delta_{ik}\delta_{jl} + \delta_{il}\delta_{jk} \right] \tag{3.89b}$$

where $D_{ijkl}$ is the inverse of $C_{ijkl}$ and is called the compliance tensor.

Another modulus of elasticity discussed here is the so-called *bulk modulus* $K$, which is introduced by a hydrostatic compression test as shown in Fig. 3.8d. In this case, $\sigma_{11} = \sigma_{22} = \sigma_{33} = p = \sigma_{kk}/3$. The bulk modulus $K$ is defined for this case as the ratio between the hydrostatic pressure $p$ and the corresponding volume change $\epsilon_{kk}$, i.e.,

$$K = \frac{p}{\epsilon_{kk}} \tag{3.90}$$

From Eq. (3.77), one gets

$$K = \lambda + \tfrac{2}{3}\mu \tag{3.91}$$

Substituting Eqs. (3.86) and (3.87) into Eq. (3.91) results in

$$K = \frac{E}{3(1-2\nu)} \tag{3.92}$$

### 3.2.4. Decomposition of Stress-Strain Relation

A neat and logical separation exists between the mean (hydrostatic or volumetric) and the shear (deviatoric) response components in an isotropic linear material. The hydrostatic response can be derived directly from Eq. (3.90) as

$$\sigma_{oct} = p = K\epsilon_{kk} \tag{3.93}$$

To derive the deviatoric response relations, we use the relation $s_{ij} = \sigma_{ij} - p\delta_{ij}$ and substitute for $\sigma_{ij}$ and $p$ from Eqs. (3.85) and (3.93), respectively, and note Eq. (3.92). This leads to

$$s_{ij} = \frac{E}{(1+\nu)} \epsilon_{ij} + \frac{\nu E}{(1+\nu)(1-2\nu)} \epsilon_{kk}\delta_{ij} - \frac{E}{3(1-2\nu)} \epsilon_{kk}\delta_{ij}$$

Substituting for $\epsilon_{ij} = e_{ij} + \frac{1}{3}\epsilon_{kk}\delta_{ij}$ and simplifying, we have the relation

$$s_{ij} = \frac{E}{1+\nu} e_{ij} = 2Ge_{ij} \tag{3.94}$$

Equations (3.93) and (3.94) give the required separation of the hydrostatic and deviatoric relations. Combining these two equations, we can write the total elastic strains $\epsilon_{ij}$ in terms of the hydrostatic and deviatoric stresses as

$$\epsilon_{ij} = \frac{1}{3} \epsilon_{kk}\delta_{ij} + e_{ij} = \frac{1}{3K} p\delta_{ij} + \frac{1}{2G} s_{ij} \tag{3.95}$$

or

$$\epsilon_{ij} = \frac{1}{9K} I_1\delta_{ij} + \frac{1}{2G} s_{ij} \tag{3.96}$$

Similarly, $\sigma_{ij}$ can be expressed in terms of the volumetric and deviatoric strains in the following form:

$$\sigma_{ij} = K\epsilon_{kk}\delta_{ij} + 2Ge_{ij} \tag{3.97}$$

### 3.2.5. Isotropic Linear Elastic Stress–Strain Relations in Matrix Form

The stress–strain relationships discussed above can be conveniently expressed in matrix form. These forms are suitable for use in solutions by numerical methods (e.g., finite-element method). In the following, matrix forms are given for various cases.

#### 3.2.5.1. THREE-DIMENSIONAL CASE

The stress and strain components are defined by the two vectors $\{\sigma\}$ and $\{\epsilon\}$, respectively, which are given by

$$\{\sigma\} = \begin{Bmatrix} \sigma_x \\ \sigma_y \\ \sigma_z \\ \tau_{xy} \\ \tau_{yz} \\ \tau_{zx} \end{Bmatrix}, \quad \{\epsilon\} = \begin{Bmatrix} \epsilon_x \\ \epsilon_y \\ \epsilon_z \\ \gamma_{xy} \\ \gamma_{yz} \\ \gamma_{zx} \end{Bmatrix} \tag{3.98}$$

Now Eq. (3.85) can be written in matrix form as

$$\{\sigma\} = [C]\{\epsilon\} \tag{3.99}$$

where the matrix $[C]$ is called the *elastic constitutive* or *elastic moduli matrix* and is given by

$$[C]=\frac{E}{(1+v)(1-2v)}\begin{bmatrix} (1-v) & v & v & 0 & 0 & 0 \\ v & (1-v) & v & 0 & 0 & 0 \\ v & v & (1-v) & 0 & 0 & 0 \\ 0 & 0 & 0 & \frac{(1-2v)}{2} & 0 & 0 \\ 0 & 0 & 0 & 0 & \frac{(1-2v)}{2} & 0 \\ 0 & 0 & 0 & 0 & 0 & \frac{(1-2v)}{2} \end{bmatrix}$$

(3.100a)

or, alternatively, substituting for $v$ and $E$ in terms of $K$ and $G$ gives

$$[C]=\begin{bmatrix} (K+\frac{4}{3}G) & (K-\frac{2}{3}G) & (K-\frac{2}{3}G) & 0 & 0 & 0 \\ (K-\frac{2}{3}G) & (K+\frac{4}{3}G) & (K-\frac{2}{3}G) & 0 & 0 & 0 \\ (K-\frac{2}{3}G) & (K-\frac{2}{3}G) & (K+\frac{4}{3}G) & 0 & 0 & 0 \\ 0 & 0 & 0 & G & 0 & 0 \\ 0 & 0 & 0 & 0 & G & 0 \\ 0 & 0 & 0 & 0 & 0 & G \end{bmatrix}$$

(3.100b)

Also, Eq. (3.82) can be written in matrix form as

$$\{\epsilon\}=[C]^{-1}\{\sigma\}=[D]\{\sigma\} \tag{3.101}$$

where the *elastic compliance* matrix, $[D]$, is given by the inverse of matrix $[C]$:

$$[D]=\frac{1}{E}\begin{bmatrix} 1 & -v & -v & 0 & 0 & 0 \\ -v & 1 & -v & 0 & 0 & 0 \\ -v & -v & 1 & 0 & 0 & 0 \\ 0 & 0 & 0 & 2(1+v) & 0 & 0 \\ 0 & 0 & 0 & 0 & 2(1+v) & 0 \\ 0 & 0 & 0 & 0 & 0 & 2(1+v) \end{bmatrix}$$

(3.102)

### 3.2.5.2. PLANE STRESS CASE

It can be shown that Eqs. (3.99) and (3.101), when reduced to the two-dimensional plane stress case ($\sigma_z = \tau_{yz} = \tau_{zx} = 0$), take the following simple forms:

$$\begin{Bmatrix} \sigma_x \\ \sigma_y \\ \tau_{xy} \end{Bmatrix} = \frac{E}{1-v^2}\begin{bmatrix} 1 & v & 0 \\ v & 1 & 0 \\ 0 & 0 & (1-v)/2 \end{bmatrix}\begin{Bmatrix} \epsilon_x \\ \epsilon_y \\ \gamma_{xy} \end{Bmatrix} \tag{3.103}$$

and

$$\left\{\begin{array}{c} \epsilon_x \\ \epsilon_y \\ \gamma_{xy} \end{array}\right\} = \frac{1}{E} \left[\begin{array}{ccc} 1 & -\nu & 0 \\ -\nu & 1 & 0 \\ 0 & 0 & 2(1+\nu) \end{array}\right] \left\{\begin{array}{c} \sigma_x \\ \sigma_y \\ \tau_{xy} \end{array}\right\} \tag{3.104}$$

It is noted that in the plane stress case, the strain component $\epsilon_z$ is nonzero, while the shear strain components $\gamma_{yz}$ and $\gamma_{zx}$ are zero. The component $\epsilon_z$ has the value

$$\epsilon_z = \frac{-\nu}{E}(\sigma_x + \sigma_y) = \frac{-\nu}{1-\nu}(\epsilon_x + \epsilon_y) \tag{3.105}$$

That is, $\epsilon_z$ is a linear function of $\epsilon_x$ and $\epsilon_y$.

The plane stress relations given above are commonly used in many practical applications. For instance, the analysis of thin, flat plates loaded in the plane of the plate ($x$-$y$ plane) are often treated as plane stress problems.

### 3.2.5.3. PLANE STRAIN CASE

The plane strain conditions ($\epsilon_z = \gamma_{yz} = \gamma_{zx} = 0$) are normally found in elongated bodies of uniform cross section subjected to uniform loading along their longitudinal axis ($z$-axis), such as in the case of tunnels, soil slopes, and retaining walls. Under the conditions of plane strain, Eqs. (3.99) and (3.101) can be reduced to the simple form

$$\left\{\begin{array}{c} \sigma_x \\ \sigma_y \\ \tau_{xy} \end{array}\right\} = \frac{E}{(1+\nu)(1-2\nu)} \left[\begin{array}{ccc} (1-\nu) & \nu & 0 \\ \nu & (1-\nu) & 0 \\ 0 & 0 & (1-2\nu)/2 \end{array}\right] \left\{\begin{array}{c} \epsilon_x \\ \epsilon_y \\ \gamma_{xy} \end{array}\right\} \tag{3.106}$$

and

$$\left\{\begin{array}{c} \epsilon_x \\ \epsilon_y \\ \gamma_{xy} \end{array}\right\} = \frac{(1+\nu)}{E} \left[\begin{array}{ccc} (1-\nu) & -\nu & 0 \\ -\nu & (1-\nu) & 0 \\ 0 & 0 & 2 \end{array}\right] \left\{\begin{array}{c} \sigma_x \\ \sigma_y \\ \tau_{xy} \end{array}\right\} \tag{3.107}$$

For this case, the stress components $\tau_{yz}$ and $\tau_{zx}$ are zero, and the stress component $\sigma_z$ has the value

$$\sigma_z = \nu(\sigma_x + \sigma_y) \tag{3.108}$$

### 3.2.5.4. AXISYMMETRIC CASE

Analysis of bodies of revolution under axisymmetric loading is similar to that for plane stress and plane strain conditions since this problem is also two-dimensional. In the usual notation, the nonzero stress components in the axisymmetric case are $\sigma_r$, $\sigma_z$, $\sigma_\theta$, and $\tau_{rz}$, and the corresponding strains are $\epsilon_r$, $\epsilon_z$, $\epsilon_\theta$, and $\gamma_{rz}$. Equations (3.99) and (3.101) can be reduced to the forms $(\tau_{z\theta} = \tau_{\theta r} = \gamma_{z\theta} = \gamma_{\theta r} = 0)$:

$$
\begin{Bmatrix} \sigma_r \\ \sigma_z \\ \sigma_\theta \\ \tau_{rz} \end{Bmatrix} = \frac{E}{(1+\nu)(1-2\nu)} \begin{bmatrix} (1-\nu) & \nu & \nu & 0 \\ \nu & (1-\nu) & \nu & 0 \\ \nu & \nu & (1-\nu) & 0 \\ 0 & 0 & 0 & (1-2\nu)/2 \end{bmatrix} \begin{Bmatrix} \epsilon_r \\ \epsilon_z \\ \epsilon_\theta \\ \gamma_{rz} \end{Bmatrix}
$$

(3.109)

and

$$
\begin{Bmatrix} \epsilon_r \\ \epsilon_z \\ \epsilon_\theta \\ \gamma_{rz} \end{Bmatrix} = \frac{1}{E} \begin{bmatrix} 1 & -\nu & -\nu & 0 \\ -\nu & 1 & -\nu & 0 \\ -\nu & -\nu & 1 & 0 \\ 0 & 0 & 0 & 2(1+\nu) \end{bmatrix} \begin{Bmatrix} \sigma_r \\ \sigma_z \\ \sigma_\theta \\ \tau_{rz} \end{Bmatrix}
$$

(3.110)

## 3.3. Nonlinear Elastic Isotropic Stress–Strain Relation

### 3.3.1. Introduction

An elastic material is characterized by its total *reversibility*. In the uniaxial case (Fig. 3.9), this means that upon loading and subsequent unloading, the material follows the same stress–strain curve, i.e., from O to A and subsequently from A to O. Therefore, following a loading cycle OAO, the state of the material is identical with that before loading. Reloading will follow the same loading path OA.

Such a reversibility implies that the mechanical work done by external loading will be regained if the load is removed statically. Thus, the work may be regarded as being stored in the deformed body in the form of energy. This stored energy is called *strain energy*.

In the uniaxial case, the *strain energy per unit volume* or the *strain energy density*, W, is represented by the area under the $\sigma$–$\epsilon$ curve shown in Fig. 3.9 and is expressed as

$$
W(\epsilon) = \int_0^\epsilon \sigma \, d\epsilon
$$

(3.111)

In the multiaxial case, the strain energy density is the sum of the contributions by all the stress components, i.e.,

$$
W(\epsilon_{ij}) = \int_0^{\epsilon_{ij}} \sigma_{ij} \, d\epsilon_{ij}
$$

(3.112)

Alternatively, the area above the $\sigma$–$\epsilon$ curve shown in Fig. 3.9, representing the *complementary energy density* (or the *complementary energy per unit*

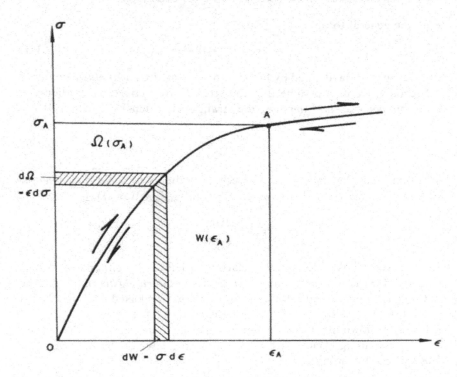

FIGURE 3.9. Strain energy density function $W$ and complementary energy density function $\Omega$.

volume) for the uniaxial case, is expressed as

$$\Omega(\sigma) = \int_0^\sigma \epsilon \, d\sigma \tag{3.113}$$

In the multiaxial case, it takes the form

$$\Omega(\sigma_{ij}) = \int_0^{\sigma_{ij}} \epsilon_{ij} \, d\sigma_{ij} \tag{3.114}$$

The strain energy density $W$ and the complementary energy density $\Omega$ are functions of strain, $\epsilon_{ij}$, and stress, $\sigma_{ij}$, respectively. It is evident that the energy functions $W$ and $\Omega$ are related by

$$W + \Omega = \sigma_{ij}\epsilon_{ij} \tag{3.115}$$

Regarding the stress–strain relationships, there are two approaches to describing the reversible behavior of elastic materials. First, we may assume that there is a one-to-one correspondence between stress and strain, or in other words, the stress $\sigma_{ij}$ is determined uniquely from the current strain

$\epsilon_{ij}$ in the general form

$$\sigma_{ij} = F_{ij}(\epsilon_{kl}) \tag{3.116}$$

The elastic material defined by Eq. (3.116) is termed *Cauchy elastic material.*
    Secondly, we may assume that the stresses are derived as gradients of the *strain potential function* (i.e., the strain energy density function $W$) as

$$\sigma_{ij} = \frac{\partial W(\epsilon_{ij})}{\partial \epsilon_{ij}} \tag{3.117}$$

or assume that the strains are derived as gradients of the *stress potential function* (i.e., the complementary energy density function $\Omega$) as

$$\epsilon_{ij} = \frac{\partial \Omega(\sigma_{ij})}{\partial \sigma_{ij}} \tag{3.118}$$

The material whose stress–strain relation is defined by either Eq. (3.117) or Eq. (3.118) is referred to as *hyperelastic* or *Green elastic material.* The isotropic nonlinear elastic stress–strain relationships based on Eqs. (3.116) to (3.118) will be discussed further in the sections that follow.
    It can be shown that Cauchy material may generate energy under certain loading–unloading cycles, thereby violating the laws of thermodynamics. This will be shown in Example 3.5.

EXAMPLE 3.3. The uniaxial stress–strain relation of a nonlinear elastic material is given by the single power-term expression

$$\epsilon = b\sigma^n \tag{3.119}$$

where $n$ is a constant. Show that the ratio of $W$ to $\Omega$ is constant in this case.

SOLUTION. Substituting for $\epsilon$ from Eq. (3.119) into the expressions for $W$ and $\Omega$, we get

$$W = \int_0^\sigma \sigma(nb\sigma^{n-1})\, d\sigma = \int_0^\sigma nb\sigma^n\, d\sigma = \frac{n}{n+1} b\sigma^{n+1} = \frac{n}{n+1}\sigma\epsilon$$

and

$$\Omega = \int_0^\sigma b\sigma^n\, d\sigma = \frac{1}{n+1} b\sigma^{n+1} = \frac{\sigma\epsilon}{n+1}$$

Thus $W/\Omega = n$; that is, the ratio $W/\Omega$ is constant in this case.

EXAMPLE 3.4. Find an expression for $\Omega$ in terms of the stress invariants $I_1$ and $J_2$ for an isotropic linear elastic material.

SOLUTION. Substituting for $\sigma_{ij} = s_{ij} + \frac{1}{3}\sigma_{kk}\delta_{ij}$ in Eq. (3.84), we have

$$\epsilon_{ij} = \frac{1+\nu}{E} s_{ij} + \frac{1-2\nu}{3E} I_1\delta_{ij} \tag{3.120}$$

where $I_1 = \sigma_{kk}$. Substituting for $\epsilon_{ij}$ from Eq. (3.120) into Eq. (3.114), we can write $\Omega$ as

$$\Omega = \frac{1+\nu}{E} \int_0^{\sigma_{ij}} s_{ij}\, d\sigma_{ij} + \frac{1-2\nu}{3E} \int_0^{\sigma_{ij}} I_1\delta_{ij}\, d\sigma_{ij}$$

which can be reduced to $(J_2 = \frac{1}{2}s_{ij}s_{ij},\ dJ_2 = s_{ij}\, ds_{ij} = s_{ij}\, d\sigma_{ij},\ dI_1 = \delta_{ij}\, d\sigma_{ij})$

$$\Omega = \frac{1+\nu}{E} \int_0^{J_2} dJ_2 + \frac{1-2\nu}{3E} \int_0^{I_1} I_1\, dI_1 \tag{3.121}$$

$$= \frac{1+\nu}{E} J_2 + \frac{1-2\nu}{6E} I_1^2 \tag{3.122a}$$

or, in terms of $G$, and $K$, we have

$$\Omega = \frac{J_2}{2G} + \frac{I_1^2}{18K} \tag{3.122b}$$

For positive values of bulk modulus $K$ and shear modulus $G$, the complementary energy density $\Omega$ in Eq. (3.122b) is a *positive definite quadratic* form in the components of stress (since both $I_1^2$ and $J_2$ are always positive and cannot be zero unless $\sigma_{ij} = 0$). For an isotropic linear elastic material, $\Omega$ is found explicitly in terms of the existing components of stress (current values of $I_1$ and $J_2$) irrespective of the loading (stress) path followed to reach these current stress components; that is, $\Omega$ in this case is *path independent*. However, in general, this is not true for Cauchy elastic materials, whether linear or nonlinear. This is illustrated further in the following example for a linear Cauchy elastic model.

EXAMPLE 3.5. In the two-dimensional principal space ($\sigma_1, \sigma_2, \epsilon_1$, and $\epsilon_2$), the behavior of a *linear Cauchy* elastic material is described by the stress–strain relations:

$$\epsilon_1 = a_{11}\sigma_1 + a_{12}\sigma_2$$
$$\epsilon_2 = a_{21}\sigma_1 + a_{22}\sigma_2 \tag{3.123}$$

where $a_{11}, a_{12}, a_{21}$, and $a_{22}$ are material constants and $a_{12} \neq a_{21}$. Consider two different stress paths 1 and 2 as shown in Fig. 3.10. Path 1 is from $(0,0)$ to $(\sigma_1^*, \sigma_2^*)$, changing first $\sigma_1$ and then $\sigma_2$. On the other hand, Path 2 is also from $(0,0)$ to $(\sigma_1^*, \sigma_2^*)$, but in this case, $\sigma_2$ is changed first and then $\sigma_1$. Calculate $\Omega$ for Paths 1 and 2. Also, find $\Omega$ for the complete cycle OACBO shown in Fig. 3.10. Comment on the results.

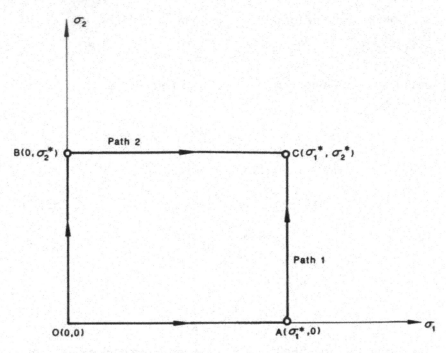

FIGURE 3.10. Two different loading paths to same final state (Example 3.5).

SOLUTION. Along Path 1, the expression for $\Omega$ in Eq. (3.114) can be written as

$$\Omega^{(1)} = \int_{(0,0)}^{(\sigma_1^*,0)} (\epsilon_1 \, d\sigma_1 + \epsilon_2 \, d\sigma_2) + \int_{(\sigma_1^*,0)}^{(\sigma_1^*,\sigma_2^*)} (\epsilon_1 \, d\sigma_1 + \epsilon_2 \, d\sigma_2)$$

Substituting for $\epsilon_1$ and $\epsilon_2$ from Eq. (3.123), and noting that $d\sigma_2 = 0$ and $d\sigma_1 = 0$ in the first and second integrands, respectively, we have

$$\Omega^{(1)} = \int_0^{\sigma_1^*} a_{11}\sigma_1 \, d\sigma_1 + \int_0^{\sigma_2^*} (a_{21}\sigma_1^* + a_{22}\sigma_2) \, d\sigma_2$$

Carrying out the indicated integrations,

$$\Omega^{(1)} = \tfrac{1}{2} a_{11}\sigma_1^{*2} + a_{21}\sigma_1^*\sigma_2^* + \tfrac{1}{2} a_{22}\sigma_2^{*2} \tag{3.124}$$

Similarly, for Path 2, it can be shown that $\Omega^{(2)}$ is given by

$$\Omega^{(2)} = \tfrac{1}{2} a_{11}\sigma_1^{*2} + a_{12}\sigma_1^*\sigma_2^* + \tfrac{1}{2} a_{22}\sigma_2^{*2} \tag{3.125}$$

Consequently, for $a_{12} \neq a_{21}$, the complementary energy density $\Omega^{(1)} \neq \Omega^{(2)}$. Thus, $\Omega$ is not unique, but depends on the loading path. Only when $a_{12} = a_{21}$ are the expressions for $\Omega^{(1)}$ and $\Omega^{(2)}$ identical. The condition $a_{12} = a_{21}$ makes the matrix of elastic coefficients in Eq. (3.123) *symmetric*. It is shown later

that the condition of symmetry of the matrix of elastic coefficients is similar
to imposing the restriction (3.117) for Green elastic material.

For the stress cycle $OACBO$, $\Omega$ is given by

$$\Omega = \int (\epsilon_1 \, d\sigma_1 + \epsilon_2 \, d\sigma_2) \tag{3.126}$$

where the integration is extended over the complete cycle. This equation
can be written as

$$\Omega = \int_{(0,0)}^{(\sigma_1^*, \sigma_2^*)} (\epsilon_1 \, d\sigma_1 + \epsilon_2 \, d\sigma_2) \qquad \text{along Path 1}$$

$$+ \int_{(\sigma_1^*, \sigma_2^*)}^{(0,0)} (\epsilon_1 \, d\sigma_1 + \epsilon_2 \, d\sigma_2) \qquad \text{along Path 2}$$

The first part yields the same expression as $\Omega^{(1)}$ in Eq. (3.124). The second
part gives the expression for $\Omega^{(2)}$ in Eq. (3.125) with a negative sign. The
net value of $\Omega$ for the complete cycle is

$$\Omega = \Omega^{(1)} - \Omega^{(2)} = (a_{21} - a_{12})\sigma_1^*\sigma_2^* \tag{3.127}$$

Depending on the values of $a_{12}$ and $a_{21}$, the net complementary energy
may be positive or negative (note that the term $\epsilon_{ij} \, d\sigma_{ij}$ in the definition of
$\Omega$ may be viewed as the rate of work done by stress increments $d\sigma_{ij}$ on
strains $\epsilon_{ij}$, and this work is regarded as energy stored in the body). Thus,
during the deformation process in the complete stress cycle, the material
model described by the stress-strain relations (3.123) may dissipate or
generate energy, the latter in violation of the laws of thermodynamics. For
a *symmetric* elastic coefficients matrix ($a_{12} = a_{21}$), the net value of $\Omega$ for the
complete cycle is zero, and full recovery of complementary energy upon
complete unloading is ensured. For isotropic linear elastic material, the
matrix $[C]$ in Eqs. (3.99) and (3.101) is *symmetric*, and thus $\Omega$ in this case
is path independent.

### 3.3.2. Nonlinear Elastic Isotropic Stress-Strain Relationships Based on Functions $W$ and $\Omega$

For an isotropic elastic material, the strain energy density $W$ [Eq. (3.112)]
can be expressed in terms of *any three independent* invariants of the strain
tensor $\epsilon_{ij}$. Choosing the three invariants $\bar{I}_1'$, $\bar{I}_2'$, and $\bar{I}_3'$ defined below, $W$
is written as

$$W = W(\bar{I}_1', \bar{I}_2', \bar{I}_3') \tag{3.128}$$

where $\bar{I}_1'$, $\bar{I}_2'$, and $\bar{I}_3'$ are given by

$$\bar{I}_1' = \epsilon_{kk}$$

$$\bar{I}_2' = \tfrac{1}{2}\epsilon_{km}\epsilon_{km} \tag{3.129}$$

$$\bar{I}_3' = \tfrac{1}{3}\epsilon_{km}\epsilon_{kn}\epsilon_{mn}$$

Then, from Eq. (3.117), we have

$$\sigma_{ij} = \frac{\partial W}{\partial \bar{I}_1'} \frac{\partial \bar{I}_1'}{\partial \epsilon_{ij}} + \frac{\partial W}{\partial \bar{I}_2'} \frac{\partial \bar{I}_2'}{\partial \epsilon_{ij}} + \frac{\partial W}{\partial \bar{I}_3'} \frac{\partial \bar{I}_3'}{\partial \epsilon_{ij}}$$

or, substituting from Eq. (3.129) for $\bar{I}_1'$, $\bar{I}_2'$, $\bar{I}_3'$ and carrying out the differentiation, we get

$$\sigma_{ij} = \alpha_1 \delta_{ij} + \alpha_2 \epsilon_{ij} + \alpha_3 \epsilon_{ik} \epsilon_{jk} \tag{3.130}$$

where

$$\alpha_i = \alpha_i(\bar{I}_j') = \frac{\partial W}{\partial \bar{I}_i'} \tag{3.131}$$

By differentiating Eqs. (3.131), the functions $\alpha_i$ (material strain functions) can be related by the three equations:

$$\frac{\partial \alpha_i}{\partial \bar{I}_j'} = \frac{\partial \alpha_j}{\partial \bar{I}_i'} \tag{3.132}$$

It should be noted that the choice of the three independent strain invariants appearing in Eqs. (3.128) and (3.129) is arbitrary. Instead, one may use the invariants $I_1'$, $I_2'$, and $I_3'$ of Section 3.1 [Eq. (3.32)], or the invariants $J_1'$, $J_2'$, and $J_3'$ of the strain deviator tensor $e_{ij}$ [see (Eq. (3.51)], or even mixed invariants such as $I_1'$, $J_2'$, and $J_3'$. The particular advantage of the choice here is the separation of the functions $\alpha_i$ in a simple, convenient manner.

At this stage, it is important to further illustrate the difference between the Cauchy and Green (hyperelastic) formulations based on the results obtained above [Eqs. (3.130) to (3.132)]. According to the *Cayley–Hamilton theorem*, all positive integer powers of any second-order tensor, such as, for example, the strain tensor $\epsilon_{ij}$, can be expressed as linear combinations of $\delta_{ij}$, $\epsilon_{ij}$, and $\epsilon_{ik} \epsilon_{kj}$ with coefficients that are polynomial functions of the three invariants of $\epsilon_{ij}$. Therefore, if the stress–strain relation of Eq. (3.116) for a Cauchy elastic material takes the form of a polynomial in $\epsilon_{ij}$ of any order, then this functional can be written in exactly the same form as Eq. (3.130) derived above based on the Green (hyperelastic) formulation. However, the $\alpha_i$ are now independent functions of the invariants of $\epsilon_{ij}$, and they are no longer restricted by the relations given in Eqs. (3.132) for Green materials. Various constitutive models (e.g., second-, third-, fourth-order) based on both Cauchy and Green formulations can thus be derived based on the general relation of Eq. (3.130) with different assumed functional forms for $\alpha_i$. The only difference is that the selected $\alpha_i$ functions are further restricted by relations (3.132) for Green-type materials.

There is no *a priori* reason for requiring that terms of a prescribed order be present in all material strain functions. Therefore, for simplicity, it may be advantageous in some cases to expand $W$ as a function of only two strain invariants, or even one strain invariant only. Also, one may not retain all the possible combinations of these invariants for a prescribed order.

In a similar procedure, one may obtain different constitutive relations from Eq. (3.118) by expanding the function $\Omega$ in terms of the stress invariants $\bar{I}_1$, $\bar{I}_2$, and $\bar{I}_3$ ($I_1$, $I_2$, and $I_3$, or $I_1$, $J_2$, and $J_3$). Thus, if we choose the following stress invariants (similar to those defined for strains),

$$\bar{I}_1 = \sigma_{kk}$$

$$\bar{I}_2 = \tfrac{1}{2}\sigma_{km}\sigma_{km} \qquad (3.133)$$

$$\bar{I}_3 = \tfrac{1}{3}\sigma_{km}\sigma_{kn}\sigma_{mn}$$

the constitutive equation is

$$\epsilon_{ij} = \phi_1 \delta_{ij} + \phi_2 \sigma_{ij} + \phi_3 \sigma_{ik}\sigma_{jk} \qquad (3.134)$$

where

$$\phi_i = \phi_i(\bar{I}_j) = \frac{\partial \Omega}{\partial \bar{I}_i} \qquad (3.135)$$

and the constraints on the material stress functions $\phi_i$ are given by the three relations:

$$\frac{\partial \phi_i}{\partial \bar{I}_j} = \frac{\partial \phi_j}{\partial \bar{I}_i} \qquad (3.136)$$

It is to be emphasized that the behavior of the isotropic models described in Eqs. (3.130) and (3.134) is *reversible* and *path independent,* as in the linear elastic models, since the state of strain (stress) is uniquely determined by the current values of stresses (strains) without regard to the loading history. Furthermore, the principal stress and strain axes always coincide in these models.

EXAMPLE 3.6. An initially unstressed and unstrained material element is subjected to a combined loading history which produces the radial straight-line path $(0, 0)$ to $(30, 10)$ ksi in $(\sigma, \tau)$ space (tension $\sigma$, shear $\tau$), as shown in Fig. 3.11. We assume that the element is of a nonlinear elastic material with a function $\Omega$ given by

$$\Omega(I_1, J_2) = aJ_2 + bI_1 J_2 \qquad (3.137)$$

where $a$ and $b$ are constants. The stress–strain relation of the material in simple tension is

$$10^3 \epsilon = \frac{\sigma}{10} + \left(\frac{\sigma}{10}\right)^2 \qquad (3.138)$$

where $\sigma$ is in ksi.

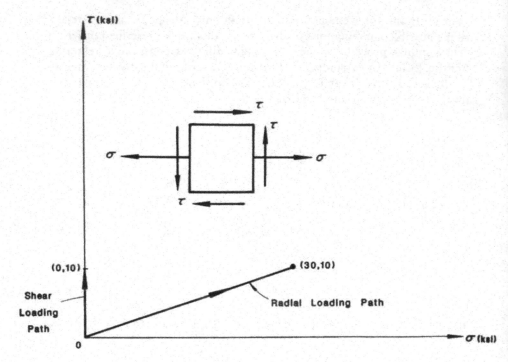

FIGURE 3.11. Loading paths in $(\sigma, \tau)$ space (Example 3.6).

(a) Determine the constants $a$ and $b$ in Eq. (3.137).

(b) Find all the components of the normal and shear strains at the end of the given stress path.

(c) Consider a shearing stress path from $(0, 0)$ to $(0, 10)$ ksi, and calculate the resulting shear strain component $\gamma_{xy}$, where the $x$–$y$ plane coincides with the plane of stress components $\sigma$ and $\tau$. What is the value of the volume change $\epsilon_{kk}$ in this case?

SOLUTION. (a) From the given expression for $\Omega$ in Eq. (3.137), we have

$$\frac{\partial \Omega}{\partial I_1} = bJ_2 \quad \text{and} \quad \frac{\partial \Omega}{\partial J_2} = (a + bI_1)$$

Since $I_1 = \sigma_{kk}$ and $J_2 = \frac{1}{2} s_{mn} s_{mn}$ (Chapter 2), then

$$\frac{\partial I_1}{\partial \sigma_{ij}} = \delta_{ij}$$

$$\frac{\partial J_2}{\partial \sigma_{ij}} = s_{mn} \frac{\partial s_{mn}}{\partial \sigma_{ij}} = s_{mn} \frac{\partial (\sigma_{mn} - \frac{1}{3} \sigma_{kk} \delta_{mn})}{\partial \sigma_{ij}} \qquad (3.139a)$$

or

$$\frac{\partial J_2}{\partial \sigma_{ij}} = s_{mn}(\delta_{im}\delta_{jn} - \tfrac{1}{3}\delta_{mn}\delta_{ij})$$

$$= s_{ij} - \tfrac{1}{3}\delta_{ij}s_{mm}$$

$$= s_{ij} \qquad (\text{since } s_{mm} = 0) \qquad\qquad (3.139b)$$

Therefore, the constitutive equations can be written as

$$\epsilon_{ij} = \frac{\partial \Omega}{\partial \sigma_{ij}} = (bJ_2)\delta_{ij} + (a + bI_1)s_{ij} \qquad\qquad (3.140)$$

In simple tension, $\sigma_{11} = \sigma$ and all other components of stress are zero. Thus,

$$I_1 = \sigma, \qquad J_2 = \tfrac{1}{3}\sigma^2, \qquad s_{11} = \tfrac{2}{3}\sigma$$

and the stress-strain relation of Eq. (3.140) reduces to

$$\epsilon = \frac{2a}{3}\sigma + b\sigma^2$$

Comparing this equation with the stress-strain relation given in Eq. (3.138), we can easily obtain the constants $a$ and $b$. The results are

$$a = \tfrac{3}{2} \times 10^{-4} \quad \text{and} \quad b = 1 \times 10^{-5}$$

(b) Substituting the above values of $a$ and $b$ into Eq. (3.140), it becomes

$$\epsilon_{ij} = (1 \times 10^{-5})J_2\delta_{ij} + (\tfrac{3}{2} \times 10^{-4} + 10^{-5}I_1)s_{ij} \qquad (3.141)$$

The values of $I_1$ and $J_2$ at the end of the given radial path are calculated from the final values $\sigma = 30$ ksi and $\tau = 10$ ksi. Therefore, we get

$$I_1 = 30$$

$$J_2 = \tfrac{1}{6}[(30)^2 + (30)^2] + (10)^2 = 400$$

Substituting these values into Eq. (3.141), we find that

$$\epsilon_{ij} = (40 \times 10^{-4})\delta_{ij} + (4.5 \times 10^{-4})s_{ij}$$

which can be used to calculate the strain components $\epsilon_{ij}$. The results are given by

$$\epsilon_{ij} = \begin{bmatrix} 130 & 45 & 0 \\ 45 & -5 & 0 \\ 0 & 0 & -5 \end{bmatrix} \times 10^{-4}$$

(c) For the shear loading path from $(0, 0)$ to $(0, 10)$ ksi, the values of $I_1$ and $J_2$ are

$$I_1 = 0, \qquad J_2 = 100$$

The constitutive equation (3.141) then becomes

$$\epsilon_{ij} = 10^{-3}\delta_{ij} + 1.5 \times 10^{-4}s_{ij}$$

Thus,

$$\gamma_{xy} = 2\epsilon_{xy} = 2\epsilon_{12} = 2 \times 1.5 \times 10^{-4}(10) = 3 \times 10^{-3}$$

$$\epsilon_{kk} = 10^{-3}(\delta_{kk}) + 1.5 \times 10^{-4}(s_{kk})$$

Since $s_{kk} = 0$, the final value of the dilatation $\epsilon_{kk}$ at the end of this shearing path is $(\delta_{kk} = 3)$

$$\epsilon_{kk} = 3 \times 10^{-3}$$

Note that unlike a linear elastic model, the nonlinear elastic model of Eq. (3.140) produces a volume increase under simple shear stress (also called *dilatation* or *dilatancy*), as illustrated in this example. This phenomenon is very important in modeling soil materials, such as dense sands and overconsolidated clays, and rocklike materials, such as concrete.

### 3.3.3. Isotropic Nonlinear Elastic Stress–Strain Relations by Modification of the Elastic Moduli

The linear elastic stress–strain relations discussed in Section 3.2.2 are *isotropic* and *reversible*. Clearly, then, a simple extension of these relations with the elastic moduli replaced by scalar functions associated with either the stress and/or the strain invariants has the properties of isotropy and reversibility also. For instance, scalar functions associated with the state of stress may include the values of the three principal stresses $\sigma_1$, $\sigma_2$, and $\sigma_3$, or equally well the three independent invariants $I_1$, $J_2$, and $J_3$. Therefore, different scalar functions such as $F(I_1, J_2, J_3)$ associated with the stress invariants, or $F(I'_1, J'_2, J'_3)$ associated with the strain invariants, may be employed to describe various nonlinear elastic constitutive models. The nonlinear stress–strain relations for each of these models reduce to the linear forms when the scalar functions are taken to be constants.

As a first example, consider the linear elastic form of Eq. (3.84) modified by replacing the reciprocal of Young's modulus $E$ by a scalar function of the invariants $I_1$, $J_2$, and $J_3$ labeled as $F(I_1, J_2, J_3)$. Thus, one has

$$\epsilon_{ij} = (1 + \nu)F(I_1, J_2, J_3)\sigma_{ij} - \nu F(I_1, J_2, J_3)\sigma_{kk}\delta_{ij} \qquad (3.142)$$

Poisson's ratio $\nu$ also may be replaced by a function of the stress invariants.

Equations (3.142) are nonlinear stress–strain relations for an isotropic elastic material which reduce to the linear forms when $F(I_1, J_2, J_3)$ is constant $(1/E)$. They represent elastic (reversible) behavior because the state of strain is determined uniquely by the current state of stress without regard to the loading history.

There is, of course, a net and logical separation between the mean response and the deviatoric or shear response of the material, exactly as for the linear elastic material. Specifically, one can obtain from Eq. (3.142)

$$\epsilon_{kk} = (1-2\nu)F(I_1, J_2, J_3)\sigma_{kk}$$
$$e_{ij} = (1+\nu)F(I_1, J_2, J_3)s_{ij} \tag{3.143}$$

where the moduli $K$ and $G$ are expressed in terms of $E$ and $\nu$ by Eqs. (3.92) and (3.83) and the reciprocal of $E$ is replaced by the scalar function $F(I_1, J_2, J_3)$. However, unlike the linear elastic relation, Eqs. (3.143) show that there is an interaction between the two responses through the change in magnitude of the scalar function $F$ with variation in the invariants $I_1, J_2$, and $J_3$. This implies that the volume change $\epsilon_{kk}$ does not depend solely on $\sigma_{kk}$. Similarly, distortions or shear deformations, $e_{ij}$, do not depend only on the stress deviation or shear stresses, $s_{ij}$. They depend on each other, and interact through the variation of the scalar function $F(I_1, J_2, J_3)$.

As a second example of the formulation of the constitutive relation of isotropic nonlinear elastic materials, consider the modification of the linear relations of Eqs. (3.93) and (3.94). The elastic bulk and shear moduli are taken as scalar functions of the stress and/or strain tensor invariants. Thus, Eqs. (3.93) and (3.94) may now be written as

$$p = K_s \epsilon_{kk} \tag{3.144}$$

$$s_{ij} = 2G_s e_{ij} \tag{3.145}$$

Thus, we have

$$\sigma_{ij} = s_{ij} + p\delta_{ij} = 2G_s e_{ij} + K_s \epsilon_{kk}\delta_{ij} \tag{3.146}$$

where $K_s$ and $G_s$ are known as the *secant bulk modulus* and the *secant shear modulus*, respectively. The scalar functional forms of $K_s$ and $G_s$ in terms of the stress and/or strain invariants are developed mainly from experimental data. In principle, any scalar function of the stress and/or the strain invariants may be used for the isotropic nonlinear elastic moduli as discussed before. Obviously, the constitutive models formulated on this basis are of the Cauchy elastic type; the state of strain is determined uniquely by the current state of stress or vice versa. For instance, for any given state of stress, $\sigma_{ij}$, the value of $F(I_1, J_2, J_3)$ and consequently the strain components, $\epsilon_{ij}$, in Eqs. (3.143) are uniquely determined without regard to the loading path. However, this does *not* imply that $W$ and $\Omega$, calculated from such stress-strain relations, are also *path independent*. Certain restrictions must be imposed on the chosen scalar functions in order to ensure the path-independent character of $W$ and $\Omega$. This in turn assures that thermodynamic laws are always satisfied and that energy is not generated during any loading-unloading cycle.

Consider the stress-strain relations of Eq. (3.146). Let $K_s$ and $G_s$ be general functions of the strain invariants $I_1', J_2'$, and $J_3'$ of the form

$K_s(I_1', J_2', J_3')$ and $G_s(I_1', J_2', J_3')$. The expression for $W$ in this case is

$$W = \int_0^{\epsilon_{ij}} \sigma_{ij}\, d\epsilon_{ij} = \int_0^{J_2'} 2G_s(I_1', J_2', J_3')\, dJ_2' + \int_0^{I_1'} \frac{1}{2} K_s(I_1', J_2', J_3')\, d(I_1')^2$$

$$(3.147)$$

in which $d(I_1')^2 = 2I_1'\, dI_1'$.

Similarly, if $K_s$ and $G_s$ are taken as functions of the stress invariants $I_1$, $J_2$, and $J_3$, it can be shown that $\Omega$ is given by

$$\Omega = \int_0^{\sigma_{ij}} \epsilon_{ij}\, d\sigma_{ij} = \int_0^{J_2} \frac{dJ_2}{2G_s(I_1, J_2, J_3)} + \int_0^{I_1} \frac{d(I_1)^2}{18K_s(I_1, J_2, J_3)} \quad (3.148)$$

As can be seen, in order for $W$ to be independent of path, the integrals in Eq. (3.147) must depend only on the current values of $I_1'$ and $J_2'$. This can always be satisfied if the moduli $K_s$ and $G_s$ are expressed as

$$K_s = K_s(I_1')$$
$$G_s = G_s(J_2')$$
$$(3.149a)$$

But since $I_1'$ and $J_2'$ are related to $\epsilon_{oct}$ and $\gamma_{oct}$, Eqs. (3.149a) may be expressed in the alternative forms

$$K_s = K_s(\epsilon_{oct})$$
$$G_s = G_s(\gamma_{oct})$$
$$(3.149b)$$

Similarly, in order to satisfy the path independence requirement for $\Omega$ in Eq. (3.148), $K_s$ and $G_s$ are taken to be functions of only $I_1$ and $J_2$, respectively; that is,

$$K_s = K_s(I_1)$$
$$G_s = G_s(J_2)$$
$$(3.150a)$$

or, in terms of octahedral stress components,

$$K_s = K_s(\sigma_{oct})$$
$$G_s = G_s(\tau_{oct})$$
$$(3.150b)$$

Furthermore, $K_s$ and $G_s$ must, of course, be positive. Consequently, the integrals in Eqs. (3.147) and (3.148) are always positive (since $I_1^2$ and $J_2$ are positive). This confirms that $W$ and $\Omega$ are *always positive definite*. The path independence for potential functions $W$ and $\Omega$ is due to the reversibility of elastic behavior, while the positive definiteness of $W$ and $\Omega$ results from the stability requirement of the material. This will be discussed in the following section.

## 3.4. Principle of Virtual Work

The principle of virtual work has proved very powerful as a technique in solving problems and in providing proofs for general theorems in solid mechanics. In the following, the virtual work equation is derived. This equation is needed for subsequent considerations of stability and uniqueness of general stress–strain relations, which may be irreversible and path dependent. In the derivation, the following assumption is made: the displacements are sufficiently *small* so that the changes in the geometry of the body are negligible and the original undeformed configuration can be used in setting up the equations for the system. This implies that nonlinear contributions in the compatibility of strains and displacement are neglected. It follows that equilibrium equations (3.70) and compatibility relations (3.71) are applicable here.

The equation of virtual work deals with two separate and unrelated sets: the *equilibrium* set and the *compatible* set. The equilibrium set and the compatible (or geometry) set are brought together, side by side but independently, in the equation of virtual work (Fig. 3.12).

equilibrium set

$$\int_A T_i u_i^* \, dA + \int_V F_i u_i^* \, dV = \int_V \sigma_{ij} \epsilon_{ij}^* \, dV \tag{3.151}$$

compatible set

Here integration is over the whole area, $A$, or volume, $V$, of the body. The quantities $T_i$ and $F_i$ are external surface and body forces, respectively. The stress field $\sigma_{ij}$ is any set of stresses, *real* or *otherwise*, in *equilibrium* with

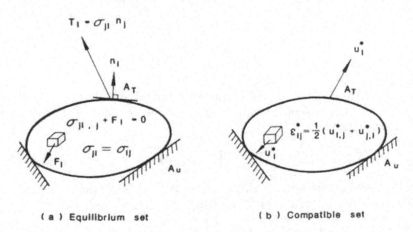

(a) Equilibrium set          (b) Compatible set

FIGURE 3.12. Two independent sets in the equation of virtual work.

body forces $F_i$ within the body and with surface forces $T_i$ on the surface where the forces $T_i$ are prescribed. Similarly, the strain field $\epsilon_{ij}^*$ represents any set of strains or deformations *compatible* with the *real* or *imagined* (virtual) displacements $u_i^*$ of the points of application of the external forces $T_i$ and $F_i$. In Fig. 3.12, the two sets (equilibrium and compatible) are shown, together with the requirements to be satisfied by each set [Eqs. (3.70) and (3.71)].

The important point to keep in mind is that neither the equilibrium set $T_i$, $F_i$, and $\sigma_{ij}$ (Fig. 3.12a) nor the compatible set $u_i^*$ and $\epsilon_{ij}^*$ (Fig. 3.12b) need be the actual state, nor need the equilibrium and compatible sets be related in any way to each other. In Eq. (3.151), asterisks are used for the compatible set to emphasize the point that these two sets are completely independent. When the actual or real states (which satisfy both equilibrium and compatibility) are substituted in Eq. (3.151), the asterisks are omitted.

### 3.4.1. Proof of Virtual Work Equation

Consider the *external* virtual work, $W_{ext}$, given by the expression on the left-hand side of Eq. (3.151). With $T_i = \sigma_{ji} n_j$ on $A$, we can write

$$W_{ext} = \int_A \sigma_{ji} n_j u_i^* \, dA + \int_V F_i u_i^* \, dV$$

The first integral can be transformed into a volume integral using the divergence theorem. Thus, we have

$$W_{ext} = \int_V (\sigma_{ji} u_i^*)_{,j} \, dV + \int_V F_i u_i^* \, dV$$

$$= \int_V (\sigma_{ji,j} u_i^* + \sigma_{ji} u_{i,j}^*) \, dV + \int_V F_i u_i^* \, dV$$

or

$$W_{ext} = \int_V [(\sigma_{ji,j} + F_i) u_i^* + \sigma_{ji} u_{i,j}^*] \, dV \tag{3.152}$$

The first term in the parentheses vanishes for the equilibrium set, which satisfies the equilibrium equations given in Eq. (3.70b). Therefore, Eq. (3.152) reduces to

$$W_{ext} = \int_V \sigma_{ij} u_{i,j}^* \, dV \tag{3.153}$$

Now consider the *internal* virtual work, $W_{int}$, given by the expression on the right-hand side of Eq. (3.151). Using Eq. (3.71a), we have

$$W_{int} = \int_V \sigma_{ij} \epsilon_{ij}^* \, dV = \int_V \frac{1}{2} \sigma_{ij} (u_{i,j}^* + u_{j,i}^*) \, dV$$

or

$$W_{int} = \int_V \left( \frac{1}{2} \sigma_{ij} u^*_{i,j} + \frac{1}{2} \sigma_{ij} u^*_{j,i} \right) dV$$

which can be written as ($i, j$ are dummy indices)

$$W_{int} = \int_V \left( \frac{1}{2} \sigma_{ij} u^*_{i,j} + \frac{1}{2} \sigma_{ji} u^*_{i,j} \right) dV$$

Finally, using the symmetry of $\sigma_{ij}$,

$$W_{int} = \int_V \sigma_{ij} u^*_{i,j} \, dV \tag{3.154}$$

Thus, from Eqs. (3.153) and (3.154), $W_{ext} = W_{int}$ and the virtual work equation (3.151) is established.

### 3.4.2. Rate Forms of Virtual Work Equations

Any equilibrium set and any compatible set may be substituted in Eq. (3.151). In particular, an increment or rate of change of external forces and interior stress $\dot{T}_i$, $\dot{F}_i$, $\dot{\sigma}_{ij}$ may be used as an equilibrium set and an increment or rate of change of displacements and strains $\dot{u}^*_i$, $\dot{\epsilon}^*_{ij}$ as a compatible set. Thus, the rate forms given below are valid as virtual work equations:

$$\int_A \dot{T}_i \dot{u}^*_i \, dA + \int_V \dot{F}_i \dot{u}^*_i \, dV = \int_V \dot{\sigma}_{ij} \dot{\epsilon}^*_{ij} \, dV \tag{3.155}$$

$$\int_A \dot{T}_i u^*_i \, dA + \int_V \dot{F}_i u^*_i \, dV = \int_V \dot{\sigma}_{ij} \epsilon^*_{ij} \, dV \tag{3.156}$$

$$\int_A T_i \dot{u}^*_i \, dA + \int_V F_i \dot{u}^*_i \, dV = \int_V \sigma_{ij} \dot{\epsilon}^*_{ij} \, dV \tag{3.157}$$

## 3.5. Drucker's Stability Postulate

Consider a material body of volume $V$ and surface area $A$, as shown in Fig. 3.13a. The applied surface and body forces are denoted by $T_i$ and $F_i$, respectively. The corresponding induced displacements, stresses, and strains are $u_i$, $\sigma_{ij}$, and $\epsilon_{ij}$, respectively. This existing system of forces, stresses, displacements, and strains satisfies both equilibrium and compatibility (geometry) conditions.

Consider now an external agency which is entirely distinct from the agency that causes the existing states of stress $\sigma_{ij}$ and strain $\epsilon_{ij}$. This external agency applies additional surface and body forces, $\dot{T}_i$ and $\dot{F}_i$, which cause the additional set of stresses $\dot{\sigma}_{ij}$, strains $\dot{\epsilon}_{ij}$, and displacements $\dot{u}_i$ as illustrated in Fig. 3.13b.

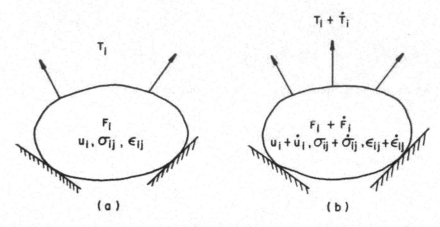

FIGURE 3.13. External agency and Drucker's stability postulate. (a) Existing system; (b) existing system and external agency.

A *stable material* is defined to be the one that satisfies the following conditions (now known as Drucker's stability postulates):

1. During the application of the added set of forces, the work done by the *external agency* on the changes in displacements it produces is positive.
2. Over the cycle of application and removal of the added set of forces, the new work performed by the *external agency* on the changes in displacements it produces is non-negative.

It is emphasized that the work referred to is only the work done by the added set of forces, $\dot{T}_i$, $\dot{F}_i$, on the *change* in displacements $\dot{u}_i$ it produces, not the work done by the total forces on $\dot{u}_i$. Mathematically, the two stability requirements can be stated as

$$\int_A \dot{T}_i \dot{u}_i \, dA + \int_V \dot{F}_i \dot{u}_i \, dV > 0 \tag{3.158}$$

$$\oint_A \dot{T}_i \dot{u}_i \, dA + \oint_V \dot{F}_i \dot{u}_i \, dV \geq 0 \tag{3.159}$$

in which $\oint$ indicates integration over a cycle of addition and removal of the additional set of forces and stresses.

The first postulate, Eq. (3.158), is called *stability in small*, while the second, Eq. (3.159), is termed *stability in cycle*. Note that these stability requirements are more restrictive than the laws of thermodynamics, which require only that the work done by the total (existing) forces $F_i$ and $T_i$ on $\dot{u}_i$ be non-negative.

Applying the *principle of virtual work* to the "added" equilibrium set, $\dot{F}_i$, $\dot{T}_i$, and $\dot{\sigma}_{ij}$, and the corresponding compatible set, $\dot{u}_i$ and $\dot{\epsilon}_{ij}$, the stability conditions in Eqs. (3.158) and (3.159) can be reduced to the following inequalities ($V$ is an arbitrary volume):

$$\dot{\sigma}_{ij}\dot{\epsilon}_{ij} > 0 \qquad \text{stability in small} \qquad (3.160)$$

$$\oint \dot{\sigma}_{ij}\dot{\epsilon}_{ij} \geq 0 \qquad \text{stability in cycle} \qquad (3.161)$$

where $\oint$ is the integral taken over a cycle of applying and removing the added stress set $\dot{\sigma}_{ij}$.

Stability conditions (3.160) may be illustrated by the uniaxial $\sigma$-$\epsilon$ curves shown in Fig. 3.14. In panels a and b of the figure, an additional stress $\dot{\sigma} > 0$ gives rise to an additional strain $\dot{\epsilon} > 0$, with the product $\dot{\sigma}\dot{\epsilon} > 0$. That is, the additional stress $\dot{\sigma}$ does positive work, which is represented by the shaded triangles in the diagram. For the *unstable material* shown in panels c and d of Fig. 3.14, however, the work done by the additional stress $\dot{\sigma}$ is always negative.

Figure 3.14 also shows that the stability postulate assures the existence of a unique inverse of the stress–strain relation. For the stable behavior shown in panels a and b, the stress is determined uniquely by a given value of strain, and vice versa. For the unstable material, however, these two strains correspond to a single value of stress (Fig. 3.14c) or two stresses correspond to a single value of strain (Fig. 3.14d).

# 3.6. Normality, Convexity, and Uniqueness for an Elastic Solid

## 3.6.1. Existence of the Potential Functions $W$ and $\Omega$

According to the concept of stable materials, useful net energy cannot be extracted from a stable material in a cycle of application and removal of the added set of forces and displacements. Furthermore, energy must be put in if only irrecoverable (permanent or plastic) deformation is to take place. For elastic materials, all deformations are recoverable and stability requires that the work done by the external agency in such a cycle be zero: that is, the integral of inequality (3.161) is always zero for elastic materials. It can be shown that this provides a *necessary* and *sufficient condition* for the *existence* of strain energy and complementary energy functions, $W$ and $\Omega$, respectively.

For example, let the existing states of stress and strain in an elastic material body be denoted by $\sigma_{ij}^{*}$ and $\epsilon_{ij}^{*}$, respectively. Consider an external agency which applies and then releases a set of stresses additional to the existing state of stress. For an elastic material, when the stress state returns back to $\sigma_{ij}^{*}$, the strain state also returns to $\epsilon_{ij}^{*}$; a strain cycle is thus completed starting and ending at $\epsilon_{ij}^{*}$. Over such a cycle, the second postulate requires

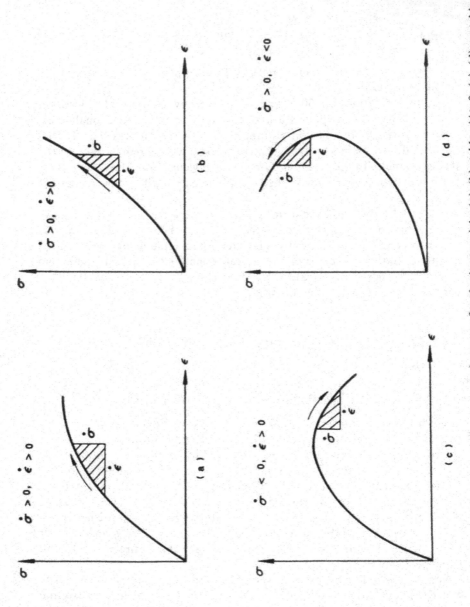

FIGURE 3.14. Stable and unstable stress–strain curves for elastic materials: (a), (b) stable, $\dot{\sigma}\dot{\epsilon} > 0$; (c), (d) unstable, $\dot{\sigma}\dot{\epsilon} < 0$.

that

$$\oint (\sigma_{ij} - \sigma_{ij}^*) \, d\epsilon_{ij} = 0$$

since no permanent (plastic) strains have occurred. Choosing the initial existing state to be stress- and strain-free, one has

$$\oint \sigma_{ij} \, d\epsilon_{ij} = 0 \qquad (3.162)$$

which must be true irrespective of the path followed during the cycle. Therefore, the integrand in Eq. (3.162) must be an exact (perfect) differential. This naturally leads to the consideration of the elastic strain energy density, $W$, written as a function of strains alone, such that

$$W(\epsilon_{ij}) = \int_0^{\epsilon_{ij}} \sigma_{ij} \, d\epsilon_{ij} \quad \text{and} \quad \sigma_{ij} = \frac{\partial W}{\partial \epsilon_{ij}}$$

These are the same relations derived previously in Section 3.3.1 as Eqs. (3.112) and (3.117).

Similarly, it can be shown that the second stability postulate leads to the existence of the elastic complementary energy density, $\Omega$, as a function of stresses alone, as given previously in Section 3.3.1 as Eqs. (3.113) and (3.118).

## 3.6.2. Normality

As we know, a function $f(x_i) = $ constant $(i = 1, 2, 3)$ represents a surface in three-dimensional Cartesian coordinate space. The outward-pointing normal to this surface at any point $x_i$ is a vector perpendicular to its tangent plane. The gradient of $f$, $\partial f / \partial x_i$, at point $x_i$ is in the direction of the normal to this surface. Thus, Eq. (3.117), $\sigma_{ij} = \partial W / \partial \epsilon_{ij}$, and Eq. (3.118), $\epsilon_{ij} = \partial \Omega / \partial \sigma_{ij}$, are the *normality conditions* in the nine-dimensional strain (stress) space. Equation (3.117) states that the outward normal to the surface $W = $ constant at a given point $\epsilon_{ij}$ represents the vector $\sigma_{ij}$ corresponding to $\epsilon_{ij}$ in the sense that its component in the direction of each of the coordinate axes of strain is proportional to the corresponding component of the normal vector $\partial W / \partial \epsilon_{ij}$. In Fig. 3.15a, the surface $W = $ constant is illustrated symbolically in the nine-dimensional strain space. The state of strain $\epsilon_{ij}$ is represented by a point in this space. The components $\sigma_{ij}$, corresponding to strains $\epsilon_{ij}$, are plotted as a *free vector* in the strain space (with $\sigma_{11}$ as the component in the $\epsilon_{11}$ direction, etc.) with its origin at the strain point $\epsilon_{ij}$. This free vector is always normal to the surface $W = $ constant at the corresponding strain point $\epsilon_{ij}$. The normality of $\epsilon_{ij}$ to the surface $\Omega = $ constant is shown in Fig. 3.15b.

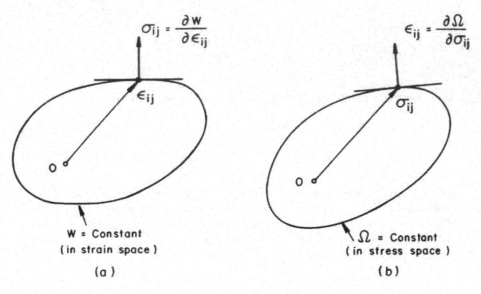

$$\sigma_{ij} = \frac{\partial W}{\partial \epsilon_{ij}} \qquad\qquad \epsilon_{ij} = \frac{\partial \Omega}{\partial \sigma_{ij}}$$

W = Constant
(in strain space)

(a)

$\Omega$ = Constant
(in stress space)

(b)

FIGURE 3.15. Normality of (a) $\sigma_{ij}$ to the surface $W = $ const. and (b) $\epsilon_{ij}$ to the surface $\Omega = $ const.

### 3.6.3. Convexity

As discussed earlier, for elastic materials, the second stability postulate implies that the constitutive relations are always of the Green (hyperelastic) type described by Eqs. (3.117) and (3.118). Moreover, these relations must satisfy the first stability requirement, inequality (3.160), which imposes additional conditions on the general form of the constitutive equations.

Consider the constitutive relations given by Eq. (3.117). The incremental stress components $\dot{\sigma}_{ij}$ can be expressed in terms of the incremental strains $\dot{\epsilon}_{ij}$ by differentiation; that is,

$$\dot{\sigma}_{ij} = \frac{\partial \sigma_{ij}}{\partial \epsilon_{kl}} \dot{\epsilon}_{kl} = \frac{\partial^2 W}{\partial \epsilon_{ij} \partial \epsilon_{kl}} \dot{\epsilon}_{kl} \tag{3.163}$$

Substituting for $\dot{\sigma}_{ij}$ from this equation into the stability condition (3.160), one obtains

$$\frac{\partial^2 W}{\partial \epsilon_{ij} \partial \epsilon_{kl}} \dot{\epsilon}_{ij} \dot{\epsilon}_{kl} > 0 \tag{3.164a}$$

That is, the quadratic form $(\partial^2 W / \partial \epsilon_{ij} \partial \epsilon_{kl}) \dot{\epsilon}_{kl} \dot{\epsilon}_{ij}$ must be positive definite for arbitrary values of the components $\dot{\epsilon}_{ij}$. The inequality (3.164a) may be rewritten in another convenient form as

$$H_{ijkl} \dot{\epsilon}_{ij} \dot{\epsilon}_{kl} > 0 \tag{3.164b}$$

where $H_{ijkl}$ is a fourth-order tensor given by

$$H_{ijkl} = \frac{\partial^2 W}{\partial \epsilon_{ij} \, \partial \epsilon_{kl}} \tag{3.165}$$

As can be easily seen from Eq. (3.165), tensor $H_{ijkl}$ satisfies the symmetry conditions ($\epsilon_{ij}$ is symmetric):

$$H_{ijkl} = H_{jikl} = H_{ijlk} = H_{jilk} = H_{klij}$$

Hence, there will be only 21 independent elements in $H_{ijkl}$.

Mathematically, the matrix of the components of $H_{ijkl} = \partial^2 W / \partial \epsilon_{ij} \, \partial \epsilon_{kl}$ is known as the *Hessian matrix* of the function $W$. When $\epsilon_{ij}$ is expressed in a vector form with six components, as defined in Eq. (3.98), then the elements of the Hessian matrix for $W$ are written as

$$[H] = \begin{vmatrix} \dfrac{\partial^2 W}{\partial \epsilon_x^2} & \dfrac{\partial^2 W}{\partial \epsilon_x \, \partial \epsilon_y} & \dfrac{\partial^2 W}{\partial \epsilon_x \, \partial \epsilon_z} & \dfrac{\partial^2 W}{\partial \epsilon_x \, \partial \gamma_{xy}} & \dfrac{\partial^2 W}{\partial \epsilon_x \, \partial \gamma_{yz}} & \dfrac{\partial^2 W}{\partial \epsilon_x \, \partial \gamma_{zx}} \\[2ex] & \dfrac{\partial^2 W}{\partial \epsilon_y^2} & \dfrac{\partial^2 W}{\partial \epsilon_y \, \partial \epsilon_z} & \dfrac{\partial^2 W}{\partial \epsilon_y \, \partial \gamma_{xy}} & \dfrac{\partial^2 W}{\partial \epsilon_y \, \partial \gamma_{yz}} & \dfrac{\partial^2 W}{\partial \epsilon_y \, \partial \gamma_{zx}} \\[2ex] & & \dfrac{\partial^2 W}{\partial \epsilon_z^2} & \dfrac{\partial^2 W}{\partial \epsilon_z \, \partial \gamma_{xy}} & \dfrac{\partial^2 W}{\partial \epsilon_z \, \partial \gamma_{yz}} & \dfrac{\partial^2 W}{\partial \epsilon_z \, \partial \gamma_{zx}} \\[2ex] & & & \dfrac{\partial^2 W}{\partial \gamma_{xy}^2} & \dfrac{\partial^2 W}{\partial \gamma_{xy} \, \partial \gamma_{yz}} & \dfrac{\partial^2 W}{\partial \gamma_{xy} \, \partial \gamma_{zx}} \\[2ex] & \text{sym.} & & & \dfrac{\partial^2 W}{\partial \gamma_{yz}^2} & \dfrac{\partial^2 W}{\partial \gamma_{yz} \, \partial \gamma_{zx}} \\[2ex] & & & & & \dfrac{\partial^2 W}{\partial \gamma_{zx}^2} \end{vmatrix} \tag{3.166}$$

and condition (3.164b) requires that $[H]$ be positive definite.

Alternatively, inequality (3.160) can be written in terms of $\Omega$ and $\sigma_{ij}$. Thus, using Eqs. (3.118) and following a similar procedure as outlined above, we finally get

$$H'_{ijkl} \dot{\sigma}_{ij} \dot{\sigma}_{kl} > 0 \tag{3.167}$$

where

$$H'_{ijkl} = \frac{\partial^2 \Omega}{\partial \sigma_{ij} \, \partial \sigma_{kl}} \tag{3.168}$$

and the elements of the Hessian matrix $[H']$ for $\Omega$ are exactly of the same form as those for $W$ in Eq. (3.166) with $W$, $\epsilon$, and $\gamma$ being replaced by $\Omega$, $\sigma$, and $\tau$, respectively.

From Eqs. (3.164) and (3.167), we conclude that the surfaces corresponding to constant $W$ and $\Omega$ in strain and stress space, respectively, are *convex*. This can be proved mathematically as follows. Consider two different stress vectors $\sigma_{ij}^a$ and $\sigma_{ij}^b$ in the nine-dimensional stress space. The difference $\Omega(\sigma_{ij}^b) - \Omega(\sigma_{ij}^a)$ can be approximated by a Taylor series expansion (neglecting higher-order terms):

$$\Omega(\sigma_{ij}^b) - \Omega(\sigma_{ij}^a) = \left(\frac{\partial \Omega}{\partial \sigma_{ij}}\right)_{\sigma_{ij}^a} \Delta\sigma_{ij} + \frac{1}{2}[H'_{ijkl}]_{\sigma_{ij}^a}\Delta\sigma_{ij}\Delta\sigma_{kl} \qquad (3.169)$$

where $\Delta\sigma_{ij} = (\sigma_{ij}^b - \sigma_{ij}^a)$ and

$$[H'_{ijkl}]_{\sigma_{ij}^a} = \text{Hessian matrix of } \Omega \text{ calculated at } \sigma_{ij}^a$$

The second term on the right-hand side of Eq. (3.169) is positive definite. [See Eq. (3.167).] Thus, we can write

$$\Omega(\sigma_{ij}^b) - \Omega(\sigma_{ij}^a) > \left(\frac{\partial \Omega}{\partial \sigma_{ij}}\right)_{\sigma_{ij}^a} (\sigma_{ij}^b - \sigma_{ij}^a) \qquad (3.170)$$

which is the condition for strict *convexity* of $\Omega(\sigma_{ij})$. Similarly, convexity of $W(\epsilon_{ij})$ can be proved.

The restrictions imposed by Drucker's material stability postulate and their implications are now summarized as follows:

1. The strain energy and complementary energy functions $W$ and $\Omega$ exist and are always *positive definite*. This follows directly from the positive definite character of their Hessian matrices, $[H]$ and $[H']$, respectively, and agrees with the requirement of the laws of thermodynamics.
2. The stress $\sigma_{ij}$ and strain $\epsilon_{ij}$ are normal to the surface $W = \text{constant}$ and $\Omega = \text{constant}$, respectively.
3. The surfaces corresponding to constant $W$ and $\Omega$ in strain and stress space, respectively, are *convex*.
4. Furthermore, the positive definiteness of $[H]$ and $[H']$ ensures that a *unique inverse* of the constitutive relations always exists. That is, for any constitutive equation $\sigma_{ij} = F(\epsilon_{ij})$ based on an assumed function for $W$, a unique inverse relation $\epsilon_{ij} = F^{-1}(\sigma_{ij})$ can always be obtained.

### 3.6.4. Uniqueness

Consider an elastic material body of volume $V$ and surface area $A$. The part of the surface area where surface tractions are prescribed is denoted by $A_T$, and that where surface displacements are prescribed is denoted by $A_u$ (see Fig. 3.7a). When the body forces $F_i$ and the surface forces $T_i$ act upon the body, the resulting stresses, strains, and displacements are given by $\sigma_{ij}$, $\epsilon_{ij}$, and $u_i$, respectively. Now suppose that we impose small changes of applied forces and displacements. These changes are characterized by the increments $dT_i$ on $A_T$, $dF_i$ in $V$, and $du_i$ on $A_u$. The problem is then to investigate whether the resulting stress and strain increments $d\sigma_{ij}$ and $d\epsilon_{ij}$, respectively, are determined uniquely by the increments of the applied forces and displacements $dT_i$, $dF_i$, and $du_i$. If not, there must then exist at least two distinct solutions corresponding to the applied changes $dT_i$, $dF_i$, and $du_i$. These two solutions are denoted by $a$ and $b$: solution $a$ with increments $d\sigma_{ij}^a$, $d\epsilon_{ij}^a$, and solution $b$ with increments $d\sigma_{ij}^b$, $d\epsilon_{ij}^b$.

Each of these solutions satisfies the equilibrium and compatibility (geometry) requirements. That is, $dT_i$, $dF_i$, and $d\sigma_{ij}^a$ constitute an equilibrium set, whereas $du_i$ and $d\epsilon_{ij}^a$ represent a compatible set. Similarly, the set $dT_i$, $dF_i$, and $d\sigma_{ij}^b$ is statically admissible and the set $du_i$ and $d\epsilon_{ij}^b$ is kinematically admissible. Because of the linearity of the equilibrium equations, Eqs. (3.70), the difference between the two statically admissible sets of solutions, $a$ and $b$, is also a statically admissible one; that is, the set $(d\sigma_{ij}^a - d\sigma_{ij}^b)$, corresponding to zero surface forces on $A_T$ and zero body forces in $V$, is an equilibrium set. Similarly, because of the linearity of the strain-displacement relations. Eqs. (3.71), the strains $(d\epsilon_{ij}^a - d\epsilon_{ij}^b)$, and the displacements $(du_i^a - du_i^b)$, which are zero on $A_u$, are kinematically admissible, and therefore constitute a compatible set. Applying the principle of virtual work to these two "difference" sets, one obtains

$$0 = \int_V (d\sigma_{ij}^a - d\sigma_{ij}^b)(d\epsilon_{ij}^a - d\epsilon_{ij}^b)\, dV \tag{3.171}$$

The "difference" state of stress $(d\sigma_{ij}^a - d\sigma_{ij}^b)$ may be considered as applied by an external agency which produces the corresponding strain $(d\epsilon_{ij}^a - d\epsilon_{ij}^b)$. The fundamental stability postulate, inequality (3.160), then gives

$$(d\sigma_{ij}^a - d\sigma_{ij}^b)(d\epsilon_{ij}^a - d\epsilon_{ij}^b) > 0 \tag{3.172}$$

That is, the integrand in Eq. (3.171) is always positive. Therefore, the integral of Eq. (3.171) can be zero if and only if the integrand is identically zero at each point in the body. Thus,

$$(d\sigma_{ij}^a - d\sigma_{ij}^b)(d\epsilon_{ij}^a - d\epsilon_{ij}^b) = 0 \tag{3.173}$$

which is satisfied when either $d\sigma_{ij}^a = d\sigma_{ij}^b$ or $d\epsilon_{ij}^a = d\epsilon_{ij}^b$. However, for stable elastic materials, the state of stress (or strain) is uniquely determined by the state of strain (or stress). Hence, $d\sigma_{ij}^a = d\sigma_{ij}^b$ implies that $d\epsilon_{ij}^a = d\epsilon_{ij}^b$, and *uniqueness* is established.

## 3.7. Incremental Stress–Strain Relations

This type of formulation is often used to describe the mechanical behavior of a class of material in which the state of stress depends on the current state of strain as well as on the stress path followed to reach this state. In general, the incremental constitutive relations are written as

$$\dot{\sigma}_{ij} = F_{ij}(\dot{\epsilon}_{kl}, \sigma_{mn}) \tag{3.174}$$

in which $\dot{\sigma}_{ij}$ and $\dot{\epsilon}_{kl}$ are the stress and strain increment tensors, respectively, and $F_{ij}$ are tensor functions. For isotropic time-independent materials, it can be shown that the expression on the right-hand side of Eq. (3.174) is a linear function of the components $d\epsilon_{kl}$ of the strain increment tensor [see the books by Chen (1982) and Chen and Saleeb (1982) for details]. Then the constitutive relations of Eq. (3.174) may be written in the incrementally linear form

$$d\sigma_{ij} = C_{ijkl}(\sigma_{mn}) \, d\epsilon_{kl} \qquad (3.175)$$

in which the material response tensor $C_{ijkl}(\sigma_{mn})$ is a function of the components of the stress tensor. The behavior described by Eqs. (3.175) is *infinitesimally* (or *incrementally*) *reversible*. This justifies the use of the prefix hypo in the term *hypoelastic* (or minimum elastic) to describe the constitutive relations (3.175).

The behavior of a hypoelastic material is in general path dependent (stress or strain history dependent). The integration of the differential equations (3.175) for different stress paths and initial conditions obviously leads to different stress-strain relations.

The tensor $C_{ijkl}$ is often called the *tangential stiffness tensor* of the material. The most general form of $C_{ijkl}$ for isotropic time-independent materials has been obtained as a polynomial function of the stress invariants with twelve material coefficients [see Chen (1982) and Chen and Saleeb (1982)]. An important characteristic exhibited is the *stress-* or *strain-induced anisotropy*. The initial isotropy of the material is destroyed, resulting in a generally anisotropic incremental stiffness. As a result of the induced anisotropy, there is a coupling between the volumetric response and deviatoric action. Also, the principal directions for the incremental stress and strain tensors do not coincide. The stress-induced anisotropy and the coupling effects are important features in modeling the behavior of real materials, such as concrete and soils, for which inelastic dilatation or compaction are dominant effects.

Recently, various special classes of incremental constitutive relations have been extensively used in modeling the nonlinear response of different engineering materials. Mainly, these models are developed on the basis of curve-fitting techniques. Examples of these can be found in the books by Chen (1982) and Chen and Saleeb (1982).

## 3.8. Summary

This chapter is concerned with the stress-strain relations for an elastic isotropic solid. The linear stress-strain relations are represented by the generalized Hooke's law, which is simple and most familiar to us, while the nonlinear elastic stress-strain relations, which are much more complicated, are in general categorized as *total* and *incremental* stress–strain relations.

There are two types of total stress–strain relations: *Cauchy type* and *Green type*. Cauchy elastic stress–strain relations take the form

$$\sigma_{ij} = F_{ij}(\epsilon_{kl})$$

which represents a one-to-one correspondence between stress $\sigma_{ij}$ and strain $\epsilon_{ij}$. Thus, the stress and strain are reversible and path independent. The most commonly used models of this type are formulated by simple modifications of the isotropic elastic stress–strain relations based on *variable secant moduli* (e.g., $E_s$, $\nu_s$, $K_s$, and $G_s$). Often, the material parameters in these models have well-defined physical relations to the observed stress–strain behavior of the material, and they can be easily determined from experimental data. However, *reversibility* and *path-independency* of the strain energy and complementary energy density functions, $W$ and $\Omega$, are not in general guaranteed. That is, thermodynamic laws may be violated since the models may generate energy for some load–unload stress paths. In order to satisfy thermodynamic laws, additional conditions must be imposed.

Green elastic stress–strain relations take the form

$$\sigma_{ij} = \frac{\partial W}{\partial \epsilon_{ij}} \quad \text{or} \quad \epsilon_{ij} = \frac{\partial \Omega}{\partial \sigma_{ij}}$$

Models of this type satisfy the laws of thermodynamics since the functions $W$ and $\Omega$ are *path independent*. Also, the stresses $\sigma_{ij}$ and strains $\epsilon_{ij}$ are reversible and path independent. Further, the *uniqueness* of stresses and strains in a boundary-value problem is satisfied if we impose the restriction of *convexity* (i.e., *positive definiteness*) on the energy functions $W$ and $\Omega$. Various functional forms for $W$ and $\Omega$ can be chosen to achieve the desired phenomena of the behavior of materials.

The incremental stress–strain relations of a *hypoelastic* material take the form

$$d\sigma_{ij} = C_{ijkl}(\sigma_{mn})\, d\epsilon_{kl}$$

The stresses and strains defined by these relations are incrementally reversible. However, the state of stress and the state of strain are load path dependent. In general, a hypoelastic model may violate the laws of thermodynamics in some load–unload cycles since it may generate energy.

## References

Chen, W.F., 1982. *Plasticity in Reinforced Concrete*, McGraw-Hill, New York.
Chen, W.F., and A.F. Saleeb, 1982. *Constitutive Equations for Engineering Materials, Volume 1: Elasticity and Modeling*, Wiley-Interscience, New York.

PROBLEMS

3.1. The relative displacement tensor $\epsilon'_{ij}$ at a point is given below:

$$\epsilon'_{ij} = \begin{bmatrix} 0.10 & 0.20 & -0.40 \\ -0.20 & 0.25 & -0.15 \\ 0.40 & 0.30 & 0.30 \end{bmatrix}$$

Determine:

(a) The strain tensor $\epsilon_{ij}$.
(b) The rotation tensor $\omega_{ij}$.
(c) The principal strains $\epsilon_1$, $\epsilon_2$, and $\epsilon_3$, and the principal directions.
(d) For a fiber element with direction $\mathbf{n} = (\frac{1}{2}, \frac{1}{2}, 1/\sqrt{2})$, find the strain vector $\overset{n}{\boldsymbol{\delta}}$, the rotation vector $\overset{n}{\boldsymbol{\Omega}}$, and the relative displacement vector $\overset{n}{\boldsymbol{\delta}}'$.

3.2. The state of strain at a point is represented by the given strain tensor $\epsilon_{ij}$.

$$\epsilon_{ij} = \begin{bmatrix} -0.005 & -0.004 & 0 \\ -0.004 & 0.001 & 0 \\ 0 & 0 & 0.001 \end{bmatrix}$$

Determine:

(a) The deviatoric strain tensor $e_{ij}$.
(b) The values of the invariants $J'_2$ and $J'_3$.
(c) The volume change per unit volume (dilatation) $\epsilon_v$.

3.3. The displacement components $u_i$ at a point in a body are given by the functional components

$$u_1 = 10x_1 + 3x_2, \qquad u_2 = 3x_1 + 2x_2, \qquad u_3 = 6x_3$$

Show that there is no rotation if the deformations are assumed to be small.

3.4. Determine the relations among the constants $a_0$, $a_1$, $b_0$, $b_1$, $c_0$, $c_1$, and $c_2$ so that the following is a possible state of strain:

$$\epsilon_x = a_0 + a_1(x^2 + y^2) + (x^4 + y^4)$$
$$\epsilon_y = b_0 + b_1(x^2 + y^2) + (x^4 + y^4)$$
$$\gamma_{xy} = c_0 + c_1 xy(x^2 + y^2 + c_2)$$
$$\epsilon_z = \gamma_{yz} = \gamma_{xz} = 0$$

3.5. Using Eq. (3.70a) and the stress-strain relations of an isotropic linear elastic material, show that the equations of equilibrium, Eqs. (3.70a), can be written in the following form

$$u_{i,jj} + \frac{1}{1-2\nu} u_{j,ji} + \frac{F_i}{G} = 0$$

in which $\nu$ and $G$ are Poisson's ratio and the shear modulus, respectively.

3.6. Prove the following relations between the elastic moduli $E$, $\nu$ and $K$, $G$:

$$E = \frac{9KG}{3K+G}; \qquad \nu = \frac{3K-2G}{2(3K+G)}$$

3.7. For an isotropic linear elastic material, the stress components, $\sigma_{ij}$, at a point are given by

$$\sigma_{ij} = \begin{bmatrix} 10 & 1 & -8 \\ 1 & -6 & 6 \\ -8 & 6 & 20 \end{bmatrix} \text{ksi}$$

The material constants are $E = 30,000$ ksi and $\nu = 0.3$. Determine the following:
(a) The strain deviator tensor, $e_{ij}$, at the given point.
(b) The values of the strain energy density, $W$, and the complementary energy density, $\Omega$, corresponding to the given state of stress.
(c) The principal strains, $\epsilon_1$, $\epsilon_2$, and $\epsilon_3$, at the same point.

3.8. An initially unstressed and unstrained material element is subjected to a combined loading history which produces the following successive straight-line paths in $(\sigma, \tau)$ space [units are in psi (tension $\sigma$, shear $\tau$)]:

Path 1: $(0, 0)$ to $(0, 10,000)$
Path 2: $(0, 10,000)$ to $(30,000, 10,000)$
Path 3: $(30,000, 10,000)$ to $(30,000, -10,000)$
Path 4: $(30,000, -10,000)$ to $(0, 0)$

The element is assumed to be of an isotropic nonlinear elastic material with the complementary energy density, $\Omega$, given as

$$\Omega = a(J_2^3 + J_3^2)$$

where $a$ is a material constant. Its stress–strain relation in simple tension is

$$10^9 \epsilon = \left(\frac{\sigma}{1000}\right)^5$$

where $\sigma$ is in psi.

(a) Give the axial extension and shear strain at the end of Path 3.
(b) Find all the components of normal and shear strains at the ends of Paths 1 and 2.
(c) Draw the curve of constant $\Omega$ in $(\sigma, \tau)$ space passing through $(30,000, 100,000)$. Demonstrate analytically that the curve $\Omega = $ const. is convex.
(d) Demonstrate pictorially that the answer to part (b) satisfies the normality condition (3.118).
(e) Convert Path 1 to a path in the principle stress space. Compute the strains at the end of this path using the stress–strain relations expressed in terms of principal stresses, and compare the answer in detail with the results obtained in terms of $\sigma$ and $\tau$, in (b).
(f) List all the paths of those given above that are straight-line paths in principal stress space.
(g) Draw Path 2 in the principal stress space.

3.9. Consider a nonlinear elastic material based on the complementary density function $\Omega$ given by

$$\Omega(I_1, J_2) = aI_1^2 + bJ_2^2$$

where $a$, and $b$ are material constants. The stress-strain relationship of the material in uniaxial tension is given by

$$10^3 \epsilon = \frac{\sigma}{10} + \frac{1}{9}\left(\frac{\sigma}{10}\right)^3$$

where $\sigma$ is in ksi.

(a) Determine the constants $a$ and $b$.

(b) Write the stress-strain relation of the material for a biaxial compression state.

(c) Write the stress-strain relation of the material for a pure shear state.

3.10. List and explain analytically and pictorially the restrictions imposed by Drucker's material stability postulate and their implications for an elastic material.

### Answers to Selected Problems

3.1. (a) $\epsilon_{ij} = \begin{bmatrix} 0.1 & 0 & 0 \\ 0 & 0.25 & 0.075 \\ 0 & 0.075 & 0.3 \end{bmatrix}$

(b) $\omega_{ij} = \begin{bmatrix} 0 & 0.2 & -0.4 \\ -0.2 & 0 & -0.225 \\ 0.4 & 0.225 & 0 \end{bmatrix}$

(c) $\epsilon_1 = 0.354$, $\epsilon_2 = 0.196$, $\epsilon_3 = 0.10$; $n_i^{(1)} = (0, \pm 0.5847, \pm 0.8113)$, $n_i^{(2)} = (0, \pm 0.8113, \mp 0.5847)$, $n_i^{(3)} = (\pm 1, 0.0)$.

(d) $\overset{n}{\delta} = (0.05, 0.178, 0.2496)$, $\overset{n}{\Omega} = (0.183, 0.259, -0.3125)$, $\overset{n}{\delta}' = (0.233, 0.437, -0.0629)$.

3.2. (a) $e_{ij} = \begin{bmatrix} -0.004 & -0.004 & 0 \\ -0.004 & 0.002 & 0 \\ 0 & 0 & 0.002 \end{bmatrix}$

(b) $J_2' = 2.8 \times 10^{-5}$, $J_3' = -4.8 \times 10^{-8}$

(c) $\epsilon_v = -0.003$.

3.4. To satisfy the compatibility conditions, we must have $a_1 + b_1 - 2c_2 = 0$, $c_1 = 4$.

3.7. (a) $e_{ij} = 10^{-4} \times \begin{bmatrix} 0.867 & 0.433 & -3.466 \\ 0.433 & -6.066 & 2.6 \\ -3.466 & 2.6 & 5.2 \end{bmatrix}$

(b) $W = \Omega = 0.01311$ in $\cdot$ k/in$^3$.

(c) $\epsilon_1 = 8.511 \times 10^{-4}$, $\epsilon_2 = 0.4592 \times 10^{-4}$, $\epsilon_3 = -5.7701 \times 10^{-4}$.

3.8. (a) $\epsilon = 43.12 \times 10^{-3}$, $\gamma = -42.336 \times 10^{-3}$

(b) At the end of Path 1, $\gamma(= \gamma_{xy}) = 2.352 \times 10^{-3}$, all other components $= 0$. At the end of Path (2); $\epsilon(= \epsilon_x) = 43.12 \times 10^{-3}$, $\epsilon_y = -20.38 \times 10^{-3}$, $\epsilon_z = -22.73 \times 10^{-3}$, $\gamma(= \gamma_{xy}) = 42.366 \times 10^{-3}$, $\gamma_{yz} = \gamma_{zx} = 0$.

(e) In principal stress space $(\sigma_1, \sigma_2)$, Path 1 is from $(0, 0)$ to $(100,000, -10,000)$ psi. At the end of this path; $\epsilon_1 = 1.176 \times 10^{-3}$, $\epsilon_3 = 0$, $\epsilon_2 = -1.176 \times 10^{-3}$.

(f) Paths 1 and 4.

# Part II:  Plastic Stress–Strain Relations

# 4
# Stress–Strain Relations for Perfectly Plastic Materials

## 4.1. Introduction

For many practical applications, a material may be idealized and assumed to have a negligible strain-hardening effect, i.e., its uniaxial stress–strain diagram beyond the yield point can be approximated by a horizontal straight line, with the constant stress level $\sigma_0$ (Fig. 4.1a). Thus, plastic deformation is assumed to occur under a constant flow stress. This behavior is called *perfectly* or *ideally plastic* behavior.

Perfectly plastic idealization can lead to a drastic simplification of the analysis of a complex structural problem. In particular, for a perfectly plastic material, the powerful upper- and lower-bound theorems of limit analysis can be established, from which simple, direct, and realistic methods for estimating the load-carrying capacity of structures in a direct manner can be developed. These bounding theorems and their applications to structural engineering problems will be given in Chapters 8 and 9. This chapter deals only with the stress–strain relations of a perfectly plastic material.

The stress–strain relation in the uniaxial case as shown in Fig. 4.1a is rather simple. However, the general behavior of the material under a complex stress state is not so straightforward, because it involves six stress and six strain components. The question therefore arises as to how the simple stress–strain relationships observed from a uniaxial stress test can be generalized to predict the behavior of the material under any general combined stress state.

This chapter is divided into three parts. The first part, Sections 4.2 through 4.6, is devoted to the classical flow theory of plasticity. The basic concepts of the flow rule and the convexity, normality, and uniqueness for elastic-perfectly plastic materials are discussed in detail. The second part, Section 4.7, provides a simple example and introduces some features of elastic-plastic behavior of a structure. The final part, Sections 4.8 through 4.11, deals with the constitutive relations for elastic-perfectly plastic materials. Specific forms of incremental stress-strain relations for different material models are presented.

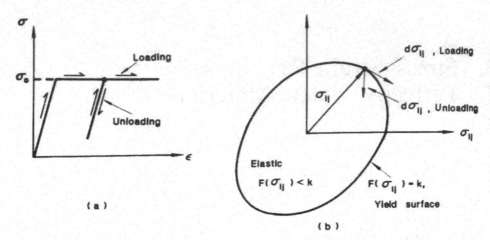

FIGURE 4.1. An elastic–perfectly plastic material. (a) Uniaxial stress–strain relation; (b) geometric representation of yield surface and criterion of loading and unloading.

### 4.1.1. Elastic Limit and Yield Function

Generalization of the *elastic limit* has been discussed previously in Chapter 2, where the elastic limit of a material under all possible combinations of stresses was defined as a yield function in terms of stress $\sigma_{ij}$ in the form

$$f(\sigma_{ij}) = F(\sigma_{ij}) - k = 0 \tag{4.1}$$

The significance of this yield function can best be interpreted geometrically as a surface in stress space. For a perfectly plastic material, the yield function is assumed to remain unchanged. Thus, the parameter $k$ in Eq. (4.1) is a constant, and the yield surface is therefore fixed in stress space (Fig. 4.1b).

### 4.1.2. Criterion for Loading and Unloading

Plastic deformation occurs as long as the stress point is on the yield surface. For the plastic flow to continue, the state of stress must remain on the yield surface. This condition is termed "loading." Otherwise, the stress state must drop below the yield value; in this case, no further plastic deformation occurs and all incremental deformations are elastic. This condition is termed "unloading."

The concept of loading and unloading for a complex stress state is clearest when $f$ is interpreted geometrically as a surface and $\sigma_{ij}$ and $d\sigma_{ij}$ as stress and stress increment vectors in stress space (Fig. 4.1b). Consider, for example, a material element in a plastic state, characterized by the stress vector $\sigma_{ij}$. If we add to the current stress state $\sigma_{ij}$ an infinitesimal increment of stress $d\sigma_{ij}$ (additional loading), will this additional stress cause further plastic deformation? For a perfectly plastic material, the stress point cannot move outside the yield surface. Plastic flow can occur only when the stress point is on the yield surface, and the additional loading $d\sigma_{ij}$ must therefore move along the tangential direction. Thus, the condition for a continuation or further plastic flow, or the criterion for loading, is

$$f(\sigma_{ij}, k) = 0 \quad \text{and} \quad df = \frac{\partial f}{\partial \sigma_{ij}} d\sigma_{ij} = 0 \tag{4.2}$$

and the criterion for unloading is

$$f(\sigma_{ij}, k) = 0 \quad \text{and} \quad df = \frac{\partial f}{\partial \sigma_{ij}} d\sigma_{ij} < 0 \tag{4.3}$$

As a result, the yield function $f(\sigma_{ij})$ also serves as the *criterion of loading* for further plastic deformation, or as the *criterion of unloading* for elastic deformation. The yield function or surface $f(\sigma_{ij})$ is also called the *loading function* or *surface*.

### 4.1.3. Elastic and Plastic Strain Increment Tensors

Since the magnitude of the plastic strain $\epsilon_{ij}^p$ is unlimited during flow, we must think therefore in terms of the strain rates $\dot{\epsilon}_{ij}$ or of infinitesimal changes of strain, or strain increments, $d\epsilon_{ij}$. The total strain increment tensor is assumed to be the sum of the elastic and plastic strain increment tensors:

$$d\epsilon_{ij} = d\epsilon_{ij}^e + d\epsilon_{ij}^p \tag{4.4}$$

Since Hooke's law or any other nonlinear elastic model (see Chapter 3) can be assumed to provide the necessary relationship between the incremental changes of stress and elastic strain, the stress–strain relation for a plastic material reduces essentially to a relation involving the current state and the incremental changes of stress and plastic strain. This latter relationship for a perfectly plastic material will be derived in detail in this chapter.

## 4.2. Plastic Potential and Flow Rule

The flow rule is the necessary kinematic assumption postulated for plastic deformation or *plastic flow*. It gives the ratio or the relative magnitudes of the components of the plastic strain increment tensor $d\epsilon_{ij}^p$. Since the increment $d\epsilon_{ij}^p$ may be represented geometrically by a vector with nine components in strain space, as shown in Fig. 4.2, the flow rule therefore also defines the direction of the plastic strain increment vector $d\epsilon_{ij}^p$ in the strain space.

We have seen in Chapter 3 that the elastic strain can be derived directly by differentiating the *elastic potential function* or complementary energy

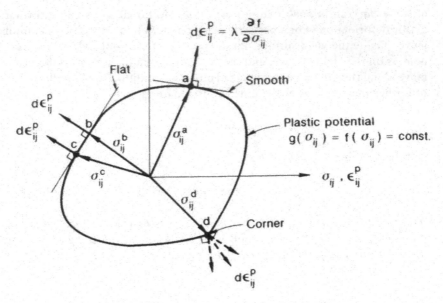

FIGURE 4.2. Geometric illustration of associated flow rule.

density function with respect to stresses $\sigma_{ij}$ [see Eq. (3.118)]. In 1928, von Mises proposed the similar concept of the *plastic potential function*, which is a scalar function of the stresses, $g(\sigma_{ij})$. Then the plastic flow equations can be written in the form

$$d\epsilon_{ij}^p = d\lambda \frac{\partial g}{\partial \sigma_{ij}} \qquad (4.5)$$

where $d\lambda$ is a *positive scalar factor of proportionality*, which is nonzero only when plastic deformations occur. The equation $g(\sigma_{ij}) = $ constant defines a surface (*hypersurface*) of plastic potential in nine-dimensional stress space. The direction cosines of the normal vector to this surface at the point $\sigma_{ij}$ on the surface are proportional to the *gradient* $\partial g/\partial \sigma_{ij}$. The relation (4.5) implies that the plastic flow vector $d\epsilon_{ij}^p$, if plotted as a free vector in stress space, is directed along the normal to the surface of plastic potential (Fig. 4.2).

Of great importance is the simplest case when the yield function and the plastic potential function coincide, $f = g$. Thus,

$$d\epsilon_{ij}^p = d\lambda \frac{\partial f}{\partial \sigma_{ij}} \qquad (4.6)$$

and plastic flow develops along the normal to the yield surface $\partial f/\partial \sigma_{ij}$ (see Fig. 4.2). Equation (4.6) is called the *associated flow rule* because the plastic flow is connected or associated with the yield criterion, while relation (4.5) with $f \neq g$ is called a *nonassociated flow rule*.

von Mises used the associated flow rule for the development of his plastic stress–strain relations for metals. It will be shown later that (1) the associated flow rule (4.6) is valid for irreversible plastic materials where work expended on plastic deformation cannot be reclaimed; (2) the stress–strain law of a material based on the associated flow rule will result in a *unique* solution of a boundary-value problem; and (3) the associated flow rule makes it possible and convenient to formulate various generalizations of the plasticity equations by considering yield and loading surfaces of more complex form.

## 4.3. Flow Rule Associated with von Mises Yield Function

We shall now take the von Mises yield function

$$f(\sigma_{ij}) = J_2 - k^2 = 0 \tag{4.7}$$

as the plastic potential. Then the flow rule has the simple form:

$$d\epsilon_{ij}^p = d\lambda \frac{\partial f}{\partial \sigma_{ij}} = d\lambda \, s_{ij} \tag{4.8}$$

where $s_{ij}$ is the deviatoric stress tensor and $d\lambda$ is a factor of proportionality with the value

$$d\lambda \begin{cases} =0 & \text{wherever } J_2 < k^2 \text{ or } J_2 = k^2, \text{ but } dJ_2 < 0 \\ >0 & \text{wherever } J_2 = k^2 \text{ and } dJ_2 = 0 \end{cases}$$

Equation (4.8) can also be expressed in terms of the components of the strain increments and stresses as

$$\frac{d\epsilon_x^p}{s_x} = \frac{d\epsilon_y^p}{s_y} = \frac{d\epsilon_z^p}{s_z} = \frac{d\gamma_{yz}^p}{2\tau_{yz}} = \frac{d\gamma_{zx}^p}{2\tau_{zx}} = \frac{d\gamma_{xy}^p}{2\tau_{xy}} = d\lambda \tag{4.9}$$

Relations (4.9) are known as the *Prandtl–Reuss equations*. It was Prandtl, in 1924, who extended the earlier Levy–von Mises equations [see Eq. (4.15)] and first proposed the stress–strain relation in the plane strain case for an elastic–perfectly plastic material. Reuss, in 1930, extended the Prandtl equations to the three-dimensional case and gave the general form of Eq. (4.9).

The relationship between the plastic strain increment $d\epsilon_{ij}^p$ and the von Mises yield function $f = J_2$ as given by Eqs. (4.8) or (4.9), or the flow rule associated with the von Mises yield condition can be shown graphically in the three-dimensional principal stress space. However, the three-dimension picture is difficult to draw and instead it is best shown by a cross section on the hydrostatic plane and by a cross section on the deviatoric plane of the three-dimensional surface as in Fig. 4.3. The normal to the yield surface as viewed along the hydrostatic axis is a radial line (Fig. 4.3b) that is parallel

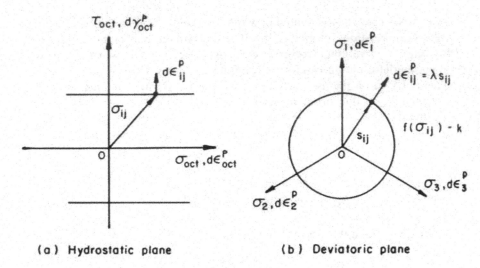

(a) Hydrostatic plane          (b) Deviatoric plane

FIGURE 4.3. Flow rule associated with von Mises yield function.

to the $\pi$-plane. Its direction is therefore parallel to the direction of the projection of the appropriate stress vector $\sigma_{ij}$ onto the $\pi$-plane, which is, of course, precisely its deviatoric stress component vector $s_{ij}$.

Equation (4.8) or (4.9) states that a small increment of plastic strain $d\epsilon_{ij}^p$ depends only on the current state of deviatoric stress $s_{ij}$, not on the stress increment $d\sigma_{ij}$ which is required to maintain the plastic flow. Also, the principal axes of stress $\sigma_{ij}$ or $s_{ij}$ and the plastic strain increment $d\epsilon_{ij}^p$ coincide. Note that these equations are only statements about the ratio or the relative magnitudes of the components of the plastic strain increment tensor; they give no direct information about its absolute magnitude.

According to Eq. (4.8), there is no plastic volumetric deformation; that is,

$$d\epsilon_{ii}^p = d\lambda \, s_{ii} = 0 \tag{4.10}$$

This can also be seen in Fig. 4.3a where the plastic strain increment vector $d\epsilon_{ij}^p$ is normal to the hydrostatic axis, and the hydrostatic strain component, $d\epsilon_{oct}^p$, is therefore zero.

The total strain increment $d\epsilon_{ij}$ is the sum of the elastic and plastic strain increments (Eq. 4.4). If Hooke's law [Eqs. (3.84) or (3.96)] is applied for the elastic component $d\epsilon_{ij}^e$ and the flow rule [Eq. (4.8)] for the plastic component $d\epsilon_{ij}^p$, we have

$$d\epsilon_{ij} = \frac{1+\nu}{E} \, d\sigma_{ij} - \frac{\nu}{E} \, d\sigma_{kk}\delta_{ij} + d\lambda \, s_{ij}$$

$$= \frac{d\sigma_{kk}}{9K} \, \delta_{ij} + \frac{ds_{ij}}{2G} + d\lambda \, s_{ij} \tag{4.11}$$

Equation (4.11) may also be separated into expressions for the volumetric and deviatoric or shear strain increments of the forms:

$$d\epsilon_{ii} = \frac{1}{3K} d\sigma_{kk}$$

$$de_{ij} = \frac{1}{2G} ds_{ij} + d\lambda \, s_{ij} \tag{4.12}$$

In practical applications, we expand Eq. (4.11) explicitly in terms of stress and strain components, giving rise to three equations for the normal strains of the form:

$$d\epsilon_x = \frac{1}{E}[d\sigma_x - \nu(d\sigma_y + d\sigma_z)] + \frac{2}{3} d\lambda \left[ \sigma_x - \frac{1}{2}(\sigma_y + \sigma_z) \right], \text{ etc. } \tag{4.13}$$

and three equations for the shear strains of the form:

$$d\gamma_{yz} = \frac{1}{G} d\tau_{yz} + 2 d\lambda \, \tau_{yz}, \text{ etc. } \tag{4.14}$$

In problems of large plastic flow, the elastic strain may be neglected. In such a case, the material can be idealized as rigid–perfectly plastic, and the total strain increment $d\epsilon_{ij}$ and the plastic strain increment $d\epsilon_{ij}^p$ are identical. The stress–strain relations for such a material may be written as

$$d\epsilon_{ij} = d\lambda \, s_{ij}$$

or

$$\frac{d\epsilon_x}{s_x} = \frac{d\epsilon_y}{s_y} = \frac{d\epsilon_z}{s_z} = \frac{d\gamma_{yz}}{2\tau_{yz}} = \frac{d\gamma_{zx}}{2\tau_{zx}} = \frac{d\gamma_{xy}}{2\tau_{xy}} = d\lambda \tag{4.15}$$

in which the superscript, $p$, of Eqs. (4.8) and (4.9) has been dropped. Equations (4.15) are known as the *Levy–von Mises equations*. In their historical development, it was St. Venant, in 1870, who first proposed that the principal axes of strain increment coincided with the principal axes of stress. These general stress–strain relations were obtained later by Levy in 1871 and independently by von Mises in 1913.

Expanding the Levy–von Mises relation in terms of stress components gives three equations for normal plastic strain increments of the form

$$d\epsilon_x = \tfrac{2}{3}d\lambda[\sigma_x - \tfrac{1}{2}(\sigma_y + \sigma_z)], \text{ etc. } \tag{4.16}$$

and three equations for shear plastic strain increments of the form

$$d\gamma_{yz} = 2\tau_{yz}d\lambda, \text{ etc. } \tag{4.17}$$

## 4.4. Flow Rule Associated with Tresca Yield Function

Now, take the Tresca yield function as the plastic potential, which in principal stress space is a right hexagonal prism consisting of six planes. The deviatoric section of the prism is shown in Fig. 4.4a. Suppose that the

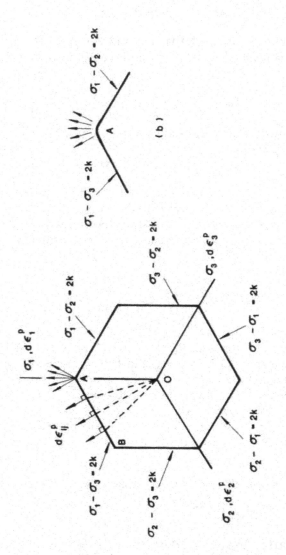

FIGURE 4.4. Flow rule associated with Tresca yield function. (a) Normality of the plastic strain increment vector; (b) vertex $A$ as a limit of a smooth surface.

ordering in magnitude of the principal stresses is $\sigma_1 > \sigma_2 > \sigma_3$; we can then write the corresponding yield function or plastic potential function in the form

$$f = F(\sigma_{ij}) - 2k = \sigma_1 - \sigma_3 - 2k = 0 \qquad (4.18)$$

According to the associated flow rule, the principal plastic strain increments, $d\epsilon_1^p$, $d\epsilon_2^p$, $d\epsilon_3^p$, satisfy the following relations:

$$d\epsilon_1^p = d\lambda \frac{\partial f}{\partial \sigma_1} = d\lambda$$

$$d\epsilon_2^p = d\lambda \frac{\partial f}{\partial \sigma_2} = 0$$

$$d\epsilon_3^p = d\lambda \frac{\partial f}{\partial \sigma_3} = -d\lambda$$

or, in a more compact form,

$$(d\epsilon_1^p, d\epsilon_2^p, d\epsilon_3^p) = d\lambda\ (1, 0, -1), \qquad d\lambda \geq 0 \qquad (4.19)$$

Similar results can be derived for the other five possible combinations of algebraic orders of magnitude of the principal stresses $\sigma_1$, $\sigma_2$, and $\sigma_3$.

The plastic strain increments can therefore be illustrated geometrically in a combined principal stress/principal strain increment space as shown in Fig. 4.4a. It is seen that anywhere on the plane $AB$ where $\sigma_1 > \sigma_2 > \sigma_3$, the directions of the plastic strain increments are parallel to each other and perpendicular to the plane $AB$ of the Tresca hexagon. Similar relationships can be developed for other planes of the hexagon.

In the special case where, for example, $\sigma_1 > \sigma_2 = \sigma_3$, the situation is more involved, because the maximum shear stress is equal to the yield value $k$ not only on the 45° shear planes parallel to the $x_2$-axis but also on the 45° planes parallel to the $x_3$-axis. We have therefore the freedom to assume that the shear slip may occur along either of the two possible maximum shear planes:

(i) $\sigma_{max} = \sigma_1$, $\sigma_{min} = \sigma_3$

$$(d\epsilon_1^p, d\epsilon_2^p, d\epsilon_3^p) = d\lambda(1, 0, -1), \qquad \text{for } d\lambda \geq 0$$

(ii) $\sigma_{max} = \sigma_1$, $\sigma_{min} = \sigma_2$

$$(d\epsilon_1^p, d\epsilon_2^p, d\epsilon_3^p) = d\mu(1, -1, 0), \qquad \text{for } d\mu \geq 0$$

In this case, we shall assume that the resulting plastic strain increment vector is a linear combination of the two increments given above, i.e.,

$$(d\varepsilon_1^p, d\epsilon_2^p, d\epsilon_3^p) = d\lambda(1, 0, -1) + d\mu(1, -1, 0), \qquad \text{for } d\lambda \geq 0, d\mu \geq 0 \quad (4.20)$$

This situation corresponds to the special case where the current stress state $\sigma_{ij}$ lies on a vertex of the hexagon. As a result, the plastic strain increment vector must lie between the directions of the normals to the two adjacent sides of the hexagon (Fig. 4.4a). This vertex or *singular point* at a potential surface can also be viewed as a limiting case of a smooth surface, and the flow rule can still be applied for a smooth surface at this corner point (Fig. 4.4b).

In general, at a singular point where several smooth yield surfaces intersect, the strain increments can generally be expressed as a linear combination of those increments given by the normals of the respective surfaces intersecting at the point, i.e.,

$$de_{ij}^p = \sum_{k=1}^{n} d\lambda_k \frac{\partial f_k}{\partial \sigma_{ij}} \tag{4.21}$$

As a result, at the vertex, the direction of the strain increment vector cannot be determined uniquely. Further, if the yield surface contains a flat part (Fig. 4.2 or Fig. 4.4a), there also exists no unique relationship between the stress and the strain increment. In general, the correspondence between the plastic strain increment vector $de_{ij}^p$ and the stress vector $\sigma_{ij}$ is not always one to one. However, it will be shown in the following example that the incremental plastic work $dW_p$ done or the rate of dissipation of energy is always uniquely determined by the magnitude of the plastic strain rate as given by

$$dW_p = \sigma_1 \, de_1^p + \sigma_2 \, de_2^p + \sigma_3 \, de_3^p = 2k \max|de^p| \tag{4.22}$$

where $\max|de^p|$ denotes the absolute value of the numerically largest principal component of the plastic strain increment vector.

EXAMPLE 4.1. Using the flow rule associated with the Tresca yield condition,

(a) show that the plastic work increment is given by expression (4.22);
(b) assuming that a material element yields at a biaxial stress state, $\sigma_1 = \sigma_0/\sqrt{3}$, $\sigma_2 = -\sigma_0/\sqrt{3}$, where $\sigma_0$ is the yield stress in uniaxial tension, and also given $de_1^p = c$, where $c$ is a constant, find the plastic strain increments and plastic work increment.

SOLUTION. (a) For a stress point on the side $AB$ with the equation $\sigma_1 - \sigma_3 = 2k$, the components of the strain increment vector are $de_2^p = 0$ and $de_3^p = -de_1^p$. The plastic work increment is therefore given by

$$dW_p = \sigma_1 \, de_1^p + \sigma_2 \, de_2^p + \sigma_3 \, de_3^p$$

$$= (\sigma_1 - \sigma_3) \, de_1^p = 2k \, de^p \tag{4.23}$$

since $\sigma_1 = \sigma_3 + 2k$ on $AB$. Note that $\max|de^p| = de_1^p$ in this case, so that $2k \, de_1^p$ can be written in the form of Eq. (4.22).

If the stress point coincides with the vertex $A$, then, $\sigma_1 = \sigma_3 + 2k$ and $\sigma_2 = \sigma_3$, and therefore we have

$$dW_p = (\sigma_3 + 2k)\, d\epsilon_1^p + \sigma_3\, d\epsilon_2^p + \sigma_3\, d\epsilon_3^p \qquad (4.24)$$

Using the *incompressibility condition*,

$$d\epsilon_1^p + d\epsilon_2^p + d\epsilon_3^p = 0$$

Equation (4.24) yields

$$dW_p = 2k\, d\epsilon_1^p \qquad (4.25)$$

Since $d\epsilon_1^p$ is the numerically largest principal component in this case, Eq. (4.25) can also be written in the form of Eq. (4.22). In a similar manner, it can be shown that Eq. (4.22) holds for every stress point on the hexagon.

(b) According to the Tresca yield condition, we have

$$\tau_{\max} = \frac{\sigma_{\max} - \sigma_{\min}}{2} = \frac{1}{2}\left(\frac{\sigma_0}{\sqrt{3}} + \frac{\sigma_0}{\sqrt{3}}\right) = k$$

Thus, $k = \sigma_0/\sqrt{3}$. The flow rule associated with this yield condition defines the increments of the plastic strain components in the $\sigma_1$ and $\sigma_2$ directions as

$$(d\epsilon_1^p,\, d\epsilon_2^p) = d\lambda\, (1, -1) = (c, -c)$$

Thus, $c$ is the largest plastic strain component and the plastic work increment is obtained as

$$dW_p = 2k\, \max|d\epsilon^p| = 2kc = \frac{2\sigma_0 c}{\sqrt{3}}$$

# 4.5. Flow Rule Associated with Mohr-Coulomb Yield Function

In the applications of limit analysis, some frictional materials such as concretes or soils are idealized as elastic-perfectly plastic materials obeying the Mohr-Coulomb yield criterion. The Mohr-Coulomb yield surface is an irregular hexagonal pyramid. Its deviatoric sections are *irregular* hexagons as shown in Fig. 4.5. The yield function takes the following form [see Eq. (2.174)]:

$$\sigma_1 \frac{1 + \sin\phi}{2c\cos\phi} - \sigma_3 \frac{1 - \sin\phi}{2c\cos\phi} = 1 \qquad (4.26)$$

where $\phi$ is the *angle of internal friction* and $c$ the *cohesion*. Equation (4.26) can also be written in compact form as [see Eq. (2.179)]

$$m\sigma_1 - \sigma_3 = f_c' \qquad \text{for } \sigma_1 \geq \sigma_2 \geq \sigma_3 \qquad (4.27)$$

where $f_c'$ is the *uniaxial compressive strength* and $m$ is the strength ratio between $f_c'$ and $f_t'$, the uniaxial tensile strength (see Section 2.3.3). To obtain

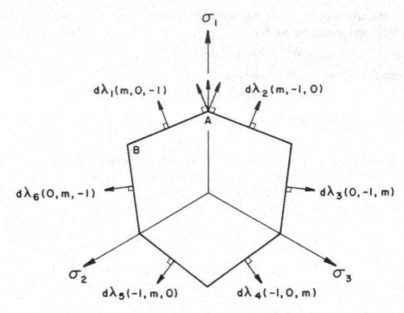

FIGURE 4.5. Flow rule associated with Mohr–Coulomb yield surface.

the expression for the plastic strain increment ($d\epsilon_1^p$, $d\epsilon_2^p$, $d\epsilon_3^p$), the following three cases must be considered separately.

*Case 1.* The yield stress point lies on the surface plane of the pyramid, say, for example, on face $AB$ (Fig. 4.5), where $\sigma_1 > \sigma_2 > \sigma_3$ and Eq. (4.27) holds. According to the associated flow rule, we have the following plastic strain increments:

$$d\epsilon_1^p = m\, d\lambda, \qquad d\epsilon_2^p = 0, \qquad d\varepsilon_3^p = -d\lambda \qquad \text{for } d\lambda \geq 0 \qquad (4.28)$$

or, in compact form,

$$(d\epsilon_1^p, d\epsilon_2^p, d\epsilon_3^p) = d\lambda\,(m, 0, -1), \qquad \text{for } d\lambda \geq 0 \qquad (4.29)$$

Similar results can be obtained for the other five possible algebraic orderings of the principal stresses $\sigma_1$, $\sigma_2$, and $\sigma_3$. These results are summarized and shown graphically in Fig. 4.5.

Notice that the plastic volumetric strain increment is

$$d\epsilon_v^p = d\epsilon_1^p + d\epsilon_2^p + d\epsilon_3^p = d\lambda\,(m - 1) \qquad (4.30)$$

Since $m = f_c'/f_t' \geq 1$, it follows that the Mohr–Coulomb material model with the associated flow rule always predicts volume dilatation except in the special case $m = 1$, which reduces to the case of the Tresca material model.

From Eq. (4.30), we can separate the sum of the principal plastic strain increments into two parts: the compressive part

$$\sum |d\epsilon_c^p| = d\lambda \qquad (4.31)$$

and the tensile part

$$\sum d\epsilon_t^p = m \, d\lambda \qquad (4.32)$$

Such a separation can be done as well for the other five planes of the pyramid. Then we have

$$\frac{\sum d\epsilon_t^p}{\sum |d\epsilon_c^p|} = m \qquad (4.33)$$

and

$$d\epsilon_v^p = \sum d\epsilon_t^p - \sum |d\epsilon_c^p| \qquad (4.34)$$

Now, consider further the plastic work increment $dW_p$. By definition, we have

$$dW_p = \sigma_1 \, d\epsilon_1^p + \sigma_2 \, d\epsilon_2^p + \sigma_3 \, d\epsilon_3^p = (\sigma_1 m - \sigma_3) \, d\lambda \qquad (4.35)$$

Using Eqs. (4.27) and (4.31), Eq. (4.35) becomes

$$dW_p = f_c' \sum |d\epsilon_c^p| \qquad (4.36)$$

or

$$dW_p = \frac{f_c'}{m} \sum d\epsilon_t^p \qquad (4.37)$$

*Case 2.* The yield stress point lies on the edges of the pyramid, say, along the edge $A$ (Fig. 4.5), where $\sigma_1 > \sigma_2 = \sigma_3$ and the two surfaces

$$m\sigma_1 - \sigma_3 = f_c'$$

and

$$m\sigma_1 - \sigma_2 = f_c'$$

intersect. In this case, Eq. (4.21) can be applied. Thus, the corresponding plastic strain increments are expressed as

$$(d\epsilon_1^p, d\epsilon_2^p, d\epsilon_3^p) = d\lambda_1(m, 0, -1) + d\lambda_2(m, -1, 0)$$
$$= [(d\lambda_1 + d\lambda_2)m, -d\lambda_2, -d\lambda_1] \qquad (4.38)$$

This strain vector lies between the directions of the normals to the two adjacent surfaces. Similar relations can be obtained for the other five edges.

The plastic volume change is obtained from Eq. (4.38) as

$$d\epsilon_v^p = m(d\lambda_1 + d\lambda_2) - (d\lambda_1 + d\lambda_2)$$

which is the sum of two parts: the compressive part

$$\sum |d\epsilon_c^p| = d\lambda_1 + d\lambda_2$$

and the tensile part

$$\sum |d\epsilon_t^p| = m(d\lambda_1 + d\lambda_2)$$

and we can see that

$$d\epsilon_v^p = \sum d\epsilon_t^p - \sum |d\epsilon_c^p| \qquad (4.39)$$

It can be seen that $d\epsilon_v^P > 0$ for $m > 1$, and that Eqs. (4.33) and (4.34) are still valid. By a similar derivation to that of Eq. (4.35), we can obtain the plastic work increment expression $dW_p$ in the following form:

$$dW_p = (\sigma_1 m - \sigma_3) \, d\lambda_1 + (\sigma_1 m - \sigma_2) \, d\lambda_2$$

$$= f_c'(d\lambda_1 + d\lambda_2) = f_c' \sum |d\epsilon_c^P| \qquad (4.40)$$

*Case 3.* The yield stress point coincides with the apex of the pyramid, where six surfaces intersect. Following the same procedure, a similar expression to Eq. (4.38) for the plastic strain $d\epsilon_i^P$ can be obtained. We can also show that Eqs. (4.34) and (4.36) are still valid. Derivations of this will be left to the reader as an exercise.

## 4.6. Convexity, Normality, and Uniqueness for Elastic-Perfectly Plastic Materials

The associated flow rule or *normality rule* discussed before has been established firmly in the mathematical theory of *metal plasticity*. It will be shown in what follows that since the condition of irreversibility of plastic deformation implies that work expended on plastic deformation in a cycle is positive, the positive plastic work leads to convexity of the yield surface and normality of the plastic flow, and that the normality condition, or the associated flow rule, guarantees the uniqueness of the solution of an elastic-plastic boundary-value problem. The normality of the plastic flow and the convexity of the yield surface are of very general nature for elastic-perfectly plastic materials as well as for materials that work harden.

### 4.6.1. Convexity of the Yield Surface and Normality of the Plastic Flow

Because of the irreversible character of plastic deformation, work expended on plastic deformation cannot be reclaimed. This means that the work of the stresses on the change of plastic strain is positive whenever a change of plastic strain occurs. In this section, we shall investigate what restrictions this *irreversibility condition* imposes on the plastic stress-strain relationship.

Consider a unit volume of material in which there is a homogeneous state of stress $\sigma_{ij}^*$ on or inside the yield surface (Fig. 4.6a). Suppose an external agency adds stresses along a path $ABC$ lying inside the surface until $\sigma_{ij}$ on the yield surface is just reached. Only elastic work has taken place so far. Now suppose that the external agency keeps the stress state $\sigma_{ij}$ on the yield surface for a short time. Plastic flow must occur, and only plastic work takes place during the flow. The external agency then releases $\sigma_{ij}$ and returns the state of stress to $\sigma_{ij}^*$ along an elastic path $DE$. As all purely elastic changes are completely reversible and independent of the

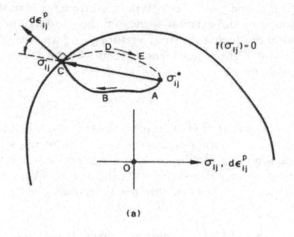

(a)

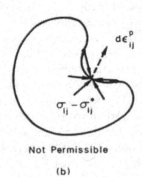

Not Permissible

(b)

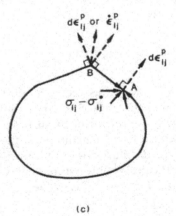

(c)

FIGURE 4.6. Convexity of the yield surface and normality of the plastic flow.

path from $\sigma_{ij}^*$ to $\sigma_{ij}$ and back to $\sigma_{ij}^*$, all the elastic energy is recovered. The plastic work done by the external agency on this loading and unloading cycle is the scalar product of the stress vector $\sigma_{ij} - \sigma_{ij}^*$ and the plastic strain increment vector $d\epsilon_{ij}^p$. The requirement that this work be positive for plastic deformation leads to

$$(\sigma_{ij} - \sigma_{ij}^*) \, d\epsilon_{ij}^p \geq 0 \tag{4.41}$$

The geometric interpretation of expression (4.41) is given below. If plastic strain coordinates are superimposed upon stress coordinates, as in Fig. 4.6, the positive scalar product requires an acute angle between the stress vector $\sigma_{ij} - \sigma_{ij}^*$ and the strain increment vector $d\epsilon_{ij}^p$. Since all possible stress vectors, $\sigma_{ij} - \sigma_{ij}^*$, must satisfy Eq. (4.41), this leads inevitably to the following consequences:

(i) Convexity: The yield surface must be convex. If not convex as shown in Fig. 4.6b, the possible directions of $d\sigma_{ij}$ cover more than 180° for some planes through $d\epsilon_{ij}^p$. Thus, the angle between $\sigma_{ij} - \sigma_{ij}^*$ and $d\epsilon_{ij}^p$ may be greater than 90°. However, Eq. (4.41) requires the angle between them to be less than 90°. Hence, the surface must be convex.

(ii) Normality: The plastic strain increment vector $d\epsilon_{ij}^p$ must be normal to the yield surface at a smooth point and lie between adjacent normals at a corner. As shown in Fig. 4.6c, if the surface is convex and smooth at point $A$, $d\epsilon_{ij}^p$ must be normal to the surface so that it makes a right angle or less with all possible $\sigma_{ij} - \sigma_{ij}^*$, and condition (4.41) is satisfied. If the surface has a corner at point $B$, there is some freedom in the direction of $d\epsilon_{ij}^p$ but the vector must lie between the normals at an adjacent point to the corner so that Eq. (4.41) is satisfied.

The irreversible character of plastic deformation requires the increment of plastic work to be positive

$$dW_p = \sigma_{ij} \, d\epsilon_{ij}^p = d\lambda \, \sigma_{ij} \frac{\partial f}{\partial \sigma_{ij}} \geq 0 \tag{4.42}$$

Since the scalar product of the radius vector $\sigma_{ij}$ on the yield surface and the exterior normal of the yield surface $\partial f / \partial \sigma_{ij}$ is non-negative (Fig. 4.2), they must make an acute angle for a convex surface. The multiplier $d\lambda$ in Eq. (4.6) is seen to be related to the magnitude of the increment of plastic work $dW_p$, and this factor $d\lambda$ must always be positive when plastic flow occurs in order to assure the irreversible nature of plastic deformation. Note that the yield function is $f = F - k = 0$; thus, $\partial f / \partial \sigma_{ij} = \partial F / \partial \sigma_{ij}$, and Eq. (4.42) can be reduced to

$$dW_p = d\lambda \, \sigma_{ij} \frac{\partial F}{\partial \sigma_{ij}} = d\lambda \, nF \tag{4.43}$$

when $F$ is a homogeneous function of degree $n$ in the stresses, as it is for most theories in metal plasticity.

## 4.6.2. Uniqueness of Solution and Normality Condition of Flow

Uniqueness of solution of a boundary-value problem for an elastic material was discussed in Section 3.6.4. In this section, we shall see that the uniqueness requirement is also satisfied for an elastic–perfectly plastic material if the normality condition is imposed on the stress–strain relation.

Let us assume that our boundary-value problem admits two solutions: $d\sigma_{ij}^{(a)}$, $d\epsilon_{ij}^{(a)}$ and $d\sigma_{ij}^{(b)}$, $d\epsilon_{ij}^{(b)}$, both corresponding to $dT_i$ on $A_T$, $du_i$ on $A_u$, and $dF_i$ in $V$. The equation of virtual work then is employed, assuming continuous $u_i$ throughout $V$,

$$\int_{A_T} dT_i^* \, du_i \, dA + \int_{A_u} dT_i^* \, du_i \, dA + \int_V dF_i^* \, du_i \, dV = \int_V d\sigma_{ij}^* \, d\epsilon_{ij} \, dV$$

$$(4.44)$$

where the starred quantities are related through equilibrium and the unstarred ones are compatible. There need be no relation between the two sets of increments. Therefore, the difference between the two assumed states $a$ and $b$ can be substituted into Eq. (4.44) although $d\sigma_{ij}^{(b)} - d\sigma_{ij}^{(a)}$ need not and often does not produce $d\epsilon_{ij}^{(b)} - d\epsilon_{ij}^{(a)}$. Substitution gives

$$0 = \int_V (d\sigma_{ij}^{(b)} - d\sigma_{ij}^{(a)})(d\epsilon_{ij}^{(b)} - d\epsilon_{ij}^{(a)}) \, dV \qquad (4.45)$$

because $dT_i^{(a)} = dT_i^{(b)}$ on $A_T$, $du_i^a = du_i^b$ on $A_u$, and $dF_i^{(a)} = dF_i^{(b)}$ in $V$.

Using the geometrical representation of the preceding section, we represent the difference of the two stress increments at a given point of the body in Eq. (4.45) by $\Delta d\sigma_{ij} = d\sigma_{ij}^{(b)} - d\sigma_{ij}^{(a)}$, the difference of the increments of elastic strain by $\Delta d\epsilon_{ij}^e$, and the difference of the increments of plastic strain by $\Delta d\epsilon_{ij}^p$. Now the integrand of the scalar product in Eq. (4.45) must vanish, i.e.,

$$dI = \Delta d\sigma_{ij} \Delta d\epsilon_{ij} = \Delta d\sigma_{ij} (\Delta d\epsilon_{ij}^e + \Delta d\epsilon_{ij}^p) = 0 \qquad (4.46)$$

Applying a stress–strain relation to Eq. (4.46), $dI$ can be expressed in a quadratic form. If we can show that $dI$ is positive definite, Eq. (4.46) would lead to $\Delta d\epsilon_{ij} = 0$ and $\Delta d\sigma_{ij} = 0$, the uniqueness is satisfied. In other words, any incremental stress–strain relation which assures that the integrand $dI$ is positive definite will therefore satisfy the condition of uniqueness.

Now $\Delta d\epsilon_{ij}^e$ is related to $\Delta d\sigma_{ij}$ by the generalized Hooke's law, and the scalar product $\Delta d\sigma_{ij} \Delta d\epsilon_{ij}^e$ is positive definite. For the scalar product $\Delta d\sigma_{ij} \Delta \epsilon_{ij}^p$, three cases must be discussed separately:

Case 1. Both solutions constitute loading at the point under consideration.

In this case, $\Delta d\sigma_{ij}$ must lie in the tangent plane to the perfectly plastic yield surface (Fig. 4.1b). It is easily seen that if the plastic strain vector $d\epsilon_{ij}^p$ is normal to the yield surface, the product $\Delta d\sigma_{ij}\,\Delta d\epsilon_{ij}^p$ will be non-negative for all vectors $\Delta d\sigma_{ij}$ which are tangent to this surface.

Case 2. Both solutions constitute unloading. In this case, $\Delta d\epsilon_{ij}^p = 0$, so that $dI$ is positive definite because $\Delta d\sigma_{ij}\,\Delta d\epsilon_{ij}^e$ is.

Case 3. One solution constitutes loading, the other unloading. If we take $d\sigma_{ij}^{(b)}$ as loading with $d\epsilon_{ij}^{p(b)}$ and $d\sigma_{ij}^{(a)}$ as unloading with $d\epsilon_{ij}^{p(a)} = 0$, the product $\Delta d\sigma_{ij}\,\Delta d\epsilon_{ij}^p$ has the form

$$(d\sigma_{ij}^{(b)} - d\sigma_{ij}^{(a)})\,d\epsilon_{ij}^{p(b)} = d\sigma_{ij}^{(b)}\,d\epsilon_{ij}^{p(b)} - d\sigma_{ij}^{(a)}\,d\epsilon_{ij}^{p(b)} \tag{4.47}$$

Since $d\sigma_{ij}^{(b)}$ constitutes loading, the stress increment vector $d\sigma_{ij}^{(b)}$ must lie in the tangent plane. If the plastic strain vector $d\epsilon_{ij}^{p(b)}$ is in the direction along the exterior normal of the yield surface (Fig. 4.2), the product $d\sigma_{ij}^{(b)}\,d\epsilon_{ij}^{p(b)}$, the first term on the right-hand side of Eq. (4.47), is zero because $d\sigma_{ij}^{(b)}$ is orthogonal to $d\epsilon_{ij}^{p(b)}$. The other stress increment vector $d\sigma_{ij}^{(a)}$ must point toward the interior of the yield surface because it constitutes unloading (Fig. 4.1b). If the plastic strain increment vector $d\epsilon_{ij}^{p(b)}$ is normal to the *convex* yield surface $f$, the stress increment vector $d\sigma_{ij}^{(a)}$ will always make an obtuse angle with $d\epsilon_{ij}^{p(b)}$. Thus, the second term on the right-hand side of Eq. (4.47) is made non-negative. In the present case, the order in which the two solutions are taken does not affect the sign of the product $\Delta d\sigma_{ij}\,\Delta d\epsilon_{ij}^p$ because both $\Delta d\sigma_{ij}$ and $\Delta d\epsilon_{ij}^p$ change sign when this order is reversed. We can therefore conclude that the associated flow rule satisfies the condition of uniqueness.

It should be noted here that although the plastic term in Eq. (4.46), $\Delta d\sigma_{ij}\,\Delta d\epsilon_{ij}^p$, may be zero, the elastic term, $\Delta d\sigma_{ij}\,\Delta d\epsilon_{ij}^e$, is always positive definite unless $\Delta d\sigma_{ij} = 0$. Uniqueness, in this sense, is established for the elastic–plastic case but not for the rigid–plastic case where the elastic term is identically zero at all times.

We are now in a position to state that the simple relation $g = f$ has a special significance in the mathematical theory of plasticity. Two immediate consequences of this are now evident. (1) The plastic strain increment vector $d\epsilon_{ij}^p$ must be normal to the yield or loading surface $f(\sigma_{ij}) = 0$. This is now known as the *normality condition*. (2) This type of plastic stress–strain relations leads to the uniqueness of the solution of a boundary-value problem. As will be seen later, the normality relation (4.6) also leads rather directly to the establishment of the powerful theorems of limit analysis of perfect plasticity (Chapter 8).

This type of normality condition is of a very general nature. In Chapter 5, it is shown that this relation is also valid for materials which work harden. The normality condition imposed on the plastic stress–strain law has strong implications with respect to uniqueness of solution for work hardening and perfectly plastic bodies. It also leads to the formulations of the variational and absolute-minimum principles as well.

## 4.7. A Simple Elastic–Plastic Problem: The Expansion of a Thick-Walled Cylinder

In this section, we shall discuss in some detail the behavior of a simple structure made of elastic–perfectly plastic material. This discussion will help us understand some basic features and useful concepts of elastic–plastic deformation of a structure. The example selected for analysis is the thick-walled tube, with closed ends, under internal pressure. The tube has inner radius $a$ and outer radius $b$ (Fig. 4.7). We shall assume that the tube is sufficiently long for end effects not to be felt in the zone which we study.

For this problem, it is best to work in cylindrical coordinates $(r, \theta, z)$; $r$ is the radial distance measured perpendicularly from the axis of the tube, $\theta$ is an angular circumferential coordinate measured from an arbitrary datum, and $z$ is the axial distance from an arbitrary datum plane parallel to the axis.

### 4.7.1. Basic Equations

The only nontrivial *equilibrium* equation is the radial one

$$\frac{d\sigma_r}{dr} - \frac{\sigma_\theta - \sigma_r}{r} = 0 \tag{4.48}$$

The *compatibility* equations express the geometrical relationships between strain and displacement. The displacement is still assumed to be small, and if $u$ is a radial displacement of a point originally at radius $r$,

$$\epsilon_r = \frac{du}{dr} \tag{4.49}$$

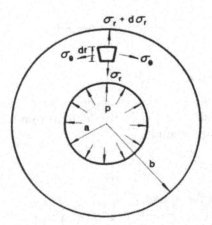

FIGURE 4.7. Transverse section of a thick-walled tube subject to interior pressure.

and, assuming symmetrical deformation,

$$\epsilon_\theta = \frac{u}{r}$$ (4.50)

In the axial direction, we can at present only state the "long tube" condition for the extension of the tube without bending:

$$\epsilon_z = \text{constant} = C$$ (4.51)

These relations are purely geometric, and thus hold irrespective of whether the strain is elastic or plastic.

The material of the tube is assumed to be elastic–perfectly plastic. In the elastic range, the behavior is described in terms of two elastic constants, Young's modulus $E$ and Poisson's ratio $\nu$. Because $r$, $\theta$, and $z$ are, by symmetry, the principal stress directions, we may write the elastic constitutive relations:

$$E\epsilon_r = \sigma_r - \nu(\sigma_\theta + \sigma_z)$$
$$E\epsilon_\theta = \sigma_\theta - \nu(\sigma_r + \sigma_z)$$ (4.52)
$$E\epsilon_z = \sigma_z - \nu(\sigma_r + \sigma_\theta)$$

The yield condition is that of Tresca, and the flow rule is associated with it by means of the normality condition.

The *boundary conditions* are especially simple:

$$\sigma_r = 0 \qquad \text{at } r = b$$ (4.53)

$$\sigma_r = -p \qquad \text{at } r = a$$ (4.54)

where $p$ is the interior gauge pressure. Lastly, in the axial direction, overall equilibrium requires

$$p\pi a^2 = \int_a^b 2\pi\sigma_z r\, dr$$ (4.55)

## 4.7.2. Elastic Solution

Elastic analysis of this problem is straightforward. First use Eq. (4.51) to eliminate $\sigma_z$ from Eq. (4.52). Then eliminate $u$ from Eqs. (4.49) and (4.50) to give a *compatibility* relation

$$\epsilon_r = \frac{d}{dr}(r\epsilon_\theta)$$ (4.56)

Into this, substitute for $\epsilon_r$ and $\epsilon_\theta$ in terms of $\sigma_\theta$, $\sigma_r$, and $C$ [Eq. (4.51)], using the relations just derived. This gives a first-order linear differential equation in $\sigma_\theta$, $\sigma_r$, $(d\sigma_r/dr)$ and $(d\sigma_\theta/dr)$, but not in fact involving $C$. Eliminate $\sigma_\theta$ and $d\sigma_\theta/dr$ using this equation and Eq. (4.48) to give a second-order differential equation in $\sigma_r$. Solve this subject to Eqs. (4.53) and (4.54) to give:

$$\sigma_r = p\left(-\frac{b^2}{r^2}+1\right)\bigg/\left(\frac{b^2}{a^2}-1\right) = \frac{pa^2(r^2-b^2)}{r^2(b^2-a^2)} \tag{4.57}$$

Substitution into Eq. (4.48) gives

$$\sigma_\theta = p\left(\frac{b^2}{r^2}+1\right)\bigg/\left(\frac{b^2}{a^2}-1\right) = \frac{pa^2(r^2+b^2)}{r^2(b^2-a^2)} \tag{4.58}$$

To find the stress $\sigma_z$, we use these results in the third relation of Eqs. (4.52) and note Eq. (4.51). This yields

$$\sigma_z = \nu(\sigma_r+\sigma_\theta) + EC = 2\nu\frac{pa^2}{b^2-a^2} + EC \tag{4.59}$$

Substitution of $\sigma_z$ in Eq. (4.55) yields $\epsilon_z = C = (1-2\nu)pa^2/[E(b^2-a^2)]$. If we assume plane strain, i.e., $\epsilon_z = 0$, then we have

$$\nu = 0.5 \quad \text{and} \quad \sigma_z = \tfrac{1}{2}(\sigma_r+\sigma_\theta) \tag{4.60}$$

Equation (4.60) implies that for the problem to satisfy both $\epsilon_z = 0$ and Eq. (4.55), $\nu$ must take the special value 0.5.

The radial displacement, $u$, is obtained from Eqs. (4.50) and the second relation of (4.52):

$$u = r\epsilon_\theta = \frac{(1+\nu)a^2p}{E(b^2-a^2)}\left[\frac{(1-2\nu)r}{(1+\nu)}+\frac{b^2}{r}\right] \tag{4.61}$$

This elastic stress distribution only applies, of course, if $p$ is sufficiently small for the stress point $(\sigma_r, \sigma_\theta, \sigma_z)$ at all radii within the wall of the tube to lie within the yield locus.

Note that, from Eq. (4.60), $\sigma_z$ always takes a value such that it is the intermediate principal stress, i.e.,

$$\sigma_\theta > \sigma_z > \sigma_r$$

Hence, the yield condition of Tresca is

$$\sigma_\theta - \sigma_r = \sigma_0 \tag{4.62}$$

where $\sigma_0$ is the yield stress in simple tension. Substitution of Eqs. (4.57) and (4.58) in Eq. (4.62) gives

$$\sigma_\theta - \sigma_r = 2p\left(\frac{b^2}{r^2}\right)\bigg/\left(\frac{b^2}{a^2}-1\right) = \sigma_0 \tag{4.63}$$

It is clear from Eq. (4.63) that if the pressure is increased steadily, the yield stress is first reached at the inner surface, $r = a$. Thus, using Eq. (4.63) with $r = a$, we find that the pressure at which the yield point is first reached is given by

$$p = p_e = \frac{\sigma_0}{2}\left(1-\frac{a^2}{b^2}\right) \tag{4.64}$$

Notice that the pressure for first yield at $r = a$ is a function of the ratio $b/a$ and not of the absolute size of the tube.

### 4.7.3. Elastic–Plastic Expansion

If the pressure is increased above the value for first yield, an enlarging plastic zone spreads outwards from the inner surface.

To analyze this partly elastic, partly plastic state of affairs, suppose that at some stage in the expansion of the tube, the elastic–plastic boundary is at radius $c$, where $a \le c \le b$, as shown in Fig. 4.8. At $r = c$, let $\sigma_r = -q$; i.e., call the radial pressure $q$ at this radius. The outer elastic zone cannot differentiate, so to speak, between pressure $q$ exerted by the plastic zone or $q$ provided by a fluid. It follows therefore that because the outer surface is not loaded, the equations we have already derived apply in the elastic region, provided the symbol $a$ is replaced throughout by $c$. In particular, because the stress must be at the yield point at $r = c$, Eq. (4.64) gives

$$q = \frac{\sigma_0}{2}\left(1 - \frac{c^2}{b^2}\right) \tag{4.65}$$

Turning now to the plastic zone, we find that the key to the situation is the *yield condition* (4.62). Substituting (4.62) into the equilibrium equation (4.48), we can integrate directly to obtain

$$\sigma_r = \sigma_0 \ln r + \text{constant} \tag{4.66}$$

The constant is determined by the boundary condition $\sigma_r = -q$ at $r = c$; using this, we find

$$\sigma_r = -q + \sigma_0 \ln \frac{r}{c}$$

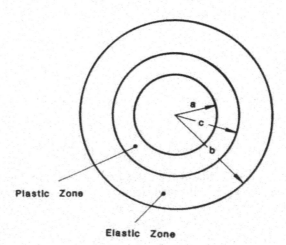

Plastic  Zone

Elastic  Zone

FIGURE 4.8. Plastic zone contained within an elastic zone.

Substituting (4.65) for $q$ and using the yield condition (4.62) gives the stresses in the yield zone as

$$\sigma_r = \sigma_0 \left[ \ln\frac{r}{c} - \frac{1}{2}\left(1 - \frac{c^2}{b^2}\right)\right]$$

$$\sigma_\theta = \sigma_0 \left[ \ln\frac{r}{c} + \frac{1}{2}\left(1 + \frac{c^2}{b^2}\right)\right]$$

(4.67)

We can now use the boundary condition $\sigma_r = -p$ at $r = a$ to obtain

$$p = q + \sigma_0 \ln\left(\frac{c}{a}\right)$$

$$= \frac{\sigma_0}{2}\left(1 - \frac{c^2}{b^2}\right) + \sigma_0 \ln\frac{c}{a}$$

(4.68)

Hence, for any value of $c$ between $a$ and $b$, the corresponding pressure may be calculated. Also for any value of $c$, $\sigma_\theta$ and $\sigma_r$ are determined throughout the tube. Figure 4.9 shows the results for a tube with $b/a = 2$ for various values of $c/a$. It is interesting to note that in the plastic zone the stresses are *statically determinate*, and, given the pressure at one boundary, the pressure at the other boundary is determined. Thus, the equations in the plastic zone, besides being simpler than those in the elastic zone, are of a different *kind*. The fact that the equilibrium equation and the yield condition can be solved directly without reference to deformation—i.e., that the situation is statically determinate—is a consequence of the *uncoupling* of stress and strain which follows from the special nonhardening form of our idealized plastic material.

### 4.7.4. Elastic–Plastic Deformation

It is noted that the radial expansion of the plastic zone is controlled by the elastic deformation of the elastic zone which entirely surrounds it. The elastic zone can be regarded as sustaining a pressure $q$ exactly as if the inner portion of the tube were filled with fluid. It follows that the pattern of strain within the tube in the elastic–plastic condition is a very simple one—there is no axial elongation, and since the material is incompressible in both the elastic and plastic ranges, the deformation may readily be expressed in terms of a single parameter. A convenient index of the deformation is the radial enlargement of the tube, $u_b$, at $r = b$.

At $r = b$, using Eq. (4.61) with $c$ substituted for $a$ and $q$ for $p$, and substituting for $q$ from Eq. (4.65), noting $\nu = \frac{1}{2}$, we have

$$\frac{u_b}{b} = \frac{3}{4}\frac{\sigma_0}{E}\left(\frac{c}{b}\right)^2$$

(4.69)

Using this in Eq. (4.68) and rearranging, we find

$$\frac{2p}{\sigma_0} = 1 - \frac{4}{3}\frac{E}{\sigma_0}\frac{u_b}{b} + \ln\left(\frac{4}{3}\frac{E}{\sigma_0}\frac{u_b}{b}\right) + 2\ln\frac{b}{a}$$

(4.70)

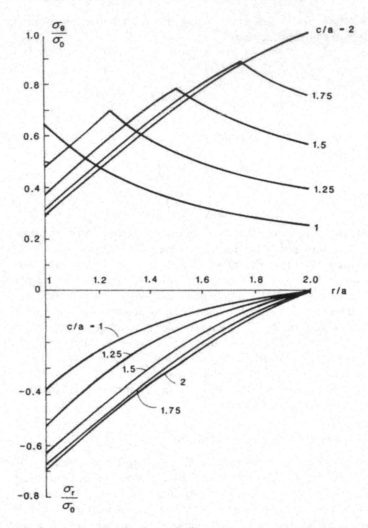

FIGURE 4.9. Successive distributions of circumferential and radial stress in the elastic–plastic expansion of a tube: $b/a = 2$.

This relationship between pressure and radial enlargement applies provided $a \le c \le b$, from which, using Eq. (4.69), we obtain

$$\frac{a^2}{b^2} \le \frac{4}{3} \frac{E}{\sigma_0} \frac{u_b}{b} \le 1 \qquad (4.71)$$

When the behavior is entirely elastic, the corresponding equation is

$$\frac{2p}{\sigma_0} = \left(\frac{b^2}{a^2} - 1\right)\left(\frac{4}{3} \frac{E}{\sigma_0} \frac{u_b}{b}\right) \qquad (4.72)$$

When the elastic–plastic boundary reaches the outer surface, $c = b$, and Eq. (4.68) becomes

$$\frac{2p_c}{\sigma_0} = 2 \ln \frac{b}{a} \tag{4.73}$$

The "full plastic" pressure $p_c = \sigma_0 \ln(b/a)$ is maintained if the tube expands further. According to the elastic–perfectly plastic assumption, it is possible for indefinitely large strains to take place in the absence of a surrounding elastic ring.

These results are plotted for $b/a = 2$ as curve $ORST$ in Fig. 4.10.

In conclusion, there are three phases of behavior for an initially stress-free tube with closed ends, made of elastic–perfectly plastic material and subject to a steadily increasing interior pressure:

 (i) An *elastic* phase, in which all the material is in the elastic range.
 (ii) An *elastic–plastic* phase in which an inner plastic zone is contained within an elastic zone. The plastic zone spreads as the pressure increases, but the deflections—which are controlled by the elastic zone—are of the same order as those in the elastic phase.

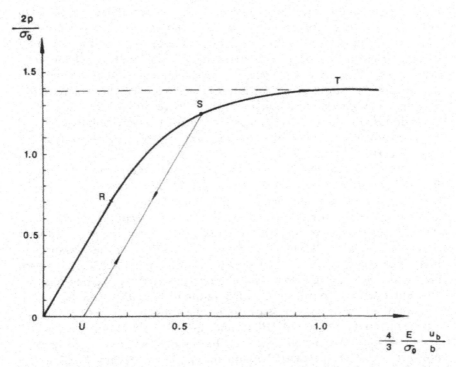

FIGURE 4.10. Elastic–plastic pressure-expansion curve showing unloading behavior.

(iii) A *full-plastic* phase in which, the outer elastic zone having vanished, the tube is free to expand by plastic deformation and achieves much larger deflections than in the elastic range. Apart from second-order effects, plastic expansion takes place at a constant pressure called the *plastic collapse pressure*. At this pressure, we predict, the tube will bulge considerably, and may *burst*.

## 4.7.5. Unloading

Suppose now that the pressure, having been raised into the elastic–plastic range, is steadily reduced until the pressure is again zero. What happens to the stresses in the tube?

For definiteness, we consider a particular case, $b = 2a$, with the pressure (applied to the stress-free tube) having risen to the value corresponding to $c = 1.5a$, i.e., by Eq. (4.68), $p = \sigma_0 [7/32 + \ln(1.5)] = 0.624\sigma_0$. The distributions of the principal stresses under these conditions are shown in Fig. 4.11 (full curves). When the pressure begins to fall, it seems likely that the material which was at the yield stress will have its stress "level" reduced, and will thus immediately reenter the elastic range. Because we now have some permanent plastic deformation in the contained plastic zone, we must regard the elastic relations (4.52) as referring to *changes* of stress and strain. As all the material is now behaving elastically, we can use results (4.57) to (4.59) to work out the *changes* in $\sigma_r$, $\sigma_\theta$, and $\sigma_z$ for negative pressure increments. For a complete removal of pressure, for example, we must subtract from the elastic–plastic stress distribution in Fig. 4.11 a stress distribution which would have occurred at the same pressure if the material had remained elastic. This is shown in Fig. 4.11 (broken curves). We must, of course, now check that the material is nowhere stressed to yield. This is easily done in the present case because—$\sigma_z$ being the intermediate principal stress—we simply have to verify that $|\sigma_\theta - \sigma_r| < \sigma_0$ everywhere; in Fig. 4.11 this is clearly so.

It is instructive to plot the stress trajectories in the $\pi$-plane, shown in Fig. 4.12. Since $\sigma_z = (\sigma_\theta + \sigma_r)/2$ everywhere (this includes the *assumption* that such is the case in the plastic zone on first loading), all points lie on a line through the origin perpendicular to the projection of the $\sigma_z$-axis. Points A, B, and C correspond to the radii a, b, and c, respectively, when $p = 0.624\sigma_0$, and A', B', and C' to the same radii when the pressure has been released. It is clear that the yield condition is not violated in the unloaded state. Having loaded the tube into the partially plastic range and then unloaded, we are thus left with a *residual stress distribution*.

If we now increase the pressure again, the stress points in Fig. 4.12 will retrace their paths between A', B', C' and A, B, C; yielding will recommence at $p = 0.624\sigma$, and at higher pressures the behavior will be exactly as if the pressure had been increased beyond this point in the first loading. The pressure–radial displacement behavior under this program of loading is

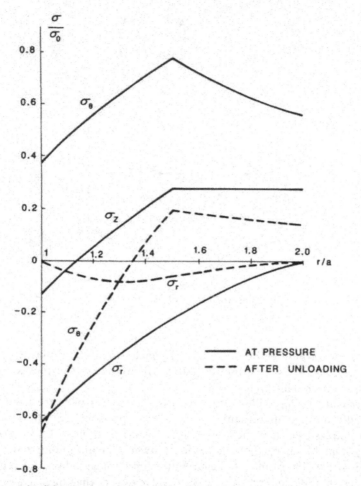

FIGURE 4.11. Distribution of circumferential, radial, and axial stress at a particular stage in the elastic–plastic expansion of a tube and after release of pressure.

shown by curve *ORSU* in Fig. 4.10; it is closely analogous to the load-extension behavior in a tensile test of a hardening material.

### 4.7.6. "Shakedown"

Another important aspect of the phenomenon of readjustment of stress distributions in structures by limited plastic flow of the ductile material is seen in structures which carry repeatedly applied and alternating loads. A possible mode of failure under these circumstances is low-cycle *fatigue* of part of the structure through cyclic plastic deformation. What tends to happen in many structures is that in the course of the first few applications

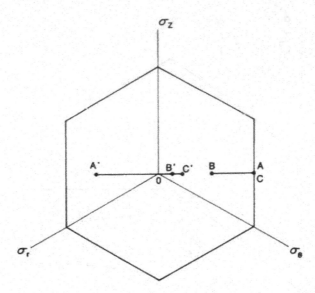

FIGURE 4.12. Stress trajectories for partly plastic tube at pressure and after release of pressure (cf. Fig. 4.11).

of loads, the structure "does its best," by means of limited plastic flow, to set up residual stress distributions that will minimize the plastic fatigue strains in subsequent cycles.

To provide a simple illustration of this, consider a pressure $p = 0.624\sigma_0$ repeatedly applied to our tube ($b/a = 2$). Supposing that the tube was initially stress-free, first yield would be reached at $p = 0.375\sigma_0$ [see Eq. (4.64)]: it might thus be thought—at least, by anyone unfamiliar with plastic analysis—that this would be the pressure limit for avoidance of repeated plasticity in repeated pressure loading. However, the analysis we have already done shows that a single application of $p = 0.624\sigma_0$ induces a residual stress pattern which enables the structure to respond to repeated pressure application up to this level by purely elastic action. We say that the structure will *shake down* to elastic behavior for repeated pressurization between $p = 0$ and $p = 0.624\sigma_0$; we draw an analogy with the behavior of a feather-filled cushion which "shakes down" when repeatedly sat upon.

Figure 4.12 suggests that our particular tube will shake down for even higher pressures: in fact, it may easily be shown that shakedown will occur for all pressures up to the plastic collapse pressure. This result however is, in a sense, a special one for sufficiently small values of $b/a$. For values of this ratio greater than about 2.2, shakedown is possible only for pressures lower than the plastic collapse pressure (Problem 4.6).

It must be admitted that the preceding example of shakedown has been extremely simple: a simple structure subject to only one kind of loading whose sign never changed. Clearly, a full discussion of shakedown should involve multiple independent loading systems with the possibility of variation of sign. This is, however, beyond the scope of the present book.

A practical application of shakedown in thick tubes is the "autofrettage" process, which has been used for many years in the manufacture of gun barrels. It is clearly desirable that the inner bore of the barrel should retain its dimensional accuracy on repeated pressurization due to firing. By subjecting the barrel to an *overpressure* before the final surface machining is done, a residual stress system is set up in the barrel which ensures that the bore never goes into the plastic range subsequently, under normal conditions.

## 4.8. Incremental Stress–Strain Relationships

In an elastoplastic analysis by a numerical approach, the most common technique is the incremental method using tangent stiffness. The constitutive relations given in Sections 4.2 to 4.5 cannot be directly applied. An incremental relationship between stress and strain is needed in forming the tangent stiffness matrix. These types of constitutive relations are studied here in this section.

As discussed before, multiaxial perfectly plastic behavior requires that the stress increment vector be tangent to the yield surface and the plastic strain increment vector be normal to the loading surface. According to the concept of perfect plasticity, the magnitude of the plastic strain increment cannot be uniquely determined by the given current stresses $\sigma_{ij}$ and stress increments $d\sigma_{ij}$. However, for given current stresses $\sigma_{ij}$ and a given plastic strain increment $d\epsilon_{ij}^p$ satisfying the flow rule, the stress increment $d\sigma_{ij}$ can be determined by the so-called *consistency condition* which ensures that the stress state remains on the yield surface.

### 4.8.1. Constitutive Relation in General Form

According to Section 4.1, the total strain increment is assumed to be the sum of the elastic strain increment and the plastic strain increment [Eq. (4.4)]:

$$d\epsilon_{ij} = d\epsilon_{ij}^e + d\epsilon_{ij}^p \tag{4.74}$$

The elastic strain increment can be obtained from Hooke's law [see Eqs. (3.89) and (3.96)]:

$$d\epsilon_{ij}^e = D_{ijkl}\, d\sigma_{kl} \tag{4.75a}$$

or

$$d\epsilon_{ij}^e = \frac{dI_1}{9K}\,\delta_{ij} + \frac{ds_{ij}}{2G} \tag{4.75b}$$

and the plastic strain increment is obtained from the flow rule, Eq. (4.6).
Then the complete strain-stress relations for an elastic-perfectly plastic
material are expressed as

$$d\epsilon_{ij} = D_{ijkl}\, d\sigma_{kl} + d\lambda\, \frac{\partial f}{\partial \sigma_{ij}} \tag{4.76a}$$

or

$$d\epsilon_{ij} = \frac{dI_1}{9K}\, \delta_{ij} + \frac{ds_{ij}}{2G} + d\lambda\, \frac{\partial f}{\partial \sigma_{ij}} \tag{4.76b}$$

where $d\lambda$ is an as yet undetermined factor with the value

$$d\lambda \begin{cases} =0 & \text{wherever } f<0 \text{ or } f=0 \text{ but } df<0 \\ >0 & \text{wherever } f=0 \text{ and } df=0 \end{cases} \tag{4.77}$$

We shall determine the form of the factor $d\lambda$ below. This can be accomplished by combining the stress-strain relations (4.76) with the *consistency
condition*

$$df = \frac{\partial f}{\partial \sigma_{ij}}\, d\sigma_{ij} = 0 \tag{4.78}$$

which ensures that the stress state $(\sigma_{ij} + d\sigma_{ij})$ existing after the incremental
change $d\sigma_{ij}$ has taken place still satisfies the yield criterion $f$

$$f(\sigma_{ij} + d\sigma_{ij}) = f(\sigma_{ij}) + df = f(\sigma_{ij}) \tag{4.79}$$

Solving Eq. (4.76) for $d\sigma_{ij}$, or directly using Hooke's law [Eq. (3.72)], the
flow rule [Eq. (4.6)], and Eq. (4.74), we can determine the stress increment
tensor

$$d\sigma_{ij} = C_{ijkl}(d\epsilon_{kl} - d\epsilon_{kl}^p) = C_{ijkl}\, d\epsilon_{kl} - d\lambda\, C_{ijkl} \frac{\partial f}{\partial \sigma_{kl}} \tag{4.80}$$

Substitute Eq. (4.80) into Eq. (4.78) and solve for $d\lambda$:

$$d\lambda = \frac{\dfrac{\partial f}{\partial \sigma_{ij}}\, C_{ijkl}\, d\epsilon_{kl}}{\dfrac{\partial f}{\partial \sigma_{rs}}\, C_{rstu}\, \dfrac{\partial f}{\partial \epsilon_{tu}}} \tag{4.81}$$

All indices in Eq. (4.81) are dummy indices, indicating the scalar character
of $d\lambda$. Therefore, if $f$ is defined for a particular material of interest and
strain increments $d\epsilon_{ij}$ are prescribed, the factor $d\lambda$ is determined uniquely.

Equation (4.81) is now substituted into Eq. (4.80); then the incremental
stress-strain relation can be expressed explicitly in the following form

$$d\sigma_{ij} = \left[ C_{ijkl} - \frac{C_{ijmn}\, \dfrac{\partial f}{\partial \sigma_{mn}}\, \dfrac{\partial f}{\partial \sigma_{pq}}\, C_{pqkl}}{\dfrac{\partial f}{\partial \sigma_{rs}}\, C_{rstu}\, \dfrac{\partial f}{\partial \sigma_{tu}}} \right] d\epsilon_{kl} \tag{4.82a}$$

in which some dummy subscripts have been properly altered. The coefficient tensor in the parentheses represents the elastic–plastic tensor of tangent moduli for an elastic-perfectly plastic material:

$$C_{ijkl}^{ep} = C_{ijkl} - \frac{C_{ijmn} \dfrac{\partial f}{\partial \sigma_{mn}} \dfrac{\partial f}{\partial \sigma_{pq}} C_{pqkl}}{\dfrac{\partial f}{\partial \sigma_{rs}} C_{rstu} \dfrac{\partial f}{\partial \sigma_{tu}}} \tag{4.82b}$$

Equation (4.82) is the most general formulation of the constitutive relation for an elastic-perfectly plastic material. It is seen that the stress increments can be determined uniquely by the yield function $f(\sigma_{ij})$ and the strain increments $d\epsilon_{ij}$. In other words, if the current stress state $\sigma_{ij}$ is known and the increments of strain $d\epsilon_{ij}$ are prescribed, the corresponding stress increments $d\sigma_{ij}$ can be found from Eq. (4.82). In general, however, if the current stress state is known and the stress increments are prescribed, the corresponding strain increments cannot be uniquely determined because the plastic strain increments can be defined only to within the indeterminate factor $d\lambda$ [see Eq. (4.76)].

## 4.8.2. Constitutive Relation in Terms of Elastic Moduli $E$ and $\nu$ or $G$ and $K$

Now we need to express the elastic tensor, $C_{ijkl}$, in the constitutive equation explicitly in terms of the elastic moduli $E$ and $\nu$ or $G$ and $K$. To this end, we substitute Eq. (3.88) for $C_{ijkl}$ into Eq. (4.81) to obtain an expression for the factor $d\lambda$:

$$d\lambda = \frac{\dfrac{\partial f}{\partial \sigma_{ij}} d\epsilon_{ij} + \dfrac{\nu}{1-2\nu} d\epsilon_{kk} \dfrac{\partial f}{\partial \sigma_{ij}} \delta_{ij}}{\dfrac{\partial f}{\partial \sigma_{rs}} \dfrac{\partial f}{\partial \sigma_{rs}} + \dfrac{\nu}{1-2\nu} \left(\dfrac{\partial f}{\partial \sigma_{rs}} \delta_{rs}\right)^2} \tag{4.83a}$$

Note that $\nu = \frac{1}{2}(3K-2G)/(3K+G)$. Thus, the above expression can be rewritten as

$$d\lambda = \frac{\dfrac{\partial f}{\partial \sigma_{ij}} d\epsilon_{ij} + \dfrac{3K-2G}{6G} d\epsilon_{kk} \dfrac{\partial f}{\partial \sigma_{ij}} \delta_{ij}}{\dfrac{\partial f}{\partial \sigma_{rs}} \dfrac{\partial f}{\partial \sigma_{rs}} + \dfrac{3K-2G}{6G} \left(\dfrac{\partial f}{\partial \sigma_{rs}} \delta_{rs}\right)^2} \tag{4.83b}$$

Also, we can substitute Eq. (3.88) for $C_{ijkl}$ into Eq. (4.80) to obtain an expression for $d\sigma_{ij}$ in terms of $E$ and $\nu$ as

$$d\sigma_{ij} = \frac{E}{1+\nu} d\epsilon_{ij} + \frac{\nu E}{(1+\nu)(1-2\nu)} d\epsilon_{kk} \delta_{ij}$$

$$- d\lambda \left[ \frac{E}{1+\nu} \frac{\partial f}{\partial \sigma_{ij}} + \frac{\nu E}{(1+\nu)(1-2\nu)} \frac{\partial f}{\partial \sigma_{mn}} \delta_{mn} \delta_{ij} \right] \tag{4.84a}$$

or in terms of $G$ and $K$ as

$$d\sigma_{ij} = 2G\, de_{ij} + K\, d\epsilon_{kk}\, \delta_{ij} - d\lambda \left[ \left( K - \frac{2}{3}G \right) \frac{\partial f}{\partial \sigma_{mn}} \delta_{mn}\delta_{ij} + 2G \frac{\partial f}{\partial \sigma_{ij}} \right] \quad (4.84b)$$

For a number of materials, the yield function is generally expressed in terms of the stress invariants $I_1$ and $J_2$ in the form

$$f(\sigma_{ij}) = F(I_1, \sqrt{J_2}) - k = 0 \quad (4.85)$$

It follows that

$$\frac{\partial f}{\partial \sigma_{ij}} = \frac{\partial f}{\partial I_1} \frac{\partial I_1}{\partial \sigma_{ij}} + \frac{\partial f}{\partial \sqrt{J_2}} \frac{\partial \sqrt{J_2}}{\partial \sigma_{ij}} \quad (4.86)$$

which reduces to

$$\frac{\partial f}{\partial \sigma_{ij}} = \frac{\partial f}{\partial I_1} \delta_{ij} + \frac{1}{2\sqrt{J_2}} \frac{\partial f}{\partial \sqrt{J_2}} s_{ij} \quad (4.87)$$

With this expression, Eq. (4.84b) becomes

$$d\sigma_{ij} = 2G\, de_{ij} + K\, d\epsilon_{kk}\, \delta_{ij} - d\lambda \left( 3K \frac{\partial f}{\partial I_1} \delta_{ij} + \frac{G}{\sqrt{J_2}} \frac{\partial f}{\partial \sqrt{J_2}} s_{ij} \right) \quad (4.88)$$

where $d\lambda$ has the form

$$d\lambda = \frac{3K\, d\epsilon_{kk}(\partial f/\partial I_1) + (G/\sqrt{J_2})(\partial f/\partial \sqrt{J_2}) s_{mn}\, de_{mn}}{9K(\partial f/\partial I_1)^2 + G(\partial f/\partial \sqrt{J_2})^2} \quad (4.89)$$

In the next two sections, we shall discuss how these equations can be used for specified yield functions.

## 4.9. Prandtl–Reuss Material Model ($J_2$ Theory)

Most of the essential features of the incremental theory of plasticity can be illustrated by the most elementary form, $F = F(J_2)$. The simplest form of $F(J_2)$ is $\sqrt{J_2}$, now known as the von Mises yield criterion. The elastic-perfectly plastic stress–strain relations derived on the basis of the von Mises yield criterion

$$f = \sqrt{J_2} - k = 0 \quad (4.90)$$

and its associated flow rule are now known as the *Prandtl–Reuss material model*. This model is probably the most widely used and perhaps the simplest elastic–perfectly plastic material model.

To find the complete stress–strain relations of Prandtl-Reuss material, we simply substitute Eq. (4.90) for the yield function $f$ into Eq. (4.89) to obtain $d\lambda$ and then substitute $d\lambda$ into Eqs. (4.76b) and (4.88) to obtain

$$d\epsilon_{ij} = \frac{ds_{ij}}{2G} + \frac{dI_1}{9K}\delta_{ij} + \frac{s_{mn}\,de_{mn}}{2k^2} s_{ij} \qquad (4.91)$$

$$d\sigma_{ij} = 2G\,de_{ij} + K\,d\epsilon_{kk}\,\delta_{ij} - \frac{Gs_{mn}\,de_{mn}}{k^2} s_{ij} \qquad (4.92)$$

When the conditions for the occurrence of plastic flow are satisfied,

$$J_2 = k^2 \quad \text{and} \quad df = \frac{\partial f}{\partial \sigma_{ij}}\,d\sigma_{ij} = s_{ij}\,ds_{ij} = 0 \qquad (4.93)$$

The quantity $s_{mn}\,de_{mn}$ in the third term of Eqs. (4.91) and (4.92) is recognized as the rate of work due to distortion. Expanding this quantity in terms of the plastic and elastic strain increments, we obtain

$$s_{mn}\,de_{mn} = s_{mn}(de^e_{mn} + de^p_{mn}) \qquad (4.94)$$

When we note that

$$de^e_{mn} = \frac{ds_{mn}}{2G} \qquad (4.95)$$

and use the fact that

$$dJ_2 = s_{mn}\,ds_{mn} = 0 \qquad (4.96)$$

Eq. (4.94) reduces to

$$s_{mn}\,de_{mn} = s_{mn}\,de^p_{mn} \qquad (4.97)$$

indicating that the rate of distortional work in the plastic range is due solely to plastic deformation. Further, from Eqs. (4.91) and (4.92) we have

$$d\epsilon_{kk} = \frac{dI_1}{3K} = d\epsilon^e_{kk} \qquad (4.98)$$

which implies that

$$d\epsilon^p_{kk} = d\epsilon_{kk} - d\epsilon^e_{kk} = 0 \qquad (4.99)$$

The volume change is purely elastic and no *plastic* volume change can occur for the Prandtl–Reuss material model. The plastic strain rate has only a deviatoric component, which is defined by the flow rule [see Eq. (4.8)]:

$$d\epsilon^p_{ij} = d\lambda\,\frac{\partial f}{\partial \sigma_{ij}} = d\lambda\,\frac{\partial J_2}{\partial \sigma_{ij}} = d\lambda\,s_{ij} \qquad (4.100)$$

The rate of plastic work can be simply derived:

$$dW_p = \sigma_{ij}\,d\epsilon^p_{ij} = d\lambda\,\sigma_{ij}s_{ij} = 2\,d\lambda\,J_2 = 2\,d\lambda\,k^2 \qquad (4.101)$$

From this, we determine the factor $d\lambda$

$$d\lambda = \frac{dW_p}{2k^2} = \frac{s_{mn}\,de^p_{mn}}{2k^2} = \frac{s_{mn}\,de_{mn}}{2k^2} \qquad (4.102)$$

When $d\lambda = 0$, Eqs. (4.91) and (4.92) reduce to Hooke's law in differential form. Since the quantity $d\lambda$ is proportional to the increment $s_{mn}\, de_{mn}$, it is evident that the strain increments $de_{ij}$ in Eq. (4.91) are not uniquely determined for a given stress state $\sigma_{ij}$; but if the strain increments $de_{ij}$ and the current stress state $\sigma_{ij}$ are given, the corresponding stress increments $d\sigma_{ij}$ are uniquely determined by Eq. (4.92).

In conclusion, the characteristics of the Prandtl–Reuss material may be summarized as follows:

1. The increments of plastic strain depend on the current values of the deviatoric stress state, not on the stress increment required to reach this state.
2. The principal axes of the stress and the plastic strain increment tensors coincide.
3. No plastic volume change can occur during plastic flow.
4. The ratios of plastic strain increments in the different directions are specified, but the actual magnitudes of the increments are determined by $d\lambda$, which is related to the amount of actual increment in the work of plastic deformation $dW_p$.

EXAMPLE 4.2. Examine the behavior of Prandtl–Reuss material under conditions of uniaxial strain.

SOLUTION. Under uniaxial strain conditions, the strain increments and stresses are given as

$$de_{ij} = [de_1, 0, 0]; \qquad de_{ij} = \tfrac{1}{3} de_1 [2, -1, -1]$$

$$\sigma_{ij} = [\sigma_1, \sigma_2, \sigma_2]; \qquad s_{ij} = [s_1, s_2, s_2] \tag{4.103}$$

and the von Mises yield criterion has the simple form

$$\sqrt{J_2} = \frac{1}{\sqrt{3}} |\sigma_1 - \sigma_2| = k \tag{4.104}$$

In the elastic range, the incremental stress–strain relations are given by

$$d\sigma_1 = \left(K + \frac{4}{3} G\right) de_1 = B\, de_1 = \frac{3K + 4G}{9K} dI_1 \tag{4.105a}$$

and

$$d\sigma_1 - d\sigma_2 = 2G\, de_1 = \frac{2G}{3K} dI_1 \tag{4.105b}$$

By substituting the yield criterion (4.104) into Eqs. (4.105), the value of the vertical stress $\sigma_1$ at yield is obtained:

$$|\sigma_1| = \frac{\sqrt{3}(3K + 4G)}{6G} k = \frac{\sqrt{3} B}{2G} k \tag{4.106}$$

where $B = K + \frac{4}{3}G$ is known as the *constrained modulus*. Thus, when $\sigma_1$ reaches the value given by Eq. (4.106), the material yields, and further increase of vertical stress results in both plastic and elastic strains as the stress state moves along the perfectly plastic yield surface. In the plastic range, the stress–strain relations (4.92) for the shear components have the form

$$ds_1 = 2G\, de_1 - \frac{G(s_1\, de_1 + 2s_2\, de_2)}{k^2}\, s_1 \tag{4.107}$$

When we use the fact that $ds_{ii} = de_{ii} = 0$ and $de_1 = 2\, d\epsilon_1/3$, Eq. (4.107) becomes

$$ds_1 = \frac{4G}{3}\, d\epsilon_1 - \frac{Gs_1^2}{k^2}\, d\epsilon_1 = 0 \tag{4.108}$$

since $k^2 = J_2 = \frac{3}{4}s_1^2$ in the plastic range. Equation (4.108) indicates that $ds_2 = 0$ also because $ds_{ii} = 0$.

Thus, in the uniaxial strain test, the stress changes beyond the initial yield are purely of the hydrostatic pressure type:

$$d\sigma_1 = ds_1 + \tfrac{1}{3}dI_1 = \tfrac{1}{3}dI_1 \tag{4.109a}$$

$$d\sigma_2 = ds_2 + \tfrac{1}{3}dI_1 = \tfrac{1}{3}dI_1 \tag{4.109b}$$

The material behaves as though it were a fluid once it has reached its limiting shear resistance, and the corresponding volume changes are purely elastic

$$\tfrac{1}{3}dI_1 = K\, d\epsilon_1 \tag{4.110}$$

Substitution of Eq. (4.110) into Eq. (4.109a) leads to the vertical stress–strain relation in the plastic range

$$d\sigma_1 = K\, d\epsilon_1 \tag{4.111}$$

Figure 4.13 depicts schematically the behavior of Prandtl–Reuss material in a uniaxial strain test. The slope of the $\sigma_1$ vs. $\epsilon_1$ curve (Fig. 4.13a) breaks, or softens, when yielding occurs and becomes equal to the bulk modulus. Accordingly, the loading slopes of the $(\sigma_1 - \sigma_2)$ vs. $(\epsilon_1 - \epsilon_2)$ curve and the $(\sigma_1 - \sigma_2)$ vs. $I_1/3$ curve become zero (Fig. 4.13b and d). Since $d\epsilon_{kk}^p = 0$, the slope of the $I_1/3$ vs. $\epsilon_{kk}$ curve remains constant (Fig. 4.13c). Once the material unloads, it follows the linear elastic relations (4.105) until it reaches the yield surface again on the opposite side of the yield surface, corresponding to

$$\frac{1}{\sqrt{3}}(\sigma_1 - \sigma_2) = -k \tag{4.112}$$

and then the material flows plastically again, according to Eqs. (4.92). This unloading behavior is also shown in Fig. 4.13.

EXAMPLE 4.3. Examine the behavior of Prandtl–Reuss material under a plane stress condition defined by $\sigma_{ij} = [\sigma_1, 0, \sigma_3]$.

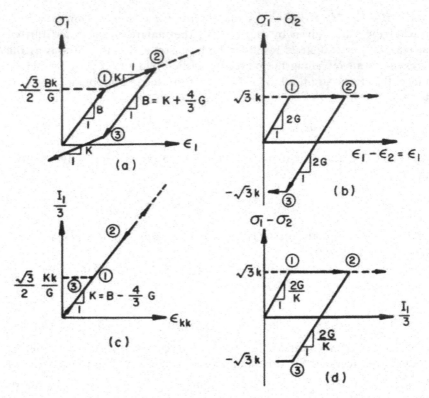

FIGURE 4.13. Behavior of Prandtl-Reuss material under conditions of uniaxial strain. (a) vertical stress-strain relation; (b) principal stress difference-strain difference relation; (c) pressure-volumetric-strain relation; and (d) principal stress difference-pressure relation (stress path).

SOLUTION. For this stress state, the material will yield when

$$J_2 = \tfrac{1}{3}(\sigma_1^2 + \sigma_3^2 - \sigma_1\sigma_3) = k^2 \qquad (4.113)$$

Equation (4.113) describes an ellipse in $(\sigma_1, \sigma_3)$ stress space (Fig. 4.14). We now consider a biaxial tension-tension test where the lateral stress $\sigma_3$ is held constant at the $k$ value while the vertical stress is increased from point $A$ to point $B$. Before reaching the yield point $B$, the behavior of the material is linearly elastic with

$$d\sigma_1 = \frac{9KG}{3K+G}\, d\epsilon_1 = E\, d\epsilon_1$$

$$d\epsilon_3 = -\frac{3K-2G}{6K+2G}\, d\epsilon_1 = -\nu\, d\epsilon_1 \qquad (4.114)$$

$$d\epsilon_2 = d\epsilon_3$$

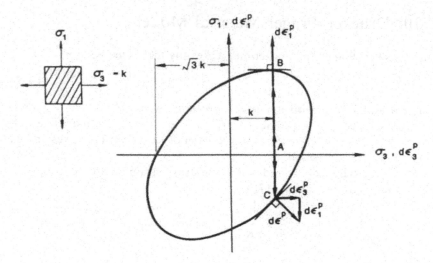

FIGURE 4.14. von Mises yield curve for special plane stress condition; $AB$ and $AC$ are stress paths; $d\epsilon^P$ = plastic strain increment vector.

At point $B$, the material yields, and unlimited plastic deformation takes place at $\sigma_1 = 2k$, $\sigma_3 = k$; the corresponding components of the plastic strain increments are

$$d\epsilon_3^p = d\lambda \frac{\partial J_2}{\partial \sigma_3} = \frac{d\lambda}{3}(2\sigma_3 - \sigma_1) = 0$$

$$d\epsilon_2^p = -d\epsilon_1^p = -k\,d\lambda \quad \text{due to incompressibility}$$

(4.115)

Note that we cannot obtain $d\epsilon_2^p$ by differentiating the yield condition (4.113) because it is a form given in $(\sigma_1, \sigma_3)$ subspace only. However, we can differentiate $J_2$ in the original form $\frac{1}{6}[(\sigma_1 - \sigma_2)^2 + (\sigma_2 - \sigma_3)^2 + (\sigma_3 - \sigma_1)^2]$ to obtain $d\epsilon_2^p = -d\lambda\,(\sigma_1 + \sigma_3)/3 = -k\,d\lambda$. If we repeat the same test and change the direction of $\sigma_1$ (a biaxial tension-compression test), we find that the material yields when $\sigma_1 = -k$ and $\sigma_3 = k$ (point $C$ in Fig. 4.14). At point $C$, the unlimited plastic flow has the value

$$d\epsilon_2^p = 0, \qquad d\epsilon_1^p = -d\epsilon_3^p = -k\,d\lambda \tag{4.116}$$

If the plastic strain increment coordinates are superimposed on the stress coordinates, as shown in Fig. 4.14, the concept of normality or the associated flow rule can be demonstrated clearly from this simple example. In the biaxial tension-tension test, $d\epsilon_3^p = 0$, and the plastic strain increment vector $d\epsilon_1^p$ is perpendicular to the yield surface at point $B$. In the biaxial tension-compression test, on the other hand, $d\epsilon_1^p = -d\epsilon_3^p$, indicating that the plastic strain increment vector is perpendicular to the yield surface at point $C$.

## 4.10. Drucker–Prager Material Model

The Drucker–Prager yield criterion $f$ takes the form (see Section 2.3.4)

$$f = \sqrt{J_2} + \alpha I_1 - k = 0 \tag{4.117}$$

where $\alpha$ and $k$ are positive material constants. As described in Chapter 2, the yield surface, $f = 0$, in principal stress space is a right-circular cone with its axis equally inclined with respect to each of the coordinate axes and its apex in the tension octant.

According to Eqs. (4.76) and (4.89), the stress-strain relation corresponding to the yield function (4.117) is

$$d\epsilon_{ij} = \frac{ds_{ij}}{2G} + \frac{dI_1}{9K}\delta_{ij} + d\lambda\left(\frac{s_{ij}}{2\sqrt{J_2}} + \alpha\delta_{ij}\right) \tag{4.118}$$

where

$$d\lambda = \frac{(G/\sqrt{J_2})s_{mn}\,de_{mn} + 3K\alpha\,de_{kk}}{G + 9K\alpha^2} \tag{4.119}$$

A very important feature of Eq. (4.118) is that the plastic rate of cubic dilatation as given by the third term on the right-hand side of this equation is

$$d\epsilon_{kk}^p = 3\alpha\,d\lambda \tag{4.120}$$

Equation (4.120) shows that plastic deformation must be accompanied by an increase in volume if $\alpha \neq 0$. This property is known as *dilatancy*; it is the consequence of the dependency of the yield function on hydrostatic pressure. For *any* yield surface open in the direction of the negative hydrostatic axis, a plastic volume expansion takes place at yield with an associated flow rule. This is perhaps easier to see from the following geometric arguments.

The meridians of the yield surface are the intersection curves between the yield surface and a plane (the meridian plane) containing the hydrostatic axis; that is, $\theta = \text{const.}$ in a general yield function. Figure 4.15 shows such a typical meridian of a Drucker–Prager yield surface open in the direction of the negative hydrostatic axis. The normality condition or associated flow rule requires that the plastic strain increment vector $d\epsilon_{ij}^p$ be perpendicular to the yield surface at the actual yield point $P$. It is therefore also perpendicular to the meridian through $P$. The vector $d\epsilon_{ij}^p$ is now decomposed into the vertical and horizontal components $d\epsilon_{ij}^{pa}$ and $d\epsilon_{ij}^{pb}$ parallel to the $\rho$ and $\xi$ axes, respectively. The horizontal component $d\epsilon_{ij}^{pb}$ represents the plastic volume change, which is always positive when the yield surface opens in the direction of the negative hydrostatic axis (Fig. 4.15). This implies that plastic flow must always be accompanied by an *increase* in volume.

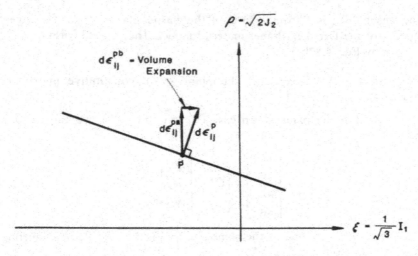

FIGURE 4.15. Plastic volume expansion associated with Drucker-Prager yield surface.

The increment of total volumetric strain $d\epsilon_{kk} = d\epsilon_{kk}^e + d\epsilon_{kk}^p$ can be determined from Eq. (4.118) and $d\lambda$ from (4.119). From Eq. (4.118), we have

$$d\epsilon_{kk} = \frac{dI_1}{3K} + 3\alpha \frac{(G/\sqrt{J_2})[\sigma_{mn}\,d\epsilon_{mn} - I_1(d\epsilon_{kk}/3)] + 3K\alpha\,d\epsilon_{kk}}{G + 9K\alpha^2} \tag{4.121}$$

Solving for $d\epsilon_{kk}$ and using Eq. (4.117), we obtain

$$d\epsilon_{kk} = \frac{\sqrt{J_2}\,dI_1}{3KGk}(G + 9K\alpha^2) + \frac{3\alpha}{k}\sigma_{mn}\,d\epsilon_{mn} \tag{4.122}$$

Substituting the yield function (4.117) into Eqs. (4.88) and (4.89), we obtain the following relationship for the stress increment tensor for the Drucker-Prager material

$$d\sigma_{ij} = 2G\,de_{ij} + K\,d\epsilon_{kk}\,\delta_{ij} - d\lambda\left(\frac{G}{\sqrt{J_2}}s_{ij} + 3K\alpha\delta_{ij}\right) \tag{4.123}$$

where $d\lambda$ is already given by Eq. (4.119). Equation (4.123) can be rewritten in the preferable form

$$d\sigma_{ij} = C_{ijmn}^{ep}\,d\epsilon_{mn} \tag{4.124}$$

for direct use in a finite-element formulation, where

$$C_{ijmn}^{ep} = 2G\delta_{im}\delta_{jn} + \left(K - \frac{2}{3}G\right)\delta_{ij}\delta_{mn}$$

$$- \frac{(G/\sqrt{J_2})s_{ij} + 3K\alpha\delta_{ij}}{G + 9K\alpha^2}\left(\frac{G}{\sqrt{J_2}}s_{mn} + 3K\alpha\delta_{mn}\right) \tag{4.125}$$

The tensor $C_{ijmn}^{ep}$ is a specific form of the elastic–plastic tensor of tangent moduli for the Drucker–Prager material model. The general form of $C_{ijmn}^{ep}$ is given by Eq. (4.82b).

EXAMPLE 4.4. Write explicitly the plane strain constitutive matrix for Drucker–Prager material.

SOLUTION. For the plane strain case ($\gamma_{yz} = \gamma_{xz} = \epsilon_z = 0$), we can write in matrix form

$$
\begin{bmatrix} d\sigma_x \\ d\sigma_y \\ d\tau_{xy} \\ d\sigma_z \end{bmatrix} = [C^{ep}] \begin{bmatrix} d\epsilon_x \\ d\epsilon_y \\ d\gamma_{xy} \end{bmatrix}
\tag{4.126}
$$

where the $z$-axis is normal to the plane and $d\gamma_{xy}$ is the so-called engineering shear strain increment

$$
d\gamma_{xy} = 2 \, d\epsilon_{xy}
\tag{4.127}
$$

and

$$
[C^{ep}] = [C] + [C^p]
$$

in which

$$
[C] = \begin{bmatrix} K + \tfrac{4}{3}G & K - \tfrac{2}{3}G & 0 \\ K - \tfrac{2}{3}G & K + \tfrac{4}{3}G & 0 \\ 0 & 0 & G \\ K - \tfrac{2}{3}G & K - \tfrac{2}{3}G & 0 \end{bmatrix}
\tag{4.128}
$$

$$
[C^p] = \frac{-1}{G + 9K\alpha^2} \begin{bmatrix} H_1^2 & H_1 H_2 & H_1 H_3 \\ H_2 H_1 & H_2^2 & H_2 H_3 \\ H_3 H_1 & H_3 H_2 & H_3^2 \\ H_4 H_1 & H_4 H_2 & H_4 H_3 \end{bmatrix}
\tag{4.129}
$$

and

$$
H_1 = 3K\alpha + \frac{G}{\sqrt{J_2}} s_x, \qquad H_2 = 3K\alpha + \frac{G}{\sqrt{J_2}} s_y
$$

$$
H_3 = \frac{G}{\sqrt{J_2}} \tau_{xy}, \qquad H_4 = 3K\alpha + \frac{G}{\sqrt{J_2}} s_z
$$

EXAMPLE 4.5. Examine the behavior of Drucker–Prager material under a uniaxial state-of-strain test:

$$
d\epsilon_{ij} = [d\epsilon_1, 0, 0]
$$

$$
de_{ij} = \tfrac{1}{3} d\epsilon_1 [2, -1, -1]
\tag{4.130}
$$

$$
\sigma_{ij} = [\sigma_1, \sigma_2, \sigma_2]
$$

SOLUTION. The elastic behavior of the material is governed by Eqs. (4.105), which can be rewritten as

$$\sigma_1 = \frac{3K+4G}{9K} I_1 \tag{4.131a}$$

and

$$\sigma_1 - \sigma_2 = \frac{2G}{3K} I_1 \tag{4.131b}$$

By using Eq. (4.131a), Eq. (4.131b) can be written as

$$\sigma_1 - \sigma_2 = \frac{6G}{3K+4G} \sigma_1 \tag{4.131c}$$

The Drucker–Prager yield condition in the case of uniaxial strain becomes

$$\alpha I_1 + \sqrt{J_2} = \alpha(\sigma_1 + 2\sigma_2) + \frac{1}{\sqrt{3}} |\sigma_1 - \sigma_2| = k \tag{4.132}$$

Substituting Eq. (4.131a) for $I_1$ and Eq. (4.131c) for $\sqrt{J_2}$ or $|\sigma_1 - \sigma_2|$ into Eq. (4.132) leads to

$$|\sigma_1| = \frac{\sqrt{3}(3K+4G)k}{6G \pm 9\sqrt{3}K\alpha} = \frac{\sqrt{3}Bk}{2G \pm 3\sqrt{3}K\alpha} \tag{4.133}$$

in which the upper sign corresponds to the case of $\sigma_1 > 0$ and the lower sign to $\sigma_1 < 0$. When $\alpha$ is equal to zero, Eq. (4.133) reduces to Eq. (4.106), corresponding to Prandtl–Reuss material. The effect of $\alpha$ in this case is to decrease the value of the vertical stress $\sigma_1$ at yield for a uniaxial tension test (upper sign) and to increase the value $\sigma_1$ at yield for a uniaxial compression test (lower sign). Further increase of the vertical stress $\sigma_1$ results in the stress state in the material moving along the yield surface undergoing both elastic and plastic deformation. The incremental relation between vertical stress and vertical strain is obtained for Eq. (4.123) in the elastoplastic range

$$d\sigma_1 = \left(K + \frac{4}{3}G\right) d\epsilon_1 - \frac{(3K\alpha \pm 2G/\sqrt{3})^2}{9K\alpha^2 + G} d\epsilon_1 \tag{4.134}$$

in which the upper sign is for the case $d\sigma_1 > 0$ while the lower sign is for $d\sigma_1 < 0$. Again, it is noted that when $\alpha$ is set equal to zero, Eq. (4.134) reduces to the corresponding equation (4.111) for Prandtl–Reuss material.

From Eq. (4.134), the slope of the $\sigma_1$ vs. $\epsilon_1$ curve during plastic flow can be obtained as

$$\frac{d\sigma_1}{d\epsilon_1} = K \frac{(1 \pm 2\sqrt{3}\alpha)^2}{1 + 9\alpha^2 K/G} \tag{4.135}$$

in which the upper sign is for the case $d\sigma_1 < 0$ while the lower sign is for $d\sigma_1 > 0$.

The stress–strain relation in a uniaxial strain–compression test is shown in Fig. 4.16 for both Prandtl–Reuss and Drucker–Prager material models. For the Prandtl–Reuss model (Fig. 4.16a), the curve is elastic until the yield condition is reached at a stress proportional to $k$ [Eq. (4.106)]. In the plastic region, the slope is simply the bulk modulus $K$. Unloading is elastic until the opposite side of the yield surface is reached and then plastic again with slope $K$. At the completion of the compressive stress cycle, a permanent (compressive) strain remains.

The case of Drucker–Prager model loaded not too far beyond the elastic range is similar (Fig. 4.16b). To see this, let's examine the slope of the $\sigma_1$ vs. $\epsilon_1$ curve in the plastic region. Since in order for the uniaxial strain–stress path to reach the yield surface in the compression test, the following condition must hold [see Eq. (4.133)]:

$$\frac{2G}{\sqrt{3}K} > 3\alpha \qquad (4.136)$$

thus from Eq. (4.135) the slope of the $\sigma_1$ vs. $\epsilon_1$ curve during the plastic flow

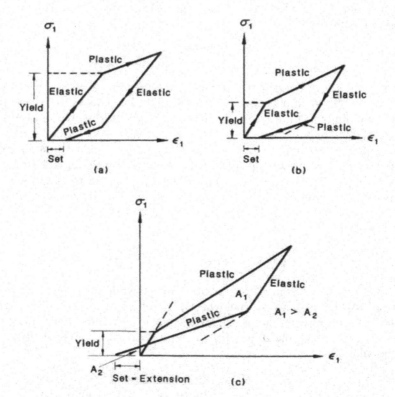

FIGURE 4.16. Uniaxial strain for Prandtl–Reuss and Drucker–Prager models. (a) Prandtl–Reuss, elastoplastic, $k$ large; (b) Drucker–Prager, stress small; (c) Drucker–Prager, stress large.

is larger than $K$ in compression loading (upper sign, for $d\sigma_1 < 0$) and smaller in unloading or reversed tension loading (lower sign, for $d\sigma_1 > 0$). The permanent strain at the end of a load–unload cycle is still compressive if the material is loaded not too far beyond yield and then unloaded, as shown in Fig. 4.16b. However, when this material is loaded well beyond the elastic range (Fig. 4.16c), the permanent set becomes an extension. This can be considered the one-dimensional analogue of the three-dimensional phenomenon of dilatancy.

To examine the volumetric strain increment under compressive loading, letting $i = j$ in Eq. (4.123) and noting that $d\lambda = d\epsilon_1 \, (2G/\sqrt{3} - 3K\alpha)/(G + 9K\alpha^2)$ for the case of $\sigma_1 < 0$, we can obtain the incremental relation between hydrostatic pressure and compressive volumetric strain for uniaxial strain tests as

$$dI_1 = \frac{9K\alpha\{[(2\sqrt{3})/3]G - 3K\alpha\}}{G + 9K\alpha^2} \, d\epsilon_{kk} + 3K \, d\epsilon_{kk} \tag{4.137}$$

When $\alpha$ is set equal to zero, Eq. (4.137) reduces to the corresponding expression for elastic material. The volumetric strain increment $d\epsilon_{kk}$ can be found from Eq. (4.137); then the increment of plastic volumetric strain is obtained by subtracting the elastic part, $d\epsilon_{kk}^e = \frac{1}{3}dI_1/K$, from $d\epsilon_{kk}$:

$$d\epsilon_{kk}^p = \frac{\alpha(2\sqrt{3}G - 9K\alpha)}{3KG(1 + 2\sqrt{3}\alpha)} \, dI_1 \tag{4.138}$$

Noting Eq. (4.136), we see that the increment of plastic volumetric strain is positive (expansion) as expected.

## 4.11. General Isotropic Material

The yield surfaces considered in the previous sections are defined in terms of only the stress invariants $I_1$ and $J_2$, and are independent of the stress invariant $J_3$, or the angle of similarity $\theta$. However, for a general isotropic material, the yield surface is a function of $I_1$, $J_2$, and $J_3$, expressed by

$$f(I_1, J_2, J_3) = 0 \tag{4.139}$$

The gradient $\partial f / \partial \sigma_{ij}$ in this case can be written as

$$\frac{\partial f}{\partial \sigma_{ij}} = \frac{\partial f}{\partial I_1} \frac{\partial I_1}{\partial \sigma_{ij}} + \frac{\partial f}{\partial J_2} \frac{\partial J_2}{\partial \sigma_{ij}} + \frac{\partial f}{\partial J_3} \frac{\partial J_3}{\partial \sigma_{ij}} \tag{4.140a}$$

or

$$\frac{\partial f}{\partial \sigma_{ij}} = B_0 \delta_{ij} + B_1 s_{ij} + B_2 t_{ij} \tag{4.140b}$$

where $B_0$, $B_1$, and $B_2$ denote the derivatives $\partial f/\partial I_1$, $\partial f/\partial J_2$, and $\partial f/\partial J_3$, respectively, and $\delta_{ij}$ is the Kronecker delta, $s_{ij}$ the deviatoric stress tensor, and $t_{ij}$ the deviation of the square of the stress deviation $s_{ij}$:

$$t_{ij} = \frac{\partial J_3}{\partial \sigma_{ij}} = s_{ik}s_{kj} - \frac{2}{3}J_2\delta_{ij} \qquad (4.141)$$

In fact, the most commonly used yield criteria of Tresca and Mohr–Coulomb belong to this type. As an illustration, recall Eq. (2.180), which is an alternative expression of the Mohr–Coulomb criterion:

$$f(\sigma_{ij}) = \frac{1}{3}I_1 \sin\phi + \sqrt{J_2}\sin\left(\theta + \frac{\pi}{3}\right) + \frac{\sqrt{J_2}}{\sqrt{3}}\cos\left(\theta + \frac{\pi}{3}\right)\sin\phi - c\cos\phi = 0 \qquad (4.142)$$

and note that

$$\cos 3\theta = \frac{3\sqrt{3}}{2}\frac{J_3}{J_2^{3/2}} \qquad (4.143)$$

Hence, we have

$$\frac{\partial\theta}{\partial J_2} = \frac{3\sqrt{3}}{4\sin 3\theta}\frac{J_3}{J_2^{5/2}} = \frac{\cot 3\theta}{2J_2}$$

$$\frac{\partial\theta}{\partial J_3} = -\frac{\sqrt{3}}{2\sin 3\theta}\frac{1}{J_2^{3/2}} = -\frac{\cot 3\theta}{3J_3} \qquad (4.144)$$

Taking the derivatives of Eq. (4.142) with respect to $I_1$, $J_2$, and $J_3$, we obtain

$$B_0 = \frac{\partial f}{\partial I_1} = \frac{\sin\phi}{3}$$

$$B_1 = \frac{\partial f}{\partial J_2} = \frac{\sin(\theta + \pi/3)}{2\sqrt{J_2}}\left\{\left[1 + \cot\left(\theta + \frac{\pi}{3}\right)\cot 3\theta\right]\right.$$

$$\left. + \frac{\sin\phi}{\sqrt{3}}\left[\cot\left(\theta + \frac{\pi}{3}\right) - \cot 3\theta\right]\right\} \qquad (4.145)$$

$$B_2 = \frac{\partial f}{\partial J_3} = \frac{\sin\left(\theta + \frac{\pi}{3}\right)\sin\phi - \sqrt{3}\cos\left(\theta + \frac{\pi}{3}\right)}{2J_2\sin 3\theta}$$

It can be seen from Eq. (4.140) that only the constants $B_i$ need be defined by the yield surface. In other words, only these three quantities have to be varied between one yield surface and the other. The constants $B_i$ are given in Table 4.1 for the four yield criteria considered in this section. Other yield functions can be expressed in the same form with equal ease.

TABLE 4.1. Constants $B_i$ defined by different yield surfaces.

| Yield surface | $B_0$ | $B_1$ | $B_2$ |
|---|---|---|---|
| von Mises [Eq. (2.143)] | 0 | 1 | 0 |
| Tresca [Eq. (2.138)] | 0 | $\left[\sin\left(\theta+\frac{\pi}{3}\right)\Big/\sqrt{J_2}\right]\left[1+\cot\left(\theta+\frac{\pi}{3}\right)\cot 3\theta\right]$ | $\left[-\sqrt{3}\cos\left(\theta+\frac{\pi}{3}\right)\Big/J_2\sin 3\theta\right]$ |
| Mohr-Coulomb [Eq. (2.180)] | $\dfrac{\sin\phi}{3}$ | $\left[\sin\left(\theta+\frac{\pi}{3}\right)\Big/2\sqrt{J_2}\right]\left\{\left[1+\cot\left(\theta+\frac{\pi}{3}\right)\cot 3\theta\right] + \sin\phi\left[\cot\left(\theta+\frac{\pi}{3}\right)-\cot 3\theta\right]\Big/\sqrt{3}\right\}$ | $\dfrac{1}{2J_2\sin 3\theta}\left[\sin\left(\theta+\frac{\pi}{3}\right)\sin\phi - \sqrt{3}\cos\left(\theta+\frac{\pi}{3}\right)\right]$ |
| Drucker-Prager [Eq. (2.185)] | $\alpha$ | $\dfrac{1}{2\sqrt{J_2}}$ | 0 |

In finite-element applications, the constitutive relation of a material is reflected by the material stiffness matrix $C_{ijkl}^{ep}$, which is used in forming the tangent stiffness. This stiffness matrix relates the strain increment with the stress increment as given by Eq. (4.82a):

$$d\sigma_{ij} = C_{ijkl}^{ep} \, d\epsilon_{kl} \tag{4.146}$$

To obtain a general form of tensor $C_{ijkl}^{ep}$, we rewrite Eq. (4.82b) as

$$C_{ijkl}^{ep} = C_{ijkl} + C_{ijkl}^{p} \tag{4.147}$$

in which $C_{ijkl}$ is the elastic tensor given by Eq. (3.88) as

$$C_{ijkl} = \frac{E}{2(1+\nu)} \left[ \frac{2\nu}{(1-2\nu)} \delta_{ij}\delta_{kl} + \delta_{ik}\delta_{jl} + \delta_{il}\delta_{jk} \right] \tag{4.148}$$

while $C_{ijkl}^{p}$ is the plastic tensor expressed as

$$C_{ijkl}^{p} = -\frac{H_{ij}H_{kl}}{h} \tag{4.149}$$

where

$$h = \frac{\partial f}{\partial \sigma_{rs}} C_{rstu} \frac{\partial f}{\partial \sigma_{tu}} \tag{4.150}$$

and

$$H_{ij} = C_{ijmn} \frac{\partial f}{\partial \sigma_{mn}} \tag{4.151}$$

Substituting Eqs. (4.148) for $C_{ijkl}$ and (4.140) for $\partial f/\partial \sigma_{ij}$ into Eqs. (4.150) and (4.151), after a lengthy but straightforward derivation [see Problem 4.13], we come up with the following expressions for $h$ and $H_{ij}$ in terms of the elastic constants $G$ and $\nu$ and the coefficients $B_0$, $B_1$, and $B_2$:

$$h = 2G \left( 3B_0^2 \frac{1+\nu}{1-2\nu} + 2B_1^2 J_2 + \frac{2}{3} B_2^2 J_2^2 + 6B_1 B_2 J_3 \right) \tag{4.152}$$

$$H_{ij} = 2G \left( B_0 \frac{1+\nu}{1-2\nu} \delta_{ij} + B_1 s_{ij} + B_2 t_{ij} \right) \tag{4.153}$$

If the stress increment tensor $d\sigma_{ij}$ and the strain increment tensor $d\epsilon_{ij}$ are expressed explicitly in vector forms as

$$\{d\sigma_{ij}\} = \{d\sigma_x, d\sigma_y, d\sigma_z, d\tau_{yz}, d\tau_{zx}, d\tau_{xy}\} \tag{4.154}$$

$$\{d\epsilon_{ij}\} = \{d\epsilon_x, d\epsilon_y, d\epsilon_z, d\gamma_{yz}, d\gamma_{zx}, d\gamma_{xy}\}$$

where $d\gamma_{xy} = 2\, d\epsilon_{xy}$, etc., are the engineering shear strains, the corresponding vector for tensor $H_{ij}$ has the form

$$[H_{ij}] = \{H_x, H_y, H_z, H_{yz}, H_{zx}, H_{xy}\} \tag{4.155}$$

_# 4.11. References   225_

where

$$H_x = 2G\left[ B_0 \frac{1+\nu}{1-2\nu} + B_1 s_x + B_2\left( s_x^2 + s_{xy}^2 + s_{xz}^2 - \frac{2}{3} J_2 \right) \right], \text{ etc.}$$

and

$$H_{yz} = 2G[B_1 s_{yz} + B_2(s_{xy}s_{xz} + s_y s_{yz} + s_{yz}s_z)], \text{ etc.}$$

Thus, the tensor $C_{ijkl}^{ep}$ can be expressed in a matrix form as

$$[C^{ep}] = [C] + [C^p] \tag{4.156}$$

where

$$[C] = \begin{bmatrix} K+\frac{4}{3}G & K-\frac{2}{3}G & K-\frac{2}{3}G & 0 & 0 & 0 \\ K-\frac{2}{3}G & K+\frac{4}{3}G & K-\frac{2}{3}G & 0 & 0 & 0 \\ K-\frac{2}{3}G & K-\frac{2}{3}G & K+\frac{4}{3}G & 0 & 0 & 0 \\ 0 & 0 & 0 & G & 0 & 0 \\ 0 & 0 & 0 & 0 & G & 0 \\ 0 & 0 & 0 & 0 & 0 & G \end{bmatrix} \tag{4.157}$$

and

$$[C^p] = -\frac{1}{h} \begin{bmatrix} H_x^2 & H_x H_y & H_x H_z & H_x H_{yz} & H_x H_{zx} & H_x H_{xy} \\ & H_y^2 & H_y H_z & H_y H_{yz} & H_y H_{zx} & H_y H_{xy} \\ & & H_z^2 & H_z H_{yz} & H_z H_{zx} & H_z H_{xy} \\ & & & H_{yz}^2 & H_{yz} H_{zx} & H_{yz} H_{xy} \\ \text{sym.} & & & & H_{zx}^2 & H_{zx} H_{xy} \\ & & & & & H_{xy}^2 \end{bmatrix} \tag{4.158}$$

# References

_Chen, W.F., 1975. _Limit Analysis and Soil Plasticity_, Elsevier, Amsterdam._

_Chen, W.F., 1982. _Plasticity in Reinforced Concrete_, McGraw-Hill, New York._

_Chen, W.F., and G.Y. Baladi, 1985. _Soil Plasticity: Theory and Implementation_, Elsevier, Amsterdam._

_Drucker, D.C., 1956. "On Uniqueness in the Theory of Plasticity," _Quarterly of Applied Mathematics._ **14**: 35–42._

_Hill, R., 1950. _The Mathematical Theory of Plasticity_, Oxford University Press, London._

_Hoffman, O., and G. Sachs, 1953. _Introduction to the Theory of Plasticity_, McGraw-Hill, New York._

_Nielsen, M.P., 1984. _Limit Analysis and Concrete Plasticity_, Prentice-Hall, Englewood Cliffs, NJ._

_Owen, D.R.J., and E. Hinton, 1980. _Finite Elements in Plasticity: Theory and Practice_, Pineridge Press Limited, Swansea, U.K._

Prager, W., 1949. "Recent Developments in the Mathematical Theory of Plasticity," *Journal of Applied Physics.* **20**: 235-241.

## PROBLEMS

4.1. A long circular thin-walled pressure vessel is subjected to an interior pressure $p$ and yielded. Find the ratio of the plastic strain increments in three principal directions according to the Prandtl-Reuss equation.

4.2. A thin-walled tube is subjected to a constant axial tension and a variable torsion. The axial stress is $\sigma_z = \frac{1}{2}\sigma_0$. According to the von Mises criterion, find the magnitude of the shear stress $\tau$ such that the tube begins to yield. Also find the ratio of the plastic strain increments $d\epsilon_{ij}^p$ when the tube is yielded.

4.3. A material element is subjected to three proportional loadings. The ratios of the principal stresses for the three loading cases are given as (1) $(2\sigma, \sigma, 0)$; (2) $(\sigma, \sigma, 0)$; (3) $(0, -\sigma, -\sigma)$. According to

(a) the von Mises yield criterion: $J_2 = k^2$;

(b) the Tresca yield criterion: $\tau_{max} = \dfrac{\sigma_{max} - \sigma_{min}}{2} = k$;

(c) the Drucker-Prager criterion: $\alpha I_1 + \sqrt{J_2} = k$;

(d) the Mohr-Coulomb criterion: $\dfrac{m\sigma_{max} - \sigma_{min}}{2} = k$,

find the magnitude of $\sigma$ for each of the above three loading cases such that the material begins to yield. Also find the principal plastic strain increment vector $(d\epsilon_1^p, d\epsilon_2^p, d\epsilon_3^p)$ during yielding based on the associated flow rule.

4.4. Show that the plastic strain increment at the apex of the Mohr-Coulomb hexagonal pyramid can be expressed as

$$d\epsilon_1^p = (d\lambda_1 + d\lambda_2)m - (d\lambda_4 + d\lambda_5)$$
$$d\epsilon_2^p = (d\lambda_5 + d\lambda_6)m - (d\lambda_2 + d\lambda_3)$$
$$d\epsilon_3^p = (d\lambda_3 + d\lambda_4)m - (d\lambda_1 + d\lambda_6)$$

Show also that Eqs. (4.34) and (4.36) are still valid in this case (see Fig. 4.5).

4.5. The modified Mohr-Coulomb yield surface is a Mohr-Coulomb surface

$$m\sigma_{max} - \sigma_{min} = f_c'$$

combined with a tension cutoff plane:

$$\sigma_{max} = f_t'$$

This yield surface consists of nine flat planes, nine edges, and seven apices. Discuss the plastic strain increment at the cutoff planes and at their relevant edges and apices. Show that

(a) the plastic strain increments satisfy

$$\frac{\sum d\epsilon_i^p}{\sum |d\epsilon_c^p|} > m$$

(b) the plastic work increment can be expressed by

$$dW_p = f'_c \sum |d\epsilon^p_c| + f_t(\sum d\epsilon^p_t - m \sum d\epsilon^p_c)$$

4.6. A thick-walled tube is first loaded into the elastic–plastic range with an internal pressure $p$, $p_e \le p \le p_c$, and then completely unloaded.

(a) Find the residual stresses.
(b) Determine the highest pressure for which the material of the tube would not yield again upon unloading.
(c) Show that if the ratio of outer and inner radii of the tube, $b/a$, is less than about 2.2, the material will shake down to elastic behavior for repeated pressurization between $p = 0$ and $p = p_c$.

4.7. A thick-walled tube of perfectly plastic material sustains the full plastic internal pressure given by Eq. (4.73). Examine the location on the Tresca yield surface of stress points for different radii, apply the normality rule to obtain information about possible plastic deformation, and verify that such deformation is compatible with a mode of collapse for the tube.

4.8. A composite tube is composed of $n$ tubes made of the same material, one inside the other. The inner and outer radii of the $n$ tubes are $(r_i, r_1)$, $(r_1, r_2), \ldots, (r_{n-1}, r_e)$, respectively. The composite tube is subjected to an internal pressure $p$. The material obeys the Tresca yield condition. Assume that yield occurs simultaneously at the inner surfaces of each tube. Show that

(a) The inner pressure for the initial yielding is given by

$$p = \frac{\sigma_0}{2} \left\{ n - \left[ \left( \frac{r_i}{r_1} \right)^2 + \left( \frac{r_1}{r_2} \right)^2 + \cdots + \left( \frac{r_{n-1}}{r_e} \right)^2 \right] \right\}$$

in which $\sigma_0$ is the yield stress in simple tension.

(b) If the ratio of the outer and inner radii of each tube is

$$\frac{r_k}{r_{k-1}} = \left( \frac{r_e}{r_i} \right)^{1/n} \qquad k = 1, 2, \ldots, n; \qquad r_0 = r_i, r_n = r_e$$

the pressure $p$ takes the maximum value for the initial yielding, and

$$(p_e)_{max} = \frac{n\sigma_0}{2} \left[ 1 - \left( \frac{r_i}{r_e} \right)^{2/n} \right].$$

(c) The full plastic pressure $p$ is given by

$$p_c = \sigma_0 \ln \frac{r_e}{r_i}$$

4.9. Consider a hollow sphere of internal radius $a$ and external radius $b$. Discuss the behavior of the sphere under an internal pressure.

(a) Find the elastic stresses and displacements. Determine the maximum pressure $p_e$ for which this elastic solution is valid.
(b) Find the elastic–plastic solution and the maximum pressure $p_c$ for which it is valid.
(c) What simplifications, if any, result if the material is assumed to be incompressible in both elastic and plastic ranges?

(d) Find the residual stresses after a complete unloading and determine the highest pressure for which shakedown occurs.

(e) Find the stresses and strain rates for uncontained plastic flow.

4.10. In a combined tension/torsion test on a thin-wall tube of circular cross section, let $\sigma$ and $\epsilon$ be the axial stress and axial strain and $\tau$ and $\gamma$ be the shear stress and shear strain, respectively. Assume that the tube is made of Prandtl–Reuss material with $\nu = \frac{1}{2}$. Calculate the stresses $\sigma$ and $\tau$ corresponding to the strain state $(\epsilon, \gamma) = (\sigma_Y/E, \sigma_Y/\sqrt{3}G)$ for the following three loading paths (Fig. P4.6):

(a) The axial strain $\epsilon$ is first increased up to the yield value $\epsilon = \sigma_Y/E$, then kept unchanged, while the shear strain $\gamma$ is increased up to its final value $\gamma = \sigma_Y/\sqrt{3}G$.

(b) Reverse of the above loading path: the shear strain $\gamma$ is first increased up to its final value $\gamma = \sigma_Y/\sqrt{3}G$, then held constant, while the axial strain $\epsilon$ is increased up to its final value $\sigma_Y/E$.

(c) Both strains $\epsilon$ and $\gamma$ are increased proportionately with a ratio of $\epsilon/\gamma = \sqrt{3}G/E = 1/\sqrt{3}$, until their final values are reached.

Hint: Noting that $\sigma_{mn}\, de_{mn} = \sigma_{mn}\, d\epsilon_{mn}$, $d\lambda$ can thus be obtained in terms of $k$, $\sigma$, $d\epsilon$, $\tau$, and $d\gamma$. Because of the incompressibility condition, Eqs. (4.91) or (4.92) will lead to a group of differential equations with respect to $\sigma(\epsilon, \gamma)$ and $\tau(\epsilon, \gamma)$. Let $d\epsilon = 0$ or $d\gamma = 0$; the differential equations can be integrated for cases (a) and (b).

4.11. Follow Example 4.3 to examine the behavior of Drucker–Prager material under a plane stress condition. Compare the results with that for Prandtl–Reuss material.

4.12. Prove Eqs. (4.145).

4.13. Prove Eqs. (4.152) and (4.153).

4.14. A long thick-walled concrete tube with open end $(\sigma_z = 0)$ is subjected to internal pressure $p$. The inner radius and outer radius are $a$ and $b$, respectively.

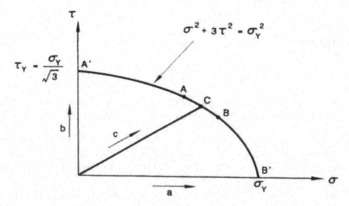

FIGURE P4.6.

Assume the concrete material follows the Rankine criterion with uniaxial tensile strength $f_t'$.

(a) Determine the elastic limit internal pressure $p_e$.
(b) Determine the relationship between the elastic–plastic boundary $r = c$ and the internal pressure $p$ for $p > p_e$.
(c) Determine the plastic limit internal pressure $p_s$.
(d) For the case of $b/a = 2$, plot $\sigma_r$ and $\sigma_\theta$ vs. $r$ curves for the elastic-plastic boundary at $c = a$, $c = \frac{1}{2}(a + b)$, and $c = b$, respectively.

4.15. A long vertical circular hole with internal radius $a$ in a half-space of rock is subjected to an internal pressure $p$ as shown in Fig. P4.15. Assume the rock material follows the Rankine criterion, with uniaxial tensile strength $f_t'$. Determine the relationship between the radius of the plastic zone and the internal pressure.

4.16. Resolve Problem 4.15 using the Tresca yield criterion. Show that the relationship between the radius of the plastic zone and the internal pressure can be obtained by letting $b \to \infty$ in Eq. (4.68).

4.17. Assuming the concrete material follows the Mohr–Coulomb criterion, resolve Problem 4.14. The uniaxial tensile and compressive strengths of the material are $f_t'$ and $f_c'$, respectively. Plot $\sigma_r$ and $\sigma_\theta$ vs. $r$ curves using $f_c'/f_t' = 10$.

4.18. Resolve Problem 4.15 using the Mohr–Coulomb criterion. Assume the uniaxial tensile and compressive strengths of the rock material are $f_t'$ and $f_c'$, respectively.

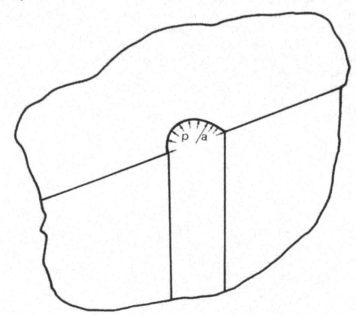

FIGURE P4.15.

**4.19.** Noting that the Tresca criterion and the Rankine criterion are special cases of the Mohr–Coulomb criterion, show that

(a) the elastic-plastic solutions of the thick-walled cylinder described in Section 4.7 and Problem 4.14 are special cases of the solution of Problem 4.17.

(b) the elastic-plastic solutions of the vertical circular hole in a half-space described in Problems 4.15 and 4.16 are special cases of the solution of Problem 4.18.

**4.20.** Derive the expression of the scalar factor $d\lambda$ for a general elastic-perfectly plastic material using the associated flow rule, such that

$$d\epsilon_{ij}^P = d\lambda \frac{\partial f}{\partial \sigma_{ij}}$$

where $f = f(\sigma_{ij})$ is the yield function. Assuming the elastic behavior of the material is linear and isotropic, express the scalar factor $d\lambda$ in terms of the two elastic constants $K$ and $G$.

**4.21.** Develop a Fortran code to calculate the material stiffness matrix $[C^{ep}]$ of Eq. (4.156) for the four yield criteria given in Table 4.1.

ANSWERS TO SELECTED PROBLEMS

**4.1.** $(d\epsilon_\theta^P, d\epsilon_z^P, d\epsilon_r^P) = d\lambda(s_\theta, s_z, s_r) = d\lambda(1, 0, -1)$.

**4.2.** $\tau_v = \sigma_0/2$; $d\epsilon_r^P : d\epsilon_\theta^P : d\epsilon_z^P : dy_{\theta z}^P = s_r : s_\theta : s_z : 2s_{\theta z} = (-1) : (-1) : (2) : (6)$.

**4.3.** (1a) $\sigma_Y = k$, $d\epsilon_i^P = d\lambda (1, 0, -1)$; (1b) $\sigma_Y = k$, $d\epsilon_i^P = d\lambda (1, 0, -1)$;
(1c) $\sigma_Y = k/(1+3\alpha)$, $d\epsilon_i^P = d\lambda (\alpha + \frac{1}{2}, \alpha, \alpha - \frac{1}{2})$;
(1d) $\sigma_Y = k/m$, $d\epsilon_i^P = d\lambda (m, 0, -1)$; (2a) $\sigma_Y = \sqrt{3}k$, $d\epsilon_i^P = d\lambda (1, 1, -2)$;
(2b) $\sigma_Y = 2k$, $d\epsilon_i^P = d\lambda_1 (1, 0, -1) + d\lambda_2 (0, 1, -1)$;
(2c) $\sigma_Y = \sqrt{3}k/(1+2\sqrt{3}\alpha)$, $d\epsilon_i^P = d\lambda (\alpha + 1/2\sqrt{3}, \alpha + 1/2\sqrt{3}, \alpha - 1/\sqrt{3})$;
(2d) $\sigma_Y = 2k/m$, $d\epsilon_i^P = d\lambda_1 (m, 0, -1) + d\lambda_2 (0, m, -1)$;
(3a) $\sigma_Y = \sqrt{3}k$, $d\epsilon_i^P = d\lambda (2, -1, -1)$;
(3b) $\sigma_Y = 2k$, $d\epsilon_i^P = d\lambda_1 (1, 0, -1) + d\lambda_2 (1, -1, 0)$;
(3c) $\sigma_Y = \sqrt{3}k/(1-2\sqrt{3}\alpha)$, $d\epsilon_i^P = d\lambda (\alpha + 1/\sqrt{3}, \alpha - 1/2\sqrt{3}, \alpha - 1/2\sqrt{3})$;
(3d) $\sigma_Y = 2k$, $d\epsilon_i^P = d\lambda_1 (m, 0, -1) + d\lambda_2 (m, -1, 0)$.

**4.6.** (a)

$$\sigma_r' = \begin{cases} \sigma_0 \ln \dfrac{r}{a} - p - \dfrac{pa^2}{b^2 - a^2}\left(1 - \dfrac{b^2}{r^2}\right), & a \leq r \leq c \\[2ex] \dfrac{\sigma_0 c^2}{2b^2}\left(1 - \dfrac{b^2}{r^2}\right) - \dfrac{pa^2}{b^2 - a^2}\left(1 - \dfrac{b^2}{r^2}\right), & c \leq r \leq b \end{cases}$$

$$\sigma_\theta' = \begin{cases} \sigma_0\left(1 + \ln \dfrac{r}{a}\right) - p - \dfrac{pa^2}{b^2 - a^2}\left(1 + \dfrac{b^2}{r^2}\right), & a \leq r \leq c \\[2ex] \dfrac{\sigma_0 c^2}{2b^2}\left(1 + \dfrac{b^2}{r^2}\right) - \dfrac{pa^2}{b^2 - a^2}\left(1 + \dfrac{b^2}{r^2}\right), & c \leq r \leq b \end{cases}$$

$$\sigma_z' = v(\sigma_r' + \sigma_\theta')$$

(b) $p \leq \sigma_0\left(1 - \dfrac{a^2}{b^2}\right) = 2p_e$.

4.10. (a) $\sigma = 0.648\sigma_Y$
$\tau = 0.440\sigma_Y$
(b) $\sigma = 0.762\sigma_Y$
$\tau = 0.374\sigma_Y$
(c) $\sigma = 0.707\sigma_Y$
$\tau = 0.408\sigma_Y$

4.13. The following formulas are provided for later use:

$$t_{ij} = s_{ik}s_{kj} - \tfrac{2}{3}J_2\delta_{ij}$$

$$t_{ii} = s_{ik}s_{ki} - 2J_2 = 0$$

$$t_{ij}s_{ij} = (s_{ik}s_{kj} - \tfrac{2}{3}J_2\delta_{ij})s_{ij} = s_{ij}s_{jk}s_{ki} = 3J_3$$

$$t_{ij}t_{ij} = s_{ik}s_{kj}s_{il}s_{lj} - \tfrac{4}{3}J_2 s_{ik}s_{ki} + \tfrac{4}{9}J_2^2\delta_{ij}\delta_{ij}$$

$$= 2J_2^2 - \tfrac{8}{3}J_2^2 + \tfrac{4}{3}J_2^2 = \tfrac{2}{3}J_2^2$$

Equations (4.153) and (4.152) are derived as follows:

$$H_{ij} = C_{ijmn}\frac{\partial f}{\partial \sigma_{mn}}$$

$$= \frac{E}{2(1+\nu)}\left[\frac{2\nu}{(1-2\nu)}\delta_{ij}\delta_{mn} + \delta_{im}\delta_{jn} + \delta_{in}\delta_{jm}\right](B_0\delta_{mn} + B_1 s_{mn} + B_2 t_{mn})$$

$$= 2G\left(\frac{\nu}{1-2\nu}\delta_{ij}\delta_{mn} + \delta_{im}\delta_{jn}\right)(B_0\delta_{mn} + B_1 s_{mn} + B_2 t_{mn})$$

$$= 2G\left(\frac{\nu B_0}{1-2\nu}\delta_{ij}\delta_{mn}\delta_{mn} + B_0\delta_{im}\delta_{jn}\delta_{mn} + \frac{\nu B_1}{1-2\nu}\delta_{ij}\delta_{mn}s_{mn} + B_1\delta_{im}\delta_{jn}s_{mn}\right.$$

$$\left. + \frac{\nu B_2}{1-2\nu}\delta_{ij}\delta_{mn}t_{mn} + B_2\delta_{im}\delta_{jn}t_{mn}\right)$$

$$= 2G\left(\frac{3\nu B_0}{1-2\nu}\delta_{ij} + B_0\delta_{ij} + 0 + B_1 s_{ij} + 0 + B_2 t_{ij}\right)$$

$$= 2G\left(B_0\frac{1+\nu}{1-2\nu}\delta_{ij} + B_1 s_{ij} + B_2 t_{ij}\right)$$

$$h = \frac{\partial f}{\partial \sigma_{ij}}C_{ijmn}\frac{\partial f}{\partial \sigma_{mn}} = \frac{\partial f}{\partial \sigma_{ij}}(2G)\left(B_0\frac{1+\nu}{1-2\nu}\delta_{ij} + B_1 s_{ij} + B_2 t_{ij}\right)$$

$$= 2G(B_0\delta_{ij} + B_1 s_{ij} + B_2 t_{ij})\left(B_0\frac{1+\nu}{1-2\nu}\delta_{ij} + B_1 s_{ij} + B_2 t_{ij}\right)$$

$$= 2G\left(B_0^2\frac{1+\nu}{1-2\nu}\delta_{ij}\delta_{ij} + B_1^2 s_{ij}s_{ij} + B_2^2 t_{ij}t_{ij} + 2B_1 B_2 s_{ij}t_{ij}\right)$$

$$= 2G\left(3B_0^2\frac{1+\nu}{1-2\nu} + 2B_1^2 J_2 + \tfrac{2}{3}B_2^2 J_2^2 + 6B_1 B_2 J_3\right)$$

# 5
# Stress–Strain Relations for Work-Hardening Materials

## 5.1. Introduction

Engineering material usually exhibits a work-hardening behavior. Increasing the stress beyond the initial yield surface and into the work-hardening range (loading) produces both plastic and elastic deformations. At each stage of plastic deformation, a new yield surface, called the *subsequent loading surface*, is established. If the state of stress is now changed such that the stress point representing it in a stress space moves inside the new yield surface (unloading), the behavior of the material is again elastic, and no plastic deformation will take place. The stress–strain behavior related to loading or unloading from new yield surface is *loading path dependent* or *loading history dependent*.

In developing constitutive equations for work-hardening materials, two basic approaches have been used. The first type of formulation is the *deformation theory* in the form of the total stress–strain relation. This theory assumes that the state of stress determines the state of strain uniquely as long as the plastic deformation continues. This is identical with the nonlinear elastic stress–strain relation of Chapter 3 as long as unloading does not occur. Thus, the most general form of this theory during loading may be written as

$$\epsilon_{ij}^{p} = \epsilon_{ij} - \epsilon_{ij}^{e} = f(\sigma_{ij}) \tag{5.1}$$

where $\epsilon_{ij}^{p}$ and $\epsilon_{ij}^{e}$ are the plastic and elastic components of the total strain $\epsilon_{ij}$, respectively. Equation (5.1) indicates a *loading-path-independent* behavior. It cannot adequately describe the phenomena associated with loading and unloading near the yield surface along a neutral loading path. Nevertheless, such theories have been used extensively in practice for the solution of elastic–plastic problems because of their comparative simplicity. However, the total stress–strain relation based on deformation theory is only valid in the case of *proportional loading*.

The other type of theory is the *incremental theory* or *flow theory*. This type of formulation relates the increment of plastic strain components $d\epsilon_{ij}^{p}$ to the state of stress, $\sigma_{ij}$, and the stress increment, $d\sigma_{ij}$. The simplest type of flow theory, as already discussed in Chapter 4, is the *theory of perfect plasticity.* A large number of the techniques used in the previous discussion on perfect plasticity carry through here with little change for work-hardening plasticity. The fundamental difference is that the yield surface is now not fixed in stress space, but rather the stress point $\sigma_{ij}$ is permitted to move outside the yield surface. The response of the material after initial yielding is described by specifying a new yield surface called the *subsequent yield surface,* and the rule that specifies this post-yield response is called the *hardening rule.*

Basic assumptions used in the development of the incremental theory of work-hardening plasticity include:

(a) The *existence* of an initial yield surface which defines the elastic limit of the material in a multiaxial state of stress. The concept of yield surface has been discussed in Chapter 2.
(b) The *hardening rule* which describes the evolution of subsequent yield surfaces. Several hardening rules have been proposed in the past and will be discussed later in this chapter.
(c) The *flow rule* which is related to a plastic potential function and defines the direction of the incremental plastic strain vector in strain space. The concept of flow rule has been discussed in some detail in Chapter 4 for perfectly plastic materials. For work-hardening materials, the associated flow rule represents a result of *Drucker's stability postulate.* This will be studied in the later part of this chapter.

This chapter is concerned with the development of the constitutive relations of work-hardening materials. The deformation theory is first introduced in Section 5.2. Then the basic concepts of the incremental theory are discussed. This later theory accounts for loading, unloading, and reloading and is suitable for describing the complete stress-history-dependent behavior of a work-hardening plastic solid. This is the main subject of the present chapter.

## 5.2. Deformation Theory of Plasticity

### 5.2.1. Deformation Theory for $J_2$-Material

The simplest and most popular deformation theory is the $J_2$ *deformation theory.* The theory is based upon the following four assumptions: (i) the material is initially isotropic; (ii) the plastic strain involves only a change in shape but no change in volume, and the elastic strain is related to the stress by Hooke's law; (iii) the principal axes of the plastic strain and the stress coincide; (iv) the principal values of the plastic strain have the same ratios to each other as the principal values of the stress deviator.

In the development of the stress–strain relation, the total strain $\epsilon_{ij}$ is decomposed into elastic and plastic components $\epsilon_{ij}^e$ and $\epsilon_{ij}^p$ by the simple superposition:

$$\epsilon_{ij} = \epsilon_{ij}^e + \epsilon_{ij}^p \tag{5.2}$$

According to assumption (ii), the elastic strain $\epsilon_{ij}^e$ is related to the stress $\sigma_{ij}$ by Hooke's law [see Eq. (3.96) of Chapter 3]

$$\epsilon_{ij}^e = \frac{s_{ij}}{2G} + \frac{\sigma_{kk}}{9K} \delta_{ij} \tag{5.3}$$

and the plastic strain $\epsilon_{ij}^p$ consists only of the component of deviatoric strain $e_{ij}^p$. Assumptions (iii) and (iv) relate this plastic strain $\epsilon_{ij}^p$ to the stress deviator $s_{ij}$ as

$$\epsilon_{ij}^p = e_{ij}^p = \phi s_{ij} \tag{5.4}$$

in which $\phi$ is a scalar, which may be considered a function of the invariant $J_2$:

$$\phi = \phi(J_2) \tag{5.5}$$

The scalar function $\phi(J_2)$ is a material property determined by experiment.

In order to calibrate the function $\phi(J_2)$ with the experimental uniaxial stress-strain curve, we introduce a stress variable called the *effective stress* $\sigma_e$, defined as

$$\sigma_e = \sqrt{3J_2} = \sqrt{\tfrac{3}{2}s_{ij}s_{ij}} \tag{5.6}$$

and a strain variable called the *effective plastic strain*, defined as

$$\epsilon_p = \sqrt{\tfrac{2}{3}\epsilon_{ij}^p \epsilon_{ij}^p} \tag{5.7}$$

It can be seen that in the uniaxial tension case with $\sigma_1 > 0$, $\sigma_2 = \sigma_3 = 0$, the effective stress $\sigma_e$ reduces to the stress $\sigma_1$. On the other hand, due to the *plastic-incompressibility condition*, in the case of uniaxial tension, we have

$$\epsilon_2^p = \epsilon_3^p = -\tfrac{1}{2}\epsilon_1^p \tag{5.8}$$

Substituting Eq. (5.8) into Eq. (5.7), we recognize that the effective strain $\epsilon^p$ reduces to the uniaxial strain $\epsilon_1^p$.

Now, we can define a single *effective stress–effective strain curve*, whose shape is governed by the simple uniaxial tension test, taking the following form

$$\sigma_e = \sigma_e(\epsilon_p) \tag{5.9}$$

Multiplying Eq. (5.4) by itself yields

$$\sqrt{\epsilon_{ij}^p \epsilon_{ij}^p} = \phi \sqrt{s_{ij}s_{ij}}$$

Using the definitions of effective stress $\sigma_e$ and effective strain $\epsilon_p$ leads to an expression for the parameter $\phi$:

$$\phi = \frac{3}{2}\frac{\epsilon_p}{\sigma_e} \tag{5.10}$$

or

$$\phi = \frac{\sqrt{3}}{2} \frac{\epsilon_p}{\sqrt{J_2}} \tag{5.11}$$

Since the effective strain $\epsilon_p$ is related to the effective stress $\sigma_e$ or the stress invariant $J_2$ through the uniaxial stress-strain relation (5.9), $\phi$ can be obtained as a function of $J_2$.

Using Eq. (5.10) for $\phi$, the constitutive equation (5.4) is now rewritten in terms of the stress and strain components explicitly as

$$\epsilon_x^p = \frac{\epsilon_p}{\sigma_e} \left[ \sigma_x - \frac{1}{2}(\sigma_y + \sigma_z) \right]$$

$$\epsilon_y^p = \frac{\epsilon_p}{\sigma_e} \left[ \sigma_y - \frac{1}{2}(\sigma_x + \sigma_z) \right]$$

$$\epsilon_z^p = \frac{\epsilon_p}{\sigma_e} \left[ \sigma_z - \frac{1}{2}(\sigma_x + \sigma_y) \right]$$

$$\gamma_{xz}^p = \frac{3\epsilon_p}{\sigma_e} \tau_{xz} \tag{5.12}$$

$$\gamma_{yz}^p = \frac{3\epsilon_p}{\sigma_e} \tau_{yz}$$

$$\gamma_{xy}^p = \frac{3\epsilon_p}{\sigma_e} \tau_{xy}$$

The stress-strain relationships of the deformational type for $J_2$-material were first formulated by Hencky in 1924 to describe a perfectly plastic behavior, and then by Nadai in 1931 to represent the behavior of a work-hardening material.

Equations (5.4) and (5.12) express the stress-strain behavior of a work-hardening material with a continuous transition from an elastic state to a plastic state, as long as the loading condition

$$dJ_2 > 0 \tag{5.13}$$

is satisfied. Otherwise, Hooke's law must be used and the plastic strain remains unchanged. In view of this, the total stress-strain relation of the deformational type is strictly valid only for or near a proportional loading path. In this case, the stress components increase in a constant ratio to each other, so that the strains can be expressed in terms of the final state of stress along this proportional loading path.

The validity of the deformation theory for loading paths other than the proportional loading path has been studied by Budiansky (1959). Assuming Drucker's postulate to constitute a criterion for physical soundness, Budiansky has shown that deformation theories are consistent with this postulate for a range of loading paths in the vicinity of proportional loading.

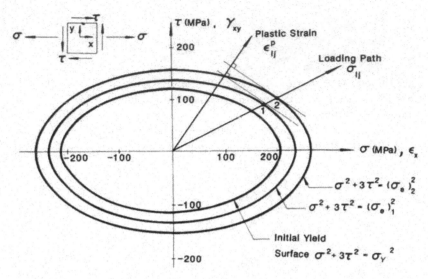

FIGURE 5.1. An illustration of $J_2$ deformation theory (Example 5.1).

EXAMPLE 5.1. An element of $J_2$-material is subjected to a proportional loading path with a stress ratio $\sigma/\tau = 2$ as shown in Fig. 5.1. The stress-strain relation in simple tension of the material is given by

$$\epsilon = \begin{cases} \dfrac{\sigma}{E} & (\sigma \le \sigma_Y) \\[2mm] \dfrac{\sigma}{E} + \dfrac{\sigma - \sigma_Y}{m} & (\sigma > \sigma_Y) \end{cases} \tag{5.14}$$

with Young's modulus $E = 207$ GPa, yield stress $\sigma_Y = 207$ MPa, constant $m = 25$ GPa, and Poisson's ratio $\nu = 0.3$. Find all the components of the normal and shear strains corresponding to the two states of stress with: (i) $\sigma = 180$ MPa, $\tau = 90$ MPa and (ii) $\sigma = 200$ MPa, $\tau = 100$ MPa.

SOLUTION. The yield condition for $J_2$-material subjected to stresses $\sigma$ and $\tau$ is expressed as

$$\sigma^2 + 3\tau^2 = \sigma_Y^2 \tag{5.15}$$

Substituting $\sigma = 180$ MPa and $\tau = 90$ MPa in Eq. (5.15) leads to

$$\sigma^2 + 3\tau^2 = (180)^2 + 3(90)^2 = (238.1)^2 > \sigma_Y^2$$

Thus, the element has yielded under the stress states (i) and (ii), and its strain included elastic and plastic parts. The elastic strain is determined by

Hooke's law [see Eqs. (3.101) and (3.102)]. At stress state (i):

$$(\epsilon_x^e)_1 = \frac{\sigma}{E} = \frac{180}{207 \times 10^3} = 8.69 \times 10^{-4}$$

$$(\epsilon_y^e)_1 = (\epsilon_z^e)_1 = -\frac{\nu}{E}\sigma = -\frac{(0.3)(180)}{207 \times 10^3} = -2.609 \times 10^{-4}$$

$$(\gamma_{xy}^e)_1 = \frac{2(1+\nu)\tau}{E} = \frac{2(1+0.3)(90)}{207 \times 10^3} = 1.130 \times 10^{-3}$$

$$(\gamma_{yz}^e)_1 = (\gamma_{xz}^e)_1 = 0$$

At stress state (ii):

$$(\epsilon_x^e)_2 = \frac{200}{207 \times 10^3} = 9.662 \times 10^{-4}$$

$$(\epsilon_y^e)_2 = (\epsilon_z^e)_2 = -\frac{(0.3)(200)}{207 \times 10^3} = -2.899 \times 10^{-4}$$

$$(\gamma_{xy}^e)_2 = \frac{2(1+0.3)(100)}{207 \times 10^3} = 1.256 \times 10^{-3}$$

$$(\gamma_{yz}^e)_2 = (\gamma_{zx}^e)_2 = 0$$

The plastic strains are calculated from Eq. (5.12), with the effective stress $\sigma_e$ obtained from Eq. (5.6):

$$\sigma_e = \sqrt{3J_2} = \sqrt{\sigma^2 + 3\tau^2}$$

For stress state (i),

$$(\sigma_e)_1 = \sqrt{(180)^2 + 3(90)^2} = 238.1 \text{ MPa}$$

For stress state (ii),

$$(\sigma_e)_2 = \sqrt{(200)^2 + 3(100)^2} = 264.6 \text{ MPa}$$

According to the given stress-strain relation (5.14), the effective strain can be expressed in terms of effective stress $\sigma_e$ as

$$\epsilon^p = \frac{\sigma_e - \sigma_Y}{m}$$

and the ratio $\epsilon_p/\sigma_e$ in Eq. (5.12) is calculated for these two stress states as

$$\left(\frac{\epsilon_p}{\sigma_e}\right)_1 = \frac{(\sigma_e)_1 - \sigma_Y}{m(\sigma_e)_1} = \frac{238.1 - 207}{(25,000)(238.1)} = 5.225 \times 10^{-6} \frac{1}{\text{MPa}}$$

and

$$\left(\frac{\epsilon_p}{\sigma_e}\right)_2 = \frac{(\sigma_e)_2 - \sigma_Y}{m(\sigma_e)_2} = \frac{264.6 - 207}{(25,000)(264.6)} = 8.707 \times 10^{-6} \frac{1}{\text{MPa}}$$

respectively. Now the plastic strains can be obtained from Eq. (5.12).

$$(\epsilon_x^p)_1 = \left(\frac{\epsilon_p}{\sigma_e}\right)_1 \sigma = (5.225 \times 10^{-6})(180) = 9.405 \times 10^{-4}$$

$$(\epsilon_y^p)_1 = (\epsilon_z^p)_1 = \left(\frac{\epsilon_p}{\sigma_e}\right)_1 \left(-\frac{\sigma}{2}\right) = (5.225 \times 10^{-6})(-90) = -4.702 \times 10^{-4}$$

$$(\gamma_{xy}^p)_1 = 3 \left(\frac{\epsilon_p}{\sigma_e}\right)_1 \tau = 3(5.225 \times 10^{-6})(90) = 1.411 \times 10^{-3}$$

$$(\gamma_{yz}^p)_1 = (\gamma_{xz}^p)_1 = 0$$

Similarly, for stress state (ii), we have

$$(\epsilon_x^p)_2 = (8.707 \times 10^{-6})(200) = 1.741 \times 10^{-3}$$

$$(\epsilon_y^p)_2 = (\epsilon_z^p)_2 = -8.707 \times 10^{-4}$$

$$(\gamma_{xy}^p)_2 = 3(8.707 \times 10^{-6})(100) = 2.612 \times 10^{-3}$$

$$(\gamma_{yz}^p)_2 = (\gamma_{xz}^p)_2 = 0$$

Finally, the total strain $\epsilon_{ij}$ is given as the sum of the elastic strain $\epsilon_{ij}^e$ and the plastic strain $\epsilon_{ij}^p$:

$$[\epsilon_{ij}]_1 = \begin{bmatrix} (\epsilon_x^e)_1 + (\epsilon_x^p)_1 & \frac{1}{2}[(\gamma_{xy}^e)_1 + (\gamma_{xy}^p)_1] & 0 \\ \frac{1}{2}[(\gamma_{xy}^e)_1 + (\gamma_{xy}^p)_1] & (\epsilon_y^e)_1 + (\epsilon_y^p)_1 & 0 \\ 0 & 0 & (\epsilon_z^e)_1 + (\epsilon_z^p)_1 \end{bmatrix}$$

$$= \begin{bmatrix} 1.810 & 1.271 & 0 \\ 1.271 & -0.731 & 0 \\ 0 & 0 & -0.731 \end{bmatrix} \times 10^{-3}$$

and similarly

$$[\epsilon_{ij}]_2 = \begin{bmatrix} 2.707 & 1.934 & 0 \\ 1.934 & -1.160 & 0 \\ 0 & 0 & -1.160 \end{bmatrix} \times 10^{-3}$$

Figure 5.1 shows the loading path and the direction of the plastic strain vector. It can be seen that the plastic strain vector is normal to the yield surface at the stress points. This has been manifested by Eq. (5.4).

## 5.2.2. Generalization of $J_2$-Theory

The stress-strain relation of Eqs. (5.4) and (5.5) is comparatively simple in structure. It involves only the stress invariant $J_2$. More elaborate stress-strain relationships have been proposed in the past. For example, Prager has formulated the following relationship between the plastic strain and stress for metals under the proportional loading condition:

$$\epsilon_{ij}^{p} = f(J_2, J_3)s_{ij} + g(J_2, J_3)t_{ij} \tag{5.16}$$

which includes also the third invariant of the stress deviator, $J_3$, and its derivatives, $t_{ij}$:

$$t_{ij} = \frac{\partial J_3}{\partial \sigma_{ij}} = s_{ik}s_{kj} - \frac{2}{3}J_2\delta_{ij} \tag{5.17}$$

For the constitutive relations (5.16), assumptions (iii) and (iv) have been eliminated but assumption (i), concerning the initial isotropy, and assumption (ii) concerning the plastic incompressibility, are still maintained. The scalar functions $f(J_2, J_3)$ and $g(J_2, J_3)$ are material properties determined by experiments. In comparison to Eq. (5.4), Eq. (5.16) provides more flexibility in fitting the experimental data.

Furthermore, if the assumption of the plastic incompressibility condition is discarded, a more general form of the stress–strain relation applicable to nonmetallic materials may be expressed as

$$\epsilon_{ij}^{p} = P(I_1, J_2, J_3)\delta_{ij} + Q(I_1, J_2, J_3)s_{ij} + R(I_1, J_2, J_3)t_{ij} \tag{5.18}$$

in which $\delta_{ij}$ is the Kronecker delta. For some nonmetallic frictional materials such as soils, concretes, and rocks, the plastic volume change is usually appreciable, and thus the effect of the first stress invariant $I_1$ must be taken into account. Since the constitutive equations (5.18) involve all three stress invariants as variables in the three scalar functions $P$, $Q$, and $R$, it follows that they are suitable for description of such materials under proportional loading.

In the following sections, we shall discuss the basic concepts of the incremental theory of plasticity for work-hardening plastic solids. This is the main topic of the present chapter.

## 5.3. Loading Surface and Hardening Rules

### 5.3.1. Loading Surface and Loading Criterion

*Loading surface* is the subsequent yield surface for an elastoplastically deformed material, which defines the boundary of the current elastic region. If a stress point lies within this region, no additional plastic deformation takes place. On the other hand, if the state of stress is on the boundary of the elastic region and tends to move out of the current loading surface, additional plastic deformations will occur, accompanied by a configuration change of the current loading surface. In other words, the current loading surface or the subsequent yield surface will change its current configuration when plastic deformation takes place. Thus, the loading surface may be generally expressed as a function of the current state of stress (or strain) and some hidden variables such that

$$f(\sigma_{ij}, \epsilon_{ij}^p, k) = 0 \qquad (5.19)$$

in which the so-called hidden variables are expressed in terms of the plastic strain $\epsilon_{ij}^p$ and a *hardening parameter k.*

To determine the nature of the subsequent loading surfaces is one of the major problems in the work-hardening theory of plasticity. The response of a material after initial yielding differs considerably in various plasticity theories. This post-yield response, called the *hardening rule,* is described by specifying the rule for the evolution of the subsequent yield surfaces or loading surfaces. Several hardening rules have been proposed in the past for use in plastic analysis. Since the configuration change of the loading surface is closely related to the "plastic loading," we shall first discuss the loading criterion for a work-hardening material. This will then be followed by the study of the hardening rules.

For a uniaxial behavior, the concepts of "loading" and "unloading" are self-evident (see Fig. 5.2a). However, this is not the case under a multiaxial stress state, and load/unload must be clearly specified. The loading surface itself is, of course, an essential part of defining loading and unloading. We shall define further here that loading or plastic flow occurs only when the stress point is on the loading surface and the additional loading or stress incremental vector $d\sigma_{ij}$ is directed outward from the current elastic region. To express the above statement more precisely, we introduce a unit vector $\mathbf{n}^f$ normal to the loading surface in stress space (Fig. 5.2b) whose components are given by

$$n_{ij}^f = \frac{\partial f/\partial \sigma_{ij}}{\left(\dfrac{\partial f}{\partial \sigma_{kl}}\dfrac{\partial f}{\partial \sigma_{kl}}\right)^{1/2}} \qquad (5.20)$$

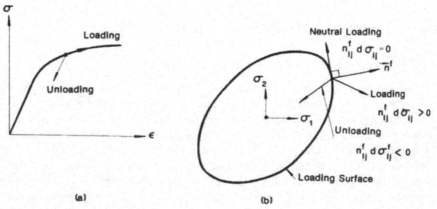

FIGURE 5.2. Loading criterion for a work-hardening material: (a) uniaxial case; (b) multiaxial case.

If the angle between the vector $d\sigma_{ij}$ and $n_{ij}^f$ is acute (Fig. 5.2b), additional plastic deformation will occur. Thus, the criterion for loading is

$$\text{if } f = 0 \text{ and } n_{ij}^f \, d\sigma_{ij} > 0, \text{ then } d\epsilon_{ij}^p \neq 0 \tag{5.21}$$

On the other hand, if the two vectors $d\sigma_{ij}$ and $n_{ij}^f$ form an obtuse angle, unloading will occur. Thus, the criterion for unloading is

$$\text{if } f = 0 \text{ and } n_{ij}^f \, d\sigma_{ij} < 0, \text{ then } d\epsilon_{ij}^p = 0 \tag{5.22}$$

In the *neutral loading* case, the additional load vector $d\sigma_{ij}$ is perpendicular to the normal vector $n_{ij}^f$, and no additional plastic deformation will occur. This condition is termed "*neutral loading.*" The criterion for "*neutral loading*" is

$$\text{if } f = 0 \text{ and } n_{ij}^f \, d\sigma_{ij} = 0, \text{ then } d\epsilon_{ij}^p = 0 \tag{5.23}$$

Recall the loading criterion for an elastic–perfectly plastic material discussed in Chapter 4 [see Eqs. (4.2) and (4.5)]; in that case, the initial yield surface becomes the *limit surface* with plastic deformation taking place only when $f = 0$ and $d\sigma_{ij}$ is tangent to the yield surface. Thus, for a perfectly plastic material, there is no neutral loading case such as satisfied by Eq. (5.23).

## 5.3.2. Hardening Rules

When an initial yield surface is known, the rule of work hardening defines its modification during the process of plastic flow. A number of hardening rules have been proposed. The most widely used rules are those of *isotropic hardening*, *kinematic hardening*, and a combination of both, i.e., the so-called *mixed hardening*. In this section, we shall discuss these three simple rules in some detail.

For clarity, the general form of the loading function of Eq. (5.19) can be written as

$$f(\sigma_{ij}, \epsilon_{ij}^p, k) = F(\sigma_{ij}, \epsilon_{ij}^p) - k^2(\epsilon_p) = 0 \tag{5.24}$$

in which the hardening parameter $k^2$ represents the size of the yield surface, while the function $F(\sigma_{ij}, \epsilon_{ij}^p)$ defines the shape of that surface. Here, the parameter $k^2$ is expressed as a function of $\epsilon_p$, called the *effective strain*, which is an integrated increasing function of the plastic strain increments but not the plastic strain itself (see Section 5.5). The value of $\epsilon_p$ depends on the loading history or the plastic strain path.

Since the work hardening of a material tends to introduce anisotropies in an initially isotropic material, it is not sufficient to represent an anisotropic yield surface in the space of principal stresses that has been used for an isotropic material. In the following, the yield surface will therefore be described in the nine-dimensional stress space of $\sigma_{ij}$. Diagrams will be drawn, however, in two dimensions, but the basic geometric ideas are readily extended to higher-dimensional spaces.

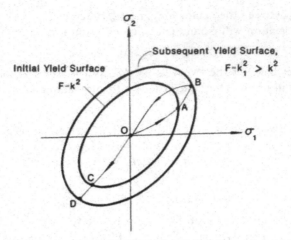

FIGURE 5.3. Subsequent yield surface for isotropic-hardening material.

## 5.3.3. Isotropic Hardening

For a perfectly plastic material, the equation for the fixed yield surface has the form $F(\sigma_{ij}) = k^2$, where $k$ is a constant. The simplest work-hardening rule is based on the assumption that the initial yield surface expands uniformly without distortion and translation as plastic flow occurs, as shown schematically in Fig. 5.3. The size of the yield surface is now governed by the value $k^2$, which depends upon plastic strain history. The equation for the subsequent yield surface or loading surface can be written in the general form

$$F(\sigma_{ij}) = k^2(\epsilon_p) \tag{5.25}$$

If, for example, the von Mises initial yield function, $F = J_2$, is used, Eq. (5.25) becomes

$$J_2 = \tfrac{1}{2} s_{ij} s_{ij} = k^2(\epsilon_p) \tag{5.26}$$

When the effective stress $\sigma_e = \sqrt{(3J_2)}$ is introduced into Eq. (5.26) as a hardening parameter, the isotropic-hardening von Mises model takes the form

$$f(\sigma_{ij}, k) = \tfrac{3}{2} s_{ij} s_{ij} - \sigma_e^2(\epsilon_p) = 0 \tag{5.27}$$

where the hardening parameter $\sigma_e(\epsilon_p)$ is related to the effective strain $\epsilon_p$ through an experimental uniaxial stress-strain curve. The effective strain $\epsilon_p$ will be defined later either as a scalar function of the work done by the plastic deformation or as the accumulated plastic strain (see Section 5.5).

### 5.3.3.1. AN ILLUSTRATIVE EXAMPLE

As an illustration, consider a material element subjected to uniaxial normal and shear stresses as shown in Fig. 5.4. Based on the von Mises yield

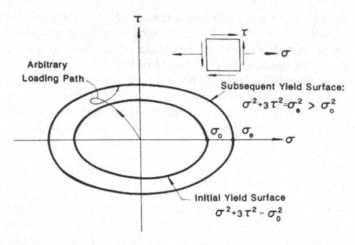

FIGURE 5.4. Hardening parameter for an element subject to normal and shear stresses.

criterion, the initial yield function is given by

$$f = \sigma^2 + 3\tau^2 - \sigma_0^2 = 0 \qquad (5.28)$$

or

$$F = \sigma^2 + 3\tau^2 = \sigma_0^2 \qquad (5.29)$$

in which $\sigma_0$ is the initial yield stress under uniaxial tension. After the initial yielding, if the material is subjected to a general loading path, according to the isotropic hardening rule, its subsequent loading surfaces are generally expressed as

$$\sigma^2 + 3\tau^2 = \sigma_e^2 \qquad (5.30)$$

in which the hardening parameter $\sigma_e^2$, characterizing the size of a loading surface, is the largest previous value of $(\sigma^2 + 3\tau^2)$ reached in the stress history. Since the recorded history of the material is represented by the hardening parameter, the material characterized by Eq. (5.30) may be regarded as a stress-hardening material.

### 5.3.3.2. THE BAUSCHINGER EFFECT

The isotropic hardening model is simple to use, but it applies mainly to monotonic loading without stress reversals. Because the loading surface expands uniformly (or isotropically) and remains self-similar with increasing plastic deformation (Fig. 5.3), it cannot account for the Bauschinger effect exhibited by most structural materials.

The term *Bauschinger effect* refers to a particular type of directional *anisotropy* induced by a plastic deformation; namely, an initial plastic deformation of one sign reduces the resistance of the material with respect to a subsequent plastic deformation of the opposite sign. The behavior predicted by the isotropic hardening rule is, in fact, contrary to this observation. The rule implies that because of work hardening, the material will exhibit an increase in the compressive yield stress equal to the increase in the tension yield stress. This is illustrated in Fig. 5.3, where the yield limits in the first loading direction $(OAB)$ and in the reversed loading direction $(OCD)$ are equal in magnitude. Since plastic deformation is an anisotropic process, it cannot be expected that the theory of isotropic hardening will lead to a realistic result when complex loading paths with stress reversal are considered.

### 5.3.4. Kinematic Hardening

The *kinematic hardening rule* assumes that during plastic deformation, the loading surface translates as a rigid body in stress space, maintaining the size, shape, and orientation of the initial yield surface. This hardening rule, due to Prager (1955, 1956), provides a simple means of accounting for the Bauschinger effect.

This rule is illustrated schematically in Fig. 5.5. As the stress point moves along its loading path from point $A$ to point $B$, the yield surface translates (no rotation) as a rigid body. Thus, the subsequent yield surface will wind up in the position indicated in Fig. 5.5 when the stress point has reached position $B$. The new position of the yield surface represents the most current yield function, whose center is denoted by $\alpha_{ij}$. Note that if the stress is unloaded from point $B$ along the initial path of loading, i.e., if $B$ now traces out path $BAO$, the material behaves elastically from point $B$ to point $C$

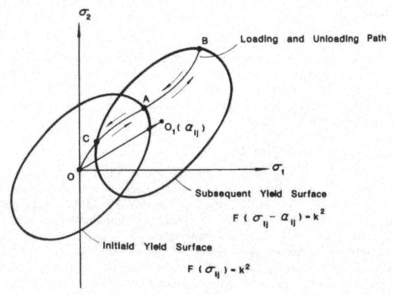

FIGURE 5.5. Subsequent yield surface for kinematic-hardening material.

but then begins to flow again before the stresses are completely relieved. In fact, the subsequent yield surface may or may not enclose the origin in stress space. As a consequence of assuming a rigid-body translation of the loading surface, the kinematic hardening rule predicts an ideal Bauschinger effect for a complete reversal of loading conditions.

For kinematic hardening, the equation of the loading surface has the general form

$$f(\sigma_{ij}, \epsilon_{ij}^p) = F(\sigma_{ij} - \alpha_{ij}) - k^2 = 0 \tag{5.31}$$

where $k$ is a constant and $\alpha_{ij}$ are the coordinates of the center of the loading surface (or the vector $OO_1$ in Fig. 5.5), which changes with the plastic deformation. As an illustration, let us consider the following simple example.

EXAMPLE 5.2. A yielded material element is subjected to a normal stress $\sigma$ and a shear stress $\tau$ as shown in Fig. 5.6. Determine the coordinate change of the center of the loading surface $d\alpha_{ij}$ due to an additional load $d\sigma_{ij} = (d\sigma, d\tau)$ which satisfies the criterion of loading. Assume $d\alpha_{ij}$ is in the direction parallel to the normal vector to the loading surface at the current yield point $A$ in the stress subspace $(\sigma, \tau)$. Assume the material satisfies the von Mises criterion.

SOLUTION. Based on the von Mises criterion, the initial yield function is given by Eq. (5.28). Due to kinematic hardening, the subsequent yield function is expressed as

$$f = (\sigma - \tilde{\sigma})^2 + 3(\tau - \tilde{\tau})^2 - \sigma_0^2 = 0 \tag{5.32}$$

in which $(\tilde{\sigma}, \tilde{\tau})$ are the coordinates of the center of the current loading surface, and the hardening parameter $\sigma_0^2$ remains constant.

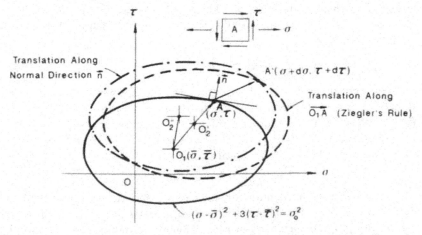

FIGURE 5.6. Subsequent yield surfaces due to different kinematic hardening rules (Examples 5.2 and 5.4).

Now, a stress increment $(d\sigma, d\tau)$ is imposed on the stress state $A(\sigma, \tau)$ which lies on the loading surface $f = 0$ and satisfies the loading condition:

$$\frac{\partial f}{\partial \sigma} d\sigma + \frac{\partial f}{\partial \tau} d\tau > 0$$

Thus, plastic strain takes place, and according to the kinematic hardening rule, the loading surface translates in the stress space. To determine the incremental translation of the center, $d\alpha_{ij}$, due to the stress increments $(d\sigma, d\tau)$, we assume that the vector $d\alpha_{ij}$ is in the direction parallel to the normal vector $\mathbf{n}$ at the contact or the current yield stress point $A$ in the *subspace of stress* $(\sigma, \tau)$. Therefore, $d\alpha_{ij}$ has only two nonzero elements, i.e.,

$$\{d\alpha_{ij}\} = \{d\tilde{\sigma}, d\tilde{\tau}\} \tag{5.33}$$

which satisfy

$$d\tilde{\sigma} = c \frac{\partial f}{\partial \sigma} = 2c(\sigma - \tilde{\sigma})$$

$$d\tilde{\tau} = c \frac{\partial f}{\partial \tau} = 6c(\tau - \tilde{\tau}) \tag{5.34}$$

where $c$ is a constant. Since the stress point $A'$ remains on the new yield surface during loading, the change in $f$, $df$ must be zero:

$$df = 2(\sigma - \tilde{\sigma})(d\sigma - d\tilde{\sigma}) + 6(\tau - \tilde{\tau})(d\tau - d\tilde{\tau}) = 0 \tag{5.35}$$

Solving Eqs. (5.34) and (5.35) for $d\tilde{\sigma}$ and $d\tilde{\tau}$, we obtain the incremental translation of the center $(d\tilde{\sigma}, d\tilde{\tau})$ as

$$d\tilde{\sigma} = (\sigma - \tilde{\sigma})[(\sigma - \tilde{\sigma})d\sigma + 3(\tau - \tilde{\tau})d\tau]/[\sigma_0^2 + 6(\tau - \tilde{\tau})^2]$$

$$d\tilde{\tau} = 3(\tau - \tilde{\tau})[(\sigma - \tilde{\sigma})d\sigma + 3(\tau - \tilde{\tau})d\tau]/[\sigma_0^2 + 6(\tau - \tilde{\tau})^2] \tag{5.36}$$

The translation $(d\tilde{\sigma}, d\tilde{\tau})$ is shown by the line $O_1 O_2''$ while the updated yield curve is shown by a dashed-dotted line in Fig. 5.6. For a given loading path, Eq. (5.36) can be integrated and the current position of the center can be determined.

### 5.3.4.1. PRAGER'S HARDENING RULE

It is seen that the key to a subsequent yield surface based on a kinematic hardening rule is the determination of the coordinates of the center, $\alpha_{ij}$. The simplest version for determining the hardening parameter $\alpha_{ij}$ is to assume a linear dependence of $d\alpha_{ij}$ on $d\epsilon_{ij}^p$. This is known as *Prager's hardening rule*, which has the simple form

$$d\alpha_{ij} = c\, d\epsilon_{ij}^p \quad \text{or} \quad \alpha_{ij} = c\epsilon_{ij}^p \tag{5.37}$$

where $c$ is the work-hardening constant, characteristic for a given material. Equation (5.37) may be taken as the definition of linear work hardening.

If we adopt the *associated flow rule*, Prager's hardening rule is equivalent to the assumption that the vector $d\alpha_{ij}$ moves in the direction parallel to the normal vector $\mathbf{n}$ at the current stress state on the yield surface in stress space.

Some inconsistencies may arise when Prager's hardening rule is used in a subspace of stress. For example, if some of the stress components are set equal to zero in Eq. (5.31), say, $\sigma''_{ij} = 0$ and $\sigma'_{ij} \neq 0$, Eq. (5.31) can be written

$$F(\sigma'_{ij} - \alpha'_{ij}, -\alpha''_{ij}) - k^2 = 0 \tag{5.38}$$

Since $d\alpha''_{ij} = c\, d\epsilon^p_{ij}$ is not necessarily zero, Eq. (5.38) no longer necessarily represents a surface which merely translates in the stress space; it may also deform as well, due to the changing values of $\alpha''_{ij}$. This can best be seen from Example 5.3 below.

It should be noted that in the last example, we have assumed that the vector $d\alpha_{ij}$ moves in the direction parallel to the normal to the yield surface at stress state $A$ in subspace $(\sigma_x, \tau_{xy})$, i.e.,

$$d\alpha_{xx} = c\frac{\partial f}{\partial \sigma_x}, \qquad d\alpha_{xy} = c\frac{\partial f}{\partial \tau_{xy}}, \qquad \text{other components} = 0 \tag{5.39}$$

which is in the direction of the projection of the vector $\partial f/\partial \sigma_{ij}$ onto the $\sigma_x$-$\tau_{xy}$ plane. Based on this assumption, the subsequent yield function holds the same form as Eq. (5.32) during hardening. However, this is not the case if Prager's rule is used to determine the parameters $\alpha_{ij}$.

EXAMPLE 5.3. Using Prager's rule, solve the same problem as in Example 5.2.

SOLUTION. Prager's rule is expressed as

$$d\alpha_{ij} = \bar{c}\, d\epsilon^p_{ij} = c\frac{\partial f}{\partial \sigma_{ij}} \tag{5.40}$$

in which the associated flow rule has been used. The general form of the subsequent yield surface of a $J_2$-material is given by

$$f = \tfrac{3}{2}(s_{ij} - \alpha_{ij})(s_{ij} - \alpha_{ij}) - \sigma_0^2 = 0 \tag{5.41}$$

Substitution of Eq. (5.41) into Eq. (5.40) leads to

$$d\alpha_{ij} = c\frac{\partial f}{\partial \sigma_{ij}} = 3c(s_{ij} - \alpha_{ij}) \tag{5.42}$$

Now the material element is only subjected to the normal stress $\sigma$ and the shear stress $\tau$, i.e.,

$$\sigma_{xx} = \sigma, \qquad \tau_{xy} = \tau, \qquad \text{other components of } \sigma_{ij} = 0$$

and

$$\alpha_{xx} = \tfrac{2}{3}\tilde{\sigma}, \qquad \alpha_{xy} = \tilde{\tau}, \qquad \text{other components of } \alpha_{ij} = 0 \tag{5.43}$$

Thus, Eq. (5.41) takes the form of Eq. (5.32):

$$f = (\sigma - \tilde{\sigma})^2 + 3(\tau - \tilde{\tau})^2 - \sigma_0^2 = 0 \tag{5.44}$$

and from Eq. (5.42), the changes of the coordinates of the center, $d\alpha_{ij}$, are obtained as

$$da_{xx} = 2c(\sigma - \tilde{\sigma}), \qquad da_{yy} = da_{zz} = -c(\sigma - \tilde{\sigma})$$

$$da_{xy} = da_{yx} = 3c(\tau - \tilde{\tau}), \qquad da_{xz} = da_{zx} = da_{yz} = da_{zy} = 0$$

(5.45)

It is seen that $da_{yy}$ and $da_{zz}$ are no longer equal to zero. Denote the updated value of the hardening parameter as

$$\tilde{\alpha}_{ij} = \alpha_{ij} + da_{ij}$$

Then the updated subsequent yield surface is expressed by

$$f = (\sigma - \tilde{a}_{xx}^2) + 3(\tau - \tilde{a}_{xy})^2 + (-\tilde{a}_{yy})^2 + (-\tilde{a}_{zz})^2 - \sigma_0^2 = 0 \qquad (5.46)$$

Comparison of Eq. (5.46) to Eq. (5.44) indicates that Prager's hardening rule leads to a subsequent yield surface which not only translates but also changes its shape during the plastic flow caused by an additional loading. Equation (5.46) does not really represent a kinematic hardening rule as described earlier.

### 5.3.4.2. ZIEGLER'S HARDENING RULE

In order to obtain a kinematic hardening rule that is also valid in subspaces, Ziegler (1959) modified Prager's hardening rule and assumed that the rate of translation takes place in the direction of the reduced-stress vector $\tilde{\sigma}_{ij} = \sigma_{ij} - \alpha_{ij}$ in the form

$$da_{ij} = d\mu(\sigma_{ij} - \alpha_{ij}) \qquad (5.47)$$

where $d\mu$ is a positive proportionality factor which depends on the history of the deformation. For simplicity, this factor can be assumed to have the simple form

$$d\mu = a\, d\epsilon_p \qquad (5.48)$$

in which $a$ is a positive constant, characteristic for a given material.

EXAMPLE 5.4. Using Ziegler's hardening rule, solve the same problem as in Example 5.2.

SOLUTION. In this case, Ziegler's rule of Eq. (5.47) is expressed as

$$d\tilde{\sigma} = da_{xx} = d\mu\,(\sigma - \tilde{\sigma})$$

$$d\tilde{\tau} = da_{xy} = da_{yx} = d\mu\,(\tau - \tilde{\tau}) \qquad (5.49)$$

$$\text{other components of } da_{ij} = 0$$

Following the same procedure as in Example 5.2, and solving Eqs. (5.49) and (5.35) for $d\tilde{\sigma}$ and $d\tilde{\tau}$, we obtain

$$d\tilde{\sigma} = \frac{1}{\sigma_0^2}(\sigma - \tilde{\sigma})[(\sigma - \tilde{\sigma})\, d\sigma + 3(\tau - \tilde{\tau})\, d\tau]$$

(5.50)

$$d\tilde{\tau} = \frac{1}{\sigma_0^2}(\tau - \tilde{\tau})[(\sigma - \tilde{\sigma})\, d\sigma + 3(\tau - \tilde{\tau})\, d\tau]$$

The translation increment $(d\tilde{\sigma}, d\tilde{\tau})$ of the center is shown by $O_1O_2'$ in Fig. 5.6, which is along the direction of the reduced-stress vector $O_1A(\sigma - \tilde{\sigma}, \tau - \tilde{\tau})$, and the updated yield surface is shown by a dashed line in the figure.

### 5.3.5. Mixed Hardening

A combination of kinematic and isotropic hardening would lead to the more general *mixed hardening rule* (Hodge, 1957):

$$f(\sigma_{ij}, \epsilon_{ij}^p, k) = F(\sigma_{ij} - \alpha_{ij}) - k^2(\epsilon_p) = 0 \tag{5.51}$$

In this case, the loading surface experiences a translation defined by $\alpha_{ij}$ and a uniform expansion measured by $k^2$; but it still retains its original shape. With the mixed hardening rule, different degrees of the Bauschinger effect can be simulated, by simply adjusting the two hardening parameters, $\alpha_{ij}$ and $k^2$.

For illustration, consider a $J_2$-material subjected to a mixed hardening rule. The general form of the subsequent loading surface is

$$f = \tfrac{1}{2}(s_{ij} - \alpha_{ij})(s_{ij} - \alpha_{ij}) - k^2(\epsilon_p) = 0 \tag{5.52}$$

If Prager's hardening rule is employed, Eq. (5.52) can be rewritten as

$$f = \tfrac{1}{2}(s_{ij} - c\epsilon_{ij}^p)(s_{ij} - c\epsilon_{ij}^p) - k^2(\epsilon_p) = 0 \tag{5.53}$$

where $c$ is a constant. In stress space, the surface moves around but does not simply expand outward, as in Fig. 5.3, or translate as in Fig. 5.5. The subsequent yield surfaces do not form a one-parameter family but intersect the previous ones, as shown in section by the dashed curve in Fig. 5.7. It is these surfaces in stress space that determine whether or not additional plastic deformations will occur in the subsequent loading.

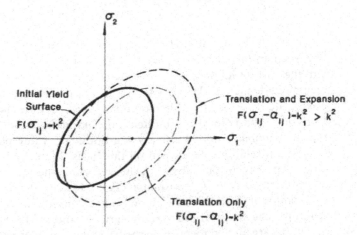

FIGURE 5.7. Subsequent yield surface for mixed-hardening $J_2$-material.

## 5.4. Flow Rule and Drucker's Stability Postulate

### 5.4.1. Flow Rules

So far, the loading surface alone has been considered, and the shape of the subsequent loading surfaces in a given loading program can be determined by the choice of a specific hardening rule. The necessary connection between the loading function $f$ and the stress–strain relation for a work-hardening material will be made here by means of a flow rule.

When the current yield surface $f$ is reached, the material is in a state of plastic flow upon further loading. Introducing the concept of a *plastic potential function* $g(\sigma_{ij}, \epsilon_{ij}^p, k)$ in analogy with ideal-fluid-flow problems, we define the flow rule

$$d\epsilon_{ij}^p = d\lambda \frac{\partial g}{\partial \sigma_{ij}} \qquad (5.54)$$

where $d\lambda > 0$ is a scalar function that will vary throughout the history of the straining process. The gradient of the plastic potential surface $\partial g/\partial \sigma_{ij}$ defines the direction of the plastic strain increment vector $d\epsilon_{ij}^p$, while the length or the magnitude of the vector is determined by the *loading parameter* $d\lambda$. Here, as in Chapter 4 for a perfectly plastic material, the flow rule is termed associated if the plastic potential surface has the same shape as the current yield or loading surface

$$g(\sigma_{ij}, \epsilon_{ij}^p, k) = f(\sigma_{ij}, \epsilon_{ij}^p, k)$$

and Eq. (5.54) takes the form

$$d\epsilon_{ij}^p = d\lambda \frac{\partial f}{\partial \sigma_{ij}} \qquad (5.55)$$

i.e., the plastic flow develops along the normal to the loading surface. Relation (5.55) is called the *associated flow rule* because the plastic flow is associated with the current loading surface. Since there is, in general, very little experimental evidence on plastic potential functions for engineering materials the associated flow rule is applied predominantly to these materials for practical reasons. Apart from its simplicity, the *normality condition* of Eq. (5.55) assures a *unique* solution for a given boundary-value problem using any stress–strain relations developed on this basis. Perhaps the most fundamental development for the subject of this section is the fact that the basic *stability postulate* or *Drucker postulate* (1951) for the definition of stable, work-hardening materials leads, among other consequences, to the normality condition. The shape of the loading surfaces and the form of the stress–strain relations are all tied together to the basic definition or postulate of a work-hardening material, as discussed below.

## 5.4.2. Drucker's Stability Postulate

Drucker's stability postulate has been discussed previously in Chapter 3 for the definition of a general *stable material*, which is expressed in terms of an external agency that adds load to the already loaded body (see Fig. 3.13). The plastic work-hardening material is a special case of the general stable inelastic materials. It also satisfies the stability postulate as given by Eqs. (3.160) and (3.161) in Chapter 3. In the following, we see that the definition of a work-hardening material as formulated by Drucker is more restrictive than the law of thermodynamics requires.

If an external agency slowly applies additional forces to a work-hardening body which is already loaded and then removes them, then

1. positive work is done by the external agency during the application of the added loads;
2. the net work performed by the external agency over a cycle of application and removal of the added loads is positive if plastic deformation has occurred in the cycle.

The work done by the added set of forces $\dot{T}_i, \dot{F}_i$ on the changes in displacement $\dot{u}$ (see Fig. 3.13) is expressed as

$$dW = \int_A \dot{T}_i \dot{u}_i \, dA + \int_V \dot{F}_i \dot{u}_i \, dV$$

Thus, the two stability requirements are stated as

$$\int_A \dot{T}_i \dot{u}_i \, dA + \int_V \dot{F}_i \dot{u}_i \, dV > 0 \tag{5.56}$$

and

$$\oint_A \dot{T}_i \dot{u}_i \, dA + \oint_V \dot{F}_i \dot{u}_i \, dV > 0 \tag{5.57}$$

in which $\oint$ indicates integration over a cycle of addition and removal of the additional set of forces, and plastic deformation is assumed to occur in this cycle.

Applying the principle of virtual work, the stability postulate can be expressed in terms of stresses and strains as follows.

$$\dot{\sigma}_{ij} \dot{\epsilon}_{ij} > 0 \quad \text{or} \quad d\sigma_{ij} \, d\epsilon_{ij}^p > 0 \qquad \text{stability in small} \tag{5.58}$$

$$\oint \dot{\sigma}_{ij} \dot{\epsilon}_{ij} > 0 \quad \text{or} \quad \oint d\sigma_{ij} \, d\epsilon_{ij}^p > 0 \qquad \text{stability in cycle in small} \tag{5.59}$$

These inequalities are illustrated geometrically in Fig. 5.8. We have assumed that the plastic strain $d\epsilon_{ij}^p \neq 0$ in Eq. (5.59). In general, we can write

$$\oint \dot{\sigma}_{ij} \dot{\epsilon}_{ij} \geq 0 \quad \text{or} \quad \oint d\sigma_{ij} \, d\epsilon_{ij}^p \geq 0 \tag{5.60}$$

The equal sign is valid if no plastic strain occurred in the cycle.

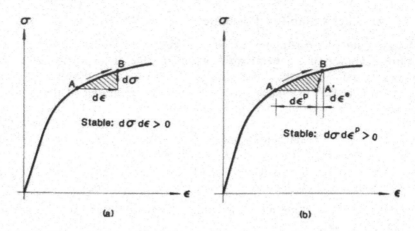

FIGURE 5.8. Stability postulate for work-hardening materials: (a) stability in small;
(b) stability in cycle in small.

Consider a material element subjected to a homogeneous state of stress
$\sigma_{ij}^*$ which is either on or inside the yield surface (Fig. 5.9). Suppose an
external agency adds stresses along a path $ABC$ with $AB$ lying inside the
yield surface and point $B$ just on it. The stresses continue to move outward
and cause the yield surface to evolute until point $C$ is reached. The external
agency then releases and returns the state of stress back to $\sigma_{ij}^*$ along an
elastic path $CDA$. As the elastic deformations are fully reversible and
independent of the path from $\sigma_{ij}^*$ to $\sigma_{ij}$ and back to $\sigma_{ij}^*$, all the elastic energy
is recovered. The plastic work done by the external agency on this loading
and unloading cycle is the scalar product of the stress vector $\sigma_{ij} - \sigma_{ij}^*$ and
the plastic strain increment vector $d\epsilon_{ij}^p$. This stability requirement of Eq.
(5.60) leads to

$$(\sigma_{ij} - \sigma_{ij}^*) \, d\epsilon_{ij}^p \geq 0 \qquad\qquad (5.61)$$

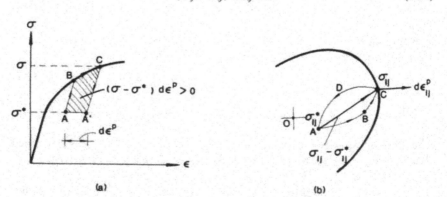

FIGURE 5.9. Stability in cycle: Existing state of stress inside the yield surface (point
$A$); stress path $ABC$ produced by external agency.

If plastic strain coordinates are superimposed upon stress coordinates, as in Fig. 5.9, Eq. (5.61) can be interpreted geometrically as the scalar product of the stress increment vector $(\sigma_{ij} - \sigma_{ij}^*)$ with the strain increment vector $d\epsilon_{ij}^p$. A positive scalar product requires an acute angle between these two vectors. The stability postulate leads therefore to the following consequences for work-hardening materials (Drucker, 1960):

Convexity: The initial yield and all the subsequent loading surfaces must be *convex*.

Normality: The plastic strain increment vector $d\epsilon_{ij}^p$ must be *normal* to the yield or loading surface $f(\sigma_{ij}, \epsilon_{ij}^p, k) = 0$ at a smooth point:

$$d\epsilon_{ij}^p = d\lambda \frac{\partial f}{\partial \sigma_{ij}} \tag{5.62}$$

and lie between adjacent normals at a corner.

Convexity and normality conditions for elastic–perfectly plastic materials have been discussed in Chapter 4. The reasoning is also found sound here for work-hardening materials.

Linearity: The plastic strain increment must be *linear* in the stress increment. Equation (5.62) indicates that the ratio of the components of plastic strain increment, $d\epsilon_{ij}^p$, are independent of the ratios of the components of stress increment, $d\sigma_{ij}$, at any smooth point on the surface. However, the magnitude of $d\epsilon_{ij}^p$, characterized by the scalar $d\lambda$, is dependent only on the projection of the stress increment, $d\sigma_{ij}$, onto the direction of the normal $\partial f/\partial \sigma_{ij}$. That is,

$$d\lambda = \bar{G}\, \partial f = \bar{G} \frac{\partial f}{\partial \sigma_{mn}} d\sigma_{mn} \tag{5.63}$$

and

$$d\epsilon_{ij}^p = \bar{G}\, \partial f \frac{\partial f}{\partial \sigma_{ij}} = \bar{G} \frac{\partial f}{\partial \sigma_{ij}} \frac{\partial f}{\partial \sigma_{mn}} d\sigma_{mn} \tag{5.64}$$

where $\bar{G}$ is a scalar function which may depend upon stress, strain, and the history of loading. But $\bar{G}$ is independent of $d\sigma_{ij}$. Note that in Eq. (5.63), the increment $\partial f$ is evaluated only with respect to increments in the stress components, i.e., with other variables unchanged [see Eq. (5.19)].

Continuity: The condition of *continuity* requires that for $d\sigma_{ij}$ tangential to the yield surface (neutral loading), no plastic increment is induced. This condition is satisfied by Eqs. (5.63) and (5.64) since for $d\sigma_{ij}$ tangential to the yield surface, we have $\partial f = (\partial f/\partial \sigma_{mn})d\sigma_{mn} = 0$.

Uniqueness: Uniqueness of the solution of a boundary-value problem for a work-hardening material can be proved directly from the stability postulate (Drucker, 1956). Suppose a body is under the action of existing surface traction $T_i$, body forces $F_i$, displacements $u_i$, stresses $\sigma_{ij}$, and strains $\epsilon_{ij}$ (see Fig. 3.7a). If small changes of the applied forces and displacements, $dT_i$ on $A_T$, $dF_i$ in $V$, $du_i$ on $A_u$, are now imposed on the body, uniqueness requires that the stress and strain changes, $d\sigma_{ij}$ and $d\epsilon_{ij}$, be uniquely determined by the changes of applied forces and displacements.

This can be proved in the usual manner. Two solutions are assumed: $d\sigma_{ij}^a$, $d\epsilon_{ij}^a$ and $d\sigma_{ij}^b$, $d\epsilon_{ij}^b$, corresponding to the same applied load and displacement increments $dT_i$ on $A_T$, $du_i$ on $A_u$, and $dF_i$ in $V$. Using the equation of virtual work, we have [Eq. (3.171)]

$$\int_V (d\sigma_{ij}^a - d\sigma_{ij}^b)(d\epsilon_{ij}^a - d\epsilon_{ij}^b)dV = 0 \tag{5.65}$$

If the integrand in Eq. (5.65) can be shown to be *positive definite*, uniqueness is proved. As a first step, the strain rates are decomposed into elastic and plastic parts, $d\epsilon_{ij} = d\epsilon_{ij}^e + d\epsilon_{ij}^p$, and the integrand is written

$$(d\sigma_{ij}^a - d\sigma_{ij}^b)(d\epsilon_{ij}^{ea} - d\epsilon_{ij}^{eb}) + (d\sigma_{ij}^a - d\sigma_{ij}^b)(d\epsilon_{ij}^{pa} - d\epsilon_{ij}^{pb}) \tag{5.66}$$

The first term is always positive definite for both linear and nonlinear elasticity (see Section 3.6, Chapter 3). Thus, if the second term can also be shown to be positive or zero, then the positive-definiteness of the integrand is proved.

To examine the second term of Eq. (5.66), consider three possibilities. If $a$ and $b$ are both elastic changes (Fig. 5.10a), both $d\epsilon_{ij}^{pa}$ and $d\epsilon_{ij}^{pb}$ vanish, so the second term is zero. If $b$ is elastic, i.e., $d\epsilon_{ij}^{pb} = 0$, and $a$ is elastic–plastic (Fig. 5.10b), then the second term is positive because both $d\sigma_{ij}^a d\epsilon_{ij}^{pa}$ and $-d\sigma_{ij}^b d\epsilon_{ij}^{pa}$ are positive. When $a$ and $b$ both represent elastic–plastic changes (Fig. 5.10c), we first note that from Eq. (5.64), the incremental plastic strain-stress relation is linear and can therefore be written in the general form:

$$d\epsilon_{ij}^p = H_{ijkl} \, d\sigma_{kl} \tag{5.67}$$

in which the coefficients $H_{ijkl}$ are functions of stress and may also depend upon the strain and the history of loading, but do not depend on the stress increment, $d\sigma_{ij}$. Now both $d\epsilon_{ij}^{pa}$, $d\sigma_{ij}^a$ and $d\epsilon_{ij}^{pb}$, $d\sigma_{ij}^b$ satisfy Eq. (5.67). Thus, the difference between $a$ and $b$ also satisfies Eq. (5.67). The stress difference $d\sigma_{ij}^a - d\sigma_{ij}^b$ may be considered then as applied by the external agency which produces the corresponding plastic strain

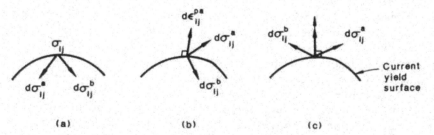

FIGURE 5.10. Proof of uniqueness: (a) both $a$ and $b$ are elastic; (b) $b$ is elastic, $a$ is elastic-plastic; (c) both $a$ and $b$ are elastic-plastic.

difference $d\epsilon_{ij}^{pa} - d\epsilon_{ij}^{pb}$. The stability postulate of Eq. (5.59) then gives

$$(d\sigma_{ij}^a - d\sigma_{ij}^b)(d\epsilon_{ij}^{pa} - d\epsilon_{ij}^{pb}) > 0 \tag{5.68}$$

Hence, Eq. (5.66) is positive definite, i.e.,

$$(d\sigma_{ij}^a - d\sigma_{ij}^b)(d\epsilon_{ij}^a - d\epsilon_{ij}^b) > 0 \tag{5.69}$$

and the uniqueness condition is therefore established.

## 5.4.3. Nonassociated Flow Rule

It has been shown that the associated flow (normality) rule and the convexity, continuity, and uniqueness conditions are all the consequences of Drucker's stability postulate. This is a fundamental unification of the theory of plasticity.

However, it is to be noted that the stability postulate is a *sufficient* but not a *necessary* criterion. In other words, this postulate may not be *necessarily* required in a general formulation of any flow rule for elastic–plastic materials (Mroz, 1963). It has been shown that for an elastic-work-hardening-plastic material, the uniqueness allows a nonassociated flow rule to occur which does not necessarily satisfy Drucker's stability postulate. Further, since when the uniqueness of stress and strain trajectories for a given loading history exists, the material can be regarded as locally stable, thus the condition of uniqueness rather than the stability postulate may be regarded as a basic criterion in establishing elastic–plastic stress–strain relationships. Based on uniqueness, certain conditions imposed on the plastic potential function have been derived by Mroz (1963).

The nonassociated flow rule can be expressed as

$$d\epsilon_{ij}^p = d\lambda \frac{\partial g}{\partial \sigma_{ij}} = \bar{G} \, \partial f \frac{\partial g}{\partial \sigma_{ij}} = \bar{G} \frac{\partial g}{\partial \sigma_{ij}} \frac{\partial f}{\partial \sigma_{mn}} d\sigma_{mn} \tag{5.70}$$

in which Eq. (5.63) has been used, and the plastic strain increment is *linear* in stress increment. The condition of continuity is also satisfied.

For some geotechnical materials, such as rocks, soils, and concretes, it has been found that the associated flow rule tends to overestimate the plastic volume expansion. The nonassociated flow rule is therefore adopted in establishing the constitutive relations. This will be discussed in Chapter 7 in the context of modeling the inelastic behavior of concrete materials.

When the plastic potential function $g$ (or $f$ for associated plastic flow) is of the most general isotropic form $g(I_1, J_2, J_3)$, Eq. (5.70) [or Eq. (5.64) using $f$ as potential function instead of $g$] leads to

$$d\epsilon_{ij}^p = \bar{G} \left( \frac{\partial g}{\partial I_1} \frac{\partial I_1}{\partial \sigma_{ij}} + \frac{\partial g}{\partial J_2} \frac{\partial J_2}{\partial \sigma_{ij}} + \frac{\partial g}{\partial J_3} \frac{\partial J_3}{\partial \sigma_{ij}} \right) \partial f \tag{5.71}$$

which can be written in the general form as

$$d\epsilon_{ij}^p = [P(I_1, J_2, J_3) \, \delta_{ij} + Q(I_1, J_2, J_3)s_{ij} + R(I_1, J_2, J_3)t_{ij}] \, \partial f \tag{5.72}$$

where $t_{ij}$, defined in Eq. (5.17), is the deviation of the square of the stress deviator $s_{ij}$. A marked similarity can be seen between the *deformation* theory [Eq. (5.18)] and the *incremental* theory [Eq. (5.72)], but the difference is extremely important. Now, when $\partial f = 0$, that is when the stress change is on the current loading surface or a neutral loading, there is no change in any component of plastic strain (*condition of continuity*).

In a true sense, such a theory is isotropic because the principal stresses may have any orientation with respect to axes fixed in the material. However, it is anisotropic in that the principal directions of the increments in plastic strain will not coincide with the stress increments. The *anisotropy* is introduced by the state of stress, but it is not *intrinsic*. Removal of the stress leaves a material isotropic in the usual sense. Similarly, rotation of the state of stress with respect to the material rotates the anisotropy. This can clearly be seen if the components of the stress–strain relation (5.72) are written out; with the stress increments appearing explicitly, they look like a highly anisotropic incremental generalized Hooke's law of the form

$$de_x = G_1 \, d\sigma_x + G_2 \, d\sigma_y + G_3 \, d\sigma_z + G_4 \, d\tau_{xy} + G_5 \, d\tau_{yz} + G_6 \, d\tau_{zx} \quad (5.73)$$

where the $G$'s are functions of the state of stress and include both the elastic and the plastic behavior. An increment of shearing stress may produce an elongation or contraction, and similarly, an increment of normal stress may cause a shearing strain. However, as previously stated, the anisotropy is produced by the existing state of stress and is not intrinsic.

## 5.5. Effective Stress and Effective Strain

For the work-hardening theory of plasticity to be of any practical use, we must relate the hardening parameters in the loading function to the experimental uniaxial stress-strain curve. To this end, we are looking for some stress variable, called *effective stress*, that is a function of the stresses and some strain variable, called *effective strain*, that is a function of the plastic strains, so that they can be plotted against each other and used to correlate the test results obtained by different loading programs. The single effective stress-effective strain curve should preferably be reduced to a uniaxial stress-strain curve for the uniaxial stress test.

### 5.5.1. Effective Stress

Since the loading function, $f(\sigma_{ij}, \epsilon_{ij}^p, k) = 0$ by definition, determines whether additional plastic flow takes place or not and is also a positively increasing function, it can be used as a truly significant stress variable to define the *effective stress*.

Consider the case of *isotropic* hardening in which the loading function takes the form of Eq. (5.25) or

$$f(\sigma_{ij}, k) = F(\sigma_{ij}) - k^2(\epsilon_p) = 0$$

The function $F(\sigma_{ij})$ is used to define the effective stress. Since the effective stress should reduce to the stress $\sigma_1$ in the uniaxial test, it follows that the loading function $F(\sigma_{ij})$ must be some constant $C$ times the effective stress $\sigma_e$ to some power

$$F(\sigma_{ij}) = C\sigma_e^n \tag{5.74}$$

For example, if we assume von Mises material, $F(\sigma_{ij}) = J_2$, then

$$J_2 = C\sigma_e^n \tag{5.75}$$

or

$$J_2 = \tfrac{1}{6}[(\sigma_1 - \sigma_2)^2 + (\sigma_2 - \sigma_3)^2 + (\sigma_3 - \sigma_1)^2] = C\sigma_e^n \tag{5.76}$$

and for the uniaxial test, $\sigma_e = \sigma_1$ and $\sigma_2 = \sigma_3 = 0$. Therefore

$$n = 2, \quad C = \tfrac{1}{3}, \quad \sigma_e = \sqrt{3J_2} \tag{5.77}$$

Similarly, for Drucker-Prager material, $F(\sigma_{ij}) = \alpha I_1 + \sqrt{J_2}$,

$$\alpha I_1 + \sqrt{J_2} = C\sigma_e^n \tag{5.78}$$

and for the uniaxial test, we have

$$\alpha\sigma_e + \frac{1}{\sqrt{3}}\sigma_e = C\sigma_e^n$$

Therefore,

$$n = 1, \quad C = \alpha + \frac{1}{\sqrt{3}}, \quad \sigma_e = \frac{\sqrt{3}\alpha I_1 + \sqrt{3J_2}}{1 + \sqrt{3}\alpha} \tag{5.79}$$

For the cases of *kinematic* hardening or *mixed* hardening, the loading function is written in the general form as Eq. (5.51)

$$f(\sigma_{ij}, \epsilon_{ij}^p, k) = F(\sigma_{ij} - \alpha_{ij}) - k^2(\epsilon_p) = 0 \tag{5.80}$$

We denote

$$\bar{\sigma}_{ij} = \sigma_{ij} - \alpha_{ij} \tag{5.81}$$

as the *reduced-stress tensor*, measured from an origin at the center of the translated yield surface. Then, the *reduced effective stress* $\bar{\sigma}_e$ is defined by the following relation similar to Eq. (5.74):

$$F(\bar{\sigma}_{ij}) = C\bar{\sigma}_e^n \tag{5.82}$$

Equations (5.77) and (5.79) are still valid for definitions of the reduced effective stress of von Mises and Drucker-Prager materials, respectively. Note that a reduced effective stress is associated with an expansion of the loading surface.

## 5.5.2. Effective Strain

The definition of *effective plastic strain*, $\epsilon_p$, is not quite as simple. Two methods are generally used. One defines the effective plastic strain increment intuitively as some simple combination of plastic strain increments which is always positive and increasing. The simplest combination of this type with the correct "dimension" is

$$d\epsilon_p = C\sqrt{d\epsilon_{ij}^p \, d\epsilon_{ij}^p} \tag{5.83}$$

For example, if we assume an $f(J_2, J_3)$ type of pressure-independent material which satisfies the plastic-incompressibility condition

$$d\epsilon_1^p + d\epsilon_2^p + d\epsilon_3^p = 0 \tag{5.84}$$

then to make the definition (5.83) agree for the uniaxial stress test, we must have

$$d\epsilon_1^p = d\epsilon_p = C\sqrt{[(d\epsilon_1^p)^2 + (\tfrac{1}{2}d\epsilon_1^p)^2 + (\tfrac{1}{2}d\epsilon_1^p)^2]}$$
$$= C\sqrt{\tfrac{3}{2}}d\epsilon_1^p \tag{5.85}$$

which leads to

$$C = \sqrt{\tfrac{2}{3}}, \qquad d\epsilon_p = \sqrt{\tfrac{2}{3}d\epsilon_{ij}^p \, d\epsilon_{ij}^p} \tag{5.86}$$

The second method defines the effective plastic strain increment in terms of the *plastic work* per unit volume in the form

$$dW_p = \sigma_e \, d\epsilon_p \tag{5.87}$$

By definiton, the increment of the plastic work is

$$dW_p = \sigma_{ij} \, d\epsilon_{ij}^p = d\lambda \, \sigma_{ij} \frac{\partial f}{\partial \sigma_{ij}} = d\lambda \, \sigma_{ij} \frac{\partial F}{\partial \sigma_{ij}} \tag{5.88}$$

in which the plastic strain increment $d\epsilon_{ij}^p$ has been related to the stresses by the flow rule of Eq. (5.62). If the function $F$ is homogeneous of degree $n$ in the stresses, as it is for many cases in metal plasticity theories, Eq. (5.88) can be further reduced to

$$dW_p = d\lambda \, nF \tag{5.89}$$

The scalar function $d\lambda$ can be obtained by squaring each of the terms in Eq. (5.62) and adding:

$$d\epsilon_{ij}^p \, d\epsilon_{ij}^p = (d\lambda)^2 \frac{\partial F}{\partial \sigma_{ij}} \frac{\partial F}{\partial \sigma_{ij}} \tag{5.90}$$

Taking the square root of both sides and substituting $d\lambda$ into Eq. (5.89) shows that $dW_p$ must be a function of $F$ and $\sqrt{(d\epsilon_{ij}^p \, d\epsilon_{ij}^p)}$

$$dW_p = \frac{\sqrt{d\epsilon_{ij}^p \, d\epsilon_{ij}^p} \, nF}{\sqrt{\partial F/\partial \sigma_{mn} \, \partial F/\partial \sigma_{mn}}} = \sigma_e \, d\epsilon_p \tag{5.91}$$

where we have used Eq. (5.87) to determine the effective plastic strain $\epsilon_p$.

If, for example, the Drucker-Prager $F = \alpha I_1 + \sqrt{J_2}$ is used, the plastic work equation (5.91) becomes

$$dW_p = \frac{\sqrt{d\epsilon_{ij}^p \, d\epsilon_{ij}^p}(1)(\alpha I_1 + \sqrt{J_2})}{\sqrt{3\alpha^2 + \frac{1}{2}}}$$

$$= \frac{\sqrt{3}(\alpha I_1 + \sqrt{J_2})}{1 + \sqrt{3}\alpha} \, d\epsilon_p \tag{5.92}$$

where $(\partial F/\partial \sigma_{ij})(\partial F/\partial \sigma_{ij}) = 3\alpha^2 + \frac{1}{2}$, $n = 1$, and $\sigma_e$ of Eq. (5.79) have been used. From Eq. (5.92), it can readily be shown that

$$d\epsilon_p = \frac{\alpha + 1/\sqrt{3}}{\sqrt{3\alpha_2 + \frac{1}{2}}} \sqrt{d\epsilon_{ij}^p \, d\epsilon_{ij}^p} \tag{5.93}$$

As for the von Mises material, for which $\alpha = 0$, Eq. (5.93) reduces to Eq. (5.86), where the effective plastic strain $\epsilon_p$ is defined by an alternative method which is rather intuitive. In general, the two definitions of effective plastic strain $\epsilon_p$—in terms of the plastic work [Eq. (5.87)] or in terms of the accumulated plastic strain [Eq. (5.83)]—will result in different scalar functions depending on the loading function. They are the same only for $F = J_2$. However, the effective plastic strain $\epsilon_p$, as defined by Eq. (5.86) for $F = J_2$ material, is found to be reasonably correct for almost any $F(J_2, J_3)$ material that is pressure independent.

### 5.5.3. Effective Stress–Effective Strain Relation

The effective stress–effective strain relation, characterizing the hardening processes of a material, is now calibrated on the uniaxial stress test, which has the general form

$$\sigma_e = \sigma_e(\epsilon_p) \tag{5.94}$$

Differentiation gives the incremental relation

$$d\sigma_e = H_p(\sigma_e) d\epsilon_p \tag{5.95}$$

where $H_p(\sigma_e)$ is a *plastic modulus* associated with the rate of expansion of the yield or loading surface

$$H_p = \frac{d\sigma_e}{d\epsilon_p} \tag{5.96}$$

with $H_p$ the slope of the uniaxial stress-plastic strain curve at the current value of $\sigma_e$.

The *strain history* for the material as recorded by the *length of the effective plastic strain path*

$$\epsilon_p = \int d\epsilon_p = \int \frac{d\sigma_e}{H_p(\sigma_e)} \tag{5.97}$$

must be a function of effective stress only. There will be a unique inverse for a work-hardening material, so that $\sigma_e$ or $F$ is a function of $\epsilon_p = \int d\epsilon_p$.

### 5.5.3.1. MIXED HARDENING

A combination of isotropic and kinematic hardening allows the yield surface to expand and to translate simultaneously in stress space. The yield and loading function takes the form of Eq. (5.80). We have defined the reduced effective stress by Eq. (5.82). Since the strain $\epsilon_p$ in Eq. (5.80) governs the isotropic expansion of the yield surface, it can be considered the *reduced effective plastic strain* and denoted by $\bar{\epsilon}_p$. Then, the loading surface of Eq. (5.80) is rewritten as

$$f = F(\bar{\sigma}_{ij}) - k^2(\bar{\epsilon}_p) = 0 \tag{5.98}$$

which is a measure of the expansion of the yield surface from an origin at the center of this surface. The rate of expansion of the yield surface is governed by the *reduced effective stress–strain relation*

$$\bar{\sigma}_e = \bar{\sigma}_e(\bar{\epsilon}_p) \tag{5.99}$$

determined by the experimental uniaxial stress–strain relation.

The total increment of plastic strain is now simply split into two collinear components

$$d\epsilon_{ij}^p = d\epsilon_{ij}^i + d\epsilon_{ij}^k \tag{5.100}$$

where $d\epsilon_{ij}^i$ is associated with the expansion of the yield surface and $d\epsilon_{ij}^k$ is associated with the translation of the yield surface. These two strain components can be written as

$$d\epsilon_{ij}^i = M \, d\epsilon_{ij}^p \tag{5.101}$$

$$d\epsilon_{ij}^k = (1 - M) \, d\epsilon_{ij}^p \tag{5.102}$$

where $M$ is the *parameter of mixed hardening*, with the range $0 < M \leq 1$.

The share of the plastic strain increment $d\epsilon_{ij}^i$ associated with the expansion of the yield surface is now used to define the *reduced effective strain* $d\bar{\epsilon}_p$ as

$$d\bar{\epsilon}_p = C\sqrt{d\epsilon_{ij}^i \, d\epsilon_{ij}^i} \tag{5.103}$$

It follows from Eq. (5.101) that the *reduced effective plastic strain* $\bar{\epsilon}_p$ associated with isotropic hardening is now related to the effective plastic strain $\epsilon_p$ by the simple relation

$$\bar{\epsilon}_p = M \int C\sqrt{d\epsilon_{ij}^p \, d\epsilon_{ij}^p} = M\epsilon_p \tag{5.104}$$

Differentiation of Eq. (5.99) gives the rate of expansion of the yield surface

$$d\bar{\sigma}_e = \bar{H}_p \, d\bar{\epsilon}_p = M\bar{H}_p \, d\epsilon_p \tag{5.105}$$

where $\bar{H}_p$ is the *plastic modulus* associated with the expansion of the yield surface.

The rate of translation of the yield surface, $d\alpha_{ij}$, as given by Eq. (5.37) or Eqs. (5.47) and (5.48), is related to the share of the plastic strain increment $d\epsilon_{ij}^k = (1-M)\, d\epsilon_{ij}^p$ that is associated with the translation. Hence, in the case of mixed hardening, $d\epsilon_{ij}^k$ and $d\epsilon_p^k = (1-M)\epsilon_p$ must replace $d\epsilon_{ij}^p$ in Eq. (5.37) and $d\epsilon_p$ in Eq. (5.48) in calculating the translation rate $d\alpha_{ij}$

$$d\alpha_{ij} = c\, d\epsilon_{ij}^k = c(1-M)\, d\epsilon_{ij}^p \qquad \text{for Prager's rule} \qquad (5.106)$$

or

$$d\alpha_{ij} = a(1-M)\, d\epsilon_p(\sigma_{ij} - \alpha_{ij}) \qquad \text{for Ziegler's rule} \qquad (5.107)$$

## 5.6. Illustrative Examples

We have discussed the hardening rules and the flow rules. Based on these important assumptions, the general stress–strain relations can now be established for a work-hardening material. The concepts of the effective stress and the effective strain allow us to calibrate the multiaxial stress–strain relation on the uniaxial stress test data. In the following, we shall present some illustrative examples.

EXAMPLE 5.5. (a) Given the uniaxial stress–plastic strain relation $\sigma_e = \sigma(\epsilon_p)$, find the hardening function $\bar{G}$ in the general stress–strain formulation of Eq. (5.64) for an isotropic-hardening von Mises material with a loading surface of the form

$$f = J_2 - k^2(\epsilon_p) = 0 \qquad (5.108)$$

(b) If the uniaxial stress–strain relation is given by the Ramberg–Osgood relation

$$\epsilon_1 = \epsilon_1^e + \epsilon_1^p = \frac{\sigma_1}{E} + a\left(\frac{\sigma_1}{b}\right)^{2n+1} \qquad (5.109)$$

in which $E$ is the initial elastic modulus and $a$, $b$, and $n$ are constants, find the incremental plastic strain–stress relation.

SOLUTION. (a) For the simplest and most commonly used isotropic-hardening von Mises model of Eq. (5.108), $\partial f/\partial\sigma_{ij} = \partial J_2/\partial\sigma_{ij} = s_{ij}$, $\partial f = dJ_2$, Eq. (5.64) takes the form:

$$d\epsilon_{ij}^p = \bar{G}s_{ij}\, dJ_2 \qquad (5.110)$$

Squaring each of the terms represented by Eq. (5.110) and adding, we have

$$d\epsilon_{ij}^p\, d\epsilon_{ij}^p = \bar{G}^2 2J_2\, (dJ_2)^2$$

Taking the square root of both sides and noting the definitions of effective stress $\sigma_e = \sqrt{(3J_2)}$ and effective strain $d\epsilon_p = \sqrt{(\tfrac{2}{3}\, d\epsilon_{ij}^p\, d\epsilon_{ij}^p)}$ yields

$$d\epsilon_p = \tfrac{2}{3}\bar{G}\sigma_e\, dJ_2 = \tfrac{4}{9}\bar{G}\sigma_e^2\, d\sigma_e \qquad (5.111)$$

in which the relation $dJ_2 = \frac{2}{3}\sigma_e \, d\sigma_e$ has been used. For a given $\sigma_e$–$\epsilon_p$ relationship, the plastic modulus $H_p = d\sigma_e/d\epsilon_p$ is known, and the hardening function $\bar{G}$ is now found to be

$$\bar{G} = \frac{9}{4}\frac{1}{H_p\sigma_e^2} = \frac{3}{4H_pJ_2} \qquad (5.112)$$

(b) Now, the uniaxial stress–strain relation is given by Eq. (5.109) in which

$$\epsilon^P = a\left(\frac{\sigma_1}{b}\right)^{2n+1}$$

Hence,

$$H_p = \frac{d\sigma_e}{d\epsilon_p} = \frac{d\sigma}{d\epsilon^P} = \frac{d\sigma_1}{d\epsilon_1^P} = \frac{1}{2n+1}\left(\frac{b}{a}\right)\left(\frac{b}{\sigma_e}\right)^{2n} \qquad (5.113)$$

Substitution of Eqs. (5.113) and (5.112) into Eq. (5.110) leads to the plastic strain–stress relation

$$d\epsilon_{ij}^P = \frac{3(2n+1)}{4}\left(\frac{a}{b}\right)\left(\frac{3J_2}{b^2}\right)^n s_{ij}\left(\frac{dJ_2}{J_2}\right) \qquad \text{for } J_2 = k^2 \text{ and } dJ_2 > 0$$

in which $dJ_2$ is linear in stress increment. This constitutive equation can be written out in component form like Eq. (5.73).

EXAMPLE 5.6. An initially unstressed and unstrained thin-walled circular tube is subjected to a combined axial-tension and twisting-moment loading history which produces the successive straight-line paths in $(\sigma, \tau)$ space shown in Fig. 5.11a. Assume that the material is elastic–plastic, and the elastic response is linear with Young's modulus $E = 210$ GPa and $\nu = 0.3$, while the plastic response is of the von Mises isotropic stress-hardening type. The stress–strain curve in simple tension is given by

$$\epsilon = \epsilon^e + \epsilon^P = \frac{\sigma}{2.1 \times 10^5} + \frac{1}{3 \times 10^6}\left(\frac{\sigma}{7}\right)^3 \qquad (5.114)$$

where $\sigma$ is in MPa. Note that the plastic strain represented by the second term takes place at the beginning of loading.

(a) Write the stress–strain relation in component form explicitly in terms of $\sigma$, $\tau$, $d\sigma$, and $d\tau$.
(b) Find the elastic and plastic components of strain at the end of each loading path.

SOLUTION. (a) For the isotropic-hardening von Mises model, from Example 5.5, we have

$$d\epsilon_{ij}^P = \bar{G}(J_2)s_{ij}\, dJ_2 \qquad (5.115)$$

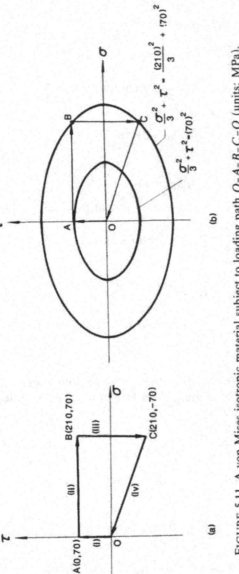

FIGURE 5.11. A von Mises isotropic material subject to loading path $O$-$A$-$B$-$C$-$O$ (units: MPa).

For the elastic strain, based on Hooke's law, we have

$$de_{ij}^e = \frac{1+\nu}{E} ds_{ij} + \frac{1-2\nu}{3E} d\sigma_{kk} \delta_{ij} \qquad (5.116)$$

Adding up the elastic and plastic strains leads to

$$d\epsilon_{ij} = \frac{1+\nu}{E} ds_{ij} + \frac{1-2\nu}{3E} d\sigma_{kk} \delta_{ij} + \bar{G} s_{ij} dJ_2 \qquad (5.117)$$

Since in $\sigma$-$\tau$ space, $\sigma_x = \sigma$, $\tau_{xy} = \tau$, other components $= 0$; we have

$$J_2 = \tfrac{1}{3}\sigma^2 + \tau^2$$

$$dJ_2 = \tfrac{2}{3}\sigma \, d\sigma + 2\tau \, d\tau$$

and

$$d\sigma_{kk} = d\sigma$$

Equation (5.117) can be expressed in component form as

$$d\epsilon_x = d\epsilon = \frac{d\sigma}{E} + \frac{2}{3} \bar{G}\sigma \left(\frac{2}{3}\sigma \, d\sigma + 2\tau \, d\tau\right)$$

$$d\epsilon_y = d\epsilon_z = -\nu \frac{d\sigma}{E} - \frac{1}{3} \bar{G}\sigma \left(\frac{2}{3}\sigma \, d\sigma + 2\tau \, d\tau\right)$$

$$(5.118)$$

$$d\gamma_{xy} = d\gamma = \frac{2(1+\nu)}{E} d\tau + 2\bar{G}\tau \left(\frac{2}{3}\sigma \, d\sigma + 2\tau \, d\tau\right)$$

$$d\gamma_{xz} = d\gamma_{yz} = 0$$

From the given simple tension data, the hardening function $\bar{G}(J_2)$ in Eq. (5.115) can be calculated by using Eqs. (5.112) and (5.113) with $a = 10^{-6}/3$, $b = 7$, and $n = 1$ as

$$\bar{G} = \frac{9}{4} \frac{1}{H_p \sigma_e^2} = \frac{9}{4} \frac{1}{\sigma_e^2 (d\sigma_e/d\epsilon_p)} = \frac{9}{4} \frac{\sigma_e^2}{\sigma_e^2 (7^3)(10^6)} = 6.56 \times 10^{-9} \quad (5.119)$$

Substituting Eq. (5.119) into (5.118) yields the formulas for the calculation of all strain increments in the plastic loading cases:

$$d\epsilon_x = d\epsilon = \left(\frac{1}{E} + 2.92 \times 10^{-9}\sigma^2\right) d\sigma + (8.75 \times 10^{-9}\sigma\tau) \, d\tau$$

$$d\epsilon_y = d\epsilon_z = -\left(\frac{\nu}{E} + 1.46 \times 10^{-9}\sigma^2\right) d\sigma - (4.38 \times 10^{-9}\sigma\tau) \, d\tau \qquad (5.120)$$

$$d\gamma_{xy} = d\gamma = (8.75 \times 10^{-9}\tau\sigma) \, d\sigma + [1.24 \times 10^{-5} + 26.25 \times 10^{-9}\tau^2] \, d\tau$$

(b) The elastic and plastic strain components at the end of each loading path (Fig. 5.11b) are found as follows.

(i) For path $OA$, plastic loading occurs. Since $\sigma = 0$, $d\sigma = 0$, we obtain the strain components from Eq. (5.120) as

$$\epsilon_{xA} = \epsilon_{yA} = \epsilon_{zA} = 0$$

and

$$\gamma_A^e = \frac{2(1+\nu)}{E} \int_0^{\tau_A} d\tau = \frac{(2)(1.3)(70)}{2.1 \times 10^5} = 8.67 \times 10^{-4}$$

$$\gamma_A^p = 26.25 \times 10^{-9} \int_0^{\tau_A} \tau^2 \, d\tau = (26.25 \times 10^{-9})\left(\frac{1}{3}\right)(70)^3 = 3 \times 10^{-3}$$

At the end of Path $OA$, the elastic and plastic strain tensors are now obtained as

$$\epsilon_A^e = \begin{bmatrix} 0 & 0.867/2 & 0 \\ 0.867/2 & 0 & 0 \\ 0 & 0 & 0 \end{bmatrix} \times 10^{-3}, \qquad \epsilon_A^p = \begin{bmatrix} 0 & \frac{3}{2} & 0 \\ \frac{3}{2} & 0 & 0 \\ 0 & 0 & 0 \end{bmatrix} \times 10^{-3}$$

and the subsequent yield surface is given as

$$J_2 = \frac{\sigma_2}{3} + \tau^2 = (70)^2 \tag{5.121}$$

which is shown in Fig. 5.11b.

(ii) Path $AB$ is also a plastic loading path (see Fig. 5.11b). Since along path $AB$, $\tau = 70$ MPa, $d\tau = 0$, the elastic plastic strain components at the end of this path are obtained from Eq. (5.120) as follows:

$$\epsilon_{xB}^e = \epsilon_{xA}^e + \int_A^B d\epsilon_x^e = 0 + \int_0^{210} \frac{d\sigma}{E} = \frac{210}{2.1 \times 10^5} = 10^{-3}$$

$$\epsilon_{yB}^e = \epsilon_{yA}^e + \int_A^B d\epsilon_y^e = 0 - \nu \int_0^{210} \frac{d\sigma}{E} = -0.3 \times 10^{-3}$$

$$\epsilon_{zB}^e = \epsilon_{yB}^e = -0.3 \times 10^{-3}$$

$$\gamma_B^e = \gamma_A^e = 8.67 \times 10^{-4}$$

$$\epsilon_{xB}^p = \epsilon_{xA}^p + \int_A^B d\epsilon_x^p = 0 + \int_0^{210} (2.92 \times 10^{-9})\sigma^2 \, d\sigma = 9.01 \times 10^{-3}$$

$$\epsilon_{yB}^p = -\tfrac{1}{2}\epsilon_{xB}^p = -4.5 \times 10^{-3}$$

$$\epsilon_{zB}^p = \epsilon_{yB}^p = -4.5 \times 10^{-3}$$

$$\gamma_B^p = \gamma_A^p + \int_A^B d\gamma^p = 3 \times 10^{-3} + \int_0^{210} (8.75 \times 10^{-9})(70)\sigma \, d\sigma = 16.5 \times 10^{-3}$$

which can be expressed in matrix form as

$$
\epsilon_B^e = \begin{bmatrix} 1 & 0.867/2 & 0 \\ 0.867/2 & -0.3 & 0 \\ 0 & 0 & -0.3 \end{bmatrix} \times 10^{-3}
$$

$$
\epsilon_B^p = \begin{bmatrix} 9.01 & 16.5/2 & 0 \\ 16.5/2 & -4.5 & 0 \\ 0 & 0 & -4.5 \end{bmatrix} \times 10^{-3}
$$

During loading, the yield surface expands and at point $B$, we have

$$
J_2 = \frac{\sigma^2}{3} + \tau^2 = \frac{(210)^2}{3} + (70)^2 \tag{5.122}
$$

which is the current yield surface as shown in Fig. 5.11b.

(iii) It can be seen from Fig. 5.11b that along path $BC$, elastic unloading occurs and the strain changes in path $BC$ are purely elastic and can be obtained by Hooke's law. Since along $BC$, $\sigma = 210$ MPa, $d\sigma = 0$, we have

$$
d\epsilon_x^e = d\epsilon_y^e = d\epsilon_z^e = 0
$$

and

$$
\int_B^C d\gamma^e = \int_{70}^{-70} \frac{2(1+\nu)}{E} \, d\tau = -1.73 \times 10^{-3}
$$

At the end of this path, the strain tensors are given by

$$
\epsilon_C^e = \begin{bmatrix} 1 & -0.867/2 & 0 \\ -0.867/2 & -0.3 & 0 \\ 0 & 0 & -0.3 \end{bmatrix} \times 10^{-3}
$$

$$
\epsilon_C^p = \epsilon_B^p = \begin{bmatrix} 9.01 & 16.5/2 & 0 \\ 16.5/2 & -4.5 & 0 \\ 0 & 0 & -4.5 \end{bmatrix} \times 10^{-3}
$$

(iv) Path $CO$ is also an elastic unloading path. The changes of the elastic strains are found by Hooke's law as

$$
\int_C^0 d\epsilon_x^e = \int_{210}^0 \frac{d\sigma}{E} = -10^{-3}
$$

$$
\int_C^0 d\epsilon_y^e = -\nu \int_{210}^0 \frac{d\sigma}{E} = 0.3 \times 10^{-3}
$$

$$
\int_C^0 d\gamma^e = \int_{-70}^0 \frac{2(1+\nu)}{E} \, d\tau = 0.867 \times 10^{-3}
$$

$$
\int_C^0 d\epsilon_z^e = 0.3 \times 10^{-3}
$$

As can be seen, the elastic strain changes during loading path $CO$ just offset the elastic strain $\epsilon_C^e$ at point $C$, and the total elastic strains for the complete cycle $O$-$A$-$B$-$O$ are zero, while the plastic strains are irrecoverable. These plastic strains are induced at the end of loading path $AB$ and remain unchanged. At the end of this loading program, we have

$$\epsilon_O^e = 0, \qquad \epsilon_O^p = \epsilon_C^p = \epsilon_B^p = \begin{bmatrix} 9.01 & 16.5/2 & 0 \\ 16.5/2 & -4.5 & 0 \\ 0 & 0 & -4.5 \end{bmatrix} \times 10^{-3}$$

Also, we have the subsequent loading surface as given by Eq. (5.122), which is a record for the complete load history of the material.

## 5.7. Incremental Stress-Strain Relationships

In the previous sections, we have discussed the basic assumptions and equations used in the development of the incremental theory of work-hardening plasticity. Based on these basic equations, a general constitutive equation for an elastic–plastic work-hardening material will be derived in this section in the form of

$$d\sigma_{ij} = C_{ijkl}^{ep} \, d\epsilon_{kl} \tag{5.123}$$

where $C_{ijkl}^{ep}$ is the *elastic–plastic stiffness tensor of tangent modulus*, which is a function of stress state and loading history. For a given stress state and loading history, Eq. (5.123) gives the stress increment $d\sigma_{ij}$ for a given strain increment $d\epsilon_{ij}$ which constitutes a plastic loading. This equation is needed in a numerical analysis of plasticity, such as finite-element analysis.

### 5.7.1. Constitutive Relation for a General Work-Hardening Material

The general expression of a yield surface or loading surface for a work-hardening material as discussed in Section 5.3 has the form

$$f(\sigma_{ij}, \epsilon_{ij}^p, k) = 0 \tag{5.124}$$

where $k = k(\epsilon_p)$ is an isotropic hardening parameter. The strain increment $d\epsilon_{ij}$ is decomposed into two parts,

$$d\epsilon_{ij} = d\epsilon_{ij}^e + d\epsilon_{ij}^p \tag{5.125}$$

in which the elastic strain increment, $d\epsilon_{ij}^e$, is related to the stress increment, $d\sigma_{ij}$, by the generalized Hooke's law as

$$d\sigma_{ij} = C_{ijkl} \, d\epsilon_{kl}^e \tag{5.126}$$

where $C_{ijkl}$ is the tensor of elastic modulus, as discussed in Chapter 3; the plastic strain increment, $d\epsilon_{ij}^p$, can be generally expressed by a nonassociated flow rule in the form

$$d\epsilon_{ij}^p = d\lambda \frac{\partial g}{\partial \sigma_{ij}} \tag{5.127}$$

where $g = g(\sigma_{ij}, \epsilon_{ij}^p, k)$, as for $f(\sigma_{ij}, \epsilon_{ij}^p, k)$, is a known plastic potential function as discussed in Section 5.4, and $d\lambda$ is a scalar function to be determined later by the consistency condition $df = 0$. Substituting the elastic strain increment, $d\epsilon_{ij}^e$, from Eq. (5.125) and the plastic strain increment, $d\epsilon_{ij}^p$, from Eq. (5.127) into Hooke's law, Eq. (5.126), we have

$$d\sigma_{ij} = C_{ijkl} \left( d\epsilon_{kl} - d\lambda \frac{\partial g}{\partial \sigma_{kl}} \right) \tag{5.128}$$

From the above equation, we see that if we know the scalar function $d\lambda$, the constitutive relation is fully determined. To obtain $d\lambda$, consider a plastic loading process. At the current state, we know the current stress state $\sigma_{ij}$ and the current plastic deformation state $\epsilon_{ij}^p$ and $k(\epsilon_p)$, and they must satisfy the current yield function, Eq. (5.124),

$$f(\sigma_{ij}, \epsilon_{ij}^p, k) = 0$$

After a small increment in total strain, $d\epsilon_{ij}$, which constitutes a plastic loading, the current state is changed to the new subsequent state, $\sigma_{ij} + d\sigma_{ij}$, $\epsilon_{ij}^p + d\epsilon_{ij}^p$, $k + dk$, and the new state must satisfy the subsequent yield function, Eq. (5.124), in the mathematical form

$$f(\sigma_{ij} + d\sigma_{ij}, \epsilon_{ij}^p + d\epsilon_{ij}^p, k + dk) = f(\sigma_{ij}, \epsilon_{ij}^p, k) + df = 0$$

or

$$df = \frac{\partial f}{\partial \sigma_{ij}} d\sigma_{ij} + \frac{\partial f}{\partial \epsilon_{ij}^p} d\epsilon_{ij}^p + \frac{\partial f}{\partial k} dk = 0 \tag{5.129}$$

This is known as the *consistency condition* for a general work-hardening material and imposes the restriction on the increments between $d\sigma_{ij}$, $d\epsilon_{ij}^p$, and $dk$. This condition assures that in a plastic loading process, the subsequent stress and deformation states remain on the subsequent yield surface. The scalar function $d\lambda$ in Eq. (5.128) can be determined directly from this condition. This is described in the following.

Consider first the increment of the isotropic hardening parameter, $dk$, in Eq. (5.129). The isotropic hardening parameter $k$ is a function of the effective plastic strain $\epsilon_p$, which can be expressed in the simple form of Eq. (5.83).

$$d\epsilon_p = C\sqrt{d\epsilon_{ij}^p d\epsilon_{ij}^p}$$

Using Eq. (5.83) for $d\epsilon_p$ and the flow rule, Eq. (5.127), we obtain

$$dk = \frac{dk}{d\epsilon_p} d\epsilon_p = \frac{dk}{d\epsilon_p} C\sqrt{\frac{\partial g}{\partial \sigma_{ij}} \frac{\partial g}{\partial \sigma_{ij}}} d\lambda \tag{5.130}$$

Substituting Eqs. (5.127) for $d\epsilon_{ij}^p$, (5.128) for $d\sigma_{ij}$, and (5.130) for $dk$ into the consistency condition (5.129), we have

$$df = \frac{\partial f}{\partial \sigma_{ij}} C_{ijkl} \, d\epsilon_{kl} - h \, d\lambda = 0 \qquad (5.131)$$

where

$$h = \frac{\partial f}{\partial \sigma_{ij}} C_{ijkl} \frac{\partial g}{\partial \sigma_{kl}} - \frac{\partial f}{\partial \epsilon_{ij}^p} \frac{\partial g}{\partial \sigma_{ij}} - \frac{\partial f}{\partial k} \frac{dk}{d\epsilon_p} C \sqrt{\frac{\partial g}{\partial \sigma_{ij}} \frac{\partial g}{\partial \sigma_{ij}}} \qquad (5.132)$$

From Eq. (5.131), the scalar function $d\lambda$ can be solved as

$$d\lambda = \frac{1}{h} \frac{\partial f}{\partial \sigma_{ij}} C_{ijkl} \, d\epsilon_{kl} = \frac{1}{h} H_{kl} \, d\epsilon_{kl} \qquad (5.133)$$

where the second-order tensor $H_{kl}$ associated with the yield function, $f$, is defined as

$$H_{kl} = \frac{\partial f}{\partial \sigma_{ij}} C_{ijkl} \qquad (5.134)$$

Similarly, we shall define the second-order tensor $H_{kl}^*$ associated with the potential function, $g$, as

$$H_{kl}^* = \frac{\partial g}{\partial \sigma_{ij}} C_{ijkl} \qquad (5.135)$$

This second-order tensor will be utilized later in connection with the development of the elastic–plastic tangent stiffness tensor $C_{ijkl}^{ep}$ below.

In Eq. (5.133), we have expressed the scalar function $d\lambda$ in terms of a given strain increment $d\epsilon_{ij}$. In the following, we shall show that the scalar function $d\lambda$ can also be expressed in terms of a given stress increment $d\sigma_{ij}$. To this end, we write the nonassociated flow rule (5.70) in the form

$$d\epsilon_{ij}^p = d\lambda \frac{\partial g}{\partial \sigma_{ij}} = \bar{G} \frac{\partial g}{\partial \sigma_{ij}} \frac{\partial f}{\partial \sigma_{kl}} \, d\sigma_{kl} = \frac{1}{\kappa} \frac{\partial g}{\partial \sigma_{ij}} \frac{\partial f}{\partial \sigma_{kl}} \, d\sigma_{k1} \qquad (5.136)$$

in which $\kappa$ is known as the *hardening modulus* and is related to the hardening function $\bar{G}$ and the scalar function $d\lambda$ by

$$\kappa = \frac{1}{\bar{G}} \qquad (5.137)$$

$$d\lambda = \frac{1}{\kappa} \partial f = \frac{1}{\kappa} \frac{\partial f}{\partial \sigma_{kl}} \, d\sigma_{kl} \qquad (5.138)$$

respectively. Since $(\partial f/\partial \sigma_{ij}) \, d\sigma_{ij} = \partial f$, $d\epsilon_{ij}^p = (\partial f/\kappa)(\partial g/\partial \sigma_{ij})$, $dk = (dk/d\epsilon_p) \, d\epsilon_p$, and

$$d\epsilon_p = C \, d\lambda \sqrt{(\partial g/\partial \sigma_{ij})/(\partial g/\partial \sigma_{ij})} = (C \, \partial f/\kappa)\sqrt{(\partial g/\partial \sigma_{ij})(\partial g/\partial \sigma_{ij})}$$

the consistency condition (5.129) can be written as

$$df = \partial f + \frac{\partial f}{\partial \epsilon_{ij}^p} \frac{\partial f}{\kappa} \frac{\partial g}{\partial \sigma_{ij}} + \frac{\partial f}{\partial k} \frac{dk}{d\epsilon_p} \frac{C \, \partial f}{\kappa} \sqrt{\frac{\partial g}{\partial \sigma_{ij}} \frac{\partial g}{\partial \sigma_{ij}}} = 0 \qquad (5.139)$$

From this, the scalar function $\kappa$ can be solved as

$$\kappa = -\frac{\partial f}{\partial \epsilon_{ij}^p} \frac{\partial g}{\partial \sigma_{ij}} - \frac{\partial f}{\partial k} \frac{dk}{d\epsilon_p} C \sqrt{\frac{\partial g}{\partial \sigma_{ij}} \frac{\partial g}{\partial \sigma_{ij}}} \tag{5.140}$$

As can be seen, the hardening modulus $\kappa$ is determined from the isotropic hardening rule. For the special case of an elastic–perfectly plastic material, the yield surface does not change during loading, i.e., $\partial f / \partial \epsilon_{ij}^p = 0$ and $dk/d\epsilon_p = 0$; hence, we have $\kappa = 0$. Using Eq. (5.140), we can also express the scalar function $h$ in Eq. (5.132) in terms of the hardening modulus $\kappa$ by

$$h = \kappa + \frac{\partial f}{\partial \sigma_{ij}} C_{ijkl} \frac{\partial g}{\partial \sigma_{kl}} = \kappa + H_{kl} \frac{\partial g}{\partial \sigma_{kl}} \tag{5.141}$$

Once the scalar function $d\lambda$ is determined, the plastic strain increment, $d\epsilon_{ij}^p$, is known from the flow rule (5.127)

$$d\epsilon_{ij}^p = d\lambda \frac{\partial g}{\partial \sigma_{ij}} = \frac{1}{h} \frac{\partial f}{\partial \sigma_{mn}} C_{mnst} \frac{\partial g}{\partial \sigma_{ij}} d\epsilon_{st} = \frac{1}{h} H_{st} \frac{\partial g}{\partial \sigma_{ij}} d\epsilon_{st} \tag{5.142}$$

and the corresponding stress increment, $d\sigma_{ij}$, can be determined from Eq. (5.128) using Eq. (5.142)

$$d\sigma_{ij} = C_{ijkl} \left( d\epsilon_{kl} - \frac{1}{h} \frac{\partial f}{\partial \sigma_{mn}} C_{mnst} \frac{\partial g}{\partial \sigma_{kl}} d\epsilon_{st} \right)$$

$$= C_{ijkl} \left( \delta_{sk}\delta_{tl} - \frac{1}{h} \frac{\partial f}{\partial \sigma_{mn}} C_{mnst} \frac{\partial g}{\partial \sigma_{kl}} \right) d\epsilon_{st}$$

$$= \left( C_{ijst} - \frac{1}{h} \frac{\partial f}{\partial \sigma_{mn}} C_{mnst} C_{ijkl} \frac{\partial g}{\partial \sigma_{kl}} \right) d\epsilon_{st}$$

$$= C_{ijst}^{ep} d\epsilon_{st} \tag{5.143}$$

or

$$d\sigma_{ij} = C_{ijkl}^{ep} d\epsilon_{kl} \tag{5.144}$$

Thus, the elastic–plastic tangent stiffness tensor can be written in the form

$$C_{ijkl}^{ep} = C_{ijkl} + C_{ijkl}^{p} \tag{5.145}$$

with

$$C_{ijkl}^{p} = -\frac{1}{h} \frac{\partial f}{\partial \sigma_{mn}} C_{mnkl} C_{ijst} \frac{\partial g}{\partial \sigma_{st}} = -\frac{1}{h} H_{kl} H_{ij}^* \tag{5.146}$$

where $C_{ijkl}^{p}$ is called the *plastic tangent stiffness tensor* and represents the degradation of the stiffness of the material due to plastic flow. It is obvious from Eq. (5.146) that the tensor $C_{ijkl}^{p}$ lacks symmetry, and so does $C_{ijkl}^{ep}$ if a nonassociated flow rule is used, i.e.,

$$\text{if } g \neq f, \text{ then } C_{ijkl}^{ep} \neq C_{klij}^{ep} \tag{5.147}$$

Furthermore, in this case, the elastic–plastic tangent stiffness $C_{ijkl}^{ep}$ may not be positive definite. In contrast, if the associated flow rule is adopted, the positive-definiteness of the tensor $C_{ijkl}^{ep}$ can be ensured.

The incremental elastic–plastic stress–strain relationship (5.144) is valid only for the case of plastic loading. Thus, before using this relationship to calculate the corresponding stress increment, $d\sigma_{ij}$, for a given strain increment, $d\epsilon_{ij}$, one must check first whether the material is in a plastic loading state corresponding to the given strain increment $d\epsilon_{ij}$. If not, the elastic stress–strain relationship

$$d\sigma_{ij} = C_{ijkl} \, d\epsilon_{kl} \tag{5.148}$$

should be used instead of the elastic–plastic stress–strain relationship. To this end, a proper loading criterion is needed. In Section 5.3, we have expressed the loading criterion in terms of stress increment. However, here we need to express the loading criterion in terms of the given strain increment, because the stress increment is unknown and is yet to be determined from the strain increment. This is probably true in most numerical analysis methods of plasticity for structures, for example, the popular finite-element method.

Refer now to the consistency condition (5.131). It can be shown that the scalar function $h$ as defined in Eq. (5.132) is always positive for a work-hardening material as well as for a strain-softening material. This will be described later in Chapter 7. The proof of this statement is outside the scope of this book. Using this fact, we shall derive the loading criterion in terms of strain increments $d\epsilon_{ij}$ as follows.

For a plastic loading, $d\lambda$ is a non-negative factor and $h \, d\lambda$ is always positive. It then follows from the consistency condition (5.131) that

$$\frac{\partial f}{\partial \sigma_{ij}} \, C_{ijkl} \, d\epsilon_{kl} > 0 \tag{5.149}$$

For a neutral loading, we have $d\epsilon_{ij}^p = 0$, or $d\lambda = 0$, and the consistency condition (5.131) leads to

$$\frac{\partial f}{\partial \sigma_{ij}} \, C_{ijkl} \, d\epsilon_{kl} = 0 \tag{5.150}$$

For an unloading from an elastic–plastic state, the stress state on the surface is moved inward, resulting in $df < 0$. Further, for this case, we have $d\lambda = 0$. Using the condition $df < 0$ and $d\lambda = 0$ in Eq. (5.131), we have

$$\frac{\partial f}{\partial \sigma_{ij}} \, C_{ijkl} \, d\epsilon_{kl} < 0 \tag{5.151}$$

In summary, the loading criterion for a given strain increment $d\epsilon_{ij}$ can be expressed as

$$\frac{\partial f}{\partial \sigma_{ij}} \, C_{ijkl} \, d\epsilon_{kl} \begin{cases} > 0, & \text{loading} \\ = 0, & \text{neutral loading} \\ < 0, & \text{unloading} \end{cases} \tag{5.152}$$

In conclusion, the complete stress–strain relationship for an elastic–plastic work-hardening material can be expressed in the general form as follows.

For $f(\sigma_{ij}, \epsilon^p_{ij}, k) = 0$, and $(\partial f/\partial \sigma_{ij})C_{ijkl} d\epsilon_{kl} > 0$, we have

$$d\sigma_{ij} = C^{ep}_{ijkl} d\epsilon_{kl} \qquad (5.153)$$

where $C^{ep}_{ijkl}$ is given in Eqs. (5.145) and (5.146).

For $f(\sigma_{ij}, \epsilon^p_{ij}, k) < 0$, or $f(\sigma_{ij}, \epsilon^p_{ij}, k) = 0$ and $(\partial f/\partial \sigma_{ij})C_{ijkl} d\epsilon_{kl} \le 0$, we have

$$d\sigma_{ij} = C_{ijkl} d\epsilon_{kl} \qquad (5.154)$$

where the elastic tangent stiffness tensor $C_{ijkl}$ is given in Chapter 3.

## 5.7.2. Constitutive Relation for a Mixed-Hardening Material

The general expression of a yield surface for a mixed-hardening material has the form

$$f(\bar{\sigma}_{ij}, k) = f(\sigma_{ij} - \alpha_{ij}, k) = 0 \qquad (5.155)$$

Here, the yield surface is expressed in terms of $\alpha_{ij}$ instead of $\epsilon^p_{ij}$ explicitly. We shall derive the incremental stress–strain relationships for a mixed-hardening material directly from the basic plasticity equations. In this development, we shall assume that the plastic strain increments $d\epsilon^p_{ij}$ can be further split into two parts, the isotropic hardening part $d\epsilon^i_{ij}$ and the kinematic hardening part $d\epsilon^k_{ij}$:

$$d\epsilon^p_{ij} = d\epsilon^i_{ij} + d\epsilon^k_{ij} \qquad (5.156)$$

in which we take

$$d\epsilon^i_{ij} = M d\epsilon^p_{ij}, \qquad d\epsilon^k_{ij} = (1 - M) d\epsilon^p_{ij} \qquad (5.157)$$

where $0 \le M \le 1$ represents the degree of mixture between the isotropic hardening and the kinematic hardening. In this case, the consistency condition becomes

$$df = \frac{\partial f}{\partial \sigma_{ij}} d\sigma_{ij} + \frac{\partial f}{\partial \alpha_{ij}} d\alpha_{ij} + \frac{\partial f}{\partial k} dk = 0 \qquad (5.158)$$

If Prager's hardening rule, Eq. (5.106), is used, we have

$$d\alpha_{ij} = c \, d\epsilon^k_{ij} = c(1 - M) \, d\epsilon^p_{ij} = c(1 - M) \frac{\partial g}{\partial \sigma_{ij}} d\lambda \qquad (5.159)$$

or if Ziegler's hardening rule, Eq. (5.107), is used, we have

$$d\alpha_{ij} = a(1 - M)(\sigma_{ij} - \alpha_{ij}) \, d\epsilon_p$$

$$= a(1 - M)(\sigma_{ij} - \alpha_{ij})C \sqrt{\frac{\partial g}{\partial \sigma_{kl}} \frac{\partial g}{\partial \sigma_{kl}}} \, d\lambda \qquad (5.160)$$

or writing these hardening rules in a general form,

$$d\alpha_{ij} = A_{ij} \, d\lambda \qquad (5.161)$$

where for Prager's rule

$$A_{ij} = c(1 - M) \frac{\partial g}{\partial \sigma_{ij}} \tag{5.162}$$

and for Ziegler's rule

$$A_{ij} = a(1 - M)(\sigma_{ij} - \alpha_{ij})C \sqrt{\frac{\partial g}{\partial \sigma_{kl}} \frac{\partial g}{\partial \sigma_{kl}}} \tag{5.163}$$

and the effective plastic strain definition

$$d\epsilon_p = C\sqrt{d\epsilon_{ij}^P \, d\epsilon_{ij}^P} = C \sqrt{\frac{\partial g}{\partial \sigma_{ij}} \frac{\partial g}{\partial \sigma_{ij}}} \, d\lambda \tag{5.164}$$

has been used in the derivation, and the parameters $a$ and $c$ have been discussed in Section 5.5. The isotropic hardening parameter $k$ is a function of the reduced effective plastic strain, $\bar{\epsilon}_p$, defined as

$$\bar{\epsilon}^p = C \int \sqrt{d\epsilon_{ij}^i \, d\epsilon_{ij}^i} = MC \int \sqrt{d\epsilon_{ij}^P \, d\epsilon_{ij}^P} = M \int d\epsilon_p = M\epsilon_p \tag{5.165}$$

$$d\bar{\epsilon}_p = M \, d\epsilon_p = MC \sqrt{\frac{\partial g}{\partial \sigma_{ij}} \frac{\partial g}{\partial \sigma_{ij}}} \, d\lambda \tag{5.166}$$

Thus, we have the relationship

$$dk = \frac{dk}{d\bar{\epsilon}_p} \, d\bar{\epsilon}_p = \frac{dk}{d\bar{\epsilon}_p} MC \sqrt{\frac{\partial g}{\partial \sigma_{kl}} \frac{\partial g}{\partial \sigma_{kl}}} \, d\lambda \tag{5.167}$$

Noting that

$$\frac{\partial f}{\partial \sigma_{ij}} = \frac{\partial f}{\partial \bar{\sigma}_{kl}} \frac{\partial \bar{\sigma}_{kl}}{\partial \sigma_{ij}} = \frac{\partial f}{\partial \bar{\sigma}_{kl}} \delta_{ik}\delta_{jl} = \frac{\partial f}{\partial \bar{\sigma}_{ij}} \tag{5.168}$$

$$\frac{\partial f}{\partial \alpha_{ij}} = \frac{\partial f}{\partial \bar{\sigma}_{kl}} \frac{\partial \bar{\sigma}_{kl}}{\partial \alpha_{ij}} = \frac{\partial f}{\partial \bar{\sigma}_{kl}} (-\delta_{ik}\delta_{jl}) = -\frac{\partial f}{\partial \bar{\sigma}_{ij}} = -\frac{\partial f}{\partial \sigma_{ij}} \tag{5.169}$$

the consistency condition (5.131) can then be rewritten as

$$df = \frac{\partial f}{\partial \sigma_{ij}} C_{ijkl} \, d\epsilon_{kl} - \bar{h} \, d\lambda = 0 \tag{5.170}$$

where

$$\bar{h} = H_{kl} \frac{\partial g}{\partial \sigma_{kl}} + A_{kl} \frac{\partial f}{\partial \sigma_{kl}} - \frac{\partial f}{\partial k} \frac{dk}{d\bar{\epsilon}_p} MC \sqrt{\frac{\partial g}{\partial \sigma_{kl}} \frac{\partial g}{\partial \sigma_{kl}}} \tag{5.171}$$

in which the tensor $H_{kl}$ is defined in Eq. (5.134). The scalar function $d\lambda$ for the mixed-hardening material is now solved from Eq. (5.170) as

$$d\lambda = \frac{1}{\bar{h}} \frac{\partial f}{\partial \sigma_{ij}} C_{ijkl} \, d\epsilon_{kl} = \frac{1}{\bar{h}} H_{kl} \, d\epsilon_{kl} \tag{5.172}$$

and the elastic–plastic tangent stiffness tensor is obtained for a mixed-hardening material as

$$C^{ep}_{ijkl} = C_{ijkl} + C^{p}_{ijkl} = C_{ijkl} - \frac{1}{h} H_{kl} H^{*}_{ij} \tag{5.173}$$

The general loading criterion, Eq. (5.152), is, of course, valid for the special mixed-hardening material. In summary, the complete stress–strain relationship for a mixed-hardening material can be expressed as follows.

For $f(\sigma_{ij} - \alpha_{ij}, k) = 0$, and $(\partial f / \partial \sigma_{ij}) C_{ijkl} \, d\epsilon_{kl} > 0$, we have

$$d\sigma_{ij} = C^{ep}_{ijkl} \, d\epsilon_{kl} \tag{5.174}$$

where the elastic–plastic tangent stiffness $C^{ep}_{ijkl}$ is given in Eq. (5.173). For $f(\sigma_{ij} - \alpha_{ij}, k) < 0$, or $f(\sigma_{ij} - \alpha_{ij}, k) = 0$, $(\partial f / \partial \sigma_{ij}) C_{ijkl} \, d\epsilon_{kl} \leq 0$, we have

$$d\sigma_{ij} = C_{ijkl} \, d\epsilon_{kl} \tag{5.175}$$

where the elastic tangent stiffness $C_{ijkl}$ is given in Chapter 3.

## 5.7.3. Illustrative Examples

The incremental stress–strain relations for a general work-hardening material and for a special mixed-hardening material have been established in the preceding sections. In the following, we shall derive the two most commonly used material models, namely, the von Mises and Drucker–Prager materials, as illustrative examples.

EXAMPLE 5.7. Derive the stress–strain equations for von Mises material with the associated mixed-hardening rule.

SOLUTION. The loading function of von Mises material corresponding to the mixed-hardening rule can be expressed as

$$f(\sigma_{ij} - \alpha_{ij}, k) = \tfrac{3}{2} \bar{s}_{ij} \bar{s}_{ij} - \bar{\sigma}^{2}_{e}(\bar{\epsilon}_{p}) = 0 \tag{5.176}$$

Here, $\bar{s}_{ij}$ denotes the deviatoric reduced stress tensor

$$\bar{s}_{ij} = \bar{\sigma}_{ij} - \tfrac{1}{3} \bar{\sigma}_{kk} \delta_{ij} \tag{5.177}$$

and $\bar{\sigma}_{ij} = \sigma_{ij} - \alpha_{ij}$ is the reduced-stress tensor; the reduced effective stress, $\bar{\sigma}_{e}$, which replaces the usual hardening parameter $k$, is defined by Eqs. (5.77) and (5.82); and the reduced effective plastic strain, $\bar{\epsilon}_{p}$, is defined by Eq. (5.103) with the constant $C = \sqrt{\tfrac{2}{3}}$.

Using the associated flow rule, $g = f$, the derivatives of $g$ and $f$ are found as

$$\frac{\partial f}{\partial \sigma_{ij}} = \frac{\partial g}{\partial \sigma_{ij}} = 3 \bar{s}_{ij} \tag{5.178}$$

For a linear-elastic isotropic material, we have the following form of Hooke's law expressed in terms of the two elastic constants $G$ and $\nu$:

$$C_{ijkl} = 2G\left(\delta_{ik}\delta_{jl} + \frac{\nu}{1-2\nu}\,\delta_{ij}\delta_{kl}\right) \tag{5.179}$$

where $G$ is the shear modulus and $\nu$ is Poisson's ratio. The tensors $H_{kl}$ and $H_{kl}^*$ as defined in Eqs. (5.134) and (5.135) are now obtained as

$$H_{kl} = H_{kl}^* = \frac{\partial f}{\partial \sigma_{ij}}\,C_{ijkl} = 6G\bar{s}_{kl} \tag{5.180}$$

Consider first Prager's hardening rule. Using Eq. (5.162) for the tensor $A_{ij}$, Eq. (5.178) for $\partial f/\partial \sigma_{ij}$ and $\partial g/\partial \sigma_{ij}$, and Eq. (5.180) for $H_{kl}$ and $H_{kl}^*$, the scalar function $\bar{h}$ as defined in Eq. (5.171) can be written as

$$\bar{h} = 18G\bar{s}_{kl}\bar{s}_{kl} + 9c(1-M)\bar{s}_{kl}\bar{s}_{kl} + 2\bar{\sigma}_e\frac{d\bar{\sigma}_e}{d\bar{\epsilon}_p}\,M\sqrt{6\bar{s}_{kl}\bar{s}_{kl}} \tag{5.181}$$

where we have replaced $k$ by $\bar{\sigma}_e$ in Eq. (5.171) and have used the yielding function of Eq. (5.176). Using the definition for the slope or the stiffness of the reduced effective stress–strain relation, $\bar{H} = d\bar{\sigma}_e/d\bar{\epsilon}_p$, the scalar function $\bar{h}$ can be reduced to the simple form

$$\bar{h} = [12G + 6c(1-M) + 4M\bar{H}_p]\bar{\sigma}_e^2 \tag{5.182}$$

In this equation, the parameter $c$ is related to the component of the kinematic hardening part and the parameter $\bar{H}_p$ is related to the component of the isotropic hardening part of the mixed-hardening model. These two parameters will be determined from a uniaxial stress–strain curve. For the uniaxial stress state, the reduced-stress tensor $\bar{\sigma}_{ij}$ can be expressed as

$$\bar{\sigma}_{11} = \sigma_{11} - \alpha_{11}, \qquad \bar{\sigma}_{22} = \bar{\sigma}_{33} = 0 - \alpha_{22} = \tfrac{1}{2}\alpha_{11}$$
$$\bar{\sigma}_{ij} = 0 \qquad \text{for } i \neq j \tag{5.183}$$

where $\alpha_{22} = \alpha_{33} = -\alpha_{11}/2$ follows directly from Prager's hardening rule, Eq. (5.106), and the plastic-imcompressibility condition of $J_2$-material. Hence, the reduced effective stress $\bar{\sigma}_e$ in the uniaxial case is obtained as

$$\bar{\sigma}_e^2 = 3\bar{J}_2 = \tfrac{1}{2}[(\bar{\sigma}_{11} - \bar{\sigma}_{22})^2 + (\bar{\sigma}_{22} - \bar{\sigma}_{33})^2 + (\bar{\sigma}_{33} - \bar{\sigma}_{11})^2]$$
$$= (\sigma_{11} - \tfrac{3}{2}\alpha_{11})^2 \tag{5.184}$$

From Eq. (5.184) and using Eq. (5.106) for $d\alpha_{11}$, we have

$$d\bar{\sigma}_e = d\sigma_{11} - \tfrac{3}{2}c(1-M)\,d\epsilon_{11}^p \qquad \text{for Prager's rule} \tag{5.185}$$

Using Eq. (5.105) for $d\bar{\sigma}_e$ and noting that $d\sigma_{11}$ and $d\epsilon_{11}^p$ are equal to $d\sigma_e$ and $d\epsilon_p$, respectively, in the uniaxial case, we obtain

$$\frac{d\sigma_e}{d\epsilon_p} = [M(\bar{H}_p - \tfrac{3}{2}c) + \tfrac{3}{2}c] = H_p \tag{5.186}$$

Here, the definition of the plastic modulus $H_p$, Eq. (5.96), has been used. Since $M$ is an arbitrary material constant and Eq. (5.162) must be valid for any value of $M$, it follows that

$$c = \tfrac{2}{3}H_p, \qquad \bar{H}_p = H_p \tag{5.187}$$

Finally, the substitution of Eq. (5.187) into (5.182) gives the scalar function $\bar{h}$ in the simple form

$$\bar{h} = 4(3G + H_p)\bar{\sigma}_e^2 \tag{5.188}$$

Therefore, the elastic–plastic stiffness tensor for von Mises material with the associated mixed hardening rule is obtained as

$$C_{ijkl}^{ep} = C_{ijkl} - \frac{36G^2}{\bar{h}}\,\bar{s}_{ij}\bar{s}_{kl} \tag{5.189}$$

It should be noted that $\bar{\sigma}_e$ and $\bar{s}_{ij}$ are reduced stress values, i.e., they are referred to the stress space with its origin at the current center of the kinematically translated yield surface.

For Ziegler's hardening rule and using Eq. (5.163) instead of Eq. (5.162) for $A_{ij}$, we obtain the scalar function $\bar{h}$ as

$$\bar{h} = [12G + 4a(1 - M)\bar{\sigma}_e + 4M\bar{H}_p]\bar{\sigma}_e^2 \tag{5.190}$$

To determine the hardening parameters $a$ and $\bar{H}_p$, we use Eq. (5.107) instead of Eq. (5.106) for $d\alpha_{11}$ in Eq. (5.184) and obtain the counterpart of Eq. (5.185)

$$d\bar{\sigma}_e = d\sigma_{11} - a(1 - M)\,d\epsilon_p\bar{\sigma}_{11} \tag{5.191}$$

In a similar manner to that of Eq. (5.186), we obtain

$$\frac{d\sigma_e}{d\epsilon_p} = [M(\bar{H}_p - a\bar{\sigma}_e) + a\bar{\sigma}_e] = H_p \tag{5.192}$$

Using the same argument as for Eq. (5.187), we can conclude from Eq. (5.192) that

$$a = \frac{H_p}{\bar{\sigma}_e}, \qquad \bar{H}_p = H_p \tag{5.193}$$

Finally, Ziegler's hardening rule leads to the same expression for $\bar{h}$, Eq. (5.188), as Prager's hardening rule for $J_2$-material.

Equations (5.188) and (5.189) define the elastic–plastic tangent stiffness tensor $C_{ijkl}^{ep}$ for the mixed-hardening von Mises material, which includes isotropic hardening, kinematic hardening, and perfectly plasticity (without hardening) as special cases. This is illustrated in the following.

(i) *Isotropic hardening case.* For this case, we have $M = 1$, $\alpha_{ij} = 0$, and $\bar{s}_{ij} = s_{ij}$. The loading function $f$, the scalar function $h$, and the elastic-plastic stiffness tensor $C_{ijkl}^{ep}$ are

$$f(\sigma_{ij}, k) = \tfrac{3}{2} s_{ij} s_{ij} - \sigma_e^2(\epsilon_p) = 0 \tag{5.194}$$

$$\bar{h} = h = 4(3G + H_p)\sigma_e^2 \tag{5.195}$$

$$C_{ijkl}^{ep} = C_{ijkl} - \frac{36G^2}{h} s_{ij} s_{kl} \tag{5.196}$$

Alternatively, using Eq. (5.141), we have

$$h = \kappa + H_{kl}\frac{\partial g}{\partial \sigma_{kl}}$$

$$= \kappa + (6Gs_{kl})(3s_{kl}) = \kappa + 12G\sigma_e^2 = \frac{1}{\bar{G}} + 12G\sigma_e^2 \tag{5.197}$$

Comparing Eq. (5.195) with Eq. (5.197), we have

$$\bar{G} = \frac{1}{4H_p\sigma_e^2} \tag{5.198}$$

This equation is similar to Eq. (5.112) given previously in Example 5.5, but differs by a factor of 9. This is because the loading function used in Eq. (5.194) differs from the loading function used in Eq. (5.108) by a factor of 3.

(ii) *Kinematic hardening case.* For this case, we have $M = 0$ and $\bar{\sigma}_e^2 = $ constant $= \sigma_0^2$. The loading function becomes

$$f(\sigma_{ij} - \alpha_{ij}) = \tfrac{3}{2} \bar{s}_{ij} \bar{s}_{ij} - \sigma_0^2 = 0 \tag{5.199}$$

The elastic-plastic tangent stiffness tensor $C_{ijkl}^{ep}$ and the expression for the scalar function $h$ take the same form as Eqs. (5.189) and (5.188), respectively, with $\bar{\sigma}_e = \sigma_0$.

Now let's examine the translation of the center, $d\alpha_{ij}$. According to Ziegler's hardening rule, Eq. (5.107), we have

$$d\alpha_{ij} = a\, d\epsilon_p\, \bar{\sigma}_{ij} = \frac{H_p\, d\epsilon_p}{\bar{\sigma}_e} \bar{\sigma}_{ij} = \frac{H_p\, d\epsilon_p}{\sigma_0} \bar{\sigma}_{ij} \tag{5.200}$$

in which Eq. (5.193) has been used. Note that in Eq. (5.200), $H_p\, d\epsilon_p = d\sigma_e = (3/2\bar{\sigma}_e)\bar{s}_{ij}\, ds_{ij}$, meaning that the incremental effective stress is evaluated only with respect to increments in the stress components, i.e., with constant $\alpha_{ij}$.

For the special case in Example 5.4, $\sigma_x = \sigma$, $\tau_{xy} = \tau$, other stress components $= 0$, $H_p\, d\epsilon_p = (1/\sigma_0)(\bar{\sigma}\, d\sigma + 3\bar{\tau}\, d\tau)$, Eq. (5.200) leads to Eq. (5.50), as it should.

(iii) *Perfectly plastic case.* For this case, we have $M = 1$, $\alpha_{ij} = 0$, $\bar{s}_{ij} = s_{ij}$, $\bar{\sigma}_e = \sigma_0 = $ constant, and $H_p = 0$. The elastic-plastic tangent stiffness takes the same form as Eq. (5.196), but the scalar function $h$ in Eq. (5.195) becomes

$$h = 12G\sigma_0^2 = 36GJ_2 \tag{5.201}$$

Thus,

$$C_{ijkl}^{ep} = C_{ijkl} - \frac{G}{J_2} s_{ij}s_{kl} \tag{5.202}$$

which leads to Eq. (4.92), the constitutive relation for a perfectly plastic von Mises material.

EXAMPLE 5.8. Derive the stress-strain equations for isotropic-hardening Drucker-Prager material with a nonassociated flow rule.

SOLUTION. The general form of the loading function of the isotropic-hardening Drucker-Prager material can be expressed as

$$f(\sigma_{ij}, \epsilon_p) = \alpha(\epsilon_p)I_1 + \sqrt{J_2} - k(\epsilon_p) = 0 \tag{5.203}$$

In Section 4.10, the stress-strain equation for a perfectly plastic material has been presented with $\alpha = $ constant and $k = $ constant. Herein, for simplicity, we shall assume that the slope of the loading surface in the $I_1$-$J_2^{1/2}$ space is a constant, $\alpha(\epsilon_p) = \alpha_1$, so that the hardening behavior of the material can be uniquely determined by a single uniaxial stress-strain relation through the hardening parameter $k(\epsilon_p)$

$$f(\sigma_{ij}, \epsilon_p) = \alpha_1 I_1 + \sqrt{J_2} + k(\epsilon_p) = 0 \tag{5.204}$$

As discussed previously in Section 4.10, the plastic deformation of Drucker-Prager material is always accompanied by a dilatation of volume if the associated flow rule is adopted. In this case, the rate of dilatation is controlled by the parameter $\alpha$, Eq. (4.120). Herein, we will use a plastic potential function similar to the loading function, Eq. (5.204),

$$g(\sigma_{ij}) = \alpha_2 I_1 + \sqrt{J_2} \tag{5.205}$$

where $0 \leq \alpha_2 \leq \alpha_1$ is a constant. The subsequent loading surfaces and the potential surface are plotted in Fig. 5.12. The derivatives of $f$ and $g$ are obtained as

$$\frac{\partial f}{\partial \sigma_{ij}} = \alpha_1 \delta_{ij} + \frac{1}{2\sqrt{J_2}} s_{ij} \tag{5.206}$$

$$\frac{\partial g}{\partial \sigma_{ij}} = \alpha_2 \delta_{ij} + \frac{1}{2\sqrt{J_2}} s_{ij} \tag{5.207}$$

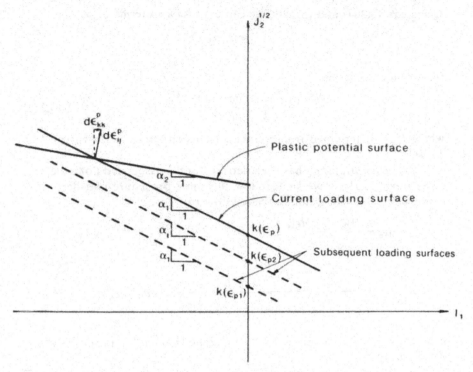

FIGURE 5.12. The loading and plastic potential surfaces for the Drucker-Prager material with a nonassociated flow rule.

Using the elastic stiffness tensor in the form

$$C_{ijkl} = (K - \tfrac{2}{3}G)\delta_{ij}\delta_{kl} + G(\delta_{ik}\delta_{jl} + \delta_{il}\delta_{jk}) \tag{5.208}$$

we obtain

$$H_{kl} = C_{ijkl}\frac{\partial f}{\partial \sigma_{ij}} = 3K\alpha_1\delta_{kl} + \frac{G}{\sqrt{J_2}}s_{kl} \tag{5.209}$$

$$H_{kl}^* = C_{ijkl}\frac{\partial g}{\partial \sigma_{ij}} = 3K\alpha_2\delta_{kl} + \frac{G}{\sqrt{J_2}}s_{kl} \tag{5.210}$$

Since the loading function is not expressed as a function of $\epsilon_{ij}^p$ explicitly, we have $\partial f/\partial \epsilon_{ij}^p = 0$. To determine the scalar function $h$ defined in Eq. (5.132), we also need to obtain $dk/d\epsilon_p$ and the parameter $C$. The effective stress $\sigma_e$ for Drucker-Prager material has been derived in Section 5.5, Eq. (5.79),

$$\sigma_e = \frac{\sqrt{3}(\alpha_1 I_1 + \sqrt{J_2})}{1 + \sqrt{3}\alpha_1} \tag{5.211}$$

Using eqs. (5.204) and (5.211), we can express $k$ in terms of $\sigma_e$

$$k = \frac{1+\sqrt{3}\alpha_1}{\sqrt{3}} \sigma_e$$

from which we obtain

$$\frac{dk}{d\epsilon_p} = \frac{1+\sqrt{3}\alpha_1}{\sqrt{3}} \frac{d\sigma_e}{d\epsilon_p} = \frac{1+\sqrt{3}\alpha_1}{\sqrt{3}} H_p \qquad (5.212)$$

where $H_p$ is determined from a uniaxial tension stress-strain curve, $d\sigma = H_p \, d\epsilon$.

The effective strain, $\epsilon_p$, has been derived for the associated flow rule case in Section 5.5. Here, we shall follow the same procedure and derive the expression of $d\epsilon_p$ for a nonassociated flow rule case.

$$d\epsilon_p = \frac{dW_p}{\sigma_e} = \frac{\sigma_{kl} \, d\epsilon_{kl}^p}{\sigma_e}$$

$$= \frac{\sigma_{kl} \dfrac{\partial g}{\partial \sigma_{kl}}}{\sigma_e} \, d\lambda = \frac{\sigma_{kl} \dfrac{\partial g}{\partial \sigma_{kl}}}{\sigma_e} \frac{\sqrt{d\epsilon_{ij}^p \, d\epsilon_{ij}^p}}{\sqrt{\dfrac{\partial g}{\partial \sigma_{st}} \dfrac{\partial g}{\partial \sigma_{st}}}} = C\sqrt{d\epsilon_{ij}^p \, d\epsilon_{ij}^p} \qquad (5.213)$$

in which we have used Eq. (5.83) as the definition of $d\epsilon_p$. From the above equation, we have

$$C = \frac{\sigma_{kl} \dfrac{\partial g}{\partial \sigma_{kl}}}{\sqrt{\dfrac{\partial g}{\partial \sigma_{st}} \dfrac{\partial g}{\partial \sigma_{st}}} \, \sigma_e} \qquad (5.214)$$

Substituting Eqs. (5.206) to (5.214) into (5.132), we can express the scalar function $h$ as

$$h = G + 9K\alpha_1\alpha_2 + \frac{\alpha_2 I_1 + \sqrt{J_2}}{3k} (1+\sqrt{3}\alpha_1)^2 H_p \qquad (5.215)$$

Finally, we obtain the elastic-plastic tangent stiffness tensor as

$$C_{ijkl}^{ep} = C_{ijkl} - \frac{1}{h} H_{ij}^* H_{kl}$$

$$= C_{ijkl} - \frac{1}{h} \left( 3K\alpha_1\delta_{kl} + \frac{G}{\sqrt{J_2}} s_{kl} \right) \left( 3K\alpha_2\delta_{ij} + \frac{G}{\sqrt{J_2}} s_{ij} \right) \qquad (5.216)$$

As noted previously in Section 5.7.2, for a nonassociated flow rule material, the stiffness $C_{ijkl}^{ep}$, as given in Eq. (5.216), is not symmetric. However, for a value of the parameter $\alpha_2$ in the range $0 \le \alpha_2 \le \alpha_1$, $C_{ijkl}^{ep}$ is still positive definite. The plastic volume change or dilatation has the value

$$d\epsilon_{kk}^p = 3\alpha_2 \, d\lambda \qquad (5.217)$$

By adjusting the value of $\alpha_2$, the rate of dilatation of volume of the material can be controlled from 0 (incompressible) up to $3\alpha_1 \, d\lambda$.

Consider the following three special cases:

Case (i): *Associated flow rule case.* For this case, $\alpha_2 = \alpha_1 = \alpha$, we have

$$h = G + 9K\alpha^2 + \left(\alpha + \frac{1}{\sqrt{3}}\right)^2 H_p \tag{5.218}$$

$$C_{ijkl}^{ep} = C_{ijkl} - \frac{1}{h}\left(3K\alpha\delta_{kl} + \frac{G}{\sqrt{J_2}}s_{kl}\right)\left(3K\alpha\delta_{ij} + \frac{G}{\sqrt{J_2}}s_{ij}\right) \tag{5.219}$$

Furthermore, if a perfectly plastic behavior is assumed, Eqs. (5.218) and (5.219) lead to Eq. (4.126) for the perfectly plastic Drucker–Prager material, as they should.

Case (ii): *Drucker–Prager surface used as the plastic potential surface for the von Mises material.* For this case, $\alpha_1 = 0$, $\alpha_2 > 0$, we have

$$h = G + \frac{\alpha_2 I_1 + \sqrt{J_2}}{3k} H_p \tag{5.220}$$

$$C_{ijkl}^{ep} = C_{ijkl} - \frac{1}{h}\left(3K\alpha_2\delta_{ij} + \frac{G}{\sqrt{J_2}}s_{ij}\right)\frac{G}{\sqrt{J_2}}s_{kl} \tag{5.221}$$

and the von Mises material is no longer plastically incompressible, and the rate of dilatation is given by Eq. (5.217).

Case (iii): *von Mises surface used as the plastic potential surface for the Drucker–Prager material.* For this case, $\alpha_1 > 0$, $\alpha_2 = 0$, we have

$$h = G + \frac{\sqrt{J_2}}{3k}(1 + \sqrt{3}\alpha_1)^2 H_p \tag{5.222}$$

$$C_{ijkl}^{ep} = C_{ijkl} - \frac{1}{h}\frac{G}{\sqrt{J_2}}s_{ij}\left(3K\alpha_1\delta_{kl} + \frac{G}{\sqrt{J_2}}s_{kl}\right) \tag{5.223}$$

and the material becomes plastically incompressible, $d\epsilon_{kk}^p = 0$.

## References

Budiansky, B., 1959. "A Reassessment of Deformation Theories of Plasticity," *Journal of Applied Mechanics*, 26:259–264.

Chen, W. F., 1982. *Plasticity in Reinforced Concrete*, McGraw-Hill, New York.

Drucker, D.C., "A More Fundamental Approach to Plastic Stress-Strain Relation," Proc. 1st National Congress of Applied Mechanics, ASME, Chicago, 1951, pp. 487–491.

Drucker, D.C., 1956. "On Uniqueness in the Theory of Plasticity," *Quarterly of Applied Mathematics* 14:35–42.

Drucker, D.C., 1960. Plasticity. In: J.N. Goodier and N.J. Hoff (Editors), *Structural Mechanics*, Pergamon Press, London, pp. 407–445.

Hill, R., 1950. *The Mathematical Theory of Plasticity*, Oxford University Press, London.

Hodge, P.G., Jr., 1957. "Discussion of [Prager (1956)]," *Journal of Applied Mechanics* 23:482–484.

Martin, J.B., 1975. *Plasticity: Fundamentals and General Results*, MIT Press, Cambridge.

Mroz, Z., 1963. "Non-Associated Flow Laws in Plasticity," *Journal de Mécanique* II(1):21–42.

Prager, W., 1955. "The Theory of Plasticity: A Survey of Recent Achievements (James Clayton Lecture)," *Institute of Mechanical Engineering* 169:41–57.

Prager, W., 1956. "A New Method of Analyzing Stress and Strains in Work-Hardening Solids," *Journal of Applied Mechanics, ASME* 23:493–496.

Ziegler, H., 1959. "A Modification of Prager's Hardening Rule," *Quarterly of Applied Mathematics* 17:55–65.

## PROBLEMS

5.1. In a deformation theory, we assume the total stress–strain relation has the form

$$\epsilon_{ij} = aI_1\delta_{ij} + (b + cJ_2^2)s_{ij}$$

(a) Determine the material constants $a$, $b$, and $c$ if the material response curves are approximated as follows.

(i) In a simple shear test ($\tau_{xy} = \tau_{yx} = \tau$, other $\sigma_{ij} = 0$, and $\gamma_{xy} = \gamma$):

$$\gamma = \frac{\tau}{28{,}000} + 27\left(\frac{\tau}{700}\right)^5, \qquad \tau \text{ in MPa}$$

(ii) In a hydrostatic stress test ($\sigma_x = \sigma_y = \sigma_z = p$, others $= 0$):

$$\epsilon_v = \frac{3}{2}\left(\frac{p}{7 \times 10^4}\right), \qquad p \text{ in MPa}$$

(b) An element of the above material is subjected to a loading history which produces the following stress state:

$$\sigma_{ij} = \begin{bmatrix} 140 & 0 & 56 \\ 0 & 70 & -42 \\ 56 & -42 & 0 \end{bmatrix} \text{MPa}$$

Predict the corresponding components of strain $\epsilon_{ij}$.

5.2. An element of von Mises kinematic-hardening material is subjected to biaxial loadings. A stress increment ($d\sigma_1$, $d\sigma_2$) is now imposed on a stress state ($\sigma_1$, $\sigma_2$) which lies on the yield surface. The stress increment satisfies the loading condition. Determine the coordinate change of the center of the loading surface, $d\alpha_{ij}$, based on Ziegler's hardening rule.

5.3. For biaxial states of stresses ($\sigma_1$, $\sigma_2$) with $\sigma_3 = 0$, write down the incremental plastic strain equation in component forms for the Tresca material in various regimes, assuming the associated flow rule.

5.4. The stress–strain curve of a metal under a simple tension test is assumed to be given by

$$\epsilon = \epsilon^e + \epsilon^p = \frac{\sigma}{E}\left[1 + \left(\frac{\sigma}{\sigma_0}\right)^{2n}\right]$$

in which $E = 2(1+\nu)G$ is the elastic modulus, $\sigma_0 = $ initial tensile yield stress, and $n = $ given material parameter. Assume that the metal is incompressible. Based on the isotropic-hardening von Mises theory, obtain the stress–strain relation

$$2G\, d\epsilon_{ij} = ds_{ij} + \frac{2n+1}{2}\left(\frac{J_2}{k^2}\right)^{n-1} s_{ij}\left(\frac{dJ_2}{k^2}\right)$$

where $k$ is the initial yield stress in pure shear.

5.5. A thin-walled circular tube is subjected to combined axial-tension and twisting-moment loadings. A stress state with $\sigma = \sigma_0$ and $\tau = \sigma_0/\sqrt{3}$ is reached in the wall of the tube. The stress–strain relation of the material in simple tension is given by

$$\epsilon = \begin{cases} \dfrac{\sigma}{E} & (\sigma < \sigma_0) \\[2mm] \dfrac{\sigma_0}{E} + \dfrac{\sigma - \sigma_0}{E_p} & (\sigma \geq \sigma_0) \end{cases}$$

which is an elastic–linear-hardening plastic stress–strain relation with constant elastic modulus $E$ and plastic modulus $E_p$, and $\sigma_0$ is the initial yield stress. Assume the material is of the isotropic-hardening von Mises type. Find the state of strain $(\epsilon, \gamma)$ corresponding to the given state of stress $(\sigma_0, \sigma_0/\sqrt{3})$ for the following loading paths (see Fig. P5.5):

(i) Normal stress $\sigma$ first increases up to $\sigma_0$ and then remains constant. Shear stress $\tau$ increases to $\sigma_0/\sqrt{3}$ (path $OCB$).
(ii) Shear stress $\tau$ first increases up to $\sigma_0/\sqrt{3}$ and then remains constant. Normal stress $\sigma$ increases to $\sigma_0$ (path $OAB$).
(iii) Stresses $\sigma$ and $\tau$ increase with a constant ratio of $\sigma/\tau = \sqrt{3}$ until $\sigma = \sigma_0$ and $\tau = \sigma_0/\sqrt{3}$ (path $OB$).

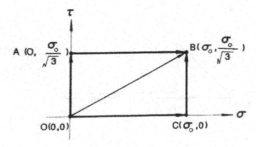

FIGURE P5.5

Use deformation theory to find $(\epsilon, \gamma)$ and compare it with the results obtained in (i), (ii), and (iii).

5.6. Continue the problem of Example 5.6:

(a) If the loading path is originally traced in reverse order: $(-iv)$, $(-iii)$, $(-ii)$, $(-i)$, that is, $O$-$C$-$B$-$A$-$O$, find the elastic and plastic strain components at the end of each loading path, and compare the permanent strains in this loading cycle with that given by Example 5.6.

(b) Draw successive yield curves at the beginning, the middle, and the end of loading path $(-iv)$, and show the normality of $d\epsilon_{ij}^p$ to these yield curves based on your numerical calculations of $d\epsilon_{ij}^p$ at the end of path $(-iv)$.

5.7. The behavior of a metal is assumed to be governed by a flow theory of plasticity based on an isotropic stress-hardening law of the $J_2$ (von Mises) type; i.e., when plastic flow occurs we have:

$$d\epsilon_{ij}^p = \bar{G}(J_2)s_{ij}\,dJ_2$$

Isotropic linear elastic behavior is assumed upon unloading. An element of this material is subjected to combined loading histories which produce the following straight-line stress paths in the $(\sigma, \tau)$ space; units are in MPa (tension, shear), as shown in Fig. P5.7:

  (i)   $(0, 0)$ to $(70, 70)$
 (ii)  $(70, 70)$ to $(70, 0)$
(iii) $(70, 0)$ to $(210, 0)$
(iv)  $(210, 0)$ to $(0, 0)$

(a) Determine the function $\bar{G}(J_2)$. The stress–strain curve of the metal on first loading in tension is approximated as follows ($\sigma$ in MPa):

$$\epsilon = \epsilon^e + \epsilon^p = \frac{\sigma}{7 \times 10^4} + \left(\frac{\sigma}{700}\right)^5$$

Poisson's ratio is assumed to be $\nu = 0.3$.

(b) Write the plastic stress–strain relation in component form explicitly in terms of $\sigma$, $\tau$, $d\sigma$, and $d\tau$.

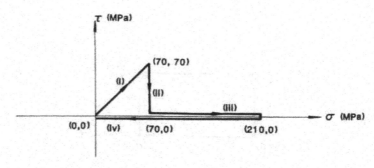

FIGURE P5.7

(c) Determine the values of the total elastic and plastic strains at the end of each stress path.

(d) Sketch the successive yield curves at the beginning and the end of each stress path. List the parts of the given stress paths which constitute (plastic) loading and those which produce unloading.

5.8. The stress–strain relation of a $J_2$-material in a pure shear test is approximated by

$$\gamma = \begin{cases} \dfrac{\tau}{G} & \tau \le \tau_y \\[2mm] \dfrac{\tau}{G} + \dfrac{\tau - \tau_y}{m} & \tau > \tau_y \end{cases}$$

where $\tau_y$ is the initial yield stress in shear. Using the deformational theory, write the complete stress-strain relationships of the material:

(i) in a simple tension state, $\sigma_x = \sigma$, $\epsilon_x = \epsilon$, other stresses are zero; and

(ii) in an equal biaxial tension–compression state, $\sigma_x = \sigma$, $\sigma_y = -\sigma$, $\epsilon_x = \epsilon$, $\epsilon_y = -\epsilon$, other stresses are zero.

5.9. The stress–strain relationship of a material in simple tension is given by

$$\epsilon = \epsilon^e + \epsilon^P = \frac{\sigma}{E}\left[1 + \left(\frac{\sigma}{\sigma_0}\right)^m\right]$$

where $\sigma_0$ is the initial yield stress in tension, and $m$ is a given constant. Using the incremental $J_2$-theory, derive the expression for $d\tau/d\gamma$ in a pure shear stress state.

5.10. A long steel thin-walled tube with diameter $D$ and wall thickness $t$ is subjected to an interior pressure $p_1$ and an external pressure $p_2$ as shown in Fig. P5.10. The ends of the tube are closed. The external pressure is assumed not to affect the axial stress component of the tube. Assuming that the material obeys the

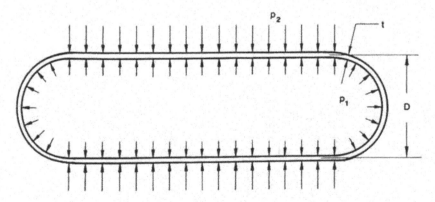

FIGURE P5.10

$J_2$ isotropic hardening theory, and using the effective stress-effective plastic strain relationship of the form

$$\epsilon_p = a\sigma_e^3$$

where $a$ is a constant, determine the plastic strains $(\epsilon_a^p, \epsilon_c^p)$ at the end of the following three loading paths using the incremental theory:

(i)   $(p_1, p_2) = (0, 0) \rightarrow (P_1, RP_1)$
(ii)  $(p_1, p_2) = (0, 0) \rightarrow (0, RP_1) \rightarrow (P_1, RP_1)$
(iii) $(p_1, p_2) = (0, 0) \rightarrow (P_1, 0) \rightarrow (P_1, RP_1)$

where $\epsilon_a^p$ and $\epsilon_c^p$ are axial and circumferential plastic strains, respectively, and $P_1$ and $R$ are constants, $R = \frac{3}{2}$. Sketch the yield surfaces and the loading paths in the $(\sigma_a, \sigma_c)$ space, and explain the result obtained in the three loading cases.

5.11. Determine the plastic strains $(\epsilon_a^p, \epsilon_c^p)$ of the thin-walled tube described in Problem 5.10 for the proportional loading path using the deformational theory for a $J_2$-material.

## ANSWERS TO SELECTED PROBLEMS

5.1. (a) $a = 2.381 \times 10^{-6}$, $b = 1.785 \times 10^{-5}$, $c = 8.032 \times 10^{-14}$

(b) $\epsilon_{ij} = \begin{bmatrix} 22.9 & 0 & 14.32 \\ 0 & 5 & 10.74 \\ 14.32 & 10.74 & -12.9 \end{bmatrix} \times 10^{-4}$

5.2. $d\alpha_1 = (\bar{\sigma}_1/2\sigma_0^2)[(2\bar{\sigma}_1 - \bar{\sigma}_2)\,d\sigma_1 + (2\bar{\sigma}_2 - \bar{\sigma}_1)\,d\sigma_2]$,
$d\alpha_2 = (\bar{\sigma}_2/2\sigma_0^2)[(2\bar{\sigma}_1 - \bar{\sigma}_2)\,d\sigma_1 + (2\bar{\sigma}_2 - \bar{\sigma}_1)\,d\sigma_2]$,  with  $\bar{\sigma}_1 = \sigma_1 - \alpha_1$, $\bar{\sigma}_2 = \sigma_2 - \alpha_2$.

5.3. (i) In ranges of $\sigma_1 > \sigma_2 > 0$ and $\sigma_1 < \sigma_2 < 0$: $d\epsilon_1^p = \bar{G}\,d\sigma_1$, $d\epsilon_2^p = 0$, $d\epsilon_3^p = -d\epsilon_1^p$;
(ii) in ranges of $\sigma_2 > \sigma_1 > 0$ and $\sigma_2 < \sigma_1 < 0$: $d\epsilon_1^p = 0$, $d\epsilon_2^p = \bar{G}\,d\sigma_2$, $d\epsilon_3^p = -d\epsilon_2^p$;
(iii) in range of $\sigma_1 > 0$, $\sigma_2 < 0$: $d\epsilon_1^p = \bar{G}(d\sigma_1 - d\sigma_2)$, $d\epsilon_2^p = -d\epsilon_1^p$, $d\epsilon_3^p = 0$;
(iv) in range of $\sigma_1 < 0$, $\sigma_2 > 0$: $d\epsilon_1^p = \bar{G}(d\sigma_1 - d\sigma_2)$, $d\epsilon_2^p = -d\epsilon_1^p$, $d\epsilon_3^p = 0$.

5.5. (i) $\epsilon = \dfrac{\sigma_0}{E} + \dfrac{\sigma_0}{E_p}\ln\sqrt{2}$

$\gamma = \dfrac{\sigma_0}{\sqrt{3}}\left[\dfrac{1}{G} + \dfrac{3}{E_p}\left(1 - \dfrac{\pi}{4}\right)\right]$

(ii) $\epsilon = \dfrac{\sigma_0}{E} + \dfrac{\sigma_0}{E_p}\left(1 - \dfrac{\pi}{4}\right)$

$\gamma = \dfrac{\sigma_0}{\sqrt{3}}\left[\dfrac{1}{G} + \dfrac{3}{E_p}\ln\sqrt{2}\right]$

(iii) $\epsilon = \dfrac{\sigma_0}{E} + \dfrac{\sigma_0}{E_p}\left(1 - \dfrac{1}{\sqrt{2}}\right)$

$\gamma = \dfrac{\sigma_0}{\sqrt{3}}\left[\dfrac{1}{G} + \dfrac{3}{E_p}\left(1 - \dfrac{1}{\sqrt{2}}\right)\right]$

The results of deformation theory are the same as in (iii).

5.6. At the end of path $(-iv)$:

$$\epsilon_{ij}^e = \begin{bmatrix} 1 & -0.867/2 & 0 \\ -0.867/2 & -0.3 & 0 \\ 0 & 0 & -0.3 \end{bmatrix} \times 10^{-3}$$

$$\epsilon_{ij}^p = \begin{bmatrix} 12 & -12/2 & 0 \\ -12/2 & -6 & 0 \\ 0 & 0 & -6 \end{bmatrix} \times 10^{-3}$$

At the end of path $(-iii)$:

$$\epsilon_{ij}^e = \begin{bmatrix} 1 & 0.867/2 & 0 \\ 0.867/2 & -0.3 & 0 \\ 0 & 0 & -0.3 \end{bmatrix} \times 10^{-3},$$

$$\epsilon_{ij}^p = \text{the same as } (-iv)$$

At the end of path $(-ii)$:

$$\epsilon_{ij}^e = \begin{bmatrix} 0 & 0.867/2 & 0 \\ 0.867/2 & 0 & 0 \\ 0 & 0 & 0 \end{bmatrix} \times 10^{-3},$$

$$\epsilon_{ij}^p = \text{the same as } (-iv)$$

At the end of path $(-i)$:

$$\epsilon_{ij}^p = \text{the same as } (-iv)$$

5.7. (a) $\bar{G}(J_2) = 2.0 \times 10^{-3} J_2$.
  (b) $d\epsilon_x^p = \frac{4}{3} \times 10^{-13}(\frac{1}{3}\sigma^2 + \tau^2)(\frac{2}{3}\sigma \, d\sigma + 2\tau\sigma \, d\tau);$ $d\epsilon_y^p = d\epsilon_z^p = -\frac{1}{2}d\epsilon_x^p;$
  $d\gamma_{xy}^p = (2 \, d\epsilon_{xy}^p) = 4 \times 10^{-13}(\frac{1}{3}\sigma^2 + \tau^2)[\frac{2}{3}\sigma\tau \, d\sigma + 2\tau^2 \, d\tau];$
  $d\gamma_{xz}^p = d\gamma_{yz}^p = 0.$
  (c) At the end of path (i):

$$\epsilon_{ij}^e = \begin{bmatrix} 1 & 2.6/2 & 0 \\ 2.6/2 & -0.3 & 0 \\ 0 & 0 & -0.3 \end{bmatrix} \times 10^{-3}$$

$$\epsilon_{ij}^p = \begin{bmatrix} 0.16 & 0.48/2 & 0 \\ 0.48/2 & -0.08 & 0 \\ 0 & 0 & -0.08 \end{bmatrix} \times 10^{-3}$$

At the end of path (ii):

$$\epsilon_{ij}^e = \begin{bmatrix} 1 & 0 & 0 \\ 0 & -0.3 & 0 \\ 0 & 0 & -0.3 \end{bmatrix} \times 10^{-3}$$

$$\epsilon_{ij}^p = \text{the same as (i)}$$

At the end of path (iii):

$$\epsilon_{ij}^{e} = \begin{bmatrix} 3 & 0 & 0 \\ 0 & -0.9 & 0 \\ 0 & 0 & -0.9 \end{bmatrix} \times 10^{-3}$$

$$\epsilon_{ij}^{p} = \begin{bmatrix} 2.27 & 0.48/2 & 0 \\ 0.48/2 & -1.135 & 0 \\ 0 & 0 & -1.135 \end{bmatrix} \times 10^{-3}$$

At the end of path (iv):

$$\epsilon_{ij}^{e} = 0, \qquad \epsilon_{ij}^{p} = \text{the same as (iii)}$$

5.8. (i) $\epsilon = \begin{cases} \dfrac{\sigma}{E}, & \sigma \le \sqrt{3}\tau_y \\[2ex] \dfrac{\sigma}{E} + \dfrac{\sigma - \sqrt{3}\tau_y}{3m}, & \sigma > \sqrt{3}\tau_y \end{cases}$

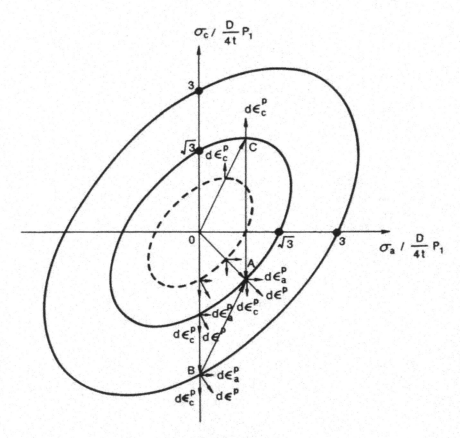

FIGURE S5.10

$$(ii) \quad \epsilon = \begin{cases} \dfrac{(1+\nu)\sigma}{E}, & \sigma \le \tau_y \\ \dfrac{(1+\nu)\sigma}{E} + \dfrac{\sigma - \tau_y}{2m}, & \sigma > \tau_y \end{cases}$$

5.9. $\dfrac{d\tau}{d\gamma} = \dfrac{1}{\dfrac{1}{G} + \dfrac{3(m+1)}{E}\left(\dfrac{\sqrt{3}\tau}{\sigma_0}\right)^m}$

5.10. See Fig. S5.10.

(i) $\epsilon_a^p = \dfrac{9a}{2}\left(\dfrac{DP_1}{4t}\right)^3$

$\epsilon_c^p = -\dfrac{9a}{2}\left(\dfrac{DP_1}{4t}\right)^3$

(ii) $\epsilon_a^p = \dfrac{27a}{2}\left(\dfrac{DP_1}{4t}\right)^3$

$\epsilon_c^p = -27a\left(\dfrac{DP_1}{4t}\right)^3$

(iii) $\epsilon_a^p = 0$

$\epsilon_c^p = \dfrac{9a}{2}\left(\dfrac{DP_1}{4t}\right)^3$

5.11. $\epsilon_a^p = \dfrac{9a}{2}\left(\dfrac{DP_1}{4t}\right)^3$

$\epsilon_c^p = -\dfrac{9a}{2}\left(\dfrac{DP_1}{4t}\right)^3$

# Part III: Metal Plasticity

# 6
# Implementation in Metals

## 6.1. Introduction

The stress–strain relations under a combined state of stresses discussed in Chapters 4 and 5 are typical for polycrystalline metals. It is well known that mild steel exhibits plastic flow under constant stress (Fig. 1.1a). This behavior can be modeled by the *theory of perfect plasticity*. The more common metals such as aluminum and copper generally fall within the category of hardening materials (Fig. 1.1b). They can best be modeled by the *work-hardening theory of plasticity*.

One of the remarkable features of metals is that the effects of hydrostatic pressure on yielding and the subsequent plastic deformation are not appreciable. These facts imply that the shear stress alone is critical for yielding and that the *plastic volume change* is negligible even for large plastic deformations. The maximum shear stress condition of Tresca and the octahedral shear stress condition of von Mises for yielding show a good agreement with experimental data, and the flow rule associated with each of these two yield functions predicts a plastic shear deformation without plastic volume change. The Tresca or von Mises model with or without work hardening is therefore generally adopted in establishing the constitutive laws of metals.

Due to the nonlinear nature of the plastic constitutive relations, analytic solutions of boundary-value problems are difficult to obtain. Up to now, only very few exact solutions of elastic–plastic boundary-value problems are available. Future prospects for exact solutions are far less bright for the case of three-dimensional processes where either redundant strain is important or geometric complexity is the characteristic feature of the problem. Thanks to the rapid development of high-speed computers and modern numerical techniques, incremental inelastic analysis of virtually any structural problems can now be carried out by the powerful *finite-element method*. This development has greatly benefited the field of plasticity. It provides the classical theory of plasticity with newer concepts and wider applications.

This chapter is concerned with the applications of the constitutive relations of metals to structural analyses by the nonlinear finite-element method. Formulations of the elastic–plastic constitutive relation are summarized in Section 6.2, in which the elastic–plastic stiffness matrix $[C^{ep}]$ based on $J_2$-theory is given in 3-D as well as 2-D forms ready for computer programming. Formulations of the finite-element method for an elastic–plastic analysis of structures are summarized in Section 6.3. In Sections 6.4 and 6.5, the algorithms for solving a general nonlinear material finite-element problem are discussed in some detail. Sections 6.6 and 6.7 deal with some further developments in metal plasticity theory. The *bounding surface theory* recently developed for modeling the behavior of metals under cyclic loadings is presented in Section 6.6, and a brief discussion of the stress–strain relations for orthotropic materials is given in Section 6.7.

## 6.2. Formulation of the Elastic–Plastic Matrix

The incremental finite-element method as applied to metal plasticity owes its success to the establishment of the elastic–plastic tangent stiffness matrix $[C^{ep}]$. Formulations of matrix $[C^{ep}]$ discussed in the preceding chapters are now summarized in what follows.

(i) The matrix $[C^{ep}]$ of a general material with the mixed work-hardening rule is given by Eqs. (5.173) and (5.171). The elastic–plastic matrix $[C^{ep}]$ is further expressed by Eqs. (5.187), (5.188), and (5.189). As can be seen, the elements of the matrices are expressed in terms of the derivatives of the yield function and the plastic potential function.

(ii) For an isotropic material, the derivatives $\partial f/\partial \sigma_{ij}$ (or $\partial g/\partial \sigma_{ij}$) can be written in terms of the derivatives with respect to the stress invariants $\partial f/\partial I_1$, $\partial f/\partial J_2$, and $\partial f/\partial J_3$ as given by Eq. (4.140) (see Section 4.11). Equations (4.156), (4.157), and (4.158) are the general expressions of the tangent stiffness matrix $[C^{ep}]$ for an elastic-perfectly plastic material (without work hardening).

(iii) For $J_2$-material that is commonly used in metal plasticity, the matrix $[C^{ep}]$ has been derived in Example 5.7. The explicit form in 3-D and 2-D cases will be given here.

### 6.2.1. 3-D Elastic–Plastic Stiffness Matrix

We rewrite the incremental stress–strain relation of Eqs. (5.143) and (5.145)

$$d\sigma_{ij} = C^{ep}_{ijkl}\, d\epsilon_{kl} = (C_{ijkl} + C^{P}_{ijkl})\, d\epsilon_{kl} \qquad (6.1)$$

in which the incremental stress and strain tensors $d\sigma_{ij}$, $d\epsilon_{ij}$ are generally expressed in vector forms:

$$\{d\sigma\}^T = \{d\sigma_x\, d\sigma_y\, d\sigma_z\, d\tau_{yz}\, d\tau_{zx}\, d\tau_{xy}\} \qquad (6.2)$$

$$\{d\epsilon\}^T = \{d\epsilon_x\, d\epsilon_y\, d\epsilon_z\, d\gamma_{yz}\, d\gamma_{zx}\, d\gamma_{xy}\} \qquad (6.3)$$

and the tensor of elastic modulus $C_{ijkl}$ is expressed in matrix form:

$$[C] = \begin{bmatrix} K+\frac{4}{3}G & K-\frac{2}{3}G & K-\frac{2}{3}G & 0 & 0 & 0 \\ K-\frac{2}{3}G & K+\frac{4}{3}G & K-\frac{2}{3}G & 0 & 0 & 0 \\ K-\frac{2}{3}G & K-\frac{2}{3}G & K+\frac{4}{3}G & 0 & 0 & 0 \\ 0 & 0 & 0 & G & 0 & 0 \\ 0 & 0 & 0 & 0 & G & 0 \\ 0 & 0 & 0 & 0 & 0 & G \end{bmatrix} \tag{6.4}$$

in which $G$ and $K$ are the shear and bulk moduli, respectively. The plastic stiffness tensor $C^P_{ijkl}$ for the $J_2$-material is given by Eq. (5.189) as

$$C^P_{ijkl} = -\frac{36G^2}{h} \bar{s}_{ij}\bar{s}_{kl} \tag{6.5}$$

in which the scalar function $h$ expressed by Eq. (5.188) is

$$h = 4(3G + H_p)\bar{\sigma}_e^2 \tag{6.6}$$

and $\bar{s}_{ij}$ is the *reduced stress deviator tensor* also expressed in vector form as

$$[\bar{S}]^T = \{\bar{s}_x, \bar{s}_y, \bar{s}_z, \bar{s}_{yz}, \bar{s}_{zx}, \bar{s}_{xy}\} \tag{6.7}$$

Thus, the plastic stiffness tensor can be expressed as a $6 \times 6$ matrix:

$$[C^P] = -\frac{1}{H} \begin{bmatrix} \bar{s}_x^2 & & & & & \text{sym.} \\ \bar{s}_y\bar{s}_x & \bar{s}_y^2 & & & & \\ \bar{s}_z\bar{s}_x & \bar{s}_z\bar{s}_y & \bar{s}_z^2 & & & \\ \bar{s}_{yz}\bar{s}_x & \bar{s}_{yz}\bar{s}_y & \bar{s}_{yz}\bar{s}_z & \bar{s}_{yz}^2 & & \\ \bar{s}_{zx}\bar{s}_x & \bar{s}_{zx}\bar{s}_y & \bar{s}_{zx}\bar{s}_z & \bar{s}_{zx}\bar{s}_{yz} & \bar{s}_{zx}^2 & \\ \bar{s}_{xy}\bar{s}_x & \bar{s}_{xy}\bar{s}_y & \bar{s}_{xy}\bar{s}_z & \bar{s}_{xy}\bar{s}_{yz} & \bar{s}_{xy}\bar{s}_{zx} & \bar{s}_{xy}^2 \end{bmatrix} \tag{6.8}$$

in which $1/H = 36G^2/h$.

## 6.2.2. Plane Strain Case

Under the condition of plane strain, $d\epsilon_z = d\gamma_{xz} = d\gamma_{yz} = 0$. Equations (6.4) and (6.8) can be reduced, and the elastic–plastic stiffness matrix is expressed in the form

$$[C^{ep}] = \begin{bmatrix} K+\dfrac{4}{3}G-\dfrac{\bar{s}_x^2}{H} & & \text{sym.} \\[2mm] K-\dfrac{2}{3}G-\dfrac{\bar{s}_y\bar{s}_x}{H} & K+\dfrac{4}{3}G-\dfrac{\bar{s}_y^2}{H} & \\[2mm] -\dfrac{\bar{s}_{xy}\bar{s}_x}{H} & -\dfrac{\bar{s}_{xy}\bar{s}_y}{H} & G-\dfrac{\bar{s}_{xy}^2}{H} \end{bmatrix} \tag{6.9}$$

## 6.2.3. Axisymmetric Case

The nonzero stress components in the axisymmetric case are $\sigma_r$, $\sigma_z$, $\sigma_\theta$, and $\tau_{rz}$, and the corresponding strains are $\epsilon_r$, $\epsilon_z$, $\epsilon_\theta$, and $\gamma_{rz}$. Equations (6.4) and (6.8) can be reduced, and the corresponding matrix $[C^{ep}]$ is given by

$$
[C^{ep}] =
\begin{bmatrix}
K + \dfrac{4G}{3} - \dfrac{\bar{s}_r^2}{H} & & & \text{sym.} \\[2ex]
K - \dfrac{2G}{3} - \dfrac{\bar{s}_z \bar{s}_r}{H} & K + \dfrac{4G}{3} - \dfrac{\bar{s}_z^2}{H} & & \\[2ex]
K - \dfrac{2G}{3} - \dfrac{\bar{s}_\theta \bar{s}_r}{H} & K - \dfrac{2G}{3} - \dfrac{\bar{s}_\theta \bar{s}_z}{H} & K + \dfrac{4G}{3} - \dfrac{\bar{s}_\theta^2}{H} & \\[2ex]
- \dfrac{\bar{s}_{rz} \bar{s}_r}{H} & - \dfrac{\bar{s}_{rz} \bar{s}_z}{H} & - \dfrac{\bar{s}_{rz} \bar{s}_\theta}{H} & G - \dfrac{\bar{s}_{rz}^2}{H}
\end{bmatrix}
\tag{6.10}
$$

## 6.2.4. Plane Stress Case

In this case, $d\sigma_z = d\tau_{yz} = d\tau_{xz} = 0$, but the strain component $d\epsilon_z$ is not zero, only the shear strain components $d\gamma_{yz}$ and $d\gamma_{zx}$ are zero. $d\epsilon_z$ must be solved from the condition $d\sigma_z = 0$ and substituted the $d\epsilon_z$ so obtained into the first and second equations of Eq. (6.1). Then the elastic–plastic stiffness matrix for the plane stress case can be obtained. However, the expression of $[C^{ep}]$ so obtained is quite involved. An alternative form proposed by Yamada (1969) is relatively simple. It has the form

$$
[C^{ep}] =
\begin{bmatrix}
\dfrac{E}{1 - \nu^2} - \dfrac{\bar{s}_1^2}{\bar{s}} & & \text{sym.} \\[2ex]
\dfrac{\nu E}{1 - \nu^2} - \dfrac{\bar{s}_1 \bar{s}_2}{\bar{s}} & \dfrac{E}{1 - \nu^2} - \dfrac{\bar{s}_2^2}{\bar{s}} & \\[2ex]
- \dfrac{\bar{s}_1 \bar{s}_6}{\bar{s}} & - \dfrac{\bar{s}_2 \bar{s}_6}{\bar{s}} & \dfrac{E}{2(1 + \nu)} - \dfrac{\bar{s}_6^2}{\bar{s}}
\end{bmatrix}
\tag{6.11}
$$

where

$$
\bar{s}_1 = \frac{E}{(1 - \nu^2)} (\bar{s}_x + \nu \bar{s}_y), \qquad
\bar{s}_2 = \frac{E}{(1 - \nu^2)} (\nu \bar{s}_x + \bar{s}_y), \qquad
\bar{s}_6 = \frac{E}{(1 + \nu)} \bar{s}_{xy}
\tag{6.12}
$$

and

$$
\bar{s} = \tfrac{4}{9} \bar{\sigma}_e H_p + \bar{s}_1 \bar{s}_x + \bar{s}_2 \bar{s}_y + 2 \bar{s}_6 \bar{s}_{xy}
\tag{6.13}
$$

Equation (6.11) is derived directly from the component form of the *flow rule*. The elastic–plastic tangent stiffness matrix $[C^{ep}]$ for an orthotropic material in the plane stress case takes a similar form. This will be given later in Section 6.7.

## 6.3. Finite-Element Formulation

In this and following sections, it is assumed that the reader has had some contact with *linear finite-element analysis*. The general governing equation of the finite-element method for a static analysis can be derived from the *principle of virtual work*, Eq. (3.157),

$$\int_V \sigma_{ij}\delta\epsilon_{ij}\,dV = \int_A T_i\delta u_i\,dA + \int_V q_i\delta u_i\,dV \tag{6.14}$$

where $\delta u_i$ and $\delta\epsilon_{ij}$ are virtual displacement increments and virtual strain increments, respectively, and they form a compatible set of deformations; $T_i$ and $q_i$ are surface traction and body force, respectively; and $\sigma_{ij}$ with $T_i$ and $q_i$ form an equilibrium set. In a matrix form. Eq. (6.14) becomes

$$\int_V \{\delta\epsilon\}^T\{\sigma\}\,dV = \int_A \{\delta u\}^T\{T\}\,dA + \int_V \{\delta u\}^T\{q\}\,dV \tag{6.15}$$

where the vectors for displacement $\{u\}$, strain $\{\epsilon\}$, and stress $\{\sigma\}$ are defined as

$$\{u\}^T = \{u_1, u_2, u_3\}, \qquad \{\delta u\}^T = \{\delta u_1, \delta u_2, \delta u_3\} \tag{6.16}$$

$$\{\epsilon\}^T = \{\epsilon_x, \epsilon_y, \epsilon_z, \gamma_{yz}, \gamma_{zx}, \gamma_{xy}\} \tag{6.17a}$$

$$\{\delta\epsilon\}^T = \{\delta\epsilon_x, \delta\epsilon_y, \delta\epsilon_z, \delta\gamma_{yz}, \delta\gamma_{zx}, \delta\gamma_{xy}\} \tag{6.17b}$$

$$\{\sigma\}^T = \{\sigma_x, \sigma_y, \sigma_z, \tau_{yz}, \tau_{zx}, \tau_{xy}\} \tag{6.18}$$

For a *geometrically linear* analysis, or a small-deformation analysis, we have

$$\{\epsilon\} = [B]\{U\}, \qquad \{\delta\epsilon\} = [B]\{\delta U\} \tag{6.19}$$

where $\{U\}$ is the displacement vector of nodal points that is related to the distributed displacement $\{u\}$ by

$$\{u\} = [N]\{U\} \tag{6.20}$$

in which $[N]$ is the matrix of the *displacement interpolation function*, or the *shape function*, and the *strain–displacement matrix* $[B]$ is a matrix defined as

$$[B] = [L][N] \tag{6.21}$$

and $[L]$ is the *differential operator matrix* defined as

$$L = \begin{bmatrix} \dfrac{\partial}{\partial x} & 0 & 0 \\[2mm] 0 & \dfrac{\partial}{\partial y} & 0 \\[2mm] 0 & 0 & \dfrac{\partial}{\partial z} \\[2mm] 0 & \dfrac{\partial}{\partial z} & \dfrac{\partial}{\partial y} \\[2mm] \dfrac{\partial}{\partial z} & 0 & \dfrac{\partial}{\partial x} \\[2mm] \dfrac{\partial}{\partial y} & \dfrac{\partial}{\partial x} & 0 \end{bmatrix} \tag{6.22}$$

such that

$$\{\epsilon\} = [L]\{u\} \tag{6.23}$$

Substituting Eqs. (6.19) and (6.20) in Eq. (6.15), we obtain the governing equation for a small-deformation analysis as

$$\int_V [B]^T\{\sigma\}\, dV = \int_A [N]^T\{T\}\, dA + \int_V [N]^T\{q\}\, dV \qquad (6.24a)$$

or

$$\int_V [B]^T\{\sigma\}\, dV = \{R\} \qquad (6.24b)$$

where $\{R\}$ is the equivalent external force acting on the nodal points,

$$\{R\} = \int_A [N]^T\{T\}\, dA + \int_V [N]^T\{q\}\, dV \qquad (6.25)$$

Furthermore, if a linear elastic stress–strain relation is assumed, we obtain the governing equation for a *linear analysis*,

$$[K]\{U\} = \{R\} \qquad (6.26a)$$

where $[K]$ is the stiffness matrix of the structure,

$$[K] = \int_V [B]^T[C][B]\, dV \qquad (6.26b)$$

in which $[C]$ is the elastic constitutive matrix.

In an elastic–plastic analysis, because of the nonlinear relationship between the stress $\{\sigma\}$ and the strain $\{\epsilon\}$, the governing equation (6.24) is a nonlinear equation of strains, and therefore, is a nonlinear function of the nodal displacement, $\{U\}$. Iterative methods are usually employed to solve Eq. (6.24) for $\{U\}$ corresponding to a given set of external forces. Moreover, since an elastic–plastic constitutive relation depends on deformation history, an incremental analysis following an actual variation of external forces should be used to trace the variation of displacement, strain, and stress along with the external forces.

In an incremental analysis, the total load $\{R\}$ acting on a structure is added in increments step by step. At the $(m+1)$th step, the load can be expressed as

$$^{m+1}\{R\} = {}^m\{R\} + {}^{m+1}\{\Delta R\} \qquad (6.27)$$

where the left superscript $m$ has been used to indicate the $m$th incremental step. Assuming that the solutions at the $m$th step, $^m\{U\}$, $^m\{\sigma\}$, $^m\{\epsilon\}$, are known, and at the $(m+1)$th step, we have, corresponding to the load increment $\{\Delta R\}$,

$$^{m+1}\{U\} = {}^m\{U\} + \{\Delta U\} \qquad (6.28)$$

$$^{m+1}\{\sigma\} = {}^m\{\sigma\} + \{\Delta\sigma\} \qquad (6.29)$$

Here, the left superscript for the increments has been dropped. Equation (6.24) becomes

$$^{m+1}\{F\} = {}^{m+1}\{R\} \tag{6.30a}$$

where $^{m+1}\{F\}$ is the equivalent force of stress acting on the nodal points,

$$^{m+1}\{F\} = \int_V [B]^{T\,m+1}\{\sigma\}\, dV \tag{6.30b}$$

or

$$\int_V [B]^T \{\Delta\sigma\}\, dV = {}^{m+1}\{R\} - \int_V [B]^{T\,m}\{\sigma\}\, dV \tag{6.30c}$$

Equation (6.30a), in fact, represents the equilibrium of the external force, $^{m+1}\{R\}$, with the internal force, $^{m+1}\{F\}$. Two types of algorithms are therefore involved in solving Eq. (6.30) for the displacement increment $\{\Delta U\}$ and stress increment $\{\Delta\sigma\}$. One is the algorithm used for solving nonlinear simultaneous equations. This will be discussed in the following section. Another is the algorithm used for determining the stress increment, $\{\Delta\sigma\}$, corresponding to a strain increment, $\{\Delta\epsilon\}$, for a given stress state and deformation history. This will be discussed later in Section 6.5.

# 6.4. Numerical Algorithms for Solving Nonlinear Equations

Many algorithms exist for solving the nonlinear simultaneous equations (6.30a). In this section, we shall present three methods of the Newton type that have been widely used in finite-element analysis.

Considering that the stress $\{\sigma\}$ is a nonlinear function of the displacement, $\{U\}$, Eq. (6.30a) can be rewritten as

$$\Psi(^{m+1}\{U\}) = {}^{m+1}\{F(^{m+1}\{U\})\} - {}^{m+1}\{R\} \tag{6.31}$$

Equation (6.31) is a nonlinear matrix equation expressed in terms of the displacement $^{m+1}\{U\}$. As noted previously in Section 6.3, this equation represents an equilibrium of the external force, $^{m+1}\{R\}$, with the internal force, $^{m+1}\{F\}$. The iterative method used to solve Eq. (6.31) is therefore called the *equilibrium iterative method*.

## 6.4.1. Newton–Raphson Method

We have already obtained the $(i-1)$th approximation, $^{m+1}\{U\}^{(i-1)}$, to the displacement $^{m+1}\{U\}$. Expanding $\Psi(^{m+1}\{U\})$ using the Taylor series expansion at $^{m+1}\{U\}^{(i-1)}$ and neglecting all higher-order terms, we obtain

$$\Psi(^{m+1}\{U\}^{(i-1)}) + \frac{\partial\Psi}{\partial U}\bigg|_{^{m+1}\{U\}^{(i-1)}} (^{m+1}\{U\} - ^{m+1}\{U\}^{(i-1)}) = 0$$

or

$$\frac{\partial F}{\partial U}\bigg|_{^{m+1}\{U\}^{(i-1)}} \{\Delta U\}^{(i)} + ^{m+1}\{F\}^{(i-1)} - ^{m+1}\{R\} = 0 \qquad (6.32)$$

where

$$\{\Delta U\}^{(i)} = ^{m+1}\{U\} - ^{m+1}\{U\}^{(i-1)} \qquad (6.33)$$

$$^{m+1}\{F\}^{(i-1)} = ^{m+1}\{F(^{m+1}\{U\}^{(i-1)})\} \qquad (6.34)$$

Recognizing that

$$^{m+1}[K]^{(i-1)} = \frac{\partial F}{\partial U}\bigg|_{^{m+1}\{U\}^{(i-1)}}$$

$$= \int_{V} [B]^{T}[C^{ep}]\bigg|_{^{m+1}\{U\}^{(i-1)}} [B]\, dV \qquad (6.35)$$

where $[C^{ep}]|_{^{m+1}\{U\}^{(i-1)}}$ is the elastic–plastic stiffness matrix corresponding to the displacement $^{m+1}\{U\}^{(i-1)}$, and $^{m+1}[K]^{(i-1)}$ is the tangential stiffness matrix of the structure, we obtain the iteration scheme of the *Newton–Raphson* algorithm as

$$^{m+1}[K]^{(i-1)}\{\Delta U\}^{(i)} = ^{m+1}\{R\} - ^{m+1}\{F\}^{(i-1)} \qquad (6.36a)$$

$$^{m+1}\{U\}^{(i)} = ^{m+1}\{U\}^{(i-1)} + \{\Delta U\}^{(i)} \qquad (6.36b)$$

$$^{m+1}\{U\}^{(0)} = ^{m}\{U\}, \qquad ^{m+1}[K]^{(0)} = ^{m}[K],$$

$$^{m+1}\{F\}^{(0)} = ^{m}\{F\} \qquad (i=1,2,\ldots) \qquad (6.36c)$$

This iteration continues until a proper convergence criterion is satisfied. The convergence criterion will be discussed later in Section 6.4.4. This iteration procedure for a one-degree-of-freedom nonlinear system is shown schematically in Fig. 6.1.

The Newton–Raphson method has a high convergence rate, and it converges quadratically. However, it should be noted from Eq. (6.36) that the tangential stiffness matrix, $^{m+1}[K]^{(i-1)}$, is evaluated and factorized at each iteration step, and such an operation can be prohibitively expensive when a large-scale system is considered. Moreover, for a perfectly plastic or a strain-softening material, the tangential stiffness matrix may become singular or ill-conditioned. This may cause difficulty in the interation procedure. Modifications of the Newton–Raphson algorithm are therefore necessary, and will be described in the following sections.

## 6.4.2. Modified Newton–Raphson Method

One of the modifications of the Newton–Raphson method is to replace the tangential stiffness matrix $^{m+1}[K]^{(i-1)}$ in Eq. (6.36a) by $^{n}[K]$, which is a

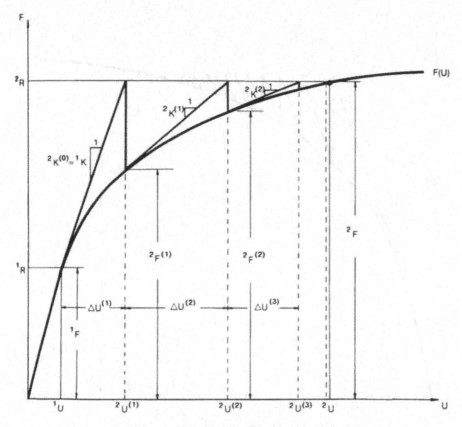

FIGURE 6.1. Newton–Raphson method.

tangential stiffness matrix evaluated at the load step $n$, $n < m+1$. If $^n[K]$ is evaluated only at the beginning of the first load step, i.e., the initial elastic stiffness matrix $^n[K] = [K]_0$ as defined in Eq. (6.26b) is used for all load steps, we obtain the so-called *initial stress method*. Often, the stiffness matrix is evaluated at the beginning of each load step, or for the $(m+1)$th step, the stiffness matrix

$$^n[K] = {}^{m+1}[K]^{(0)} = {}^m[K] \tag{6.37}$$

is used. The iteration scheme for the *modified Newton–Raphson* algorithm is expressed as

$$^n[K]\{\Delta U\}^{(i)} = {}^{m+1}\{R\} - {}^{m+1}\{F\}^{(i-1)} \tag{6.38a}$$

$$^{m+1}\{U\}^{(i)} = {}^{m+1}\{U\}^{(i-1)} + \{\Delta U\}^{(i)} \tag{6.38b}$$

$$^{m+1}\{U\}^{(0)} = {}^m\{U\}, \qquad {}^{m+1}\{F\}^{(0)} = {}^m\{F\} \qquad (i = 1, 2, \ldots) \tag{6.38c}$$

Again, the iteration continues until a proper convergence criterion is satisfied. This modified iteration procedure for a one-degree-of-freedom nonlinear system is shown schematically in Fig. 6.2.

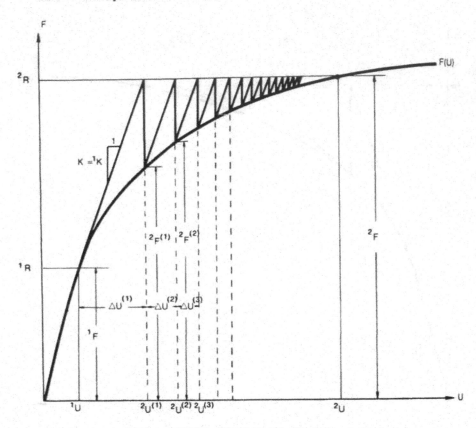

FIGURE 6.2. Modified Newton-Raphson method.

The modified Newton-Raphson algorithm involves fewer stiffness matrix evalution and factorization steps. As a result, the computational effort for one iteration cycle is much less than that for the Newton-Raphson algorithm for a large-scale system. However, the modified Newton-Raphson method converges linearly and, in general, more slowly than the Newton-Raphson method, i.e., for a given problem, more iterations are needed to reach a convergence when the modified Newton-Raphson method is used. In some situations, such as in the analysis of a strain-softening material, it becomes prohibitively slow. The convergence rate of the modified Newton-Raphson algorithm depends to a large extent on the number of times the stiffness matrix is updated. The more frequently the stiffness matrix is updated, the fewer iterations are needed to reach convergence. Moreover, the problem that the stiffness matrix may be singular or ill-conditioned still exists.

Another problem associated with the modified Newton-Raphson method is that if a change in the external load causes an unloading, i.e., the stress state is unloaded from a plastic state to an elastic state, this method may not lead to a convergent iteration, unless the structural stiffness matrix is updated for this situation. This in turn increases the complexity of coding in the numerical implementation of the modified method.

### 6.4.3. The Quasi-Newton Method

A compromise between the Newton–Raphson algorithm and the modified Newton–Raphson algorithm is the *quasi-Newton algorithm*. The Newton–Raphson method requires an evaluation and factorization of the stiffness matrix of a structure in each iteration, and this requires a large amount of computational time. On the other hand, the modified Newton–Raphson method keeps the same stiffness matrix during several iterations and, as a result, has a poorer convergence rate. Unlike these two methods, the quasi-Newton method employs a lower-rank matrix to update the inverse of the stiffness matrix, $^{m+1}[K^{-1}]^{(i-1)}$, and this procedure provides a secant approximation to the marix $^{m+1}[K^{-1}]^{(i)}$. It belongs to a class of methods known as *matrix update methods*. A Broyden–Fletcher–Goldfarb–Shanno (BFGS) rank 2 method commonly used with the quasi-Newton algorithm (Bathe, 1982) will be presented here.

Define the displacement increment $\{\delta\}$ as

$$\{\delta\}^{(i)} = {}^{m+1}\{U\}^{(i)} - {}^{m+1}\{U\}^{(i-1)} \tag{6.39}$$

and the out-of-balance force vector $\{R\}$ and its increment $\{\gamma\}$ as

$$\{R\}^{(i)} = {}^{m+1}\{R\} - {}^{m+1}\{F\}^{(i)} \tag{6.40}$$

$$\{\gamma\}^{(i)} = \{R\}^{(i-1)} - \{R\}^{(i)} \tag{6.41}$$

such that the updated matrix $^{m+1}[K]^{(i)}$ satisfies the quasi-Newton equation

$$^{m+1}[K]^{(i)}\{\delta\}^{(i)} = \{\gamma\}^{(i)} \tag{6.42}$$

For a symmetric positive-definite stiffness matrix, the recurrence formula of the inverse of the matrix is

$$^{m+1}[K^{-1}]^{(i)} = [A]^{(i-1)T}\,{}^{m+1}[K^{-1}]^{(i-1)}[A]^{(i-1)} \tag{6.43}$$

in which $[A]$ is the modification matrix defined as

$$[A]^{(i-1)} = [I] + \{V\}^{(i-1)}\{W\}^{(i-1)T} \tag{6.44}$$

where $[I]$ is a unit matrix with the same dimension as $[K]$; $\{V\}^{(i-1)}$ and $\{W\}^{(i-1)}$ are vectors expressible in terms of vectors $\{\delta\}$, $\{R\}$, and $\{\gamma\}$.

The iteration procedure for an iteration step $i$ $(i = 1, 2, \ldots)$ is divided into two steps:

Step 1: Evaluate the displacement increment $\{\Delta U\}$

$$\{\Delta U\} = {}^{m+1}[K^{-1}]^{(i-1)}\{R\}^{(i-1)}$$

$$= [A]^{(i-1)T} \cdots [A]^{(1)T}\,{}^n[K^{-1}][A]^{(1)} \cdots [A]^{(i-1)}\{R\}^{(i-1)} \tag{6.45a}$$

and

$$\{\delta\}^{(i)} = \{\Delta U\} \qquad (6.45b)$$

$$^{m+1}\{U\}^{(i)} = {}^{m+1}\{U\}^{(i-1)} + \{\Delta U\} \qquad (6.45c)$$

The actual computation in this step involves the vector inner product, the vector scalar product, and the solution of linear simultaneous equations with an already factorized coefficient matrix $^{m}[K]$.

Step 2: Compute the correction vectors $\{V\}^{(i)}$ and $\{W\}^{(i)}$ that will be used in the next iteration step

$$\{V\}^{(i)} = -c^{(i)} \, {}^{m+1}[K]^{(i-1)}\{\delta\}^{(i)} - \{\gamma\}^{(i)} \qquad (6.46a)$$

$$= \{R\}^{(i)} - (1 + c^{(i)})\{R\}^{(i-1)}$$

$$\{W\}^{(i)} = \frac{\{\delta\}^{(i)}}{\{\delta\}^{(i)T}\{\gamma\}^{(i)}} = \frac{\{\delta\}^{(i)}}{G(0) - G(1)} \qquad (6.46b)$$

where $c^{(i)}$ is the condition number for the modification matrix $[A]$,

$$c^{(i)} = \left( \frac{\{\delta\}^{(i)T}\{\gamma\}^{(i)}}{\{\delta\}^{(i)T} \, {}^{m+1}[K]^{(i-1)} \{\gamma\}^{(i)}} \right)^{1/2}$$

$$= \frac{G(0) - G(1)}{G(0)} \qquad (6.47)$$

To avoid the numerically dangerous updating, the updating will be performed only when $c^{(i)}$ is less than a preset tolerance, say, $10^5$. $G(x)$ denotes the dot product of vectors defined as

$$G(x) = G({}^{m+1}\{U\}^{(i-1)} + x\{\Delta U\})$$

$$= \{\Delta U\}^T [{}^{m+1}\{R\} - {}^{m+1}\{F({}^{m+1}\{U\}^{(i-1)} + x\{\Delta U\})\}] \qquad (6.48)$$

It should be noted that in the second step, the computation of the equivalent force for the stress state corresponding to the displacement $^{m+1}\{U\}^{(i)}$ is required. This iteration continues until a proper convergence criterion is met. The iteration procedure for a one-degree-of-freedom non-linear system is shown schematically in Fig. 6.3.

The computation required in the quasi-Newton algorithm for one iteration step is more than that in the modified Newton–Raphson algorithm, but is much less than that in the Newton–Raphson algorithm. However, this method has a better convergence property than the modified Newton–Raphson method, and its convergence rate lies somewhere between the linear and the quadratic rate. Furthermore, the stiffness matrix in this method is less important than in the other two methods, when using the matrix updating approach. In fact, the initial elastic stiffness matrix of the structure can even be used for all incremental steps without losing much efficiency. As a result, this method is suitable for the analysis of elastic–plastic solids exhibiting work hardening, strain softening, or perfect plasticity. No difficulties arise in the case of unloading. This method provides therefore a safe

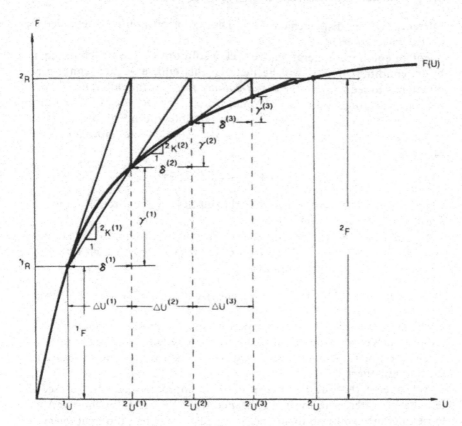

FIGURE 6.3. Quasi-Newton method.

and efficient procedure in solving nonlinear simultaneous equations (6.30) for a general elastic–plastic material. It is the best algorithm available at the present time.

### 6.4.4. Convergence Criteria

A properly defined convergence criterion to terminate the equilibrium iteration is an essential part of an efficient incremental solution strategy. At the end of each iteration, the solution obtained must be checked against a selected tolerance to see whether convergence has occurred.

For a finite-element displacement analysis, the computed displacement should approach the true displacement. Since the true displacement is not known in advance, an approximation of this criterion can be expressed as

$$\|\{\Delta U\}^{(i)}\|_2 \leq \epsilon_D \|^{m+1}\{U\}^{(i)} - {}^m\{U\}\|_2 \qquad (6.49)$$

where $\{\Delta U\}^{(i)}$ is the displacement increment obtained in the $i$th iteration, $\| \ \|_2$ is used to denote the Euclidean norm of a vector, and $\epsilon_D$ is a prescribed tolerance for the displacement $\{U\}$. This criterion is therefore called the *displacement criterion*.

For an equilibrium iteration, we seek a solution $\{U\}$ at which the equilibrium condition (6.30a) will be met. To this end, a second convergence criterion is to require that the *out-of-balance force*, or the difference between the internal force and the external force, $^{m+1}\{R\} - {}^{m+1}\{F\}$, vanishes. However, in a numerical process, it is impossible and needless to reach the state of zero out-of-balance force. Thus, we introduce an approximation of the form

$$\| {}^{m+1}\{R\} - {}^{m+1}\{F\}^{(i)} \|_2 \leq \epsilon_F \| {}^{m+1}\{R\} - {}^{m}\{F\} \|_2 \qquad (6.50)$$

where $\epsilon_F$ is a prescribed tolerance for the out-of-balance force. This criterion is called the *force criterion*.

A third criterion is to provide a measure of how close both displacement and force are to their equilibrium values. It is called the *internal energy criterion* and can be expressed as

$$\{\Delta U\}^{(i)T} ({}^{m+1}\{R\} - {}^{m+1}\{F\}^{(i)}) \leq \epsilon_E \{\Delta U\}^{(1)T} ({}^{m+1}\{R\} - {}^{m}\{F\}) \quad (6.51)$$

where the left-hand side of the inequality represents the work done by the out-of-balance force on the displacement increment, and the right-hand side is the initial value of the same work; $\epsilon_E$ is a prescribed tolerance for the internal energy.

Any one of these three criteria or their combinations can be used to terminate an iteration, but the tolerance must be carefully chosen. A too loose tolerance will lead to an inaccurate result, while a too tight tolerance may lead to wasteful computation to obtain a needless accuracy.

To compare the efficiency of these three equilibrium iteration algorithms, an example is presented in Fig. 6.4 using three convergence criteria. A typical elastic–perfectly plastic problem, a thick-walled cylinder subjected to an internal pressure as shown in Fig. 6.4a, has been analyzed using these three iterations algorithms. In a load step, step 2, the internal pressure is increased from $0.5p_s$ to $0.99p_s$, where $p_s$ is the plastic limit pressure of the cylinder. In Fig. 6.4b, the norm of the accumulated displacement increment in this load step, $\| {}^2\{U\}^{(i)} - {}^1\{U\} \|_2$, is plotted against the iteration step, Itera. In Figs. 6.4c and 6.4d, the norm of the out-of-balance force, the left-hand side of Eq. (6.50), and the internal energy, the left-hand side of Eq. (6.51), are plotted against Itera, respectively. All curves are normalized with respect to their respective maximum values.

For this problem, we can see that the fastest convergence is by the Newton–Raphson method. The quasi-Newton method iterates two more times than the Newton–Raphson method. However, the CPU time used by these two methods is almost the same. The slowest one is the modified Newton–Raphson method, which iterates 57 times to reach a convergence

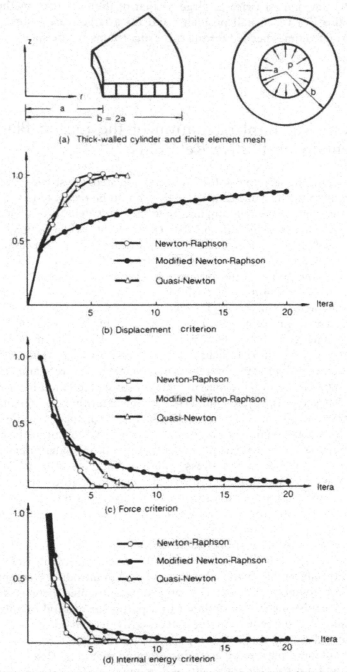

(a) Thick-walled cylinder and finite element mesh

(b) Displacement criterion

(c) Force criterion

(d) Internal energy criterion

FIGURE 6.4. A thick-walled cylinder subjected to internal pressure.

with a CPU time almost twice as large as that of the other two methods. Note that this is a very small problem, and for a large-scale system, the ratios of the CPU times needed for the three algorithms can be significantly different.


## 6.5. Numerical Implementation of the Elastic–Plastic Incremental Constitutive Relations

As already shown in Section 6.4, for each iteration step, the stress $^{m+1}\{\sigma\}^{(i)}$ corresponding to the displacement $^{m+1}\{U\}^{(i)}$ should be computed using an elastic–plastic constitutive relation, and then the equivalent force of stress, $^{m+1}\{F\}^{(i)}$, is calculated using Eq. (6.30b). The integration of Eq. (6.30b) is then performed over each element of the structure, using the popular Gaussian integration technique. Thus, the stress will be computed at all Gaussian sampling points of the structure in each iteration step.

The incremental constitutive relation for a general elastic–plastic material has been presented in Chapter 5. This relationship relates an infinitesimal stress increment to an infinitesimal strain increment corresponding to a given stress state and plastic deformation history. However, in a finite-element analysis, since a finite load increment instead of an infinitesimal one is applied in a load step, the relevant increments of stress and strain have finite sizes. Therefore, the incremental constitutive relation presented in Chapter 5 has to be integrated numerically. The algorithms to implement this numerical integration play an important role and, together with the algorithms for solving nonlinear simultaneous equations, constitute the core of an elastic–plastic finite-element analysis. An improper algorithm may lead not only to an inaccurate stress solution, but may also delay the convergence of the equilibrium iteration, and even lead to divergence of the iteration. Since stress computation consumes usually a significant share of the total computing time, the efficiency of an algorithm is therefore essential.

In this section, we shall first rewrite the incremental constitutive relation for a general elastic–plastic material in matrix form and then discuss in some detail the algorithms involved. Finally, we shall introduce a complete procedure for computing stress. The procedure presented here is quite general. It is neither necessarily related to a particular method of solving nonlinear simultaneous equations nor necessarily related to a particular type of material model. In fact, the procedure is suitable for a perfectly plastic, a hardening plastic, or a strain-softening material. However, for simplicity and for clarity, smooth yield and potential surfaces and isotropic hardening behavior are assumed here. It is not difficult to extend this procedure to include cases without these restrictions.

### 6.5.1. General Description

In matrix form, the stress increment, $\{d\sigma\}$, can be expressed in terms of the elastic strain increment, $\{d\epsilon^e\}$, or the total strain increment, $\{d\epsilon\}$, as

$$\{d\sigma\} = [C]\{d\epsilon^e\} = [C](\{d\epsilon\} - \{d\epsilon^p\}) \qquad (6.52a)$$

$$\{d\sigma\} = [C^{ep}]\{d\epsilon\} \qquad (6.52b)$$

The plastic strain increment, $\{d\epsilon^p\}$, is expressed, using a nonassociated flow rule, as

$$\{d\epsilon^p\} = d\lambda \left\{\frac{\partial g}{\partial\{\sigma\}}\right\} \qquad (6.53)$$

where $\{\partial g/\partial\{\sigma\}\}$ is the gradient vector of the plastic potential function, $g(\sigma_{ij}, k)$. The scalar function $d\lambda$, Eq. (5.133), is expressed as

$$d\lambda = \frac{L}{h} \qquad (6.54)$$

where $L$ is the loading criterion function defined in Eq. (5.152),

$$L = \left\{\frac{\partial f}{\partial\{\sigma\}}\right\}^T [C]\{d\epsilon\} \qquad (6.55)$$

where $\{\partial f/\partial\{\sigma\}\}$ is the gradient of the yield function, $f(\sigma_{ij}, k)$. Here, the yield function is not expressed explicitly as a function of $\epsilon_{ij}^p$. The positive scalar function $h$ defined in Eq. (5.132) becomes

$$h = \left\{\frac{\partial f}{\partial\{\sigma\}}\right\}^T [C]\left\{\frac{\partial g}{\partial\{\sigma\}}\right\} - n\frac{\partial f}{\partial k} \qquad (6.56)$$

and

$$n = \frac{dk}{d\epsilon} C\sqrt{\left\{\frac{\partial g}{\partial\{\sigma\}}\right\}^T \left\{\frac{\partial g}{\partial\{\sigma\}}\right\}} \qquad (6.57)$$

Finally, the elastic–plastic stiffness matrix $[C^{ep}]$, Eqs. (5.145) and (5.146), is expressed as

$$[C^{ep}] = [C] - \frac{1}{h}[C]\left\{\frac{\partial g}{\partial\{\sigma\}}\right\}\left\{\frac{\partial f}{\partial\{\sigma\}}\right\}^T [C] \qquad (6.58)$$

It is clear that the matrix $[C^{ep}]$ is not symmetric when a nonassociated flow rule is used.

The computation of stress will be done for all Gaussian sampling points. In the following, the computation for only one Gaussian point will be considered. In a typical load step, say the $(m+1)$th step, we have already known the stress and strain, $^m\{\sigma\}$, $^m\{\epsilon\}$, and the hardening parameters, say,

$^m\epsilon_p$, $^mk$, at the end of the $m$th load step in which the equilibrium iteration has converged. For a typical iteration step, say, $i$, of the $(m+1)$th load step, the $i$th approximation of the displacement, $^{m+1}\{U\}^{(i)}$, has been obtained. Then, using Eq. (6.19), the corresponding strain and strain increment at a Gaussian point are

$$^{m+1}\{\epsilon\}^{(i)} = [B]^{m+1}\{U\}^{(i)} \tag{6.59a}$$

$$\{\Delta\epsilon\} = {}^{m+1}\{\epsilon\}^{(i)} - {}^m\{\epsilon\} \tag{6.59b}$$

We define a trial stress increment $\{\Delta\sigma^e\}$, assuming an elastic behavior,

$$\{\Delta\sigma^e\} = [C]\{\Delta\epsilon\} \tag{6.60}$$

Assume that at the end of the $m$th load step, the stress state at a Gaussian point is in an elastic state satisfying $f(^m\{\sigma\}, {}^mk) < 0$ and enters into an elastic–plastic state in the $(m+1)$th step, $f(^m\{\sigma\} + \{\Delta\sigma^e\}, {}^mk) > 0$. Therefore, there exists a *scaling factor* $r$ such that $f(^m\{\sigma\} + r\{\Delta\sigma^e\}, {}^mk) = 0$. The strain is then divided in two parts, $r\{\Delta\epsilon\}$ and $(1-r)\{\Delta\epsilon\}$. The first part corresponds to a pure elastic response, while the second part corresponds to an elastic–plastic response. Hence, the stress increment can be integrated as

$$
\begin{aligned}
\{\Delta\sigma\} &= \int_{^m\{\epsilon\}}^{^{m+1}\{\epsilon\}^{(i)}} [C](\{d\epsilon\} - \{d\epsilon^p\}) \\
&= \int_{^m\{\epsilon\}}^{^m\{\epsilon\}+r\{\Delta\epsilon\}} [C](\{d\epsilon\}) + \int_{^m\{\epsilon\}+r\{\Delta\epsilon\}}^{^m\{\epsilon\}+\{\Delta\epsilon\}} [C](\{d\epsilon\} - \{d\epsilon^p\}) \\
&= r\{\Delta\sigma^e\} + \int_{^m\{\epsilon\}+r\{\Delta\epsilon\}}^{^m\{\epsilon\}+\{\Delta\epsilon\}} [C](\{d\epsilon\} - \{d\epsilon^p\})
\end{aligned}
\tag{6.61a}
$$

or

$$\{\Delta\sigma\} = r\{\Delta\sigma^e\} + \int_{^m\{\epsilon\}+r\{\Delta\epsilon\}}^{^m\{\epsilon\}+\{\Delta\epsilon\}} [C^{ep}]\{d\epsilon\} \tag{6.61b}$$

Finally, we obtain the stress corresponding to $^{m+1}\{U\}^{(i)}$ as

$$^{m+1}\{\sigma\}^{(i)} = {}^m\{\sigma\} + \{\Delta\sigma\} \tag{6.62}$$

In the following sections, the determination of the scaling factor $r$ and the technique employed to perform the integration will be described in detail.

## 6.5.2. Determination of the Loading State

The first step in a stress computation is to determine the loading state of a Gaussian point, i.e., whether it is in a plastic loading state, an elastic loading state, or an unloading state, corresponding to the strain increment, $\{\Delta\epsilon\}$. Only for the case that constitutes a plastic loading will the elastic-plastic constitutive relation be used to determine the corresponding stress increment. To this end, we shall discuss this procedure for two separate cases. One is that the Gaussian point is in an elastic state at the end of the $m$th load step, and the other is that it is in an elastic-plastic state at the end of the $m$th load step.

If the Gaussian point is in an elastic state at the end of the $m$th step, $f(^m\{\sigma\}, {}^mk) < 0$, the trial stress increment $\{\Delta\sigma^e\}$ defined in Eq. (6.60) could be used to check whether an elastic-plastic state will result in the $(m+1)$th step. If $f(^m\{\sigma\} + \{\Delta\sigma^e\}, {}^mk) \leq 0$, this Gaussian point will remain in an elastic state in the $(m+1)$th step, and the elastic stress-strain relationship leads to

$$\{\Delta\sigma\} = \{\Delta\sigma^e\} \tag{6.63}$$

If $f(^m\{\sigma\} + \{\Delta\sigma^e\}, {}^mk) > 0$, the Gaussian point enters into an elastic-plastic state in this load step. As a result, there exists a scaling factor $r$ such that

$$f(^m\{\sigma\} + r\{\Delta\sigma^e\}, {}^mk) = 0 \tag{6.64}$$

This is shown schematically in Fig. 6.5. Equation (6.64) is usually nonlinear in the factor $r$. We can solve Eq. (6.64) for $r$ analytically or numerically. If

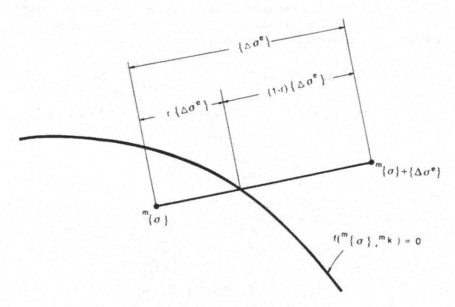

FIGURE 6.5. Schematic illustration of entering an elastic-plastic state from an elastic state at a Gaussian point.

the yield function is expressed in a simple form of stress invariants, the factor $r$ could be solved analytically. For example, the yield function of the von Mises isotropic hardening material can be expressed as

$$f(\{\sigma\}, k) = \tfrac{1}{2}\{s\}^T\{s\} - k^2(\epsilon_p) = 0 \tag{6.65}$$

where $\{s\}$ is the vector of the deviatoric stress defined as

$$\{s\}^T = \{s_x, s_y, s_z, s_{yz}, s_{zx}, s_{xy}\} \tag{6.66a}$$

We also define the increment of $\{s\}$, $\{\Delta s\}$, as

$$\{\Delta s\}^T = \{\Delta s_x, \Delta s_y, \Delta s_z, \Delta s_{yz}, \Delta s_{zx}, \Delta s_{xy}\} \tag{6.66b}$$

Equation (6.64) in this case becomes

$$f(^m\{\sigma\} + r\{\Delta\sigma^e\}, {}^mk) = \tfrac{1}{2}(^m\{s\} + r\{\Delta s\})^T({}^m\{s\} + r\{\Delta s\}) - {}^mk^2 = 0 \tag{6.67}$$

or

$$\tfrac{1}{2}r^2\{\Delta s\}^T\{\Delta s\} + r\,{}^m\{s\}^T\{\Delta s\} + \tfrac{1}{2}{}^m\{s\}^T\,{}^m\{s\} - {}^mk^2 = 0 \tag{6.68}$$

The scaling factor $r$ can then be solved as

$$r = \frac{-{}^m\{s\}^T\{\Delta s\} + \sqrt{({}^m\{s\}^T\{\Delta s\})^2 + \{\Delta s\}^T\{\Delta s\}(2\,{}^mk^2 - {}^m\{s\}^T\,{}^m\{s\})}}{\{\Delta s\}^T\{\Delta s\}} \tag{6.69}$$

On the other hand, Eq. (6.64) may lead to a highly nonlinear form of $r$ for some yield functions, and recourse must be made to a numerical method. The simplest method is to expand Eq. (6.64) in the Taylor series form, neglect all higher-order terms, and obtain

$$f(^m\{\sigma\} + r\{\Delta\sigma^e\}, {}^mk) = f(^m\{\sigma\}, {}^mk) + \left\{\frac{\partial f}{\partial\{\sigma\}}\right\}\Bigg|_{{}^m\{\sigma\}}^T r\{\Delta\sigma^e\} = 0 \tag{6.70}$$

which leads to an approximation of $r$ as (Nayak and Zienkiewicz, 1972)

$$r = \frac{-f(^m\{\sigma\}, {}^mk)}{\left\{\dfrac{\partial f}{\partial\{\sigma\}}\right\}\Bigg|_{{}^m\{\sigma\}}^T \{\Delta\sigma^e\}} \tag{6.71}$$

We can also keep all quadratic terms in the Taylor series expansion and obtain a quadratic equation in $r$:

$$f(^m\{\sigma\} + r\{\Delta\sigma\}, {}^mk) = f(^m\{\sigma\}, {}^mk) + \left\{\frac{\partial f}{\partial\{\sigma\}}\right\}\Bigg|_{{}^m\{\sigma\}}^T r\{\Delta\sigma^e\}$$
$$+ r^2\{\Delta\sigma^e\}^T \left(\frac{\partial^2 f}{\partial\{\sigma\}^2}\right)\Bigg|_{{}^m\{\sigma\}} \{\Delta\sigma^e\} = 0 \tag{6.72}$$

The above equation can be solved to obtain a better approximation of $r$.

For the case that the Gaussian point is in an elastic–plastic state at the end of the $m$th load step, $f(^m\{\sigma\}, {}^mk) = 0$, we can use the load criterion function $L$ defined in Eq. (6.55) to determine the loading state. Assuming the small load step is proportional, Eq. (6.55) becomes

$$L = \left\{\frac{\partial f}{\partial\{\sigma\}}\right\}^T [C]\{\Delta\epsilon\} \tag{6.73}$$

As discussed previously for the loading criterion, if $L \leq 0$, i.e., the Gaussian point is in an unloading or neutral loading state, an elastic constitutive relation should be used, or

$$\{\Delta\sigma\} = \{\Delta\sigma^e\} \tag{6.74}$$

If $L > 0$, i.e., the Gaussian point is in a plastic loading state, the integration of Eq. (6.61) must be performed numerically with $r = 0$ to obtain $\{\Delta\sigma\}$. This will be described in the following section.

### 6.5.3. Integrating Techniques

The algorithms employed to perform the integration in Eq. (6.61a) or Eq. (6.61b) can be grouped into two categories: those based on an explicit technique and those based on an implicit technique. For both types, the strain increment that constitutes an elastic–plastic response is further divided into a sufficient number, say, $m$, of subincrements, $\{\Delta\bar{\epsilon}\}$,

$$\{d\epsilon\} = \{\Delta\bar{\epsilon}\} = (1 - r)\{\Delta\epsilon\}/m \tag{6.75}$$

in order to achieve the required accuracy of the integration process. If an explicit technique is employed, such as the Euler forward method, the stress is calculated forward one strain subincrement by one strain subincrement. If an implicit technique is employed, such as the Euler backward method, the stress at the end of each subincrement is calculated iteratively. Therefore, in this case, there are two iteration cycles in solving the nonlinear equations (6.30): one is the equilibrium iteration cycle discussed in Section 6.4, and the other is during the integration (6.62) to evaluate an accurate stress.

In this section, we shall only discuss the explicit method in some detail. Equation (6.62) can be rewritten in the form of differential equations as

$$\{d\sigma\} = [C](\{d\epsilon\} - \{d\epsilon^P\}) \tag{6.76}$$

and

$$\{d\epsilon^P\} = d\lambda \left\{\frac{\partial g}{\partial\{\sigma\}}\right\} = \frac{L}{h}\left\{\frac{\partial g}{\partial\{\sigma\}}\right\} \tag{6.77a}$$

with

$$L = \left\{\frac{\partial f}{\partial\{\sigma\}}\right\}^T [C]\{\Delta\bar{\epsilon}\} \tag{6.77b}$$

or

$$\{d\epsilon^P\} = [P]\{d\epsilon\} \tag{6.77c}$$

with

$$[P] = P(\{\epsilon\}, \{\epsilon^p\}, \epsilon_p) = \frac{1}{h} \left\{ \frac{\partial g}{\partial\{\sigma\}} \right\} \left\{ \frac{\partial f}{\partial\{\sigma\}} \right\}^T [C] \qquad (6.77d)$$

with the initial conditions

$$\{\epsilon\} = {}^m\{\epsilon\} + r\{\Delta\epsilon\} \qquad (6.78a)$$

$$\{\sigma\} = {}^m\{\sigma\} + r\{\Delta\sigma^e\} \qquad (6.78b)$$

$$\{\epsilon^p\} = {}^m\{\epsilon^p\}, \qquad \epsilon_p = {}^m\epsilon_p \qquad (6.78c)$$

where Eqs. (6.54) and (6.55) have been used, and $r$ is the scaling factor discussed in the previous section.

For each strain subincrement, $\{\Delta\bar{\epsilon}\}$, the explicit technique involves the following steps:

Step 1: Determine the plastic strain subincrement, $\{\Delta\bar{\epsilon}^p\}$, using Eq. (6.77) with a proper algorithm, and the effective plastic strain subincrement, $\Delta\bar{\epsilon}_p$.
Step 2: Compute the stress subincrement, $\{\Delta\bar{\sigma}\}$, using Eq. (6.76):

$$\{\Delta\bar{\sigma}\} = [C](\{\Delta\bar{\epsilon}\} - \{\Delta\bar{\epsilon}^p\})$$

Step 3: Update stress, strains, and hardening parameters:

$$\{\sigma\} \leftarrow \{\sigma\} + \{\Delta\bar{\sigma}\}$$

$$\{\epsilon\} \leftarrow \{\epsilon\} + \{\Delta\bar{\epsilon}\}$$

$$\{\epsilon^p\} \leftarrow \{\epsilon^p\} + \{\Delta\bar{\epsilon}^p\}$$

$$\epsilon_p \leftarrow \epsilon_p + \Delta\bar{\epsilon}_p, \qquad k \leftarrow k(\epsilon_p)$$

In this procedure, the accuracy of the stress so obtained depends mainly on the accuracy of the plastic strain subincrement computed. Denoting

$$[P_i] = P(\{\epsilon\} + r_i\{\Delta\bar{\epsilon}\}, \{\epsilon^p\} + r_i\{\Delta\bar{\epsilon}^p\}_{i-1}, \epsilon_p + r_i(\Delta\bar{\epsilon}_p)_{i-1}) \qquad (6.79)$$

where the matrix $[P]$ has been defined in Eq. (6.77c), three algorithms for computing $\{\Delta\bar{\epsilon}^p\}$ are expressed as follows:
The Euler forward method:

$$\{\Delta\bar{\epsilon}^p\} = [P_1]\{\Delta\bar{\epsilon}\} \qquad (6.80a)$$

$$r_1 = 0 \qquad (6.80b)$$

The second-order Runge–Kutta method:

$$\{\Delta\bar{\epsilon}^p\} = w_1\{\Delta\bar{\epsilon}_1^p\} + w_2\{\Delta\bar{\epsilon}_2^p\} \qquad (6.81a)$$

$$\{\Delta\bar{\epsilon}_i^p\} = [P_i]\{\Delta\bar{\epsilon}\} \qquad (6.81b)$$

$$r_1 = 0, \qquad r_2 = 1, \qquad w_1 = w_2 = \tfrac{1}{2} \qquad (6.81c)$$

The fourth-order Runge–Kutta method:

$$\{\Delta \tilde{\epsilon}^P\} = w_1\{\Delta \tilde{\epsilon}_1^P\} + w_2\{\Delta \tilde{\epsilon}_2^P\} + w_3\{\Delta \tilde{\epsilon}_3^P\} + w_4\{\Delta \tilde{\epsilon}_4^P\} \qquad (6.82a)$$

$$\{\Delta \tilde{\epsilon}_i^P\} = [P_i]\{\Delta \tilde{\epsilon}\} \qquad (6.82b)$$

$$r_1 = 0, \qquad r_2 = r_3 = \tfrac{1}{2}, \qquad r_4 = 1$$

$$w_1 = w_4 = \tfrac{1}{6}, \qquad w_2 = w_3 = \tfrac{1}{3} \qquad (6.82c)$$

In an actual coding, it is not necessary to form the matrix $[P]$; instead, Eq. (6.77a) can be used directly to compute $\{\Delta \tilde{\epsilon}^P\}$.

It should be noted that for the three algorithms described previously, the accuracy and the computational effort needed for one subincrement both increase in the order in which they were presented. For a given problem, to obtain the same accuracy, the use of a method with a higher accuracy will require a lesser number of subincrements. The choice of a particular algorithm and the number of subincrements, $m$, required depend on the problem type. It should also be noted that, in fact, in the procedure for computing the stress subincrement, Eq. (6.61a) has been used instead of Eq. (6.61b). Equation (6.61b) has been widely employed by many authors. The reason for this change is that using Eq. (6.61a), we can formulate different algorithms relatively easily, and for a work-hardening material, the computation of the plastic strain is a necessary step. The elastic–plastic stiffness matrix, $[C^{ep}]$, is not involved in this stress computation procedure, and $[C^{ep}]$ is evaluated only when the stiffness matrix of the structure is required.

## 6.5.4. Forcing the Increments to Satisfy the Consistency Condition

As we have already discussed in Chapters 4 and 5, the consistency condition $df = 0$ must be met in a plastic loading process. However, since many approximations have been made in the numerical implementation of the incremental constitutive relation, the consistency condition is often invalid. Adding a strain subincrement at the subsequent state leads to

$$f(\{\sigma\}, \epsilon_p) \neq 0$$

or, in other words, the stress will not stay on the subsequent yield surface. Such a departure of stress from the subsequent yield surface is accumulative and may lead to a very significant error in solving the nonlinear equation (6.30). A correction of the stress vector has to be made to meet the consistency condition. Such a correction is often achieved by adding to the stress vector a correction vector in the direction normal to the yield surface,

$$\{\delta\sigma\} = a \left\{ \frac{\partial f}{\partial \{\sigma\}} \right\} \qquad (6.83)$$

where $a$ is a small scalar to be determined, such that the yield condition is satisfied at the subsequent position,

$$f(\{\sigma\}+\{\delta\sigma\}, \epsilon_p) = f\left(\{\sigma\}+a\left\{\frac{\partial f}{\partial \{\sigma\}}\right\}, \epsilon_p\right) = 0 \qquad (6.84)$$

Equation (6.84) is a nonlinear equation for the scalar $a$. Here, as for the scaling factor $r$, $a$ can be solved from Eq. (6.84) analytically or numerically. If a Taylor series expansion is used for Eq. (6.84), and all higher-order terms are neglected, we can obtain the scalar $a$ as

$$a = \frac{-f(\{\sigma\}, \epsilon_p)}{\left\{\frac{\partial f}{\partial \{\sigma\}}\right\}^T \left\{\frac{\partial f}{\partial \{\sigma\}}\right\}} \qquad (6.85)$$

and the correction vector has the form

$$\{\delta\sigma\} = \frac{-f(\{\sigma\}, \epsilon_p)}{\left\{\frac{\partial f}{\partial \{\sigma\}}\right\}^T \left\{\frac{\partial f}{\partial \{\sigma\}}\right\}} \left\{\frac{\partial f}{\partial \{\sigma\}}\right\} \qquad (6.86)$$

Finally, the corrected stress vector is obtained as

$$\{\sigma\} \leftarrow \{\sigma\} + a\left\{\frac{\partial f}{\partial \{\sigma\}}\right\}$$

## 6.5.5. General Procedure for Stress Computation

In short, a typical procedure for stress computation will be summarized here. In this procedure, the symbol IPEL is used to indicate the state at a Gaussian point under consideration. IPEL = 0 indicates the Gaussian point is in an elastic state, and IPEL = 1 indicates it is in an elastic–plastic state.

Step 1: Compute the strain increment $\{\Delta\epsilon\}$ and the trial stress increment $\{\Delta\sigma^e\}$ assuming an elastic behavior.

$$\{\Delta\epsilon\} = {}^{m+1}\{\epsilon\}^{(i)} - {}^m\{\epsilon\}$$

$$\{\Delta\sigma^e\} = [C]\{\Delta\epsilon\}$$

Step 2: Determine the loading state
    If IPEL = 1, the Gaussian point is in an elastic–plastic state previously.
    Compute the loading criterion function $L$, Eq. (6.73).
        If $L > 0$, $r \leftarrow 0$, plastic loading.
        If $L \leq 0$, $r \leftarrow 1$, IPEL $\leftarrow 0$, unloading or neutral loading.
    If IPEL = 0, the Gaussian point is in an elastic state previously.
    Compute the yield function $f$:

$$f \leftarrow f({}^m\{\sigma\} + \{\Delta\sigma^e\}, \epsilon_p)$$

    If $f \leq 0$, $r \leftarrow 1$, remains in the elastic state. Go to Step 5.

If $f > 0$, IPEL $\leftarrow 1$, enters into an elastic-plastic state.
Determine $r$ such that $f(^m\{\sigma\} + r\{\Delta\sigma^e\}, \epsilon_p) = 0$

$$\{\sigma\} \leftarrow {}^m\{\sigma\} + r\{\Delta\sigma\}$$

Step 3: Compute the subincrement of strain

$$\{\Delta\bar{\epsilon}\} = \frac{(1-r)}{m}\{\Delta\epsilon\}$$

Step 4: Integrate numerically, loop from 1 to $m$.
Determine the plastic strain subincrement $\{\Delta\bar{\epsilon}^p\}$ and $\Delta\bar{\epsilon}_p$

$$\{\Delta\bar{\sigma}\} = [C](\{\Delta\bar{\epsilon}\} - \{\Delta\bar{\epsilon}^p\})$$
$$\{\sigma\} \leftarrow \{\sigma\} + \{\Delta\bar{\sigma}\}, \qquad \epsilon_p \leftarrow \epsilon_p + \Delta\bar{\epsilon}_p$$

Check the subsequent yield condition.
If $|f(\{\sigma\}, \epsilon_p)| > \epsilon_f$, where $\epsilon_f$ is a prescribed tolerance for a yield function, determine the correction stress vector $\{\delta\sigma\}$ and $\{\sigma\} \leftarrow \{\sigma\} + \{\delta\sigma\}$.
Step 5:

$$^{m+1}\{\sigma\}^{(i)} \leftarrow \{\sigma\}$$

If necessary, calculate the elastic-plastic matrix, $[C^{ep}]$.

# 6.6. Bounding Surface Theory

The classical models of isotropic or kinematic hardening plasticity are simple and reasonably good for simple loading histories. For complex loading histories, such as cyclic loadings in the plastic range, these models are incapable of describing the observed hysteretic behavior shown in Fig. 6.6b. An evident feature of cyclic behavior is the noncoincidence of the subsequent yield stress ($A$) and subsequent loading stress ($A'$). However, the idealized stress-strain curve of Fig. 6.6a implies that the loading cycle is completely reversible, and thus no plastic deformation is recorded if no reversed yielding has occurred. Additional plastic strains can occur only upon reloading to a stress state beyond point $A$, and the subsequent behavior is identical to that which would have been obtained if unloading had never occurred. Such an inadequacy of the classical models has led to the development of alternative plasticity models.

The *bounding surface theory* proposed by Dafalias and Popov (1975) and Krieg (1975) is an attempt to generalize the conventional flow theory to account for the cyclic behavior of materials.

## 6.6.1. General Considerations

Before proceeding to the discussion of the *bounding surface theory*, let us recall the constitutive relations of Eqs. (5.125), (5.136), and (5.140) derived

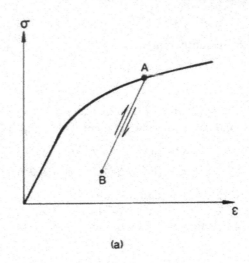

(a)

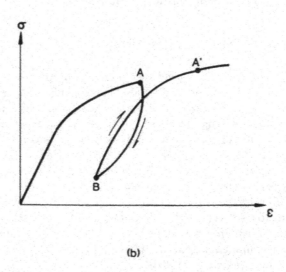

(b)

FIGURE 6.6. Unloading and reloading from a plastic state: (a) idealized stress–strain curve; (b) hysteretic loop in a cycle of loading.

in Chapter 5 based on the plastic flow theory:

$$de_{ij} = de_{ij}^e + de_{ij}^p = \left( D_{ijkl} + \frac{1}{\kappa} \frac{\partial g}{\partial \sigma_{ij}} \frac{\partial f}{\partial \sigma_{kl}} \right) d\sigma_{kl} \tag{6.87}$$

and

$$d\sigma_{ij} = \left( C_{ijkl} - \frac{C_{ijtu}(\partial g/\partial \sigma_{tu})(\partial f/\partial \sigma_{pq}) C_{pqkl}}{\kappa + (\partial f/\partial \sigma_{mn}) C_{mnrs}(\partial g/\partial \sigma_{rs})} \right) d\epsilon_{kl} \tag{6.88}$$

These tensor expressions can now be rewritten in matrix forms:

$$\{d\epsilon\} = \{d\epsilon\}^e + \{d\epsilon\}^p = \left( [D] + \frac{1}{\kappa} \{n_g\}\{n_f\}^T \right)\{d\sigma\} \tag{6.89}$$

and

$$\{d\sigma\} = \left( [C] - \frac{[C]\{n_g\}\{n_f\}^T[C]}{\bar{\kappa} + \{n_f\}^T[C]\{n_g\}} \right)\{d\epsilon\} \tag{6.90}$$

in which $[C]$ is the tensor of *elastic moduli* and $[D]$ is the inverse of $[C]$, i.e., the *elastic compliance tensor*, $\{n_g\}$ and $\{n_f\}$ are the unit normals of the *plastic potential surface g* and the *loading surface f*, respectively, namely,

$$\{n_g\} = \frac{\{\partial g/\partial \sigma\}}{(\{\partial g/\partial \sigma\}^T\{\partial g/\partial \sigma\})^{1/2}} \tag{6.91}$$

$$\{n_f\} = \frac{\{\partial f/\partial \sigma\}}{(\{\partial f/\partial \sigma\}^T\{\partial f/\partial \sigma\})^{1/2}} \tag{6.92}$$

and $\bar{\kappa}$ may be referred to as the *plastic modulus* and is related to the *hardening modulus* $\kappa$ by

$$\bar{\kappa} = \frac{\kappa}{(\{\partial g/\partial \sigma\}^T\{\partial g/\partial \sigma\})^{1/2}(\{\partial f/\partial \sigma\}^T\{\partial f/\partial \sigma\})^{1/2}} \tag{6.93}$$

while the hardening modulus $\kappa$ is related to the hardening parameters and the plastic modulus $H_p$ in a uniaxial test, which is the slope of the uniaxial-stress–plastic-strain curve.

From Eqs. (6.89) through (6.93) it is noted that the definition of the constitutive relation is complete if, for any stress state, we can specify

1. the historical parameters, e.g., the hardening parameters,
2. the direction $\{n_f\}$ of loading,
3. the direction $\{n_g\}$ of flow,
4. the elastic modulus tensor $[C]$, and
5. the plastic modulus $\bar{\kappa}$.

There are many possible ways to make such specifications. These include classical plasticity, bounding surface plasticity, and many as yet untried variants.

The basic idea of the boundary surface model by Dafalias and Popov (1975) may be outlined as follows:

1. The concept of a boundary surface is introduced by assuming the existence of such a surface that encloses all the yield and the subsequent loading surfaces in stress space. The definition of the bounding surface is identical to that of a yield surface in the classical theory of plasticity. In general, both the bounding surface and the yield surface may deform and translate in stress space if plastic loading takes place.
2. A rule is established for associating the stress point $\{\sigma\}$, which is within the bounding surface and on the yield or subsequent loading surface, to a point $\{\bar{\sigma}\}$ on the bounding surface. Then, the distance between these two points, $\delta$, is defined.
3. The model assumes that the variation of the plastic modulus, $E^p$, (i.e., $H_p$ in our nomenclature) may be separated into three regions. The first is associated with the end of the elastic behavior, where $E^p = \infty$; the second region occurs beyond initial yield, where $E^p$ is a smoothly decreasing function of the distance, $\delta$, until it reaches the third region, where $E^p$ takes on a constant value $E_0$.

## 6.6.2. Uniaxial Loading

Figure 6.7a shows a schematic sketch of the uniaxial stress–strain response. The material exhibits a linear elastic response up to point $A$. Within this elastic range, the plastic modulus $E^p$ is infinity. After passing $A$, plastic strains occur and $E^p$ starts assuming finite values, which change gradually and approach an ultimate value associated with the bounding lines $XX'$ and $YY'$. In Fig. 6.7a, $OA$ represents the elastic part, and $ABD$ the elastic-plastic part. Elastic unloading is assumed to occur at point $D$ along $DD'$ and plastic reloading occurs along $D'D''F$. The region $FF'$ represents the new elastic unloading, which is followed by the plastic reloading along $F'F''X$. After point $B$, the curve followed during plastic reloading is the same as that during the first plastic loading for the purely kinematic hardening. In general, the bounds can change with loading history.

In an elastic region such as $FF'$, the plastic modulus $E^p$ is infinite. The value of $E^p$ varies during the plastic behavior in regions such as $F'F''$. Along $FF'$, $E^p$ is infinite, but right after $F'$, it assumes a finite value which changes as the process goes to $F''$. The zone $F''X$ represents a plastic behavior during which $E^p$ is assumed to remain constant. The stress–strain response is thus bounded by the two lines $XX'$ and $YY'$. In multiaxial cases, the projections of the points on the bounds are generalized into a bounding surface, hence the name *bounding surface* model.

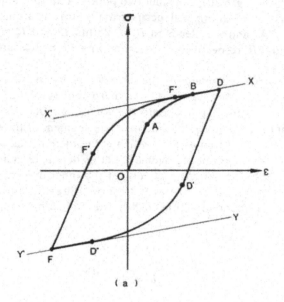

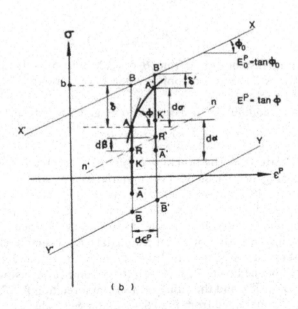

FIGURE 6.7. Schematic of bounding surface (line) model in (a) $\sigma-\epsilon$ and (b) $\sigma-\epsilon^P$ spaces (Dafalias and Popov, 1975).

The model may be explained by the stress–plastic strain curve for the uniaxial case (Fig. 6.7b). Consider two points $A$ and $A'$ on the stress–plastic strain curve. Unloading will occur from point $A$, along $A\bar{A}$. Extension of $A\bar{A}$ to $XX'$ and $YY'$ leads to $B$, $\bar{B}$. Points $K$ and $R$ are the midpoints of $A\bar{A}$ and $B\bar{B}$, respectively. Corresponding nomenclature applies to the state $A'$.

Under a stress increment $d\sigma$, the state of stress moves from $A$ to $A'$. Then the elastic zone $A\bar{A}$ moves to $A'\bar{A}'$ and $B\bar{B}$ moves to $B'\bar{B}'$. Now we define $d\alpha$ as the projection of $KK'$ onto the stress axis. $KK'$ is the incremental movement of the center $K$ of the inner line segment or the elastic zone $A\bar{A}$, and $d\beta$ is the projection of $RR'$, which is the incremental movement of the center $R$ of the outer line segment $B\bar{B}$. Let us denote by $S$ and $dS$ the length and increment of length, respectively, of the inner line segment and by $E_0^p$ the plastic modulus associated with the bounding line (surface). Then from geometric considerations (Fig. 6.7b), the following expressions can be written:

$$d\sigma = E^p \, d\epsilon^p \tag{6.94}$$

$$d\alpha = d\sigma - \frac{dS}{2} \tag{6.95}$$

$$d\beta = E_0^p \, d\epsilon^p \tag{6.96}$$

$$\delta' - \delta = d\delta = \left( \frac{E_0^p}{E^p} - 1 \right) d\sigma \tag{6.97}$$

where $\delta$ is the distance of $A$ from $B$ on $XX'$ and $\delta'$ is the distance of $A'$ from $B'$ on $XX'$. The above applies when the two bounds are assumed to be straight lines.

The plastic modulus $E^p$ is assumed to be given by a relation of the form

$$E^p = \hat{E}^p(\delta, W_p) \tag{6.98}$$

with $E_0^p = \hat{E}^p(0, W_p)$. The function $\hat{E}^p$ is: (i) an increasing function of $\delta$, (ii) a decreasing function of $W_p$ if the material softens, and (iii) an increasing function of $W_p$ if the material hardens. $W_p$ is the plastic work during the half-cycle preceding the current plastic state. $S$ is usually assumed to be a function of plastic work $W_p$; hence, $d\delta$ can be determined.

Now, the incremental process for a uniaxial case is completely determined from the above set of equations. Indeed, given $d\sigma$, $W_p$, and $\delta$, the $d\epsilon^p$, $d\alpha$, and $d\beta$ are obtained from Eqs. (6.94), (6.95), and (6.96) using the value of $E^p$ from Eq. (6.98), and the value of $dS$ from a relation for $S$. For the next increment, $\delta'$ is obtained from Eq. (6.97) and a new value of $E^p$ is obtained from Eq. (6.98). Then the process is repeated.

The projections of the two line segments $A\bar{A}$ and $B\bar{B}$ onto the $\sigma$-axis move during plastic loading. The inner segment moves due to the stress increment $d\sigma$. The outer segment moves in a similar manner but at a slower rate. Hence, the inner segment can eventually reach the outer one, and then both move together. During the motion, $E^p$ changes with $\delta$ and approaches $E_0^p$ when $\delta = 0$. It is possible that $E^p$ becomes $E_0^p$ for $\delta \neq 0$. Then both segments move at the same rate without being in contact.

### 6.6.3. Multiaxial Loading

The foregoing description for the uniaxial loading case can easily be generalized to a multiaxial loading case. In a multiaxial case, the elastic region represented by the inner line segments becomes a region bounded by the yield or the subsequent loading surface, shown for simplicity as a circle with its center at $k$ (Fig. 6.8). This corresponds to the inner line segment on the $\sigma$-axis of the uniaxial case. The outer line segment is represented by another surface in the stress space, shown again as a circle with its center at $r$, enclosing the yield surface. This second surface in the stress space will be called the *bounding surface*, from the fact that the yield surface is constrained to move always within this surface, which also moves in the stress space.

If a state of stress $\{\sigma\}$, represented by point $a$, lies on the loading surface and continues to move outwards, plastic loading occurs. Assuming the associated flow rule, i.e., $\{n_g\} = \{n_f\}$, according to Eq. (6.89), one has

$$\{d\epsilon\}^p = \frac{1}{\bar{\kappa}}\{n_f\}(\{n_f\}^T\{d\sigma\}) = \frac{1}{\bar{\kappa}}\,d\sigma\{n_f\} \qquad (6.99)$$

in which $d\sigma = \{n_f\}^T\{d\sigma\}$, is a scalar representing the projection of the stress increment $\{d\sigma\}$ on the unit normal $\{n_f\}$, and $\bar{\kappa}$ is the plastic modulus, given by a relation similar to Eq. (6.98):

$$\bar{\kappa} = \hat{\kappa}(\delta, W_p) \qquad (6.100)$$

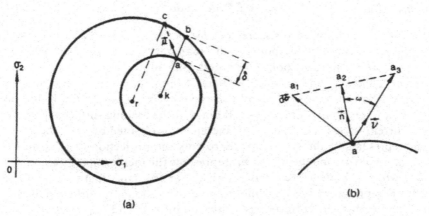

FIGURE 6.8. Schematic representation of a loading surface and a bounding surface (a) and illustration of their motions (b).

where the function $\hat{\kappa}$ has the same properties as the function $\hat{E}^P$ for the uniaxial case. The plastic work now is defined as $W_p = \int \sigma_{ij} \, d\epsilon_{ij}^p$, where the integration is carried out in the strain space along the loading path during the plastic deformation prior to the elastic deformation preceding the current plastic state.

Now, a question arises in the multiaxial case as to how to determine the distance $\delta$. Recall that in the uniaxial case, $\delta = AB$, where $A$ is the state of stress and $B$ is a point on line $X'X$. Accordingly, in Fig. 6.8a, point $a$ is taken on the yield surface, and point $b$ on the bounding surface. In this manner, the distance $\delta = ab$. As for the determination of $b$, many assumptions are possible. For example, if the yield surface and the bounding surface are congruent, $b$ can be the point corresponding to $a$ with respect to the congruency; or $b$ can be obtained as the intersection of the normal to the loading surface at $a$ with the bounding surface; or simply $b$ can be the intersection of $ka$ with the bounding surface as shown in Fig. 6.8a. The distance $\delta$ may be given by the usual Euclidean metric, i.e., if $\{\bar{\sigma}\}$ is the stress state at $b$, then

$$\delta = [(\{\bar{\sigma}\} - \{\sigma\})^T (\{\bar{\sigma}\} - \{\sigma\})]^{1/2} \qquad (6.101)$$

The loading surface can translate and possibly deform in the stress space according to any hardening rule considered appropriate. Simultaneously, the bounding surface may translate in the stress space. The bounding surface may also deform; for example an expansion of the bounding surface would correspond to an increase in the distance between the two bounding lines $XX'$ and $YY'$ in the uniaxial case.

The rule for translation of the bounding surface is obtained from the movement $d\beta$ of its center. In the uniaxial case, Eq. (6.96) for $d\beta$ can be rewritten in the following form:

$$d\beta = E_0^p \, d\epsilon^p = E^p \, d\epsilon^p - (E^p - E_0^p) \, d\epsilon^p = d\sigma - (E^p - E_0^p) \, d\epsilon^p \qquad (6.102)$$

In this equation, the stress increment $d\sigma$ is the incremental displacement in the stress space of the point of the loading surface that represents the current stress state. It should be noted that this quantity has a different meaning than $d\sigma$ in Eq. (6.99), where it applies to the multiaxial case. Figure 6.8b illustrates the quantity corresponding to $d\sigma$ of Eq. (6.102) in the multidimensional stress space. Point $a$ of the loading surface represents the current stress state. If the stress increment is denoted by $\{d\sigma\}$, the vector $aa_1$ represents it; the vector $aa_2$ represents the projection of $\{d\sigma\}$ on the unit normal $\{n\}$; and the vector $aa_3$ represents the incremental displacement of $a$ which is in the direction of a unit vector $\{\nu\}$. Note that the new stress state is represented by $a_1$, while point $a$ of the loading surface moves to $a_3$. The vector $aa_3$ corresponds to the quantity $d\sigma$ of Eq. (6.102) in the uniaxial case. Within a second-order approximation, it can be asserted that $aa_2 = \{n\}^T \{d\sigma\}$ is the projection on $\{n\}$ of both $aa_1$ and $aa_3$. Therefore, if

$\omega$ is the generalized angle between $\{n\}$ and $\{\nu\}$, and since $\{n\}^T\{\nu\} = n_{ij}\nu_{ij} = \cos \omega$,

$$aa_3 = \frac{d\sigma}{\cos \omega}\{\nu\} \tag{6.103}$$

For generalizing the part $(E^p - E_0^p) \, d\epsilon^p$ in Eq. (6.102), the idea proposed by Mroz (1967) is used. Denote by $c$ in Fig. 6.8a the point on the bounding surface where the outward normal has the same direction as the normal to the loading surface at $a$. Then, denoting by $\{\mu\}$ the unit vector along $ac$, the incremental translation of the bounding surface along $\{\mu\}$ would satisfy the requirement that the point of contact is at the same time the current stress state. Using Eq. (6.99), Eq. (6.103), and the above assumption, one gets

$$\{d\beta\} = \frac{d\sigma}{\cos \omega}\{\nu\} - (\bar{\kappa} - \bar{\kappa}_0)\sqrt{\{d\epsilon^p\}^T\{d\epsilon^p\}}\{\mu\}$$

$$= d\sigma\left[\frac{1}{\cos \omega}\{\nu\} - \left(1 - \frac{\bar{\kappa}_0}{\bar{\kappa}}\right)\{\mu\}\right] \tag{6.104}$$

This equation corresponds to Eq. (6.96). Relations corresponding to Eqs. (6.95) and (6.97) of the uniaxial case require specific formulation of the loading and bounding surfaces. Observe that if the loading surface moves as a rigid body, the term $d\sigma\{\nu\}/\cos \omega$ is also the incremental displacement $\{d\alpha\}$ of the center of the loading surface. In such a case, when the two surfaces come in contact, the point in common has $\delta = 0$ and $\bar{\kappa} = \bar{\kappa}_0$, and Eq. (6.104) yields $\{d\beta\} = \{d\alpha\}$, i.e., the surfaces move together at the same rate. As in the uniaxial case, it is also possible that $\bar{\kappa}$ takes the value $\bar{\kappa}_0$ before the loading surface reaches the bounding surface, and then the two surfaces move at the same rate and direction without being in contact.

It can be shown that Prager's and Ziegler's kinematic hardening models are special cases of the bounding surface model. A more general formulation of the bounding surface plasticity with internal variable formalism is given in papers by Dafalias and Popov (1976, 1977).

## 6.6.4. Other Bounding Surface Models

According to the basic idea of the bounding surface theory, many variant constitutive models can be developed for metals as well as for geological materials. Since the definition of the shapes and the changes of the bounding surface and the loading surface can differ from one material to the other, so does the definition of the stress point $\{\bar{\sigma}\}$ on the bounding surface corresponding to the stress point $\{\sigma\}$ on the loading surface, and the rules for obtaining the values of $\{n_g\}$, $\{n_f\}$, $\bar{\kappa}$, and $\delta$ can be of varying complexity.

As summarized by Zienkiewicz (1982), some typical models of this sort for geomechanics include the two-surface anisotropic hardening model, the bounding surface model with degradation, the infinite-surface model, and the modified infinite-surface model (see Mroz, Norris, and Zienkiewicz, 1978, 1979). A special bounding surface model for soils was also developed by Dafalias and Herrmann (1982).

## 6.7. Extension to Anisotropic Case

Some materials, such as sheet metals and fabricated forms, manifest predominant directional properties. In studying the plastic flow of such materials, it is necessary to consider the effect of *anisotropy*. One of the first treatments of the plastic flow of an anisotropic material was given by Hill (1950). Hill proposed a yield criterion with a quadratic expression of the stresses which reduces to the von Mises criterion when the degree of anisotropy is null [see Eq. (2.198), Section 2.4]. Similar stress–strain relations for anisotropic materials with strain hardening were proposed by Jackson, Smith, and Lankford (1948), Dorn (1949), and Hu (1956). A more general theory that combines the isotropic and the kinematic hardening theory was presented by Baltov and Sawczuk (1965).

Based on the classical developments, Yamada (1968, 1969) reformulated constitutive relations for an orthotropic material with work hardening suitable for applications to finite-element analysis.

### 6.7.1. Yield Criterion

Following Hill's theory, we define the yield or loading function, $f$, in a form similar to that of Eq. (2.198):

$$f = \frac{1}{2(F+G+H)}[F(\sigma_y - \sigma_z)^2 + G(\sigma_z - \sigma_x)^2 + H(\sigma_x - \sigma_y)^2$$
$$+ 2L\tau_{yz}^2 + 2M\tau_{zx}^2 + 2N\tau_{xy}^2] = k \tag{6.105}$$

in which $F$, $G$, $H$, $L$, $M$, and $N$ are material parameters. It should be noted that the coordinate axes $x$, $y$, and $z$ are taken to be the principal axes of anisotropy of the material in Eq. (6.105). The hardening parameter $k$, representing the size of the yield surface, is expressed in terms of the effective stress $\bar{\sigma}_e$, which will be specified later. Equation (6.105) can alternatively be expressed by the following implicit function $Q$;

$$Q = f - k = 0 \tag{6.106}$$

### 6.7.2. Flow Rule and Plastic Strain Increments

We shall use the *associated flow rule* to determine the plastic strain increments, where the *potential function* g is assumed to be the same as the yield or loading function $f$ given in Eq. (6.105):

$$g = f \tag{6.107}$$

Then, the plastic strain increments are given by

$$d\epsilon_{ij}^P = d\lambda \frac{\partial g}{\partial \sigma_{ij}} = d\lambda \frac{\partial f}{\partial \sigma_{ij}} \tag{6.108}$$

In matrix form,

$$\{d\epsilon^P\} = d\lambda \left\{\frac{\partial f}{\partial \sigma}\right\} \tag{6.109}$$

where

$$\{d\epsilon^P\}^T = \{d\epsilon_x^P \; d\epsilon_y^P \; d\epsilon_z^P \; d\gamma_{yz}^P \; d\gamma_{zx}^P \; d\gamma_{xy}^P\} \tag{6.110}$$

$$\left\{\frac{\partial f}{\partial \sigma}\right\}^T = \left\{\frac{\partial f}{\partial \sigma_x} \; \frac{\partial f}{\partial \sigma_y} \; \frac{\partial f}{\partial \sigma_z} \; \frac{\partial f}{\partial \tau_{yz}} \; \frac{\partial f}{\partial \tau_{zx}} \; \frac{\partial f}{\partial \tau_{xy}}\right\} \tag{6.111}$$

and $d\lambda$ is a positive scalar of proportionality. Note that the shear components defined here are engineering plastic strain increments, i.e.,

$$d\gamma_{yz}^P = 2 \, d\epsilon_{yz}^P, \ldots \tag{6.112}$$

Using the yield function $f$ in Eq. (6.105), we can obtain the expressions of plastic strain increments as follows:

$$d\epsilon_x^P = d\lambda \frac{\partial f}{\partial \sigma_x} = \frac{d\lambda}{F+G+H}[H(\sigma_x - \sigma_y) + G(\sigma_x - \sigma_z)] = \sigma_x' \, d\lambda$$

$$d\epsilon_y^P = d\lambda \frac{\partial f}{\partial \sigma_y} = \frac{d\lambda}{F+G+H}[F(\sigma_y - \sigma_z) + H(\sigma_y - \sigma_x)] = \sigma_y' \, d\lambda$$

$$d\epsilon_z^P = d\lambda \frac{\partial f}{\partial \sigma_z} = \frac{d\lambda}{F+G+H}[G(\sigma_z - \sigma_x) + F(\sigma_z - \sigma_y)] = \sigma_z' \, d\lambda$$

$$\tag{6.113}$$

$$d\gamma_{yz}^P = d\lambda \frac{\partial f}{\partial \tau_{yz}} = \frac{d\lambda}{F+G+H}[2L\tau_{yz}] = 2\tau_{yz}' \, d\lambda$$

$$d\gamma_{zx}^P = d\lambda \frac{\partial f}{\partial \tau_{zx}} = \frac{d\lambda}{F+G+H}[2M\tau_{zx}] = 2\tau_{zx}' \, d\lambda$$

$$d\gamma_{xy}^P = d\lambda \frac{\partial f}{\partial \tau_{xy}} = \frac{d\lambda}{F+G+H}[2N\tau_{xy}] = 2\tau_{xy}' \, d\lambda$$

in which

$$\sigma_x' = \frac{\partial f}{\partial \sigma_x} = \frac{H(\sigma_x - \sigma_y) + G(\sigma_x - \sigma_z)}{F+G+H}$$

$$\sigma_y' = \frac{\partial f}{\partial \sigma_y} = \frac{F(\sigma_y - \sigma_z) + H(\sigma_y - \sigma_x)}{F+G+H}$$

$$\sigma'_z = \frac{\partial f}{\partial \sigma_z} = \frac{G(\sigma_z - \sigma_x) + F(\sigma_z - \sigma_y)}{F + G + H}$$

$$\tau'_{yz} = \frac{1}{2}\frac{\partial f}{\partial \tau_{yz}} = \frac{L\tau_{yz}}{F + G + H} \tag{6.114}$$

$$\tau'_{zx} = \frac{1}{2}\frac{\partial f}{\partial \tau_{zx}} = \frac{M\tau_{zx}}{F + G + H}$$

$$\tau'_{xy} = \frac{1}{2}\frac{\partial f}{\partial \tau_{xy}} = \frac{N\tau_{xy}}{F + G + H}$$

Equation (6.113) is expressed in matrix forms as follows:

$$\{d\epsilon^P\} = \{\sigma'\}\, d\lambda \tag{6.115}$$

where

$$\{\sigma'\} = \left\{\frac{\partial f}{\partial \sigma}\right\} \tag{6.116}$$

$$\{\sigma'\}^T = \{\sigma'_x\ \sigma'_y\ \sigma'_z\ 2\tau'_{yz}\ 2\tau'_{zx}\ 2\tau'_{xy}\} \tag{6.117}$$

The stresses $\sigma'_x, \ldots, \tau'_{yz}, \ldots$ can be reduced to the deviatoric stresses $s_x, \ldots, s_{yz}, \ldots$, respectively, for an isotropic material when

$$F = G = H = 3L = 3M = 3N$$

The material parameters $F$, $G$, $H$, $L$, $M$, and $N$, may be functions of the state of stresses and plastic deformation. For simplicity, they are also assumed to be constant during plastic flow.

## 6.7.3. Stress–Strain Relations

In the theory of plasticity, we assume that the total strain increment consists of plastic components and elastic components:

$$\{d\epsilon\} = \{d\epsilon^P\} + \{d\epsilon^e\} \tag{6.118}$$

where

$$\{d\epsilon\}^T = \{d\epsilon_x\ d\epsilon_y\ d\epsilon_z\ d\gamma_{yz}\ d\gamma_{zx}\ d\gamma_{xy}\} \tag{6.119}$$

$$\{d\epsilon^e\}^T = \{d\epsilon^e_x\ d\epsilon^e_y\ d\epsilon^e_z\ d\gamma^e_{yz}\ d\gamma^e_{zx}\ d\gamma^e_{xy}\} \tag{6.120}$$

The elastic strain increments $\{d\epsilon^e\}$ can be related to the stress increments through an anisotropic elastic stress–strain relationship:

$$\{d\epsilon^e\} = [D^e]\{d\sigma\} \tag{6.121}$$

where

$$\{d\sigma\}^T = \{d\sigma_x\ d\sigma_y\ d\sigma_z\ d\tau_{yz}\ d\tau_{zx}\ d\tau_{xy}\} \tag{6.122}$$

$$[D^e] = \begin{bmatrix} d_{11} & d_{12} & d_{13} & 0 & 0 & 0 \\ d_{12} & d_{22} & d_{23} & 0 & 0 & 0 \\ d_{13} & d_{23} & d_{33} & 0 & 0 & 0 \\ 0 & 0 & 0 & d_{44} & 0 & 0 \\ 0 & 0 & 0 & 0 & d_{55} & 0 \\ 0 & 0 & 0 & 0 & 0 & d_{66} \end{bmatrix} \qquad (6.123)$$

It should be noted that the elastic anisotropy, with the same principal axes as the plastic anisotropy, has been incorporated in the matrix $[D^e]$. The symmetric nature of the matrix $[D^e]$ implies the existence of the elastic-strain-energy function. The inverse of Eq. (6.121) is represented by

$$\{d\sigma\} = [C^e]\{d\epsilon^e\} \qquad (6.124)$$

where $[C^e]$ is the inverse of $[D^e]$, expressed by

$$[C^e] = [D^e]^{-1} = \begin{bmatrix} c_{11} & c_{12} & c_{13} & 0 & 0 & 0 \\ c_{12} & c_{22} & c_{23} & 0 & 0 & 0 \\ c_{13} & c_{23} & c_{33} & 0 & 0 & 0 \\ 0 & 0 & 0 & c_{44} & 0 & 0 \\ 0 & 0 & 0 & 0 & c_{55} & 0 \\ 0 & 0 & 0 & 0 & 0 & c_{66} \end{bmatrix} \qquad (6.125)$$

Substituting Eqs. (6.115) and (6.121) into Eq. (6.118), we obtain the expression for the total strain increments as:

$$\{d\epsilon\} = \{d\epsilon^P\} + \{d\epsilon^e\}$$
$$= \{\sigma'\} \, d\lambda + [D^e]\{d\sigma\} \qquad (6.126)$$

### 6.7.4. Effective Stress and Effective Strain

Similarly to the von Mises theory for the isotropic case, we define the effective (or equivalent) stress $\bar{\sigma}_e$ as follows:

$$\bar{\sigma}_e^2 = \frac{3}{2} \frac{F(\sigma_y - \sigma_z)^2 + G(\sigma_z - \sigma_x)^2 + H(\sigma_x - \sigma_y)^2 + 2L\tau_{yz}^2 + 2M\tau_{zx}^2 + 2N\tau_{xy}^2}{F + G + H}$$

$$(6.127)$$

By definition, the plastic work increment $dW_p$ is represented by

$$dW_p = \sigma_x \, d\epsilon_x^P + \cdots + \tau_{yz} \, d\gamma_{yz}^P + \cdots \qquad (6.128)$$

Using Eq. (6.113) to substitute the plastic strain increments in Eq. (6.128), we obtain

$$dW_p = \tfrac{2}{3}\bar{\sigma}_e^2 \, d\lambda \qquad (6.129)$$

An alternative way to obtain Eq. (6.129) is to apply Euler's theorem on homogeneous functions. Comparing Eq. (6.105) and Eq. (6.127), we have

$$f = \tfrac{1}{3}\bar{\sigma}_e^2$$

Then we obtain

$$dW_p = \sigma_{ij}\, d\epsilon_{ij}^p = \sigma_{ij}\frac{\partial f}{\partial \sigma_{ij}}\, d\lambda = 2f\, d\lambda = \frac{2}{3}\bar{\sigma}_e^2\, d\lambda$$

since the yield function $f$ given in Eq. (6.105) is a homogeneous polynomial in stress components of degree 2.

We define the effective (or equivalent) plastic strain increment $d\epsilon_p$ as

$$dW_p = \bar{\sigma}_e\, d\epsilon_p \qquad (6.130)$$

From Eq. (6.129), the effective plastic strain increment can be expressed by

$$d\epsilon_p = \tfrac{2}{3}\bar{\sigma}_e\, d\lambda \qquad (6.131)$$

In order to express the effective plastic strain increment in terms of plastic strain increments, $d\epsilon_x^p, \ldots, d\gamma_{yz}^p, \ldots$, the following manipulations of Eq. (6.113) will be made:

$$G\, d\epsilon_y^p - H\, d\epsilon_z^p = \frac{d\lambda}{F+G+H}(FG+GH+HF)(\sigma_y - \sigma_z) \qquad (6.132)$$

Then

$$(\sigma_y - \sigma_z) = \frac{G\, d\epsilon_y^p - H\, d\epsilon_z^p}{FG+GH+HF}\frac{F+G+H}{d\lambda}$$

Similarly,

$$(\sigma_z - \sigma_x) = \frac{H\, d\epsilon_z^p - F\, d\epsilon_x^p}{FG+GH+HF}\frac{F+G+H}{d\lambda}$$

$$(\sigma_x - \sigma_y) = \frac{F\, d\epsilon_x^p - G\, d\epsilon_y^p}{FG+GH+HF}\frac{F+G+H}{d\lambda} \qquad (6.133)$$

and also

$$\tau_{yz} = \frac{d\gamma_{yz}^p}{2L}\frac{F+G+H}{d\lambda}$$

$$\tau_{zx} = \frac{d\gamma_{zx}^p}{2M}\frac{F+G+H}{d\lambda}$$

$$\tau_{xy} = \frac{d\gamma_{xy}^p}{2N}\frac{F+G+H}{d\lambda}$$

Substituting Eqs. (6.133) into Eq. (6.127), we obtain the expression for $\bar{\sigma}_e^2$ in terms of plastic strain increments as follows:

$$\bar{\sigma}_e^2 = \frac{3}{2} \frac{F+G+H}{(d\lambda)^2}$$

$$\times \left[ \frac{F(G\,d\epsilon_y^p - H\,d\epsilon_z^p)^2 + G(H\,d\epsilon_z^p - F\,d\epsilon_x^p)^2 + H(F\,d\epsilon_x^p - G\,d\epsilon_y^p)^2}{(FG+GH+HF)^2} \right.$$

$$\left. + \frac{(d\gamma_{yz}^p)^2}{2L} + \frac{(d\gamma_{zx}^p)^2}{2M} + \frac{(d\gamma_{xy}^p)^2}{2N} \right] \tag{6.134}$$

Substituting Eq. (6.134) into Eq. (6.131), we then obtain the expression for the effective plastic strain increment $d\epsilon_p$ in terms of plastic strain increments:

$$d\epsilon_p = \sqrt{\frac{2}{3}(F+G+H)}$$

$$\times \left[ \frac{F(G\,d\epsilon_y^p - H\,d\epsilon_z^p)^2 + G(H\,d\epsilon_z^p - F\,d\epsilon_x^p)^2 + H(F\,d\epsilon_x^p - G\,d\epsilon_y^p)^2}{(FG+GH+HF)^2} \right.$$

$$\left. + \frac{(d\gamma_{yz}^p)^2}{2L} + \frac{(d\gamma_{zx}^p)^2}{2M} + \frac{(d\gamma_{xy}^p)^2}{2N} \right]^{1/2} \tag{6.135}$$

It may be noted that the hardening parameter $k$ is related to the effective stress $\bar{\sigma}_e$ in Eq. (6.127) by

$$k = \tfrac{1}{3}\bar{\sigma}_e^2 \tag{6.136}$$

and that the definitions of $\bar{\sigma}_e$ and $d\epsilon_p$ in Eqs. (6.127) and (6.135) can be reduced to those of the von Mises theory if the anisotropy is made vanishingly small.

## 6.7.5. Consistency Condition

From Eq. (6.136), we can rewrite Eq. (6.106) as follows:

$$Q = f - \tfrac{1}{3}\bar{\sigma}_e^2 = 0 \tag{6.137}$$

The *consistency condition*, which requires the state of stress to remain on the yield or loading surface during plastic flow, is expressed by

$$dQ = 0 \tag{6.138}$$

From Eq. (6.137),

$$dQ = df - \tfrac{2}{3}\bar{\sigma}_e\,d\bar{\sigma}_e = 0 \tag{6.139}$$

If we assume a unique relationship between the effective stress ($\bar{\sigma}_e$) and the effective plastic strain ($\int d\epsilon_p$), Eq. (6.139) can be written as

$$dQ = df - \frac{2}{3}\bar{\sigma}_e \frac{d\bar{\sigma}_e}{d\epsilon_p} d\epsilon_p$$

$$= df - \frac{2}{3}\bar{\sigma}_e\,d\epsilon_p H_p = 0 \tag{6.140}$$

where $H_p = d\bar{\sigma}_e/d\epsilon_p$ corresponds to the slope of the equivalent stress $(\bar{\sigma}_e)$ versus plastic strain $(\int d\epsilon_p)$ curve.

Recalling Eq. (6.131), the consistency condition can be expressed by

$$df - \tfrac{4}{9}\bar{\sigma}_e^2 H_p \, d\lambda = 0 \tag{6.141}$$

in which

$$df = \frac{\partial f}{\partial \sigma_{ij}} \, d\sigma_{ij}$$

$$= \sigma_x' \, d\sigma_x + \sigma_y' \, d\sigma_y + \sigma_z' \, d\sigma_z + 2\tau_{yz}' \, d\tau_{yz} + 2\tau_{zx}' \, d\tau_{zx} + 2\tau_{xy}' \, d\tau_{xy} \tag{6.142}$$

where $\sigma_x', \ldots, \tau_{yz}', \ldots$ are defined in Eq. (6.114).

Hence, the consistency condition of Eq. (6.141) is expressed in matrix form as

$$\{\sigma'\}^T \{d\sigma\} - \tfrac{4}{9}\bar{\sigma}_e^2 H_p \, d\lambda = 0 \tag{6.143}$$

where $\{\sigma'\}$ and $\{d\sigma\}$ are defined in Eqs. (6.117) and (6.122), respectively.

## 6.7.6. Elastic–Plastic Stress–Strain Matrix $[C^{ep}]$

So far, the plastic strain increments are expressed in terms of the current stress state $\sigma_{ij}$ including, however, a positive scalar $d\lambda$. In order to derive the incremental elastic–plastic stress–strain relationships, we have to determine the expression for $d\lambda$. The explicit expression of $d\lambda$ can be obtained through the consistency conditions (6.141) through (6.143) by substituting for $\{d\sigma\}$.

The expressions for the stress increments are obtained by inverting Eq. (6.126),

$$\{d\sigma\} = [C^e](\{d\epsilon\} - \{d\epsilon^P\})$$

$$= [C^e](\{d\epsilon\} - \{\sigma'\} \, d\lambda) \tag{6.144}$$

$$= [C^e]\{d\epsilon\} - \{S\} \, d\lambda \tag{6.145}$$

where

$$\{S\} = [C^e]\{\sigma'\} \tag{6.146}$$

$$= \{s_1 \ s_2 \ s_3 \ s_4 \ s_5 \ s_6\}^T \tag{6.147}$$

in which

$$s_1 = c_{11}\sigma_x' + c_{12}\sigma_y' + c_{13}\sigma_z', \qquad s_4 = 2c_{44}\tau_{yz}'$$

$$s_2 = c_{12}\sigma_x' + c_{22}\sigma_y' + c_{23}\sigma_z', \qquad s_5 = 2c_{55}\tau_{zx}' \tag{6.148}$$

$$s_3 = c_{13}\sigma_x' + c_{23}\sigma_y' + c_{33}\sigma_z', \qquad s_6 = 2c_{66}\tau_{xy}'$$

Substituting the stress increments (6.145) into the consistency condition (6.143), we obtain

$$\{\sigma'\}^T[C^e]\{d\epsilon\} - \{\sigma'\}^T\{S\} \, d\lambda - \tfrac{4}{9}\bar{\sigma}_e^2 H_p \, d\lambda = 0 \qquad (6.149)$$

The symmetry of $[C^e]$ and the use of Eq. (6.146) lead to

$$\{S\}^T\{d\epsilon\} - \{\sigma'\}^T\{S\} \, d\lambda - \tfrac{4}{9}\bar{\sigma}_e^2 H_p \, d\lambda = 0 \qquad (6.150)$$

Solving for $d\lambda$, we obtain

$$d\lambda = \frac{\{S\}^T\{d\epsilon\}}{s} \qquad (6.151)$$

$$= \frac{s_1 \, d\epsilon_x + s_2 \, d\epsilon_y + s_3 \, d\epsilon_z + s_4 \, d\gamma_{yz} + s_5 \, d\gamma_{zx} + s_6 \, d\gamma_{xy}}{s} \qquad (6.152)$$

where

$$s = \tfrac{4}{9}\bar{\sigma}_e^2 H_p + \{\sigma'\}^T\{S\} \qquad (6.153)$$

$$= \tfrac{4}{9}\bar{\sigma}_e^2 H_p + s_1\sigma'_x + s_2\sigma'_y + s_3\sigma'_z + 2s_4\tau'_{yz} + 2s_5\tau'_{zx} + 2s_6\tau'_{xy} \qquad (6.154)$$

Finally, substituting $d\lambda$ from Eq. (6.151) into Eq. (6.145) leads to

$$\{d\sigma\} = [C^{ep}]\{d\epsilon\} \qquad (6.155)$$

where

$$[C^{ep}] = [C^e] - \frac{1}{s}\{S\}\{S\}^T \qquad (6.156)$$

or

$$[C^{ep}] = \begin{bmatrix}
c_{11} - \dfrac{s_1^2}{s} & & & & & \text{sym.} \\[2ex]
c_{12} - \dfrac{s_1 s_2}{s} & c_{22} - \dfrac{s_2^2}{s} & & & & \\[2ex]
c_{13} - \dfrac{s_1 s_3}{s} & c_{23} - \dfrac{s_2 s_3}{s} & c_{33} - \dfrac{s_3^2}{s} & & & \\[2ex]
-\dfrac{s_1 s_4}{s} & -\dfrac{s_2 s_4}{s} & -\dfrac{s_3 s_4}{s} & c_{44} - \dfrac{s_4^2}{s} & & \\[2ex]
-\dfrac{s_1 s_5}{s} & -\dfrac{s_2 s_5}{s} & -\dfrac{s_3 s_5}{s} & -\dfrac{s_4 s_5}{s} & c_{55} - \dfrac{s_5^2}{s} & \\[2ex]
-\dfrac{s_1 s_6}{s} & -\dfrac{s_2 s_6}{s} & -\dfrac{s_3 s_6}{s} & -\dfrac{s_4 s_6}{s} & -\dfrac{s_5 s_6}{s} & c_{66} - \dfrac{s_6^2}{s}
\end{bmatrix}$$

### 6.7.7. Elastic–Plastic Matrix $[C^{ep}]$ for Plane Stress Case

In the case of the plane stress condition, $\sigma_z = \tau_{zx} = \tau_{yz} = 0$; then $d\sigma_z = d\tau_{zx} = d\tau_{yz} = 0$, but the strain component $d\epsilon_z$ is not zero. Recalling Eq. (6.144),

$$\{d\sigma\} = [C^e](\{d\epsilon\} - \{\sigma'\} \, d\lambda) \qquad (6.158)$$

Expanding Eq. (6.158),

$$d\sigma_x = c_{11}(d\epsilon_x - \sigma'_x \, d\lambda) + c_{12}(d\epsilon_y - \sigma'_y \, d\lambda) + c_{13}(d\epsilon_z - \sigma'_z \, d\lambda)$$

$$d\sigma_y = c_{12}(d\epsilon_x - \sigma'_x \, d\lambda) + c_{22}(d\epsilon_y - \sigma'_y \, d\lambda) + c_{23}(d\epsilon_z - \sigma'_z \, d\lambda)$$

$$d\sigma_z = c_{13}(d\epsilon_x - \sigma'_x \, d\lambda) + c_{23}(d\epsilon_y - \sigma'_y \, d\lambda) + c_{33}(d\epsilon_z - \sigma'_z \, d\lambda)$$

$$d\tau_{yz} = c_{44}(d\gamma_{yz} - 2\tau'_{yz} \, d\lambda) \tag{6.159}$$

$$d\tau_{zx} = c_{55}(d\gamma_{zx} - 2\tau'_{zx} \, d\lambda)$$

$$d\tau_{xy} = c_{66}(d\gamma_{xy} - 2\tau'_{xy} \, d\lambda)$$

Since $d\sigma_z = 0$,

$$(d\epsilon_z - \sigma'_z \, d\lambda) = -\frac{c_{13}}{c_{33}}(d\epsilon_x - \sigma'_x \, d\lambda) - \frac{c_{23}}{c_{33}}(d\epsilon_y - \sigma'_y \, d\lambda) \tag{6.160}$$

Substituting Eq. (6.160) into the expressions for $d\sigma_x$ and $d\sigma_y$ in Eq. (6.159), we obtain

$$d\sigma_x = \left(c_{11} - \frac{c_{13}c_{13}}{c_{33}}\right)(d\epsilon_x - \sigma'_x \, d\lambda) + \left(c_{12} - \frac{c_{13}c_{23}}{c_{33}}\right)(d\epsilon_y - \sigma'_y \, d\lambda)$$

$$= \bar{c}_{11} \, d\epsilon_x + \bar{c}_{12} \, d\epsilon_y - \bar{s}_1 \, d\lambda \tag{6.161a}$$

$$d\sigma_y = \left(c_{12} - \frac{c_{23}c_{13}}{c_{33}}\right)(d\epsilon_x - \sigma'_x \, d\lambda) + \left(c_{22} - \frac{c_{23}c_{23}}{c_{33}}\right)(d\epsilon_y - \sigma'_y \, d\lambda)$$

$$= \bar{c}_{12} \, d\epsilon_x + \bar{c}_{22} \, d\epsilon_y - \bar{s}_2 \, d\lambda \tag{6.161b}$$

Also, $d\tau_{yz} = d\tau_{zx} = 0$, and

$$d\tau_{xy} = c_{66}(d\gamma_{xy} - 2\tau'_{xy} \, d\lambda)$$

$$= \bar{c}_{66} \, d\gamma_{xy} - \bar{s}_6 \, d\lambda \tag{6.161c}$$

where

$$\bar{c}_{11} = c_{11} - \frac{c_{13}c_{13}}{c_{33}}, \qquad \bar{c}_{12} = c_{12} - \frac{c_{13}c_{23}}{c_{33}}, \qquad \bar{c}_{22} = c_{22} - \frac{c_{23}c_{23}}{c_{33}}, \qquad \bar{c}_{66} = c_{66} \tag{6.162}$$

$$\bar{s}_1 = \bar{c}_{11}\sigma'_x + \bar{c}_{12}\sigma'_y, \qquad \bar{s}_2 = \bar{c}_{12}\sigma'_x + \bar{c}_{22}\sigma'_y, \qquad \bar{s}_6 = 2\bar{c}_{66}\tau'_{xy} \tag{6.163}$$

From the above expressions, we can show that

$$\{\overline{d\sigma}\} = [\bar{C}^e]\{\overline{d\epsilon}^e\}$$

$$= [\bar{C}^e](\{\overline{d\epsilon}\} - \{\overline{d\epsilon}^P\})$$

$$= [\bar{C}^e](\{\overline{d\epsilon}\} - \{\bar{\sigma}'\} \, d\lambda)$$

$$= [\bar{C}^e]\{\overline{d\epsilon}\} - \{\bar{S}\} \, d\lambda \tag{6.164}$$

where

$$\{\overline{d\sigma}\}^T = \{d\sigma_x \; d\sigma_y \; d\tau_{xy}\} \qquad (6.165)$$

$$[\bar{C}^e] = \begin{bmatrix} \bar{c}_{11} & \bar{c}_{12} & 0 \\ \bar{c}_{12} & \bar{c}_{22} & 0 \\ 0 & 0 & \bar{c}_{66} \end{bmatrix} \qquad (6.166)$$

$$\{\overline{d\epsilon}\}^T = \{d\epsilon_x \; d\epsilon_y \; d\gamma_{xy}\} \qquad (6.167)$$

$$\{\overline{d\epsilon^e}\}^T = \{d\epsilon_x^e \; d\epsilon_y^e \; d\gamma_{xy}^e\} \qquad (6.168)$$

$$\{\overline{d\epsilon^p}\}^T = \{d\epsilon_x^p \; d\epsilon_y^p \; d\gamma_{xy}^p\} \qquad (6.169)$$

and

$$\{\bar{\sigma}'\}^T = \{\sigma_x' \; \sigma_y' \; 2\tau_{xy}'\} \qquad (6.170)$$

and

$$\{\bar{S}\}^T = \{\bar{s}_1 \; \bar{s}_2 \; \bar{s}_6\} \qquad (6.171)$$

in which

$$\{\bar{S}\} = [\bar{C}^e]\{\bar{\sigma}'\} \qquad (6.172)$$

In the above notations, vectors and matrices are written with an overbar in order to distinguish the plane stress case from the 3-D case.

The consistency condition for the plane stress case is expressed by

$$\{\bar{\sigma}'\}^T\{\overline{d\sigma}\} - \tfrac{4}{9}\bar{\sigma}_e^2 H_p \, d\lambda = 0 \qquad (6.173)$$

Substituting Eq. (6.164) into the first term, we obtain

$$\{\bar{\sigma}'\}^T\{d\bar{\sigma}\} = \{\bar{\sigma}'\}^T[\bar{C}^e]\{\overline{d\epsilon}\} - \{\bar{\sigma}'\}^T\{\bar{S}\} \, d\lambda$$
$$= \{\bar{S}\}^T\{\overline{d\epsilon}\} - \{\bar{\sigma}'\}^T\{\bar{S}\} \, d\lambda \qquad (6.174)$$

Then the consistency condition leads to

$$\{\bar{S}\}^T\{\overline{d\epsilon}\} - \{\bar{\sigma}'\}^T\{\bar{S}\} \, d\lambda - \tfrac{4}{9}\bar{\sigma}_e^2 H_p \, d\lambda = 0 \qquad (6.175)$$

Solving for $d\lambda$, we obtain

$$d\lambda = \frac{\{\bar{S}\}^T\{\overline{d\epsilon}\}}{\bar{s}} = \frac{\bar{s}_1 \, d\epsilon_x + \bar{s}_2 \, d\epsilon_y + \bar{s}_6 \, d\gamma_{xy}}{\bar{s}} \qquad (6.176)$$

where

$$\bar{s} = \tfrac{4}{9}\bar{\sigma}_e^2 H_p + \{\bar{\sigma}'\}^T\{\bar{S}\}$$
$$= \tfrac{4}{9}\bar{\sigma}_e^2 H_p + \bar{s}_1\sigma_x' + \bar{s}_2\sigma_y' + 2\bar{s}_6\tau_{xy}' \qquad (6.177)$$

Substituting $d\lambda$ from Eq. (6.176) into Eq. (6.164), we obtain

$$\{\overline{d\sigma}\} = [\bar{C}^e]\{\overline{d\epsilon}\} - \frac{1}{\bar{s}}\{\bar{S}\}\{\bar{S}\}^T\{\overline{d\epsilon}\}$$

$$= ([\bar{C}^e] - \frac{1}{\bar{s}}\{\bar{S}\}\{\bar{S}\}^T)\{\overline{d\epsilon}\}$$

$$= [\bar{C}^{ep}]\{\overline{d\epsilon}\} \qquad (6.178)$$

where $[\bar{C}^{ep}]$ is the elastic-plastic matrix for the plane stress condition:

$$[\bar{C}^{ep}] = \begin{bmatrix} \bar{c}_{11} - \dfrac{\bar{s}_1^2}{\bar{s}} & & \text{sym.} \\[2ex] \bar{c}_{12} - \dfrac{\bar{s}_1 \bar{s}_2}{\bar{s}} & \bar{c}_{22} - \dfrac{\bar{s}_2^2}{\bar{s}} & \\[2ex] -\dfrac{\bar{s}_1 \bar{s}_6}{\bar{s}} & -\dfrac{\bar{s}_2 \bar{s}_6}{\bar{s}} & \bar{c}_{66} - \dfrac{\bar{s}_6^2}{\bar{s}} \end{bmatrix} \qquad (6.179)$$

On the other hand, the strain increment in the $z$ direction, $d\epsilon_z$, can be obtained as follows:

$$d\epsilon_z = d\epsilon_z^p + d\epsilon_z^e$$

where

$$d\epsilon_z^p = \sigma_z' \, d\lambda = -(\sigma_x' + \sigma_y') \frac{\bar{s}_1 \, d\epsilon_x + \bar{s}_2 \, d\epsilon_y + \bar{s}_6 \, d\gamma_{xy}}{\bar{s}}$$

$$d\epsilon_z^e = d_{13} \, d\sigma_x + d_{23} \, d\sigma_y = d_{13}(\bar{c}_{11} \, d\epsilon_x + \bar{c}_{12} \, d\epsilon_y) + d_{23}(\bar{c}_{12} \, d\epsilon_x + \bar{c}_{22} \, d\epsilon_y)$$

where $d_{13}$ and $d_{23}$ are components of the matrix $[D^e]$ given in Eq. (6.123). If we assume that the material is elastically isotropic, then

$$\bar{c}_{11} = \frac{E}{1-\nu^2}, \qquad \bar{c}_{12} = \frac{\nu E}{1-\nu^2}, \qquad \bar{c}_{22} = \frac{E}{1-\nu^2}, \qquad \bar{c}_{66} = \frac{E}{2(1+\nu)} \qquad (6.180)$$

In this case, Eq. (6.179) can be expressed as

$$[\bar{C}^e] = \begin{bmatrix} \dfrac{E}{1-\nu^2} - \dfrac{\bar{s}_1^2}{\bar{s}} & & \text{sym.} \\[2ex] \dfrac{\nu E}{1-\nu^2} - \dfrac{\bar{s}_1 \bar{s}_2}{\bar{s}} & \dfrac{E}{1-\nu^2} - \dfrac{\bar{s}_2^2}{\bar{s}} & \\[2ex] -\dfrac{\bar{s}_1 \bar{s}_6}{\bar{s}} & -\dfrac{\bar{s}_2 \bar{s}_6}{\bar{s}} & \dfrac{E}{2(1+\nu)} - \dfrac{\bar{s}_6^2}{\bar{s}} \end{bmatrix} \qquad (6.181)$$

where

$$\bar{s} = \tfrac{4}{9} \bar{\sigma}_e^2 H_p + \bar{s}_1 \sigma_x' + \bar{s}_2 \sigma_y' + 2\bar{s}_6 \tau_{xy}' \qquad (6.182)$$

$$\bar{s}_1 = \frac{E}{1-\nu^2}(\sigma_x' + \nu\sigma_y'), \qquad \bar{s}_2 = \frac{E}{1-\nu^2}(\nu\sigma_x' + \sigma_y'), \qquad \bar{s}_6 = \frac{E}{1+\nu}\tau_{xy}' \qquad (6.183)$$

Stresses $\sigma_x'$, $\sigma_y'$, and $\tau_{xy}'$ are defined by Eq. (6.114).

### 6.7.8. Radial Drawing of a Blank—An Example (Yamada, 1969)

For plane problems, the material parameters involved in Eq. (6.105) are $F$, $G$, $H$, and $N$. These parameters can be determined by tension tests of plate specimens as follows.

Suppose that $x$ and $y$ are the principal axes of orthotropy of the material and that a tensile plate specimen is cut with angle $\alpha$ from the $x$-axis. If the specimen is subjected to a tensile stress $\sigma$ (Fig. 6.9), the stress components with respect to the $x$- and $y$-axes are

$$\sigma_x = \sigma \cos^2 \alpha, \qquad \sigma_y = \sigma \sin^2 \alpha, \qquad \tau_{xy} = \sigma \sin \alpha \cos \alpha$$

Using Eq. (6.113), and considering $\sigma_z = \tau_{yz} = \tau_{zx} = 0$, we can calculate the plastic strain increments as follows:

$$d\epsilon_x^p = \sigma_x' \, d\lambda = \frac{(G+H)\sigma_x - H\sigma_y}{F+G+H} \, d\lambda$$

$$= \frac{d\lambda \, \sigma}{F+G+H} [(G+H) \cos^2 \alpha - H \sin^2 \alpha]$$

$$d\epsilon_y^p = \sigma_y' \, d\lambda = \frac{(F+H)\sigma_y - H\sigma_x}{F+G+H} \, d\lambda$$

$$= \frac{d\lambda \, \sigma}{F+G+H} [(F+H) \sin^2 \alpha - H \cdot \cos^2 \alpha]$$

$$d\epsilon_z^p = \sigma_z' \, d\lambda = \frac{-G\sigma_x - F\sigma_y}{F+G+H} \, d\lambda$$

$$= \frac{d\lambda \, \sigma}{F+G+H} [-F \sin^2 \alpha - G \cos^2 \alpha]$$

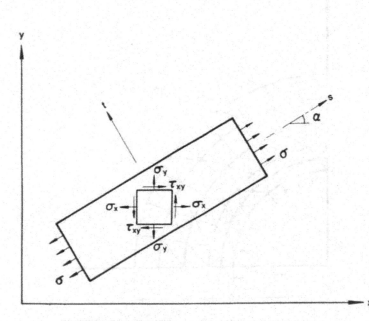

FIGURE 6.9. Uniaxial tension of orthotropic plate.

$$d\gamma_{xy}^p = 2\tau_{xy}' \, d\lambda = \frac{2N\tau_{xy}}{F+G+H} \, d\lambda$$

$$= \frac{d\lambda \, \sigma}{F+G+H} [2N \sin \alpha \cos \alpha]$$

The plastic strain increment in the direction of $t$ can then be obtained by the transformation of plastic strain components:

$$d\epsilon_t^p = d\epsilon_x^p \sin^2 \alpha + d\epsilon_y^p \cos^2 \alpha - d\gamma_{xy}^p \sin \alpha \cos \alpha$$

$$= \frac{d\lambda \, \sigma}{F+G+H} [-H - (2N - F - G - 4H) \sin^2 \alpha \cos^2 \alpha]$$

Then, the ratio of transverse to through-thickness plastic strain increments is given by

$$r = \frac{d\epsilon_t^p}{d\epsilon_z^p} = \frac{H + (2N - F - G - 4H) \sin^2 \alpha \cos^2 \alpha}{F \sin^2 \alpha + G \cos^2 \alpha} \tag{6.184}$$

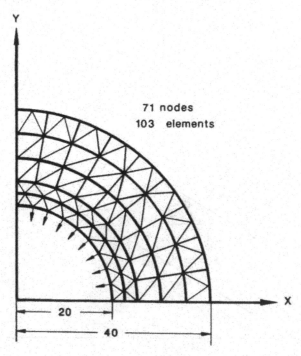

FIGURE 6.10. Radial drawing of annular plate and element division.

This ratio is often referred to as *Lankford's r-value*. Putting $\alpha = 0°$, $90°$, and $45°$, and denoting the corresponding $r$-values by $r_x$, $r_y$, and $r_{45}$, we obtain

$$\frac{G}{H} = \frac{1}{r_x}, \qquad \frac{F}{H} = \frac{1}{r_y}, \qquad \frac{N}{H} = \left(r_{45} + \frac{1}{2}\right)\left(\frac{1}{r_x} + \frac{1}{r_y}\right) \qquad (6.185)$$

Auxiliary useful relations are

$$\frac{1}{X^2} : \frac{1}{Y^2} : \frac{1}{T^2} = (G+H) : (F+H) : \left(\frac{F+G+2N}{4}\right) \qquad (6.186)$$

where $X$, $Y$, and $T$ are the tensile yield stresses in the $0°$, $90°$, and $45°$ directions.

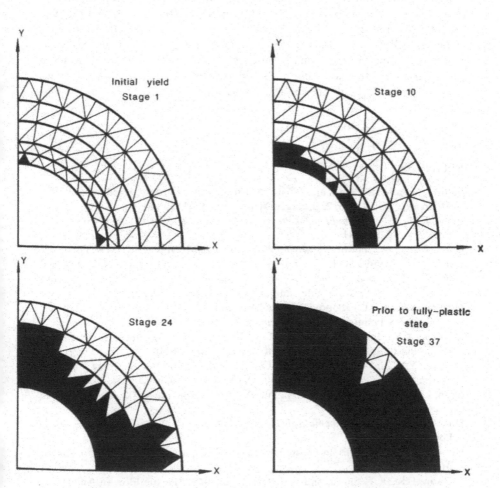

FIGURE 6.11. Spread of plastic region at typical stages of radial drawing.

The radial drawing problem is shown in Fig. 6.10. Relevant material properties adopted in the calculation are:

Young's modulus: $E = 21,000 \, kg/mm^2$, Poisson's ratio: $\nu = 0.3$,
Lankford's $r$-values: $r_x = 2.5$, $r_y = 2.0$, and $r_{45} = 1.0$
Yield stress in 90° direction: $Y = 25 \, kg/mm^2$, and
$X : Y : T = 1.035 : 1 : 1.291$ from Eq. (6.186)
Work-hardening rate: $H_p = d\bar{\sigma}/d\epsilon^p = 0.001 E$.

It is noted that all elastic properties are assumed to be isotropic.

The plastic regions at typical stages of drawing are illustrated in Fig. 6.11; stage 1 corresponds to the initial yield and stage 37 is just prior to the fully plastic state. It can be seen from Fig. 6.11 that plastic regions develop most rapidly in the $x$-direction, and the elements along the 45° direction are the last part to yield. The outer annulus of a drawn blank is essentially in the state of circumferential compressive stress, which predominates the radial tensile stress. Therefore, the results of Fig. 6.11 are consistent with the assumed order of magnitude of yield stresses $Y < X < T$ in three directions. Also, the calculated radial displacements predict, in conformity with the order of yield stresses and Lankford's $r$-values, ears at the 0° and 90° positions.

# References

Baltov, A., and A. Sawczuk, 1965. "A Rule of Anisotropic Hardening," *Acta Mechanica*, Vol. 1, No. 2, Springer, Wien–New York, pp. 81–92.

Bathe, K.-J., 1982. *Finite Element Procedures in Engineering Analysis*, Prentice-Hall, Englewood Cliffs, NJ.

Chen, W.F., 1982. *Plasticity in Reinforced Concrete*, McGraw-Hill, New York.

Dafalias, Y.F., and L.R. Herrmann, 1982. "Bounding Surface Formulation of Soil Plasticity," in *Soils Under Cyclic and Transient Load* (G.N. Pande and O.C. Zienkiewicz, eds.), John Wiley and Sons, Chichester, England.

Dafalias, Y.F., and E.P. Popov, 1975. "A Model for Nonlinearly Hardening Materials for Complex Loading," *Acta Mechanica* 21(3): 173–192.

Dafalias, Y.F., and E.P. Popov, 1976. "Plastic Internal Variables Formalism of Cyclic Plasticity," *Journal of Applied Mechanics* 43: 645–651.

Dafalias, Y.F., and E.P. Popov, 1977. "Cyclic Loading for Materials with a Vanishing Elastic Region," *Nuclear Engineering Design* 41(2): 293–302.

Dorn, J.E., 1949. "Stress–Strain Relations for Anisotropic Flow," Journal of Applied Physics, 20: 15–20.

Hill, R., 1950. *The Mathematical Theory of Plasticity*, Oxford University Press, London.

Hu, L.W., 1956. "Studies on Plastic Flow of Anisotropic Metals," *Journal of Applied Mechanics* 12: 444.

Jackson, L.R., K.F. Smith, and W.T. Lankford, 1948. "Plastic Flow in Anisotropic Sheet Metal," *Metal Technology*, T.P., 2440.

Krieg, R.D., 1975. "A Practical Two-Surface Plasticity Theory," *Journal of Applied Mechanics* **42**: 641–646.

Mroz, Z., 1967. "On the Description of Anisotropic Work-Hardening," *Journal of the Mechanics and Physics of Solids* **15**: 163–175.

Mroz, Z., V.A. Norris, and O.C. Zienkiewicz, 1978. "An Anisotropic Hardening Model for Soils and Its Application to Cyclic Loading," *International Journal of Numerical and Analytical Methods in Geomechanics* **2**: 203–221.

Mroz, Z., V.A. Norris, and O.C. Zienkiewicz, 1979. "Application of an Anisotropic Hardening Model in the Analysis of Elasto–Plastic Deformation of Soils," *Geotechnique* **29**(1): 1–34.

Nayak, G.C., and O.C. Zienkiewicz, 1972. "Convenient Form of Stress Invariants for Plasticity," *Journal of the Structural Division, ASCE* Vol. 98, No. ST4, April; 949–953.

Owen, D.R.J., and E. Hinton, 1980. *Finite Elements in Plasticity: Theory and Practice*, Pineridge Press, Swansea, U.K.

Yamada, Y., 1969. "Recent Japanese Developments in Matrix Displacement Method for Elasto–Plastic Problems," Japan–U.S. Seminar on Matrix Methods of Structural Analysis and Design, Tokyo, Japan, August, 1969.

Yamada, Y., N. Yoshimura, and T. Sakurai, 1968. "Plastic Stress–Strain Matrix and Its Application for the Solution of Elastic–Plastic Problems by Finite Element Method," *International Journal of Mechanical Science* **10**: 343–354.

Zienkiewicz, O.C., 1982. "Generalized Plasticity and Some Models for Geomechanics," *Applied Mathematics and Mechanics* (Published by Techmodern Business Promotion Center, Hong Kong), English edition, Vol. 3, No. 3.

# Part IV: Concrete Plasticity

# 7
# Implementation in Concretes

## 7.1. Introduction

### 7.1.1. Background

In past years, the methods of analysis and design for concrete structures were mainly based on elastic analysis combined with various classical procedures as well as on empirical formulas developed on the basis of a large amount of experimental data. Such approaches are still necessary and desirable and continue to be the most convenient and effective methods for ordinary design. However, the rapid development of modern numerical analysis techniques and high-speed digital computers has provided structural engineers with a powerful tool for a complete *nonlinear analysis* of concrete structures. By using the finite-element method and performing an incremental inelastic analysis, deformational and failure characteristics of concrete structures can be assessed with some degree of accuracy. For example, some complex behaviors of reinforced concrete, such as multiaxial nonlinear stress–strain properties, cracking, aggregate interlocking, bond slip, and other effects previously ignored or treated in a very approximate manner can now be modeled and studied more rationally. In addition, as the quantitative information on the load–deformation behavior of concrete develops and computing capability expands, the scope of nonlinear analysis can be broadened to include triaxially loaded concrete structures, such as nuclear power reactors, floating vessels, offshore platforms, arch dams, etc., for which this type of analysis is of particular value because large-scale experimental studies of these special types of structures are often prohibitively expensive.

The first attempt to apply the finite-element method to a reinforced concrete structure was made by Ngo and Scordelis in 1967. Since then, a rapid development has taken place. It has been well recognized that the incompleteness of material models for reinforced concrete is the major factor limiting the further expansion of the capability to analyze concrete structures. Therefore, most efforts made in the past have been directed toward improving the mathematical description of the constitutive relations of concrete materials as well as the modeling of interactions between reinforcement and concrete. To date, a variety of modeling approaches have been proposed, including nonlinear elasticity, plasticity, endochronic theory, and damage theory. This chapter is concerned only with the plasticity modeling of concrete materials.

## 7.1.2. Features of Concrete Behavior

Concrete is a composite material. It consists of coarse aggregate and a continuous matrix of mortar, which itself comprises a mixture of cement paste and smaller aggregate particles. Its physical behavior is very complex, being largely determined by the structure of the composite material, such as the ratio of water to cement, the ratio of cement to aggregate, the shape and size of aggregate, and the kind of cement used. Our discussion is confined to the stress–strain behavior of an average ordinary concrete. The structure of the material is ignored and the rules of material behavior are developed on the basis of a homogeneous *continuum*. Also, the material is customarily assumed to be initially isotropic.

Concrete is a sort of brittle material. Its stress–strain behavior is affected by the development of micro- and macrocracks in the material body. Particularly, concrete contains a large number of microcracks, especially at interfaces between coarse aggregates and mortar, even before the application of external load. These initial microcracks are caused by segregation, shrinkage, or thermal expansion in the cement paste. Under applied loading, further microcracking may occur at the aggregate–cement paste interface, which is the weakest link in the composite system. The progression of these cracks, which are initially invisible, to become visible cracks occurs with the application of external loads and contributes to the generally obtained nonlinear stress–strain behavior.

### 7.1.2.1. NONLINEAR STRESS-STRAIN BEHAVIOR

A typical stress–strain curve in a uniaxial compression test is shown in Fig. 7.1. There are three deformation stages observed in this simple test (Kotsovos and Newman, 1977). The first stage corresponds to a stress in the region up to 30% of the maximum compressive stress $f'_c$. At this stage, the cracks existing in concrete before loading remain nearly unchanged. Hence, the stress–strain behavior is linearly elastic. Therefore, $0.3f'_c$ is usually proposed as the *limit of elasticity*. Beyond this limit, the stress–strain curve begins to deviate from a straight line. Stress between 30% and about 75% of $f'_c$ characterizes the second stage, in which bond cracks start to increase in length, width, and number, and later some cracks at nearby aggregate surfaces start to bridge in the form of mortar cracks. With the significant cracking developed, material nonlinearity becomes more evident. However, the crack propagation at this stage is still *stable* until the stress reaches the level of about 75% of $f'_c$. Hence, $0.75f'_c$ is generally termed the *onset of*

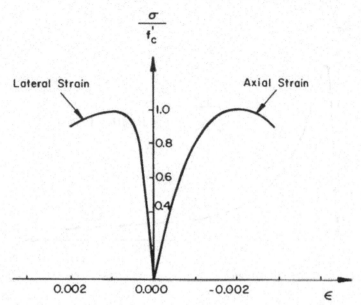

FIGURE 7.1. Typical uniaxial compressive stress–strain curve (Kupfer et al., 1969).

*unstable fracture propagation.* Further increase of the load eventually results in *unstable fracture,* and in the third stage, the progressive failure of concrete is primarily caused by cracks through the mortar. These cracks join bond cracks at the surface of nearby aggregates and form crack zones or *internal damage.* Then a smoothly varying deformation pattern may change, and further deformations may be localized. Finally, major cracks form parallel to the direction of applied load, causing failure of the specimen.

Although the above discussion is concerned only with the uniaxial compression case, three deformation stages can also be identified qualitatively in other loading cases, i.e., the linear elastic stage, the inelastic stage, and the localized stage, and these have to be considered in modeling concrete behavior.

### 7.1.2.2. DIFFERENT RESPONSES IN TENSION AND COMPRESSION

Figure 7.2 shows a typical uniaxial tension stress–elongation curve. In general the limit of elasticity is observed to be about 60 to 80% of the ultimate tensile strength. Above this level, the bond microcracks start to grow. As the uniaxial tension state of stress tends to arrest the cracks much less frequently than the compressive stage of stress, one can expect the interval of stable crack propagation to be quite short, and the unstable crack propagation to start very soon. That is why the deformational behavior of concrete in tension is quite brittle in nature. In addition, the aggregate–mortar interface has a significantly lower tensile strength than mortar. This is the primary reason for the low tensile strength of concrete materials.

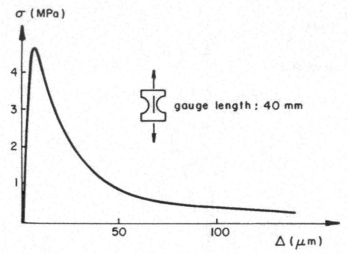

FIGURE 7.2. Uniaxial tensile stress–elongation curve (Peterson, 1981).

### 7.1.2.3. MULTIAXIAL COMPRESSIVE LOADING

A typical stress–strain behavior of concrete under multiaxial loading conditions is shown in Fig. 7.3 (Palaniswamy and Shah, 1974). The results are obtained from tests of cylindrical specimens. The concrete cylinders are submitted to constant lateral pressures, $\sigma_2 = \sigma_3$. The axial pressure $\sigma_1$ is increased up to failure. Figure 7.3 shows the axial stress–axial strain $(\sigma_1 - \epsilon_1)$

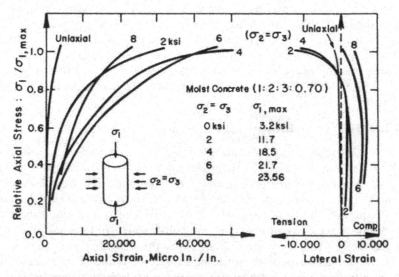

FIGURE 7.3. Stress–strain curves under multiaxial compression (Palaniswamy and Shah, 1974). Stress and strain, positive in compression.

and the axial stress–lateral strain ($\sigma_1$–$\epsilon_2$) curves for different values of confining (lateral) pressure. It is observed that the confining pressure has significant effects on the deformation behavior of the specimen. At first, the axial and lateral strains at failure increase with increasing confining pressure. But, for lateral stresses beyond a certain value, e.g., 4 ksi, increasing lateral stresses will decrease the values of axial strains at failure. However, compared to the uniaxial compression case, much larger strains occur in the confined concrete specimens. It can be seen that under compressive loadings with confining pressure, concrete exhibits a certain degree of ductility before failure.

Similar to the uniaxial case, qualitatively, there are also three deformational stages for a moderately confined concrete, i.e., linearly elastic, inelastic, and localized.

### 7.1.2.4. VOLUME EXPANSION UNDER COMPRESSIVE LOADING

The volumetric strain plotted against the stress in biaxial compression tests is shown in Fig. 7.4. Initially, the strain decreases up to about 0.75 to 0.90 of the ultimate stresses. Then the tendency is reversed with increasing stress. It was shown by Shah and Chandra (1968) that the cement paste itself does not expand under the compression load. The paste specimen continues to consolidate up to failure. Volumetric expansion is observed only when the cement paste is mixed up with aggregates, this indicates that the *composite*

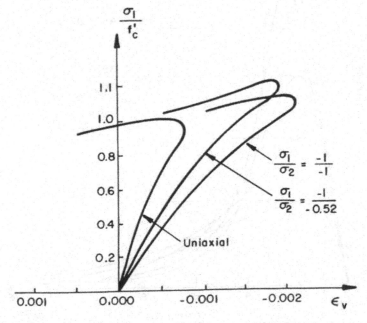

FIGURE 7.4. Volumetric strain under biaxial compression (Kupfer et al., 1969).

nature of concrete is primarily responsible for the volume dilatation. It is noted that the stress at which volume begins to increase is related to a noticeable increase of microcracks through the mortar, i.e., to the beginning of an unstable crack propagation.

### 7.1.2.5. STRAIN SOFTENING

Many engineering materials such as concretes, rocks, and soils exhibit a significant strain-softening behavior beyond the peak of failure stress. Figure 7.5 shows typical uniaxial compressive stress-strain curves obtained from strain-controlled tests. Each of these curves has a sharp descending branch beyond the peak of failure stress.

However, as mentioned above, even before the peak stress, strain localization has started to occur. It becomes more significant after the peak stress. Strain distribution in the specimen is no longer continuous. Such questions therefore arise as "what does the softening branch of a stress–strain curve

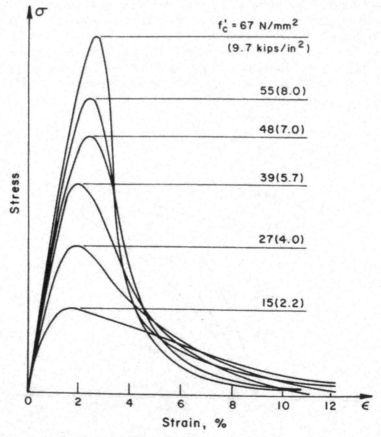

FIGURE 7.5. Uniaxial compressive stress–strain curves for concrete (Wischers, 1978).

mean?" and "can the post-peak test results be interpreted in terms of stress and strain?"

It is generally agreed now that the softening branch of a stress–strain curve does not reflect a material property, but rather represents the response of the structure formed by the specimen together with its complete loading system (van Mier, 1984). This argument is supported by compression tests of specimens with different heights. The test results in terms of stress and strain are shown in Fig. 7.6. As can be seen, the descending branches of the stress–strain curves are not identical but have slopes decreasing with increasing specimen heights. On the other hand, however, if the post-peak displacement rather than the strain is plotted against the stress, the stress–displacement curves are almost identical, independent of the specimen heights (van Mier, 1984). This phenomenon can be explained as follows. Since the post-peak strain is localized in a small region of the specimens, it would result in the same post-peak displacement for all the specimens. However, when we calculate the strains for each specimen, we are using different heights to divide the same value of displacement. This will result in different strain values. These values are not real ones measured from experiments, but average ones over the heights of the specimens.

Consequently, as the post-peak deformation is localized, the descending branch of the stress–strain curve cannot be considered a material property. Rather, it has to be referred to as a structural property.

### 7.1.2.6. STIFFNESS DEGRADATION

Figure 7.7 shows a typical uniaxial compressive stress–strain curve of concrete under cyclic loading. As can be seen, the unloading-reloading

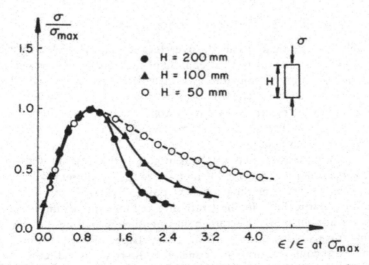

FIGURE 7.6. Influence of specimen height on uniaxial stress–strain curve (van Mier, 1984).

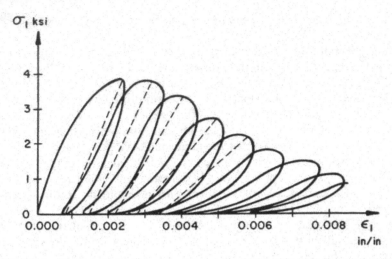

FIGURE 7.7. Cyclic uniaxial compressive stress–strain curve (Sinha et al., 1964).

curves are not straight-line segments but loops of changing size with decreasing average slopes. If we assume that the average slope is the slope of a straight line connecting the turning points of one cycle and that the material behavior upon unloading and reloading is linearly elastic (dotted line in Fig. 7.7), then the elastic modulus (or the slope) degrades with increasing straining. This stiffness degradation behavior is related to some kinds of damage (say, microvoids and microcracks). This damage becomes much more significant in the post-peak range.

## 7.1.3. Constitutive Modeling of Concrete Materials

The intensive investigations made over recent years have led to a better understanding of the constitutive behavior of concrete under various loading conditions. A variety of constitutive models have been proposed. Most are of the phenomenal type. The aim of a phenomenal model is to reproduce mathematically the macroscopic stress–strain relations for different loading conditions, neglecting the microscopic mechanism of the behavior. The plasticity approach falls into this category.

The classical theory of plasticity was originally developed for metals, and the deformational mechanisms of metals are quite different from those of concrete. However, from a macroscopic point of view, they still have some similarities, particularly in the prefailure regime. For example, concrete exhibits a nonlinear stress–strain behavior during loading and has a significant irreversible strain upon unloading. Especially under compressive loadings with confining pressure, concrete may show some ductile behavior. The irreversible deformations of concrete are induced by microcracking and slip and may be treated by the theory of plasticity.

Any plasticity model must involve three basic assumptions:

(i) An *initial yield surface* in stress space defines the stress level at which plastic deformation begins.

(ii) A *hardening rule* defines the change of the loading surface as well as the change of the hardening properties of the material during the course of plastic flow.

(iii) A *flow rule*, which is related to a *plastic potential function*, gives an incremental plastic stress–strain relation.

In plasticity modeling of concrete, the ultimate strength condition, i.e., the failure condition which sets the upper bound of the attainable states of stress, has to be assumed, in addition to the above three assumptions. Since the classical theory of plasticity was developed originally for metals, some of its fundamental assumptions do not hold for other engineering materials such as concrete. Thus, considerable modifications are necessary in applying this classical theory to concrete materials.

In this chapter, the development of a plasticity-based model in the prefailure range is first discussed in some detail. Emphasis is placed on the underlying concepts of the yield surface, the hardening rule, and the flow rule peculiar to concrete materials.

A proper constitutive model for analysis of concrete structures, however, requires a complete description of the behavior of the material, not only in its prefailure (hardening) range, but also including its postfailure (softening) behavior. From a macroscopic point of view, the postfailure behavior is characterized by a descending branch of the stress–strain relation as well as a degradation of the elastic modulus upon unloading. In the framework of *strain-space plasticity*, the strain-softening behavior can generally be described, and further, the elastic degradation coupled with plastic deformation can also be formulated by the *plastic-fracturing theories*. Therefore, the later part of this chapter is devoted to a discussion of the strain-space plasticity formulation. Since the model for the postfailure behavior of concrete is still under active development, the discussion is limited only to the basic concepts of the general theory rather than a specific model of concrete.

In the following, we shall start with a discussion of the failure criteria that form the basis for almost all concrete constitutive models.

## 7.2. Failure Criteria

### 7.2.1. Characteristics of Failure Surface of Concrete

Since the assumption of isotropy is reasonable for concrete materials, the general form of a failure surface can be expressed as (see Chapter 2)

$$f(I_1, J_2, J_3) = 0 \tag{7.1a}$$

or

$$f(\xi, \rho, \theta) = 0 \tag{7.1b}$$

Alternatively, using *octahedral* stresses $\sigma_{oct}$ and $\tau_{oct}$ to replace stress invariants $I_1$ and $J_2$, Eq. (7.1a, b) may be written as

$$f(\sigma_{oct}, \tau_{oct}, \theta) = 0 \tag{7.1c}$$

The explicit form of the failure function is defined by experimental data. Strength tests of plain concrete are well documented in the literature. To name a few, the reports by Kupfer et al. (1969) and Tasuji et al. (1978) nearly cover the full area of the biaxial stress section. For triaxial stress states, we mention the results of Mills and Zimmerman (1970), Launay and Gachon (1970), and Gerstle et al. (1978), among others. The available experimental data clearly indicate the essential features of a failure surface. This is illustrated in Fig. 7.8.

Being hydrostatic-pressure-dependent materials, concretes have a failure surface with curved meridians, indicating that the hydrostatic pressure produces effects of increasing the shearing capacity of the material (Fig. 7.8a). The meridians start from the hydrostatic tensile failure point and open in the direction of the negative hydrostatic axis. A pure hydrostatic loading cannot cause failure. A failure curve along the compressive meridian up to $I_1 = -79f'_c$ has been determined experimentally by Chinn and Zimmerman (1965) without observing any tendency of this meridian to approach the hydrostatic axis. The tensile meridian $\rho_t$, compressive meridian $\rho_c$, and shear meridian $\rho_s$ corresponding to $\theta = 0°$, $\theta = 60°$, and $\theta = 30°$, respectively, satisfy $\rho_t < \rho_s < \rho_c$. The value of $\rho_t / \rho_c$ increases with increasing hydrostatic pressure. It is about 0.5 near the $\pi$-plane and reaches a high of about 0.8 near the hydrostatic pressure $\xi = -7f'_c$.

As shown in Fig. 7.8b, the cross sections of the failure surface in the deviatoric planes have 120° period and 60° symmetry because of isotropy. The shape of the traces changes from nearly triangular for tensile and small compressive stresses to a bulged shape (near circular) for higher compressive stresses. The deviatoric sections are convex and $\theta$-dependent (Fig. 7.8b).

Based on our knowledge concerning the shape of the failure surface of concrete materials, a variety of failure criteria have been proposed. Most of these criteria were discussed in the book by Chen (1982), where they were classified by the number of material constants appearing in the expression as one-parameter through five-parameter models. Some of the most popular models are shown in Fig. 7.9.

One-parameter models, including the von Mises or Tresca type of failure surface for ductile metals, are generally used in early finite-element analyses for concrete under compressive stresses. This type of pressure-independent yield surface corresponds to a pure shear dependence. To account for the limited tensile capacity of concrete, the von Mises or Tresca surface usually is augmented by the *maximum principal stress surface* or *tension cutoff surface*.

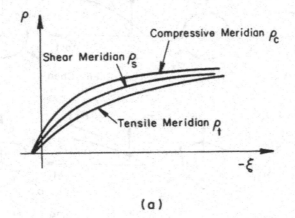

(a)

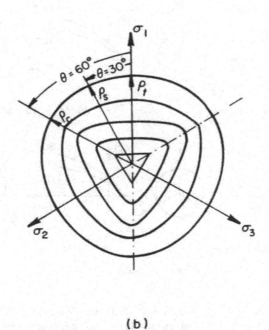

(b)

FIGURE 7.8. Basic features of failure surface: (a) meridians of the failure surface; (b) sections in deviatoric plane.

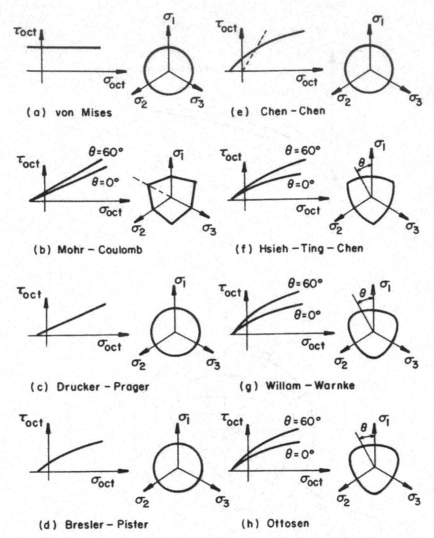

FIGURE 7.9. Failure models.

Among two-parameter models, the Drucker-Prager and Mohr-Coulomb surfaces are probably the simplest types of pressure-dependent failure criteria. In these models, the pure shear or octahedral shear $\tau_{oct}$ depends linearly on the hydrostatic stress $\sigma_m$ or octahedral normal stress $\sigma_{oct}$. The Mohr-Coulomb surface has frequently been used as the failure surface for concrete, while the Drucker-Prager surface has most frequently been used for soils. The Drucker-Prager surface has two basic shortcomings in connection with concrete modeling: the linear relationship between $\tau_{oct}$ and $\sigma_{oct}$ and the independence of the angle of similarity $\theta$ (see Fig. 7.9c). As mentioned above, the $\tau_{oct}$-$\sigma_{oct}$ relation has been experimentally shown to be curved, and the trace of the failure surface on deviatoric sections is not circular. Two-parameter models with straight lines as meridians are therefore inadequate for describing the failure of concrete in the high-compression range.

The generalized Drucker–Prager surface proposed by Bresler and Pister (1958) is a three-parameter model which assumes a parabolic dependence of $\tau_{oct}$ on $\sigma_{oct}$, while the deviatoric sections are independent of $\theta$. On the other hand, the early version of the three-parameter surface developed by Willam and Warnke retains the linear $\tau_{oct}$-$\sigma_{oct}$ relation, but deviatoric sections exhibit $\theta$-dependence. The four-parameter models of Ottosen (1977) and Hsieh et al. (1982) and the refined five-parameter model of Willam and Warnke (1975) have both a parabolic $\tau_{oct}$-$\sigma_{oct}$ relation and $\theta$-dependence. These refined models (Fig. 7.9f, g, h) reproduce all the important features of the triaxial failure surface and give a close estimate of the relevant experimental data. The simple one- and two-parameter models have been discussed in Chapter 2. The refined models will be introduced in the following sections.

## 7.2.2. Ottosen Four-Parameter Model

To meet the geometric requirements of the failure surface for concrete materials, Ottosen (1977) suggested the following criterion involving all three stress invariants $I_1$, $J_2$, and $\theta$:

$$f(I_1, J_2, \theta) = aJ_2 + \lambda \sqrt{J_2} + bI_1 - 1 = 0 \qquad (7.2a)$$

where $\lambda$ is a function of $\cos 3\theta$:

$$\lambda = \begin{cases} k_1 \cos\left[\dfrac{1}{3} \cos^{-1}(k_2 \cos 3\theta)\right] & \text{for } \cos 3\theta \geq 0 \\[3mm] k_1 \cos\left[\dfrac{\pi}{3} - \dfrac{1}{3} \cos^{-1}(-k_2 \cos 3\theta)\right] & \text{for } \cos 3\theta \leq 0 \end{cases} \qquad (7.2b)$$

In Eq. (7.2a, b), $a$, $b$, $k_1$, and $k_2$ are constants. For convenience in later discussions, we shall assume all the stresses and stress invariants appearing in a failure criterion to be normalized by $f'_c$, the uniaxial compressive strength of concrete, e.g., $I_1$, $J_2$ in Eq. (7.2a) represent $I_1/f'_c$ and $J_2/f'^2_c$, respectively.

Equations (7.2a, b) define a failure surface with curved meridians and noncircular cross sections on the deviatoric planes. The meridians described by Eq. (7.2a) are quadratic parabolas which are convex if $a > 0$ and $b > 0$. The cross sections have the geometric properties of symmetry and convexity, and have changing shapes from nearly triangular to nearly circular with increasing hydrostatic pressure. The model encompasses several earlier models as special cases, e.g., the von Mises model for $a = b = 0$ and $\lambda = $ constant, and the Drucker–Prager model for $a = 0$ and $\lambda = $ constant.

The four parameters in the failure criterion may be determined on the basis of the following two typical uniaxial concrete tests ($f'_c$ and $f'_t$) and two typical biaxial and triaxial data:

1. Uniaxial compressive strength $f'_c$ ($\theta = 60°$).
2. Uniaxial tensile strength $f'_t$ ($\theta = 0°$).
3. Biaxial compressive strength ($\theta = 0°$). In particular, we choose $\sigma_1 = \sigma_2 = -1.16f'_c$, $\sigma_3 = 0$, corresponding to the tests of Kupfer et al. (1969); that is $f'_{bc} = 1.16f'_c$.
4. The triaxial stress state $(\xi/f'_c, \rho/f'_c) = (-5, 4)$ on the compressive meridian ($\theta = 60°$) which gives the best fit to the test results of Balmer (1949) and Richart et al. (1928).

The values obtained for the parameters from these data and their dependence on $\bar{f}'_t = f'_t/f'_c$ are presented in Table 7.1.

The failure stresses estimated by the Ottosen criterion with the parameters of Table 7.1, $\bar{f}'_t = 0.1$, will now be compared with experimental tests. Figure 7.10 shows the comparison of the failure criterion with triaxial data in meridian planes, indicating that the best fit to the relevant part ($\xi/f'_c > -5$) of the compressive meridian as determined by Balmer (1949) and Richart et al. (1928) is achieved by the chosen failure stress point $(\xi/f'_c, \rho/f'_c) = (-5, 4)$, as should be the case. For $\xi/f'_c < -5$, the criterion seems to give a conservative estimate. It should be noted that if another region is considered more essential for a good fit, another point on the compressive meridian should be chosen in order to have a closer fit for that region. Along the tensile meridian, the criterion is seen to give a good fit with the biaxial strength point marked $f'_{bc}$ on the curve, which was used as a fitting point corresponding to the tests of Kupfer et al. (1969). The intersection point for the meridians with the $\xi/f'_c$ axis (corresponding to hydrostatic tension) is found in the range 0.14 to 0.22, depending on the $f'_t/f'_c$ value, and it is obvious that a relatively large uncertainty in this value influences the tensile meridian very little.

The ability of the Ottosen criterion to represent the experimental biaxial results of Kupfer et al. (1969) is shown in Fig. 7.11. The agreement is considered satisfactory, the largest difference occurring in compression when $\sigma_1/\sigma_2 \approx 0.5$ (shear meridian $\theta = 30°$), where Kupfer et al. reported $1.27f'_c$ as the mean value of tests. The failure criterion with the parameters of Table 7.1 gives $1.35f'_c$ ($\bar{f}'_t = 0.08$), $1.38f'_c$ ($\bar{f}'_t = 0.1$), and $1.41f'_c$ ($\bar{f}'_t = 0.12$).

TABLE 7.1. Parameter values and their dependence on $\bar{f}'_t = f'_t/f'_c$.

| $\bar{f}'_t = f'_t/f'_c$ | $a$ | $b$ | $k_1$ | $k_2$ |
|---|---|---|---|---|
| 0.08 | 1.8076 | 4.0962 | 14.4863 | 0.9914 |
| 0.10 | 1.2759 | 3.1962 | 11.7365 | 0.9801 |
| 0.12 | 0.9218 | 2.5969 | 9.9110 | 0.9647 |

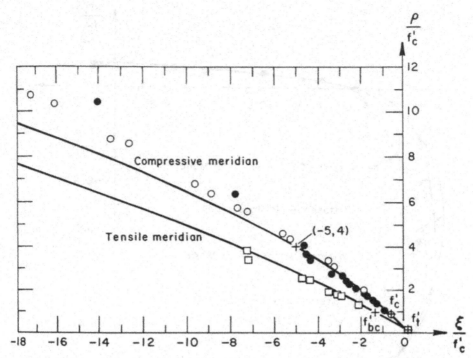

FIGURE 7.10. Comparison of Ottosen criterion with triaxial data in meridian planes: open circles, Balmer (1949), compressive; solid circles, Richart et al. (1928), compressive; squares, Richart et al. (1928), tensile; crosses, Kupfer et al. (1969), tensile.

In general, the four-parameter failure criterion is valid for a wide range of stress combinations. It is in a mathematical form suitable for computer applications. However, the expression for the λ-function is quite involved. Hsieh et al. (1982) proposed a simpler form which can also fit the experimental data very well.

### 7.2.3. Hsieh–Ting–Chen Four-Parameter Model

Hsieh et al. proposed a λ-function with the simple form $\lambda(\theta) = b \cos \theta + c$ for $|\theta| \le 60°$, where $b$ are $c$ are constants. Replacing $\lambda$ in Eq. (7.2b) by this expression and using Haigh–Westergaard coordinates yields a failure function of the following form:

$$f(\xi, \rho, \theta) = a\rho^2 + (b \cos \theta + c)\rho + d\xi - 1 = 0 \qquad (7.3)$$

in which $a$, $b$, $c$, and $d$ are material constants. Noting that $\rho \cos \theta = (\sqrt{3/2}\sigma_1 - I_1/\sqrt{6})$ [see Eq. (2.123)], we can rewrite Eq. (7.3) in terms of the stress invariants $I_1$, $J_2$, $J_3$ with the four new material constants $A$, $B$, $C$, $D$ as

$$AJ_2 + B\sqrt{J_2} + C\sigma_1 + DI_1 - 1 = 0 \qquad (7.4)$$

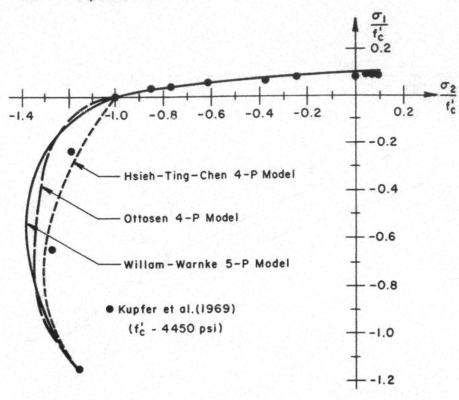

FIGURE 7.11. Comparison of failure criteria with biaxial tests ($f'_c = 30.7$ MPa).

Note that all stresses in the failure criterion expressions have been normalized by $f'_c$. Hence, $f'_c$ does not appear explicitly in Eqs. (7.3) and (7.4). Further, it is interesting to note that the functional form of Eq. (7.4) appears to be a linear combination of three well-known failure criteria, namely, the von Mises, the Drucker–Prager, and the Rankine criteria (see Chapter 2).

To determine the four material parameters, $A$, $B$, $C$, and $D$, use was made of the biaxial tests of Kupfer et al. (1969) and of the triaxial tests of Mills and Zimmerman (1970). The parameters are determined from the following four failure states:

1. Uniaxial compressive strength $f'_c$.
2. Uniaxial tensile strength $f'_t = 0.1f'_c$.
3. Equally biaxial compressive strength $f'_{bc} = 1.15f'_c$.
4. The stress state $(\sigma_{oct}/f'_c, \tau_{oct}/f'_c) = (-1.95, 1.6)$ on the compressive meridian ($\theta = 60°$) which seems to give the best fit to the test results of Mills and Zimmerman.

The values of the four constants $A$, $B$, $C$, $D$ obtained were:

$$A = 2.0108, \qquad B = 0.9714, \qquad C = 9.1412, \qquad D = 0.2312$$

The relations between the two sets of material constants $(a, b, c, d)$ and $(A, B, C, D)$ can be obtained by comparing Eqs. (7.3) and (7.4) and using some formulas given in the Appendix, yielding

$$a = A/2 = 1.0054 \qquad b = \sqrt{2/3}\,C = 7.4638$$
$$c = B/\sqrt{2} = 0.6869 \qquad d = \sqrt{3}\,(D + b/\sqrt{6}) = 5.678$$

In Fig. 7.12, the test results of Mills and Zimmerman are shown in the $\sigma_{oct}/f'_c$, $\tau_{oct}/f'_c$ coordinate system. Only the compressive $(\theta = 60°)$ and the tensile $(\theta = 0°)$ meridians are shown; $\sigma_{oct}$ and $\tau_{oct}$ have been normalized by the uniaxial compressive strength $f'_c > 0$. The point $(\sigma_{oct}/f'_c, \tau_{oct}/f'_c) = (-1.95, 1.6)$ used to determine the parameters of the failure criterion is also marked on the compressive meridian. The agreement is considered satisfactory. The failure criterion and the experiments of Launay and Gachon (1970) are compared in the deviatoric plane in Fig. 7.13. Close agreement can be observed in the low-pressure regime for $I_1/f'_c = 3\sigma_{oct}/f'_c = -1$ and $-3$. In the high-compression regime, $I_1/f'_c \leq -5$, the criterion seems to give a conservative estimate. This is expected, since the tests of Launay and Gachon are known to give a very high biaxial compressive strength $(f'_{bc} = 1.8f'_c)$. Figure 7.11 shows the biaxial failure envelope of the four-parameter model. The comparison with test data from Kupfer et al. (1969) indicates a close agreement.

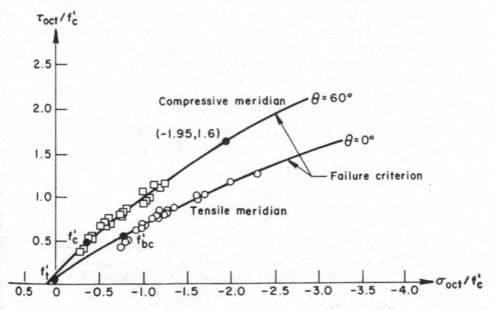

FIGURE 7.12. Comparison of Hsieh-Ting-Chen criterion with test results of Mills and Zimmerman (1970) in octahedral shear and normal-stress plane. Tests (open circles for $\theta = 0°$, open squares for $\theta = 60°$), determination of parameter (dots for the four failure states used).

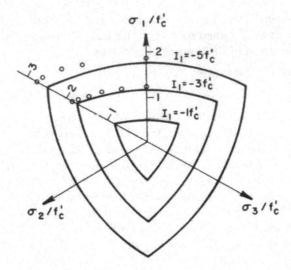

FIGURE 7.13. Comparison of Hsieh–Ting–Chen criterion with triaxial data in deviatoric plane; open circles, Launay and Gachon (1970).

Although the four-parameter criterion satisfies the convexity requirement for all stress conditions, it still has edges along compressive meridians (Fig. 7.13) where continuous derivatives along the edges do not exist. Continuous derivatives used in a general constitutive relation would result in a better convergence during iteration in a numerical analysis. Thus, smoothness everywhere of the yield surface is a desirable property.

### 7.2.4. Willam–Warnke Five-Parameter Model

The Willam–Warnke five-parameter model is illustrated in Fig. 7.14. This model has curved tensile and compressive meridians expressed by quadratic parabolas of the form

$$\sigma_m = a_0 + a_1\rho_t + a_2\rho_t^2 \tag{7.5a}$$

$$\sigma_m = b_0 + b_1\rho_c + b_2\rho_c^2 \tag{7.5b}$$

where $\sigma_m = I_1/3$ is the mean stress, $\rho_t$ and $\rho_c$ are the stress components perpendicular to the hydrostatic axis at $\theta = 0°$ and $\theta = 60°$, respectively, and $a_0, a_1, a_2, b_0, b_1,$ and $b_2$ are material constants. All the stresses have been normalized by $f'_c$, i.e., $\sigma_m, \rho_t,$ and $\rho_c$ in Eqs. (7.5a, b) represent $\sigma_m/f'_c, \rho_t/f'_c,$ and $\rho_c/f'_c$, respectively.

Since the two meridians must intersect the hydrostatic axis at the same point, it follows that

$$a_0 = b_0 \tag{7.6}$$

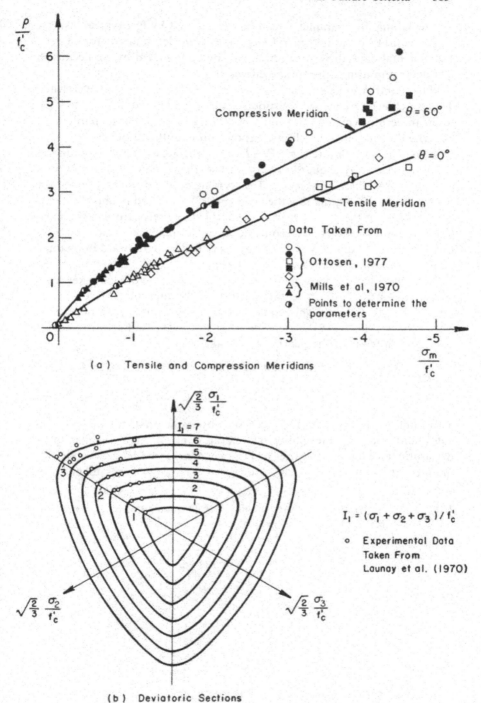

(a)  Tensile and Compression Meridians

(b)  Deviatoric Sections

FIGURE 7.14. Willam–Warnke five-parameter model.

The remaining five parameters can be determined by five typical tests (see Table 7.2). Once the two meridians have been determined from a set of experimental data, the cross section can be constructed by connecting the meridians and using appropriate curves.

Willam and Warnke's failure curves are convex and smooth everywhere. This is achieved by using a portion of an elliptic curve (Fig. 7.15a). Due to the threefold symmetry, it is only necessary to consider the part $0° \le \theta \le 60°$. The derivation of the elliptic expression is outlined below.

In Fig. 7.15b, the failure curve $P_1 - P - P_2$, with $(\rho, \theta)$ as polar coordinates, is approximated by a quarter of an ellipse $P_1 - P - P_2 - P_3$ with $(x, y)$ as principal axes (half-axes $a$, $b$). The symmetry conditions at $\theta = 0°$ and $\theta = 60°$ (Fig. 7.15) require that the position vectors $\rho_t$ and $\rho_c$ must be normal to the ellipse at the points $P_1(0, b)$ and $P_2(m, n)$, respectively. The minor $y$-axis is therefore chosen to coincide with the position vector $\rho_t$ so that the normality condition at $P_1$ is always satisfied. The outward normal unit vector to the ellipse at $P_2(m, n)$ must form an angle of 30° with the $x$-axis.

Using the conditions given above, we can determine the half-axes $a$ and $b$ in terms of the position vectors $\rho_t$ and $\rho_c$ and then obtain the standard equation of the ellipse. This equation is finally transformed in terms of polar coordinates $(\rho, \theta)$. After some algebra, the radius $\rho(\theta)$ can be expressed in terms of the parameters $\rho_t$ and $\rho_c$:

$$\rho(\theta) = \frac{2\rho_c(\rho_c^2 - \rho_t^2) \cos \theta + \rho_c(2\rho_t - \rho_c)[4(\rho_c^2 - \rho_t^2) \cos^2 \theta + 5\rho_t^2 - 4\rho_t\rho_c]^{1/2}}{4(\rho_c^2 - \rho_t^2) \cos^2 \theta + (\rho_c - 2\rho_t)^2}$$

(7.7)

Two limiting cases of Eq. (7.7) can be observed. First, for $\rho_t/\rho_c = 1$ (or, equivalently, $a = b$), the ellipse degenerates into a circle (similar to the deviatoric trace of the von Mises or Drucker–Prager models). Second, when the ratio $\rho_t/\rho_c$ approaches the value $\frac{1}{2}$ (or $a/b$ approaches infinity), the

TABLE 7.2. Test points to determine the material constants (five-parameter model).

| Test | Failure point | | |
| --- | --- | --- | --- |
|  | $\sigma_m$ | $\rho$ | $\theta$ |
| Uniaxial tension | $\frac{1}{3}f_t'$ | $\frac{\sqrt{2}}{\sqrt{3}}f_t'$ | 0° |
| Uniaxial compression | $-\frac{1}{3}f_c'$ | $\frac{\sqrt{2}}{\sqrt{3}}f_c'$ | 60° |
| Biaxial compression | $-\frac{2}{3}f_{bc}'$ | $\frac{\sqrt{2}}{\sqrt{3}}f_{bc}'$ | 0° |
| Confined biaxial compression with $(\sigma_1 > \sigma_2 = \sigma_3)$ | $\sigma_{mt}$ | $\rho_t$ | 0° |
| Confined biaxial compression with $(\sigma_1 = \sigma_2 > \sigma_3)$ | $\sigma_{mc}$ | $\rho_c$ | 60° |

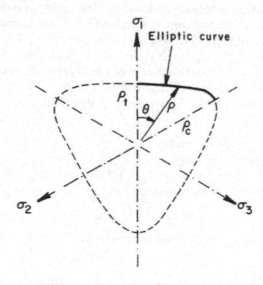

(a) Deviatoric Section

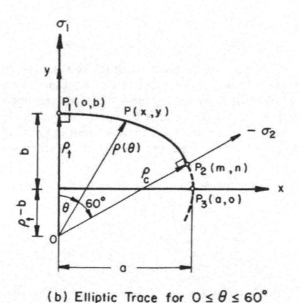

(b) Elliptic Trace for $0 \leq \theta \leq 60°$

FIGURE 7.15. Trace of the deviatoric section of the Willam–Warnke five-parameter failure surface.

deviatoric trace becomes nearly triangular (similar to that of Fig. 2.20c for the maximum-tensile-stress criterion). The triangular deviatoric curve corresponding to $\rho_t/\rho_c = \frac{1}{2}$ has corners at the compressive meridians. Therefore, both convexity and smoothness of the failure curve (Fig. 7.15) can be assured for the ratio $\rho_t/\rho_c$ in the range $\frac{1}{2} < \rho_t/\rho_c \leq 1$. Equations (7.5a, b) and (7.7) completely define the failure criterion of the Willam–Warnke five-parameter model. Based on Kupfer's biaxial tests and other triaxial tests, the five parameters of the failure function are now determined by the following five failure states:

1. Uniaxial compressive strength $f'_c$.
2. Uniaxial tensile strength $f'_t = 0.1f'_c$.
3. Biaxial compressive strength $f'_{bc} = 1.15f'_c$.
4. Confined biaxial compression strength with $\sigma_1 > \sigma_2 = \sigma_3$:

$$(\sigma_{mt}, \rho_t) = (-1.95f'_c, 2.77f'_c)$$

5. Confined biaxial compression strength with $\sigma_1 = \sigma_2 > \sigma_3$:

$$(\sigma_{mc}, \rho_c) = (-3.9f'_c, 3.461f'_c)$$

The five failure points expressed in terms of coordinates $(\sigma_m/f'_c, \rho/f'_c, \theta)$ are listed in Table 7.3. The values of the constants $a_0$, $a_1$, $a_2$, $b_1$, and $b_2$ obtained are

$$a_0 = 0.1025, \qquad a_1 = -0.8403, \qquad a_2 = -0.0910$$

$$b_1 = -0.4507, \qquad b_2 = -0.1018$$

Comparison of the model predictions with the test data is shown in Fig. 7.14a, b, in which the five test points used to determine the material constants are indicated by half-open circles. As can be seen, the agreement is satisfactory along both the meridians and the deviatoric sections.

The three models discussed above are compared in Fig. 7.11, where the two-dimensional failure envelopes in the $\sigma_1$–$\sigma_2$ plane are plotted. It can be

TABLE 7.3. Determination of material constants (five-parameter model).[a]

| Failure point | | | Material constants | |
|---|---|---|---|---|
| $\sigma_m$ | $\rho$ | $\theta$ | | |
| −0.3333 | 0.81650 | 60° | $a_0$ | 0.10250 |
| −1.93330* | 2.77600 | 60° | $u_1$ | −0.84030 |
| 0.03333 | 0.081650 | 0° | $a_2$ | −0.09100 |
| −0.76700 | 0.93900 | 0° | $b_1$ | −0.45070 |
| −3.90000* | 3.46100 | 0° | $b_2$ | −0.10180 |

[a] Data taken from Kupfer et al. (1969) except* taken from other triaxial tests.

seen that the three curves match each other quite well except in a region around the shear meridian where the smooth Willam–Warnke five-parameter model and the Ottosen four-parameter model predict a higher failure load than the Hsieh–Ting–Chen four-parameter model, which gives lower and safer predictions. Considering the scatter of the test data, however, these three models are all good in representing the failure surface of concrete.

## 7.2.5. A General Formulation of the Failure Surface

In the following formulation, the failure surface of concrete is generally expressed in the form:

$$f(\rho, \sigma_m, \theta) = \rho - \rho_f(\sigma_m, \theta) = 0 \qquad |\theta| \le 60° \qquad (7.8)$$

where $\rho = \sqrt{(2J_2)}$ is the stress component perpendicular to the hydrostatic axis, and $\rho_f(\sigma_m, \theta)$ defines the failure envelope on deviatoric planes and is given by different expressions in different failure models. Equation (7.8) can also be rewritten as

$$f(\tau_0, \sigma_0, \theta) = \tau_0 - \tau_{0f}(\sigma_0, \theta) = 0 \qquad |\theta| \le 60° \qquad (7.9)$$

where $\tau_0$ and $\sigma_0$ are octahedral shear and normal stresses, respectively, and $\tau_{0f}(\sigma_0, \theta)$ is the failure octahedral shear stress. Noting that $\sigma_0 = \sigma_m$ and $\tau_0 = (1/\sqrt{3})\rho$, it can be seen that Eqs. (7.8) and (7.9) are basically the same.

The expressions for $\rho_f(\sigma_m, \theta)$ in the three failure models considered above can be derived as follows.

(i) *Ottosen four-parameter model*: Solving Eq. (7.2a) for $\sqrt{(2J_2)} = \rho_f$ leads to the following expression:

$$\rho_f(\sigma_m, \theta) = \frac{1}{2a}[-\sqrt{2}\lambda + \sqrt{2\lambda^2 - 8a(3b\sigma_m - 1)}] \qquad (7.10)$$

where $\lambda$ is a function of $\cos 3\theta$ defined by Eq. (7.2b).

(ii) *Hsieh–Ting–Chen four-parameter model*: Similarly, from Eq. (7.3), one gets

$$\rho_f(\sigma_m, \theta) = \frac{1}{2a}[-(b\cos\theta + c) + \sqrt{(b\cos\theta + c)^2 - 4a(\sqrt{3}\,d\sigma_m - 1)}] \qquad (7.11)$$

in which the coordinate $\xi$ has been replaced by $\sqrt{3}\sigma_m$ and $\sigma_m$ is the mean stress.

(iii) *Willam–Warnke five-parameter model*: The right-hand side of Eq. (7.7) may be written as

$$\rho_f(\sigma_m, \theta) = \frac{s+t}{v} \qquad (7.12)$$

where

$$s = s(\sigma_m, \theta) = 2\rho_c(\rho_c^2 - \rho_t^2) \cos \theta \qquad (7.12a)$$

$$t = t(\sigma_m, \theta) = \rho_c(2\rho_t - \rho_c)u^{1/2} \qquad (7.12b)$$

$$u = u(\sigma_m, \theta) = 4(\rho_c^2 - \rho_t^2) \cos^2 \theta + 5\rho_t^2 - 4\rho_t\rho_c \qquad (7.12c)$$

$$v = v(\sigma_m, \theta) = 4(\rho_c^2 - \rho_t^2) \cos^2 \theta + (\rho_c - 2\rho_t)^2 \qquad (7.12d)$$

From Eq. (7.5a, b), $\rho_c$ and $\rho_t$ can be obtained as functions of $\sigma_m$:

$$\rho_c = -\frac{1}{2b_2}[b_1 + \sqrt{b_1^2 - 4b_2(b_0 - \sigma_m)}] \qquad (7.13a)$$

$$\rho_t = -\frac{1}{2a_2}[a_1 + \sqrt{a_1^2 - 4a_2(a_0 - \sigma_m)}] \qquad (7.13b)$$

## 7.3. Plasticity Modeling: Hardening Behavior

One of the better-known plasticity models of concrete is the one proposed by Chen and Chen (1975), which in fact sets up a general framework for this type of modeling. Several other plasticity-based models proposed in the past can be mentioned: these include the models by Chen and Schnobrich (1981), Hsieh, Ting, and Chen (1982), Fardis, Alibe, and Tassoulas (1983), Vermeer and De Borst (1984), Han and Chen (1985), and Chen and Buyukozturk (1985), among others. These models differ from each other in the shape of the failure and loading surfaces, in the hardening rule, and in the flow rule. In what follows, we shall present the formulation of the *nonuniform hardening plasticity* model proposed recently by Han and Chen (1985, 1987) in some detail. This model is used here as an illustrative example to demonstrate the general techniques employed in the plasticity modeling of concrete materials.

The model is outlined and illustrated in the hydrostatic plane in Fig. 7.16. The failure surface encloses all the loading surfaces and serves as a *bounding surface*, which is assumed to remain unchanged during loading. The shapes of the deviatoric sections of the loading surfaces are assumed to be similar to those of the failure surface, but their meridians are different. The initial yield surface has a closed shape. During hardening, the loading surface expands and changes its shape from the initial yield surface to the final shape that matches with the failure surface. Each loading surface is characterized by a hardening parameter $k_0$. A nonassociated flow rule is assumed. Then, the incremental stress–strain relationships can be established according to the classical theory of plasticity.

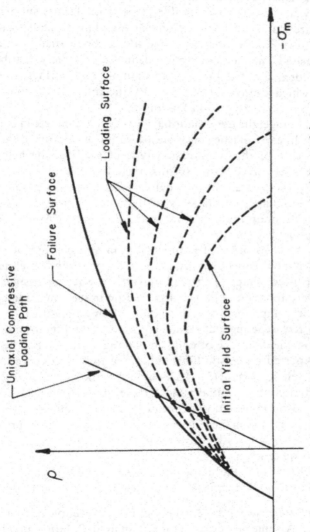

FIGURE 7.16. A nonuniform hardening plasticity model.

## 7.3.1. Yield Criterion

Since the yield stress of concrete is not easy to measure experimentally, the yield criterion of concrete is usually assumed on the basis of the known failure criterion. For example, in some early plasticity models, the yield stress was taken as a proportionally reduced value of the failure stress. This implies that the yield surface has a similar shape to the failure surface but with a reduced size. Later, this assumption was found to be inadequate for concrete: first, the yield surface so assumed has an open shape, which is unreasonable. Second, such an assumption defines a uniformly distributed elastoplastic (hardening) zone between the initial yield surface and the failure surface, which cannot reflect correctly the different responses of concrete to tension and compression loadings.

In the prefailure range, the deformational behavior of concrete in tension loading is almost linearly elastic: only elastic strain occurs up to failure. However, in the case of the compression type of loading, the behavior becomes nonlinear: relatively large strains, including reversible (elastic) and irreversible (plastic) strains, occur before failure. This is particularly true in the cases of compressive loading with confining pressure. In such cases, the irreversible strain could be quite large, and concrete would exhibit to some extent a ductile behavior.

Since concrete behaves differently under tension and compression, the initial yield surface should not be simply assumed to have a similar, although proportionally reduced, shape to the failure surface. Such an assumption can lead to an overestimation of plastic deformation in tension loading and underestimation in compression loading with confining pressure.

There are very few experimental results reported in the literature on the shape of the initial yield surface of concrete. Launay and Gachon (1972), among others, reported an elastic limit and crack initiation curve in the hydrostatic stress plane (Fig. 7.17). This curve may be considered a first qualitative description of the initial yield surface of concrete materials. It reveals that the elastic limit almost coincides with the failure curve, and the hardening zone vanishes in the tensile and very-low-hydrostatic-pressure region, whereas in the compressive region with high confining pressure, the hardening zone is quite large.

Based on this observation, the shape of the meridians of the yield surface may be assumed to be as shown in Fig. 7.18, which consists of four parts:

1. In the tension zone, i.e., $\sigma_m \geq \xi_t$, the yield surface coincides with the failure surface. Assume no plastic deformation up to failure, representing brittle behavior.
2. In the compression–tension mixed zone, i.e., $\xi_t > \sigma_m \geq \xi_c$, a plastic-hardening zone gradually evolves.
3. In the compression zone with low confining pressure, i.e., $\xi_c > \sigma_m \geq \xi_k$, the meridian represents a proportionally reduced size of the failure surface.

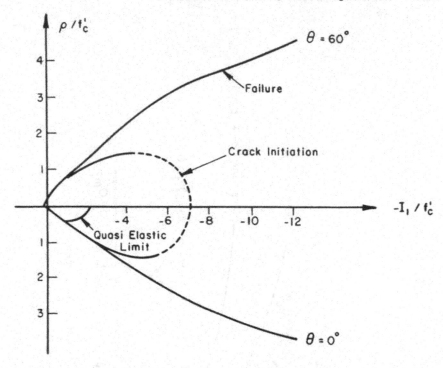

FIGURE 7.17. Experimentally obtained failure and crack initiation curves (Launay and Gachon, 1972).

4. In the compression zone with a relatively high confining pressure, i.e., $\sigma_m < \xi_k$, the yield surface gradually closes up at the hydrostatic axis and a wide plastic-hardening region is generated.

To identify the zones as tension–tension, tension–compression, compression–tension, or compression–compression, the following triaxial zoning criterion may be used (Fig. 7.18):

Tension–tension:

$$\sqrt{J_2} - \frac{1}{\sqrt{3}} I_1 > 0$$

Tension–compression:

$$\sqrt{J_2} - \frac{1}{\sqrt{3}} I_1 \leq 0 \quad \text{and} \quad I_1 \geq 0$$

Compression–tension:

$$\sqrt{J_2} + \frac{1}{\sqrt{3}} I_1 \geq 0 \quad \text{and} \quad I_1 < 0$$

Compression–compression:    $\sqrt{J_2} + \frac{1}{\sqrt{3}} I_1 < 0$

Note that the surface $\sqrt{J_2} + I_1/\sqrt{3} = 0$ passes through the uniaxial compression state, while the surface $\sqrt{J_2} - I_1/\sqrt{3} = 0$ passes through the uniaxial

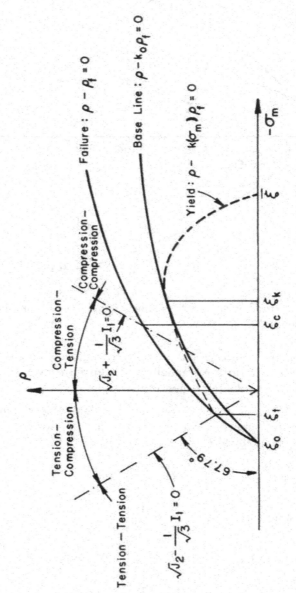

FIGURE 7.18. Construction of the yield surface.

tension state. This is why these two surfaces are used as the boundary surfaces.

In choosing the values of parameters $\xi_t$, $\xi_c$, and $\xi_k$, it is not necessary to follow exactly the above zoning criterion. For simplicity, we may take $\xi_t = 0$, $\xi_c = \xi_k = -f'_c/3$.

With the failure function defined by Eq. (7.8), the initial yield function can be formulated by introducing a *shape factor k* into the failure function, yielding an expression of the form

$$\rho - k\rho_f = 0 \qquad (7.14)$$

The shape factor $k$ is a function of the hydrostatic stress, $\sigma_m$, which modifies the failure surface so as to give a proper shape for the initial yield surface.

## 7.3.2. Hardening Rule and Subsequent Yield Surface

As can be seen from Fig. 7.16, the initial yield surface so defined has a different shape from that of the failure surface, and during hardening, both the size and the shape of the subsequent yield surfaces must vary continuously from the initial yield shape to the final failure shape. Such a hardening rule is nonuniform with respect to the hydrostatic stress. This is quite different from the rule of isotropic hardening.

According to the nonuniform hardening rule, the subsequent yield surface can be expressed in a form similar to Eq. (7.14) as

$$f = \rho - k(k_0, \sigma_m)\rho_f(\sigma_m, \theta) = 0 \qquad (7.15)$$

in which the shape factor $k$ is defined as a function of $\sigma_m$ as well as a hardening parameter $k_0$. A proposed functional form of $k(k_0, \sigma_m)$ is given in the Appendix.

The parameter $k_0$ indicates the hardening level, which can take a value between $k_y$ and 1, i.e.,

$$k_y \le k_0 \le 1$$

such that $k_0 = k_y$ corresponds to the initial yield surface, while $k_0 = 1$ indicates that the ultimate stress state has been reached and that the loading surface has finally met the failure surface. For this reason, we let

$$\bar{\xi} = A/(1 - k_0) \qquad (7.16)$$

where $\bar{\xi}$ is the intersection of the loading surface with the hydrostatic axis and $A$ is a constant. According to Eq. (7.16), the intersection point $\bar{\xi}$ would approach infinity as $k_0 \to 1$.

The hardening parameter $k_0$ is related to a plastic modulus $H_b^p$, which is the modulus defined from the uniaxial compressive stress–plastic strain curve. The relationship between $k_0$ and $H_b^p$ can be found by determining the intersection of the uniaxial compression loading path with the loading surface (see Fig. 7.16). The intersection point gives the uniaxial compressive stress corresponding to a given $k_0$, and also a plastic modulus, which is the slope of the experimental uniaxial compressive stress–plastic strain curve at the given stress level. In this way, each loading surface of parameter $k_0$ is related to a base plastic modulus $H_b^p$ explicitly and also to the plastic work $W_p$ implicitly.

The plastic modulus $H_b^p$ obtained from the experimental uniaxial compression test is called the *base plastic modulus*, distinct from the *plastic modulus* $H^p$ used in the constitutive equation. Since $H_b^p$ is determined from a uniaxial compression stress state, it is only suitable for cases in which the mean stress $\sigma_m$ is about $(-1/3)f_c'$. In cases of compression with high confining pressure, concrete becomes more ductile, so the plastic modulus must be modified. To account for the hydrostatic pressure sensitivity and lode angle dependence, a *modification factor* $M(\sigma_m, \theta)$ may be introduced, and the plastic modulus $H^p$ can be generally expressed as

$$H^p = M(\sigma_m, \theta)H_b^p \qquad (7.17)$$

where $\sigma_m$ is the mean stress and $\theta$ the lode angle. A proposed functional form of $M(\sigma_m, \theta)$ is also given in the Appendix.

### 7.3.3. Nonassociated Flow Rule and Dilatancy Factor

The nonlinear volume change during hardening is a prominent feature of concrete materials. Experimental results indicate that under compressive loadings, inelastic volume contraction occurs at the beginning of yielding and volume dilatation occurs at about 75 to 90% of the ultimate stress. Inflection points are usually observed (see Fig. 7.4). This sort of behavior generally violates the associated flow rule. A plastic potential other than the loading function is therefore needed to define the flow rule. For simplicity, we may choose a functional form of the Drucker–Prager type:

$$g = \alpha I_1 + \sqrt{J_2} \qquad (7.18)$$

Then the flow rule becomes

$$d\epsilon_{ij}^p = d\lambda \frac{\partial g}{\partial \sigma_{ij}} = d\lambda \left( \alpha \delta_{ij} + \frac{s_{ij}}{2\sqrt{J_2}} \right) \qquad (7.19)$$

According to the flow rule, the incremental plastic volume change, $d\epsilon_v^p = d\epsilon_{ii}^p$, is given by $3\alpha\, d\lambda$; thus, $\alpha$ represents a measure of plastic volume dilatation. To reflect the nonlinear volume change, a functional form of $\alpha$ may be defined according to the available experimental data. Herein, we shall assume $\alpha$ to be a linear function of the hardening parameter $k_0$, taking a negative value at the beginning of yielding ($k_0 = k_y$) and a positive value at failure ($k_0 = 1$).

## 7.3.4. Incremental Stress-Strain Relationships

### 7.3.4.1. GENERAL CONSTITUTIVE RELATIONS

Based on the classical theory of plasticity (see Chapters 4 and 5), the incremental stress-strain relation can generally be derived in the following manner. The total strain increment is taken to be the sum of the elastic and plastic increments

$$d\epsilon_{ij} = d\epsilon_{ij}^e + d\epsilon_{ij}^p \tag{7.20}$$

According to Hooke's law, the stress increment is determined by

$$d\sigma_{ij} = C_{ijkl} d\epsilon_{kl}^e = C_{ijkl}(d\epsilon_{kl} - d\epsilon_{kl}^p) \tag{7.21}$$

where

$$C_{ijkl} = G\left(\delta_{ik}\delta_{jl} + \delta_{il}\delta_{jk} + \frac{2\nu}{1-2\nu}\delta_{ij}\delta_{kl}\right) \tag{7.22}$$

is an *isotropic tensor* with elastic constants $G$ and $\nu$. While plastic flow takes place, the *consistency condition*

$$df = 0 \tag{7.23}$$

must hold. In nonuniform hardening, the loading function of Eq. (7.15) is relevant. The total differential of function $f$ is

$$df = \frac{\partial f}{\partial \sigma_{ij}} d\sigma_{ij} + \frac{\partial f}{\partial \tau}\frac{d\tau}{d\epsilon_p} d\epsilon_p = 0 \tag{7.24}$$

where $\tau$ is the *effective stress*. Substituting Eq. (7.21) for $d\sigma_{ij}$ into Eq. (7.24) and noting that

$$\frac{d\tau}{d\epsilon_p} = H^p \tag{7.25}$$

where $H^p$ is the *plastic modulus*, we have

$$df = \frac{\partial f}{\partial \sigma_{ij}} C_{ijkl}(d\epsilon_{kl} - d\epsilon_{kl}^p) + \frac{\partial f}{\partial \tau} H^p d\epsilon_p = 0 \tag{7.26}$$

The effective plastic strain $d\epsilon_p$ can be related to $d\lambda$ as

$$d\epsilon_p = \phi \, d\lambda \tag{7.27}$$

where $\phi$ is a scalar function of stress state (see Section 7.3.5). Using the flow rule (7.19) and the effective plastic strain definition (7.27) in Eq. (7.26), and solving for $d\lambda$, we have

$$d\lambda = \frac{1}{h}\frac{\partial f}{\partial \sigma_{pq}} C_{pqkl} d\epsilon_{kl} \tag{7.28}$$

where

$$h = \frac{\partial f}{\partial \sigma_{mn}} C_{mnpq}\frac{\partial g}{\partial \sigma_{pq}} - H^p \frac{\partial f}{\partial \tau}\phi \tag{7.29}$$

The quantities $\phi$ and $\partial f/\partial \tau$ will be given later in Section 7.3.5. Substitution of Eq. (7.28) into Eq. (7.19) yields the expression for plastic strain increment. Finally, using Eq. (7.21) leads to the constitutive equation

$$d\sigma_{ij} = (C_{ijkl} + C_{ijkl}^p)\, d\epsilon_{kl} \tag{7.30}$$

in which the plastic stiffness tensor $C_{ijkl}^p$ has the form

$$C_{ijkl}^p = -\frac{1}{h}\, H_{ij}^* H_{kl} \tag{7.31}$$

and

$$H_{ij}^* = C_{ijmn}\frac{\partial g}{\partial \sigma_{mn}} \tag{7.32}$$

$$H_{kl} = \frac{\partial f}{\partial \sigma_{pq}}\, C_{pqkl} \tag{7.33}$$

Now, if the loading function $f$ and plastic potential function $g$ are given, the constitutive relation can be directly derived from Eqs. (7.29) through (7.33) in a rather straightforward manner.

### 7.3.4.2. CONSTITUTIVE EQUATIONS BASED ON THE ASSOCIATED FLOW RULE

The associated flow rule assumes that $g = f$, with $f$ given by Eq. (7.15). Its derivatives can generally be expressed by [see Eq. (4.140), Section 4.11]

$$\frac{\partial g}{\partial \sigma_{ij}} = \frac{\partial f}{\partial \sigma_{ij}} = B_0\delta_{ij} + B_1 s_{ij} + B_2 t_{ij} \tag{7.34}$$

where the coefficients $B_0$, $B_1$, and $B_2$ are the derivatives of the loading function $f$ with respect to the stress invariants. Expressions for $B_0$, $B_1$, and $B_2$ for three different forms of failure surfaces are listed in the Appendix. Now, we substitute Eq. (7.34) into Eqs. (7.29), leading to

$$h = 2G\left[3B_0^2\frac{1+\nu}{1-2\nu} + 2B_1^2 J_2 + 6B_1 B_2 J_3 - \frac{2}{3}B_2^2 J_2^2\right] - \phi H^p\frac{\partial f}{\partial \tau} \tag{7.35}$$

In the derivation, we have used the relation $s_{ik}s_{kj}s_{il}s_{lj} = 2J_2^2$.

Also, using Eq. (7.34) in Eqs. (7.32) and (7.33), we have

$$H_{ij} = H_{ij}^* = 2G\left[B_0\frac{1+\nu}{1-2\nu}\delta_{ij} + B_1 s_{ij} + B_2 t_{ij}\right] \tag{7.36}$$

Then, the plastic stiffness tensor has the symmetric form

$$C_{ijkl}^p = -\frac{1}{h}\, H_{ij}H_{kl} \tag{7.37}$$

### 7.3.4.3. CONSTITUTIVE EQUATIONS BASED ON A NONASSOCIATED FLOW RULE

The Drucker–Prager surface, Eq. (7.28), is used here as the plastic potential. We substitute Eqs. (7.19) and (7.34) into Eqs. (7.29) and (7.32), leading to

$$h = 2G\left(3B_0\alpha\frac{1+\nu}{1-2\nu} + B_1\sqrt{J_2} + \frac{3B_2}{2\sqrt{J_2}}J_3\right) - \phi H^p \frac{\partial f}{\partial \tau} \tag{7.38}$$

and

$$H_{ij}^* = 2G\left(\alpha\frac{1+\nu}{1-2\nu}\delta_{ij} + \frac{1}{2\sqrt{J_2}}s_{ij}\right) \tag{7.39}$$

The plastic stiffness tensor $C_{ijkl}^p$ is expressed by Eq. (7.31), in which the tensors $H_{ij}$ and $H_{ij}^*$ are now given by Eqs. (7.36) and Eq. (7.39), respectively. Obviously, the stiffness tensor is *not* symmetric.

To implement the incremental stress–strain relationship in a finite-element program, it is convenient to write Eqs. (7.36) and (7.39) explicitly as

$$H_{xx} = 2G\left[B_0\frac{1+\nu}{1-2\nu} + B_1s_{xx} + B_2\left(s_{xx}^2 + s_{xy}^2 + s_{xz}^2 - \frac{2J_2}{3}\right)\right], \text{ etc.} \tag{7.40}$$

$$H_{yz} = 2G[B_1s_{yz} + B_2(s_{xy}s_{xz} + s_{yy}s_{yz} + s_{yz}s_{zz})], \text{ etc.} \tag{7.41}$$

and

$$H_{xx}^* = 2G\left(\alpha\frac{1+\nu}{1-2\nu} + \frac{1}{2\sqrt{J_2}}s_{xx}\right), \text{ etc.} \tag{7.42}$$

$$H_{yz}^* = \frac{G}{\sqrt{J_2}}s_{yz}, \text{ etc.} \tag{7.43}$$

Further, the stress-strain relation is written as

$$\{d\sigma\} = ([C] + [C]^p)\{d\epsilon\} \tag{7.44}$$

where

$$\{d\sigma\} = (d\sigma_x, d\sigma_y, d\sigma_z, d\tau_{yz}, d\tau_{xz}, d\tau_{xy})^T$$
$$\{d\epsilon\} = (d\epsilon_x, d\epsilon_y, d\epsilon_z, d\gamma_{yz}, d\gamma_{xz}, d\gamma_{xy})^T$$

and

$$[C]^p = -\frac{1}{h}[H^*]^T[H] \tag{7.45}$$

in which

$$[H] = [H_{xx}, H_{yy}, H_{zz}, H_{yz}, H_{xz}, H_{xy}] \tag{7.46}$$
$$[H^*] = [H_{xx}^*, H_{yy}^*, H_{zz}^*, H_{yz}^*, H_{xz}^*, H_{xy}^*] \tag{7.47}$$

## 7.3.5. Effective Stress and Effective Strain

Based on the loading function given by Eq. (7.15), the effective stress $\tau$ and effective strain increment $d\epsilon_p$ are defined for the multiaxial stress state such that the single $\tau$-$\epsilon_p$ curve can be calibrated against a uniaxial compression stress–plastic strain curve.

In the uniaxial compression test, i.e., $(0, 0, -\tau)$, we have $\rho = \sqrt{2/3}\,\tau$, $\sigma_m = -\tau/3$, $\rho_f = \rho_c$, and the loading function reduces to

$$f = \sqrt{2/3}\,\tau - k\rho_c = 0 \tag{7.48}$$

Now we define the effective stress $\tau$ as

$$\tau = \sqrt{3/2}\,\rho_c k \tag{7.49}$$

Then, the corresponding effective plastic strain increment $d\epsilon_p$ can be defined in terms of the plastic work per unit volume in the form

$$dW_p = \tau\, d\epsilon_p \tag{7.50}$$

On the other hand, we have

$$dW_p = \sigma_{ij}\, d\epsilon_{ij}^p = \sigma_{ij}\, d\lambda\, \frac{\partial g}{\partial \sigma_{ij}} \tag{7.51}$$

Thus, $d\epsilon_p$ can be derived from Eqs. (7.50) and (7.51) as

$$d\epsilon_p = \phi\, d\lambda \tag{7.52}$$

where

$$\phi = \frac{1}{\tau}\, \frac{\partial g}{\partial \sigma_{ij}}\, \sigma_{ij} \tag{7.53}$$

For the case of the associated flow rule, we use Eq. (7.34) for $\partial g/\partial \sigma_{ij}$ and Eq. (7.49) for $\tau$ and obtain

$$\phi = \frac{\sqrt{2}(B_0 I_1 + 2B_1 J_2 + 3B_2 J_3)}{\sqrt{3}\,\rho_c k} \tag{7.54}$$

For the case of a nonassociated flow rule, we use Eq. (7.19) for $\partial g/\partial \sigma_{ij}$ and Eq. (7.49) for $\tau$ and obtain

$$\phi = \frac{\sqrt{2}(\alpha I_1 + \sqrt{J_2})}{\sqrt{3}\,\rho_c k} \tag{7.55}$$

To find an expression for the quantity $\partial f/\partial \tau$, we rewrite the consistency equation (7.24) as

$$df = \frac{\partial f}{\partial \sigma_{ij}}\, d\sigma_{ij} + \frac{\partial f}{\partial \tau}\, d\tau = 0 \tag{7.56}$$

Note that in uniaxial compressive loading, the only nonzero stress component is $d\sigma_{33}$, and according to the definition of the effective stress $\tau$, we have $d\tau = -d\sigma_{33}$. Equation (7.56) then leads to

$$\frac{\partial f}{\partial \tau} = \frac{\partial f}{\partial \sigma_{33}} \tag{7.57}$$

Now we differentiate Eq. (7.48), noting that $\tau = -\sigma_{33}$, $\rho = -\sqrt{2/3}\sigma_{33}$, and $\sigma_m = \sigma_{33}/3$, and obtain

$$\frac{\partial f}{\partial \tau} = \frac{\partial f}{\partial \sigma_{33}} = -\left( \frac{\sqrt{2}}{\sqrt{3}} + \frac{k}{3} \frac{d\rho_c}{d\sigma_m} + \frac{\rho_c}{3} \frac{dk}{d\sigma_m} \right) \tag{7.58}$$

## 7.3.6. Parameters and Model Predictions

The nonuniform hardening plasticity model incorporates many aspects of the properties of concrete materials, including brittle failure in tension, ductile behavior in compression, hydrostatic sensitivities, and inelastic volume dilatation. It thus requires some tests to determine the material constants. For the failure behavior, we need four or five strength tests to determine the material constants in Eqs. (7.11) and (7.13). For the hardening behavior, we need a uniaxial compression test to give the stress–strain curve up to failure, and we also need a general knowledge of the deformational behavior of concrete to determine the modification factor $M(\sigma_m, \theta)$. For the plastic flow behavior, we need to define the initial and final values of the dilatancy factor $\alpha$. A specific form of $M(\sigma_m, \theta)$ and some typical values of $\alpha$ have been proposed and incorporated in a computer program, EPM1. It has been found that the modification factor does not change much from one type of concrete to another. Thus, the input parameters consist of only four or five material constants for the failure surface and a uniaxial compressive stress–strain curve.

Since in structural analysis, various stress states could occur, the performance of the proposed model should be examined in some detail before it is applied to any finite-element computer program for a general structural analysis. A good constitutive model should be able to predict reasonably well the response of the material under all possible stress combinations. The program EPM1 has been developed for this purpose. In this program, the Willam–Warnke five-parameter model is used as the failure surface. Details of this development are given in the companion book by Chen and Zhang (1988).

Four sets of test results have been used to compare test data with the model predictions. These include the best-known biaxial tests of Kupfer et al. (1969), the triaxial compression tests of Schickert and Winkler (1977), the biaxial and triaxial compression tests on low-strength concrete (Traina et al., 1983), and the cyclic loading tests of the University of Colorado (Scavuzzo et al., 1983). Typical results are shown in Figs. 7.19 through 7.21. Good agreement is generally observed.

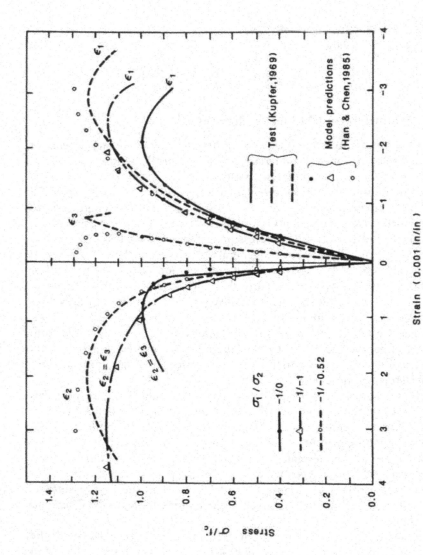

FIGURE 7.19. Comparisons of model predictions with the test data by Kupfer et al (1969) for biaxial compressive loading cases.

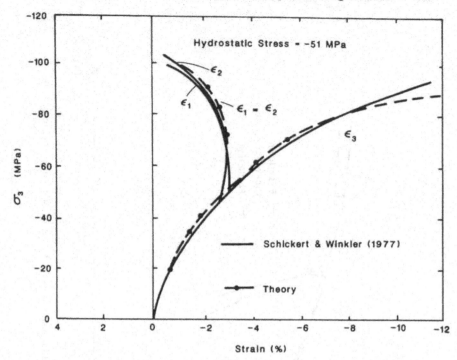

FIGURE 7.20. Comparison of model predictions for triaxial compression loading with data of Schickert and Winkler (1977).

## 7.3.7. Summary

Based on the above discussion, the characteristics of a concrete plasticity model may be summarized as follows.

(i) The initial yield surface should not be simply assumed to be proportionally reduced in shape from the failure surface, because such an assumption generally leads to a uniformly distributed plasticity zone between the yield surface and the failure surface. This can lead to an unreasonable estimation of plastic deformation in the cases of tension and high compression loadings. Such an assumption cannot reflect the properties of concrete, which behaves quite differently in tension and in compression.

(ii) Since the initial yield surface has a different shape than the failure surface, both the size and shape of the subsequent loading surfaces must vary continuously during hardening from the initial yield shape to the final failure shape. This is why the proposed nonuniform hardening rule is relatively complicated. The nonuniform hardening rule is not isotropic.

(iii) To account for the behavior of volumetric compaction/dilatation during plastic deformation, the associated flow rule may not be used

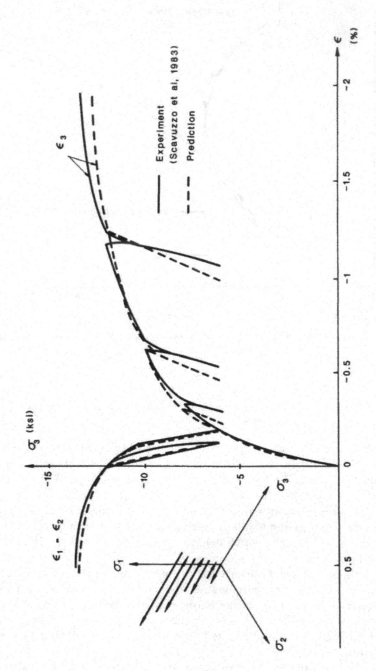

FIGURE 7.21. Comparison of model predictions and experimental data for cyclic loading in deviatoric plane $\sigma_m = 6$ ksi along compressive meridian: $\sigma_1 = \sigma_2 > \sigma_3$ ($\theta = 60°$).

because it can lead to an exaggerative volume expansion for concrete materials. A nonassociated flow rule with a variable dilatancy factor is therefore suggested in the present formulation of the constitutive stress–strain relation of concrete.

# 7.4. Plasticity Modeling: Softening Behavior

As discussed in Section 7.1, axial compression tests on concrete specimens exhibit, in general, the softening behavior of the material in the postfailure regime. The strain-softening behavior, i.e., the negative slope of the load–deformation curve, will be considered in the following as a material property and will be treated by the strain–space plasticity formulation. Before we do this, we shall first examine some material behaviors shown in Fig. 7.22.

## 7.4.1. Types of Material Behaviors

### 7.4.1.1. ELASTIC-PLASTIC SOLIDS

Figure 7.22a shows a stress–strain diagram of a hardening–softening solid, in which the unloading–reloading lines follow straight lines that are parallel to the initial tangent of the stress–strain curve, i.e., the slope of the unloading–reloading line does not change with plastic deformation. This is a typical behavior of an *elastic–plastic solid.*

### 7.4.1.2. PROGRESSIVELY FRACTURING SOLIDS

The behavior described in the previous section is not the case for many engineering materials such as concrete. For example, the elastic modulus or the stiffness usually decreases with increasing straining. This sort of behavior is considered to be due to microcracking or fracturing. Thus, on the other extreme, an ideal material model, called a *progressively fracturing solid* and shown in Fig. 7.22b, was proposed by Dougill (1975). This ideal material is perfectly elastic. Upon unloading, the material returns to its initial stress- and strain-free state; no permanent (plastic) strain occurs.

Since the stiffness degradation behavior is due mainly to fracturing (microcracking), which is different from slip, it cannot be satisfactorily interpreted within the framework of plasticity. Recognizing the difference between fracturing and plastic flow, Dougill (1975, 1976) proposed a theory called *fracturing theory.* This idea is further realized in the development of the more recent so-called *damage theory.*

### 7.4.1.3. PLASTIC-FRACTURING SOLIDS

A material exhibiting both plastic deformation and stiffness degradation behaviors is shown in Fig. 7.22c. Concretes fall into this category, particularly in their softening range. To account for both behaviors, a combined

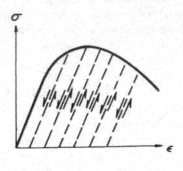

(a) Elasto-Plastic Solid

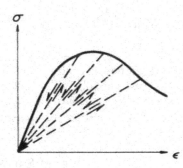

(b) Progressively Fracturing Solid

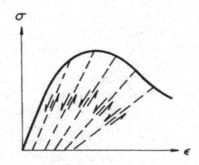

(c) Plastic-Fracturing Solid

FIGURE 7.22. Typical material behaviors.

theory called plastic-fracturing theory was proposed by Bazant and Kim (1979). In this theory, plastic deformation is defined by the flow theory of plasticity in a traditional manner, while the stiffness degradation is modeled by the fracturing theory of Dougill. This approach encounters some difficulties in the definition of the loading criterion, because it involves two loading surfaces—the yield surface specified in stress space and the fracturing surface specified in strain space. To avoid this problem, a strain–space plasticity approach can be used in formulating the plastic-fracturing behavior (Han and Chen, 1986). This later formulation presents a consistent form of the constitutive equations for an elastic–plastic material with stiffness degradation in the ranges of work hardening as well as strain softening.

### 7.4.1.4. Strain Softening and Strain-Space Formulation

The one-dimensional softening behavior shown in Fig. 7.23 is now generalized to a multiaxial state of stress and strain in a similar manner as for hardening behavior. We first discuss the softening behavior in stress space. This will then lead up to a discussion of strain-space formulation. In a stress-space formulation, a state of stress is represented by a point in stress space (Fig. 7.24a). If the state $A$ is on the loading surface $f = 0$ but the material is still in the range of work hardening, a stress increment $d\sigma$ must point outward in order to produce a plastic as well as an elastic increment of strain. A stress increment pointing inward would cause elastic strain only. The outward motion of the stress point $A$ carrying with it the yield surface corresponds to a hardening or ascending branch of the stress–strain curve for increasing stress in the one-dimensional case. On the other hand, if the material is in the range of strain softening, plastic deformation causes the yield surface to contract or move inward at the current stress point $C$

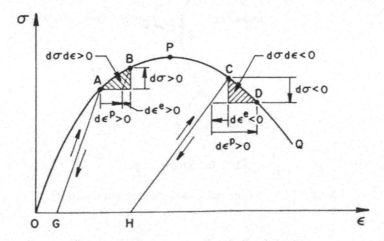

FIGURE 7.23. Features of softening behavior.

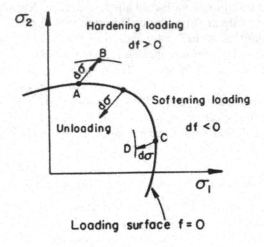

### (a) In Stress Space

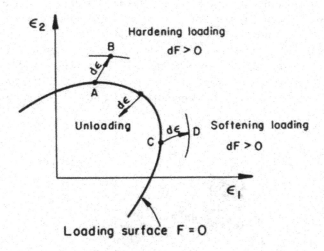

### (b) In Strain Space

FIGURE 7.24. Loading surfaces defined in stress and strain space.

(Fig. 7.24a). This inward motion corresponds to a softening or descending branch of the stress-strain curve for increasing strain in the one-dimensional case. For elastic unloading, too, the stress increment $d\sigma$ points inward from the loading surface. Hence, the stress-space formulation presents difficulties in distinguishing between a reduction of stress which causes additional plastic deformation and one due to elastic unloading.

Referring to point $A$ and point $C$ in Fig. 7.23, however, the strain increment $d\epsilon$ is always positive for a plastic loading and negative for an elastic unloading along either path $AG$ or path $CH$. A generalization to the multidimensional case is shown in Fig. 7.24b, where the loading surface $F = 0$ is a function of strains. For any strain point on the loading surface ($A$ or $C$, for example), a strain increment $d\epsilon$ pointing outward represents a plastic loading case and one pointing inward represents an elastic unloading case. There is no ambiguity. It is clear that if strains are used as independent variables in formulating the plasticity constitutive relation, hardening and softening behavior can be studied simultaneously.

In this section, the discussion is concerned only with modeling techniques for the typical material behaviors shown in Fig. 7.22, rather than a specific model of concrete. A general form of the strain-space formulation of plasticity is first introduced to describe the constitutive behavior of an elastic-plastic solid (Fig. 7.22a). This formulation is then extended to a plastic-fracturing solid (Fig. 7.22c). The modeling method for a progressively fracturing solid (Fig. 7.22b) will not be discussed here, since it involves the concepts of damage theory, which is outside the scope of this book.

## 7.4.2. Plasticity Formulation in Strain Space

The strain-space plasticity formulation has been discussed in the literature (Naghdi and Trapp, 1975; Yoder and Iwan, 1981; Qu and Yin, 1981; Casey and Naghdi, 1983). The following formulation is based upon a *weak stability criterion* that relaxes the requirements of Drucker's stability postulates and allows for an *unstable* behavior. It will be seen that many of the familiar features of stress-space plasticity can be carried over to the strain-space formulation.

### 7.4.2.1. BASIC RELATIONSHIPS

Consider the typical stress-strain diagram shown in Fig. 7.25. In the traditional formulation of plasticity, the stresses play the role of independent variables. The total strain is found by adding the elastic strain, $\epsilon_{ij}^e$, which would arise from the given stress, to the plastic strain, $\epsilon_{ij}^p$, as indicated in Fig. 7.25. Thus,

$$\epsilon_{ij} = \epsilon_{ij}^e + \epsilon_{ij}^p \tag{7.59}$$

An alternative approach would take the strains as independent variables in describing the material state. To find the stress, one could first compute

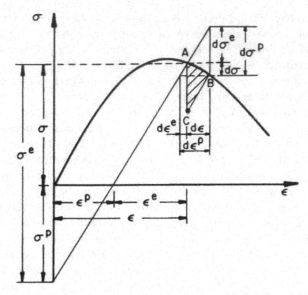

FIGURE 7.25. Schematic description of plasticity formulation based on Il'yushin's postulate.

the elastic stress, denoted by $\sigma_{ij}^e$, which is the elastic response of the current strain, $\epsilon_{ij}$, and then subtract the amount, $\sigma_{ij}^p$, by which the stress has been relaxed on account of the plastic effect (see Fig. 7.25):

$$\sigma_{ij} = \sigma_{ij}^e - \sigma_{ij}^p \tag{7.60}$$

The elastic response $\sigma_{ij}^e$ is related to the total strain $\epsilon_{ij}$ by the generalized Hooke's law as

$$\sigma_{ij}^e = C_{ijkl}\epsilon_{kl} \tag{7.61}$$

where $C_{ijkl}$ is the isotropic tensor of elastic moduli. It has the form, in the usual notation,

$$C_{ijkl} = \frac{\nu E}{(1+\nu)(1-2\nu)}\delta_{ij}\delta_{kl} + \frac{E}{2(1+\nu)}(\delta_{ik}\delta_{jl} + \delta_{il}\delta_{jk}) \tag{7.62}$$

The relaxation stress due to the plastic effect, $\sigma_{ij}^p$, is related to the plastic strain $\epsilon_{ij}^p$ by an equation similar to Eq. (7.61):

$$\sigma_{ij}^p = C_{ijkl}\epsilon_{kl}^p \tag{7.63}$$

In Eq. (7.59), the strain $\epsilon_{ij}^e$ is the elastic response to the total stress $\sigma_{ij}$ and can therefore be expressed as

$$\epsilon_{ij}^e = D_{ijkl}\sigma_{kl} \tag{7.64}$$

where the elastic compliance tensor $D_{ijkl}$ is the inverse of $C_{ijkl}$ and has the form

$$D_{ijkl} = -\frac{\nu}{E}\delta_{ij}\delta_{kl} + \frac{1+\nu}{2E}(\delta_{ik}\delta_{jl} + \delta_{il}\delta_{jk}) \tag{7.65}$$

The relations of the quantities $\sigma_{ij}$, $\sigma^e_{ij}$, $\sigma^p_{ij}$, $\epsilon_{ij}$, $\epsilon^e_{ij}$, and $\epsilon^p_{ij}$ are shown schematically in Fig. 7.25 for the one-dimensional case and are summarized in the following two equations:

$$\sigma_{ij} = \sigma^e_{ij} - \sigma^p_{ij} = C_{ijkl}\epsilon_{kl} - \sigma^p_{ij} \tag{7.66}$$

$$\epsilon_{ij} = \epsilon^e_{ij} + \epsilon^p_{ij} = D_{ijkl}\sigma_{kl} + \epsilon^p_{ij} \tag{7.67}$$

### 7.4.2.2. THE LOADING SURFACE IN STRAIN SPACE AND THE FLOW RULE

In the strain-space formulation, it is required that a loading surface in strain space be defined such that the instantaneous strain state always lies inside or on this surface and the stress relaxation will occur only if the strain lies on the surface and if the strain increment is directed outward from it. This surface may undergo translation and/or distortion. As for the loading function in stress space (see Chapter 5), we may assume that the loading function $F$ takes the following form

$$F(\epsilon_{ij}, \epsilon^p_{ij}, k) = 0 \tag{7.68}$$

in which $\epsilon_{ij}$ is the tensor of current strain, $\epsilon^p_{ij}$ is the plastic strain tensor (see Fig. 7.25), and $k$ is a parameter indicating the loading history, which may be taken as the plastic work $W_p$ (see Fig. 7.25) or the accumulated plastic strain $\epsilon_p$. The surface given by Eq. (7.68) is also called the *relaxation surface* (Yoder and Iwan, 1981) since only when the strain state lies on this surface will it be possible for stress relaxation, which accompanies plastic deformation, to occur.

With the loading function $F$ known, the loading criterion can be given as follows:

$$\text{if } F = 0 \text{ and } \frac{\partial F}{\partial \epsilon_{ij}} d\epsilon_{ij} < 0, \text{ unloading, } d\sigma^p_{ij} = 0$$

$$\text{if } F = 0 \text{ and } \frac{\partial F}{\partial \epsilon_{ij}} d\epsilon_{ij} = 0, \text{ neutral loading, } d\sigma^p_{ij} = 0 \tag{7.69}$$

$$\text{if } F = 0 \text{ and } \frac{\partial F}{\partial \epsilon_{ij}} d\epsilon_{ij} > 0, \text{ loading, } d\sigma^p_{ij} \neq 0$$

In the case of three-dimensional loading, it is necessary to assign a direction to the stress relaxation increment, $d\sigma^p_{ij}$. The *weak stability criterion* proposed by Il'yushin (1961) may be used for this purpose.

Consider now a closed cycle of deformation $A - B - C$ illustrated in Fig. 7.25. In the case of three-dimensional loading, this deformation cycle corresponds to a strain path in which the strain moves out from the relaxation surface, advances incrementally, and then returns to the original state. *Il'yushin's postulate* states that the work done by the external forces in a closed cycle of deformation of an elastic–plastic material is non-negative, i.e., the work is positive if plastic deformation, and thus stress relaxation, takes place and it is zero if only elastic deformation occurs. The shaded area in Fig. 7.25, $dW$, represents the work done in a deformation cycle $A - B - C$. According to Il'yushin's postulate, we have

$$dW = \oint d\sigma_{ij}^p \, d\epsilon_{ij} \geq 0 \qquad (7.70)$$

from which the normality rule or flow rule follows:

$$d\sigma_{ij}^p = d\lambda \, \frac{\partial F}{\partial \epsilon_{ij}} \qquad (7.71)$$

where $d\lambda$ is a scalar greater than 0. The normality rule for an elastoplastic coupling material has also been discussed by Dafalias (1977a, b) and Yin and Qu (1982).

Relating to the loading criterion (7.69), we may assume

$$d\lambda = \frac{\partial F}{h} = \frac{1}{h} \frac{\partial F}{\partial \epsilon_{kl}} d\epsilon_{kl} \qquad (7.72)$$

in which

$$\partial F = F(\epsilon_{ij} + d\epsilon_{ij}, \epsilon_{ij}^p, k) - F(\epsilon_{ij}, \epsilon_{ij}^p, k) = \frac{\partial F}{\partial \epsilon_{kl}} d\epsilon_{kl} \qquad (7.72a)$$

and $h$ is a scalar greater than 0. The stress relaxation increment $d\sigma_{ij}^p$ is then expressed as

$$d\sigma_{ij}^p = \frac{1}{h} \frac{\partial F}{\partial \epsilon_{ij}} \frac{\partial F}{\partial \epsilon_{kl}} d\epsilon_{kl} \qquad (7.73)$$

### 7.4.2.3. INCREMENTAL CONSTITUTIVE RELATIONS

The incremental stress–strain relations are now directly obtained by substituting Eq. (7.73) into the incremental form of Eq. (7.66):

$$d\sigma_{ij} = \left( C_{ijkl} - \frac{1}{h} \frac{\partial F}{\partial \epsilon_{ij}} \frac{\partial F}{\partial \epsilon_{kl}} \right) d\epsilon_{kl} \qquad (7.74)$$

As can be seen, the strain-space plasticity gives the stiffness tensor directly in terms of the strain state, and the tangent stiffness tensor is now represented by

$$C_{ijkl}^{ep} = C_{ijkl} + C_{ijkl}^p = C_{ijkl} - \frac{1}{h} \frac{\partial F}{\partial \epsilon_{ij}} \frac{\partial F}{\partial \epsilon_{kl}} \qquad (7.75)$$

To find the compliance tensor, one can multiply Eq. (7.74) by $(\partial F/\partial \epsilon_{mn}) D_{mnij}$:

$$\frac{\partial F}{\partial \epsilon_{mn}} D_{mnij} d\sigma_{ij} = \frac{\partial F}{\partial \epsilon_{mn}} D_{mnij} C_{ijkl} d\epsilon_{kl} - \frac{\partial F}{\partial \epsilon_{mn}} D_{mnij} \frac{\partial F}{\partial \epsilon_{ij}} \left( \frac{1}{h} \frac{\partial F}{\partial \epsilon_{kl}} d\epsilon_{kl} \right) \quad (7.76)$$

Substituting Eq. (7.72) in the right-hand side of Eq. (7.76) and then solving for $d\lambda$, we obtain

$$d\lambda = \frac{\dfrac{\partial F}{\partial \epsilon_{pq}} D_{pqkl} d\sigma_{kl}}{h + \dfrac{\partial F}{\partial \epsilon_{mn}} D_{mnrs} \dfrac{\partial F}{\partial \epsilon_{rs}}} \quad (7.77)$$

Inverting Eq. (7.63), substituting it into Eq. (7.67), and writing the result in incremental form, we obtain

$$d\epsilon_{ij} = d\epsilon^e_{ij} + d\epsilon^p_{ij} = D_{ijkl} d\sigma_{kl} + D_{ijtu} d\sigma^p_{tu} \quad (7.78)$$

Using the relaxation rule (7.71) in Eq. (7.78) and substituting Eq. (7.77) for $d\lambda$ leads to the following form of the constitutive relations:

$$d\epsilon_{ij} = \left( D_{ijkl} + \frac{D_{ijtu} \dfrac{\partial F}{\partial \epsilon_{tu}} \dfrac{\partial F}{\partial \epsilon_{pq}} D_{pqkl}}{h + \dfrac{\partial F}{\partial \epsilon_{mn}} D_{mnrs} \dfrac{\partial F}{\partial \epsilon_{rs}}} \right) d\sigma_{kl} \quad (7.79)$$

The expression in parentheses represent the *compliance tensor*.

Equations (7.74) and (7.79) are the general incremental constitutive relations for an elastoplastic solid with loading function $F$ in strain space. These equations are suitable for both the hardening and softening ranges but it is undefined in the case of perfect plasticity. Up to now, the scalar parameter $h$ or $d\lambda$ has not been determined yet. This is given in what follows.

### 7.4.2.4. CONSISTENCY CONDITION

With the loading surface defined by Eq. (7.68), the scalar $h$ or $d\lambda$ can be determined from the consistency condition, which states that during stress relaxation, each strain increment leads from one plastic state to another. Equation (7.68) holds both before and after strain increment. Differentiating Eq. (7.68) yields

$$dF = \frac{\partial F}{\partial \epsilon_{ij}} d\epsilon_{ij} + \frac{\partial F}{\partial \epsilon^p_{ij}} d\epsilon^p_{ij} + \frac{\partial F}{\partial k} \frac{\partial k}{\partial \epsilon^p_{ij}} d\epsilon^p_{ij} = 0 \quad (7.80)$$

Inverting the incremental form of Eq. (7.63) and recalling Eq. (7.71), one can express $d\epsilon_{ij}^p$ in terms of $d\lambda$ and then solve Eq. (7.80) for $d\lambda$:

$$d\lambda = \frac{1}{h}\frac{\partial F}{\partial \epsilon_{ij}}d\epsilon_{ij} \tag{7.81}$$

in which

$$h = -\frac{\partial F}{\partial \epsilon_{mn}^p}D_{mnpq}\frac{\partial F}{\partial \epsilon_{pq}} - \frac{\partial F}{\partial k}\frac{\partial k}{\partial \epsilon_{mn}^p}D_{mnpq}\frac{\partial F}{\partial \epsilon_{pq}} \tag{7.82}$$

It is seen that $h$ relies on the *evolution rule* of the yield surface in strain space. As soon as the functional form of $F$ is given, the parameter $h$ or $d\lambda$ can finally be determined.

As can be seen, derivation of the stress–strain relations in strain space is parallel to that in stress space (see Section 5.7). The correspondence between the stress-space and strain-space formulations is shown in Table 7.4.

### 7.4.3. Plastic-Fracturing Formulation in Strain Space

A consistent form of the constitutive relation for a plastic-fracturing material in the ranges of work hardening as well as strain softening is given in this section.

#### 7.4.3.1. BASIC RELATIONSHIPS

Consider the typical stress–strain diagram shown in Fig. 7.26. The stress increment is assumed to comprise three components:

$$d\sigma_{ij} = d\sigma_{ij}^e - d\sigma_{ij}^p - d\sigma_{ij}^f \tag{7.83}$$

in which $d\sigma_{ij}^e$ is the elastic response to the total strain increment, $d\epsilon_{ij}$,

$$d\sigma_{ij}^e = C_{ijkl}\,d\epsilon_{kl} \tag{7.84}$$

and $d\sigma_{ij}^p$ is the *relaxation stress increment* in relation to the plastic strain increment, $d\epsilon_{ij}^p$,

$$d\sigma_{ij}^p = C_{ijkl}\,d\epsilon_{kl}^p \tag{7.85}$$

while $d\sigma_{ij}^f$ is the relaxation stress increment due to stiffness degradation (see Fig. 7.26a) and is related to the fracturing strain increment, $d\epsilon_{ij}^f$, as

$$d\sigma_{ij}^f = C_{ijkl}\,d\epsilon_{kl}^f \tag{7.86}$$

In Eqs. (7.84) through (7.86), $C_{ijkl}$ is the tensor of current elastic moduli.

We may further define an elastic strain increment, $d\epsilon_{ij}^e$, as the elastic response to the total stress increment, $d\sigma_{ij}$; that is,

$$d\epsilon_{ij}^e = D_{ijkl}\,d\sigma_{kl} \tag{7.87}$$

TABLE 7.4. Stress- and strain-space formulation of plasticity.

| | Stress space | Strain space |
|---|---|---|
| Independent variables | $\sigma_{ij}$ | $\epsilon_{ij}$ |
| Variables to be determined | $\epsilon_{ij} = \epsilon^e_{ij} + \epsilon^p_{ij}$ | $\sigma_{ij} = \sigma^e_{ij} - \sigma^p_{ij}$ |
| | $d\epsilon_{ij} = d\epsilon^e_{ij} + d\epsilon^p_{ij}$ | $d\sigma_{ij} = d\sigma^e_{ij} - d\sigma^p_{ij}$ |
| Hooke's law | $\epsilon^e_{ij} = D_{ijkl}\sigma_{kl}$ | $\sigma^e_{ij} = C_{ijkl}\epsilon_{kl}$ |
| Basic relationship | $d\epsilon_{ij} = D_{ijkl}\,d\sigma_{kl} + d\epsilon^p_{ij}$ | $d\sigma_{ij} = C_{ijkl}\,d\epsilon_{kl} - d\sigma^p_{ij}$ |
| Yield function | $f(\sigma_{ij}, \epsilon^p_{ij}, k) = 0$ | $F(\epsilon_{ij}, \epsilon^p_{ij}, k) = 0$ |
| Loading criterion | $f = 0$ and $\dfrac{\partial f}{\partial \sigma_{ij}}\,d\sigma_{ij} > 0$ | $F = 0$ and $\dfrac{\partial F}{\partial \epsilon_{ij}}\,d\epsilon_{ij} > 0$ |
| Postulate and normality rule | Drucker's postulate | Il'yushin's postulate |
| | $\displaystyle\oint d\sigma_{ij}\,d\epsilon^p_{ij} \geq 0$ | $\displaystyle\oint d\sigma^p_{ij}\,d\epsilon_{ij} \geq 0$ |
| | $d\epsilon^p_{ij} = d\lambda\,\dfrac{\partial f}{\partial \sigma_{ij}}$ | $d\sigma^p_{ij} = d\lambda\,\dfrac{\partial F}{\partial \epsilon_{ij}}$ |
| Linearity | $d\epsilon^p_{ij} = \dfrac{1}{\kappa}\dfrac{\partial f}{\partial \sigma_{ij}}\dfrac{\partial f}{\partial \sigma_{kl}}\,d\sigma_{kl}$ | $d\sigma^p_{ij} = \dfrac{1}{h}\dfrac{\partial F}{\partial \epsilon_{ij}}\dfrac{\partial F}{\partial \epsilon_{kl}}\,d\epsilon_{kl}$ |
| | or | or |
| | $d\lambda = \dfrac{\partial f}{\kappa} = \dfrac{1}{\kappa}\dfrac{\partial f}{\partial \sigma_{kl}}\,d\sigma_{kl}$ | $d\lambda = \dfrac{\partial F}{h} = \dfrac{1}{h}\dfrac{\partial F}{\partial \epsilon_{kl}}\,d\epsilon_{kl}$ |
| Constitutive relation | | |
| Stiffness tensor $C^{ep}_{ijkl}$ | $C_{ijkl} - \dfrac{C_{ijtu}\dfrac{\partial f}{\partial \sigma_{tu}}\dfrac{\partial f}{\partial \sigma_{pq}}C_{pqkl}}{\kappa + \dfrac{\partial f}{\partial \sigma_{mn}}C_{mnrs}\dfrac{\partial f}{\partial \sigma_{rs}}}$ | $C_{ijkl} - \dfrac{1}{h}\dfrac{\partial F}{\partial \epsilon_{ij}}\dfrac{\partial F}{\partial \epsilon_{kl}}$ |
| Compliance tensor $D^{ep}_{ijkl}$ | $D_{ijkl} + \dfrac{1}{\kappa}\dfrac{\partial f}{\partial \sigma_{ij}}\dfrac{\partial f}{\partial \sigma_{kl}}$ | $D_{ijkl} + \dfrac{D_{ijtu}\dfrac{\partial F}{\partial \epsilon_{tu}}\dfrac{\partial F}{\partial \epsilon_{pq}}D_{pqkl}}{h + \dfrac{\partial F}{\partial \epsilon_{mn}}D_{mnrs}\dfrac{\partial F}{\partial \epsilon_{rs}}}$ |

in which $D_{ijkl}$ is the tensor of current compliance, i.e., the inverse of the tensor $C_{ijkl}$.

Solving $d\sigma_{ij}$ from Eq. (7.87) and substituting it together with Eqs. (7.84) through (7.86) into Eq. (7.83), we obtain the following relation for the strain increments $d\epsilon_{ij}$, $d\epsilon^e_{ij}$, $d\epsilon^p_{ij}$, and $d\epsilon^f_{ij}$:

$$d\epsilon_{ij} = d\epsilon^e_{ij} + d\epsilon^p_{ij} + d\epsilon^f_{ij} \qquad (7.88)$$

which implies that the total strain increment $d\epsilon_{ij}$ comprises three parts: the elastic strain increment, $d\epsilon^e_{ij}$, which is reversible in an incremental sense; the plastic strain increment, $d\epsilon^p_{ij}$, which is the permanent strain increment;

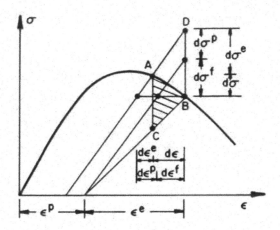

(a) Stress and Strain Increments

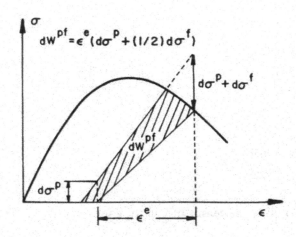

(b) Plastic-Fracturing Work Increment

FIGURE 7.26. Schematic description of the combined formulation.

and the fracturing strain increment, $d\epsilon_{ij}^f$, which is recoverable only if the stress is fully released.

In contrast to the incremental strains, it should be noted that the total strain $\epsilon_{ij}$ comprises only two parts: $\epsilon_{ij}^p$, the plastic (permanent) strain, and $\epsilon_{ij}^e$, the recoverable strain or elastic strain; that is,

$$\epsilon_{ij} = \epsilon_{ij}^e + \epsilon_{ij}^p \tag{7.89}$$

and the elastic strain $\epsilon_{ij}^e$ is related to the total stress $\sigma_{ij}$ by the tensor of current elastic moduli

$$\sigma_{ij} = C_{ijkl}\epsilon_{ij}^e \tag{7.90}$$

### 7.4.3.2. Relaxation Surface and Flow Rule

Herein, we shall assume the relaxation surface in strain space to have a form similar to that of Eq. (7.68), but the parameter $k$ is replaced by $W^{pf}$:

$$F(\epsilon_{ij}, \epsilon_{ij}^p, W^{pf}) = 0 \tag{7.91}$$

where $W^{pf}$ is the *plastic-fracturing work*, which is the *total energy dissipation* during loading and unloading (Fig. 7.26b). The relaxation surface is also the loading surface since only if the strain state lies on this surface will it be possible for stress relaxation, which accompanies plastic deformation as well as stiffness degradation, to occur. Thus, we have the following loading criterion:

$$\text{if } F = 0 \text{ and } \frac{\partial F}{\partial \epsilon_{ij}} d\epsilon_{ij} < 0, \text{ unloading, } d\sigma_{ij}^{pf} = 0$$

$$\text{if } F = 0 \text{ and } \frac{\partial F}{\partial \epsilon_{ij}} d\epsilon_{ij} = 0, \text{ neutral loading, } d\sigma_{ij}^{pf} = 0 \tag{7.92}$$

$$\text{if } F = 0 \text{ and } \frac{\partial F}{\partial \epsilon_{ij}} d\epsilon_{ij} > 0, \text{ loading, } d\sigma_{ij}^{pf} \neq 0$$

where $d\sigma_{ij}^{pf}$ is the incremental stress relaxation, equal to the sum of the plastic stress increment, $d\sigma_{ij}^p$, and the fracturing stress increment, $d\sigma_{ij}^f$, i.e.,

$$d\sigma_{ij}^{pf} = d\sigma_{ij}^p + d\sigma_{ij}^f \tag{7.93}$$

Il'yushin's postulate requires that the work done in a deformation cycle, $dW$, be non-negative. In Fig. 7.26a, $dW$ is shown by the shaded area. Thus, we have

$$dW = \oint d\sigma_{ij}^{pf} d\epsilon_{ij} \geq 0 \tag{7.94}$$

from which the *normality rule* (or *flow rule*) is represented as

$$d\sigma_{ij}^{pf} = d\lambda \frac{\partial F}{\partial \epsilon_{ij}} \tag{7.95}$$

The normality rule for an elastic–plastic coupling material has been discussed by Dafalias (1977b) and Yin and Qu (1982).

### 7.4.3.3. RATE OF ENERGY DISSIPATION AND PARTITION OF $d\sigma_{ij}^{pf}$

By definition, the rate of energy dissipation per unit volume, $D = dW^{pf}$, for a plastic-fracturing solid consists of two parts: one is due to plastic deformation, and the other is due to stiffness degradation. Thus we have

$$D = dW^{pf} = \sigma_{ij}\,d\epsilon_{ij}^{p} - \tfrac{1}{2}\,dC_{ijkl}\epsilon_{ij}^{e}\epsilon_{kl}^{e} \tag{7.96}$$

In view of the basic relations (7.85) and (7.90), the first term of Eq. (7.96) can be rewritten in terms of the plastic stress component $d\sigma_{ij}^{p}$ and the elastic strain increment $\epsilon_{ij}^{e}$ as $\epsilon_{ij}^{e}\,d\sigma_{ij}^{p}$. [This relationship can be easily seen by simply multiplying Eq. (7.85) by $\epsilon_{ij}^{e}$ and using Eq. (7.90).] Further, noting that the fracturing stress component, $d\sigma_{ij}^{f}$, depends on the stiffness degradation as

$$d\sigma_{ij}^{f} = -dC_{ijkl}\,\epsilon_{kl}^{e} \tag{7.97}$$

the second term of Eq. (7.96) can be expressed in terms of $d\sigma_{ij}^{f}$ as $\tfrac{1}{2}\,d\sigma_{ij}^{f}\,\epsilon_{ij}^{e}$ and Eq. (7.96) becomes

$$D = dW^{pf} = \epsilon_{ij}^{e}(d\sigma_{ij}^{p} + \tfrac{1}{2}\,d\sigma_{ij}^{f}) \tag{7.98}$$

This, in fact, represents the shaded area of Fig. 7.26b. We may assume that the elastic stiffness tensor $C_{ijkl}$ is a function of the plastic-fracturing work, $W^{pf}$, i.e.,

$$C_{ijkl} = C_{ijkl}(W^{pf}) \tag{7.99}$$

Then the rate of stiffness degradation can be expressed as

$$C'_{ijkl} = \frac{dC_{ijkl}}{dW^{pf}} \tag{7.100}$$

Noting the definition of the energy dissipation rate of Eq. (7.98), we obtain the stiffness degradation $dC_{ijkl}$ as

$$dC_{ijkl} = \frac{dC_{ijkl}}{dW^{pf}}\,dW^{pf} = C'_{ijkl}\epsilon_{mn}^{e}\left(d\sigma_{mn}^{p} + \frac{1}{2}\,d\sigma_{mn}^{f}\right) \tag{7.101}$$

Substitution of Eq. (7.101) into (7.97) leads to

$$d\sigma_{ij}^f = -C'_{ijkl}\epsilon_{kl}^e\epsilon_{mn}^e(d\sigma_{mn}^p + \tfrac{1}{2} d\sigma_{mn}^f)$$
(7.102)

After some tensor manipulations of Eq. (7.102), the relation between fracturing stress increment $d\sigma_{ij}^f$ and total inelastic stress increment $d\sigma_{ij}^{pf}$ can be obtained in the form

$$d\sigma_{ij}^f = T'_{ijkl} \, d\sigma_{kl}^{pf}$$
(7.103)

where $T'_{ijkl}$ can be viewed as a transformation tensor and is expressed by

$$T'_{ijkl} = \bar{M}_{ijmn} N_{mnkl}$$
(7.104)

in which the tensor $\bar{M}_{ijmn}$ is the inverse of tensor $\bar{M}_{ijmn}$ and

$$\bar{M}_{ijmn} = \delta_{im}\delta_{jn} - \tfrac{1}{2}C'_{ijpq}\epsilon_{pq}^e\epsilon_{mn}^e$$
(7.105)

while $N_{mnkl}$ is defined as

$$N_{mnkl} = -C'_{mnpq}\epsilon_{pq}^e\epsilon_{kl}^e$$
(7.106)

From Eq. (7.93), $d\sigma_{ij}^p$ can be related to $d\sigma_{ij}^{pf}$ by

$$d\sigma_{ij}^p = T'^p_{ijkl} \, d\sigma_{kl}^{pf}$$
(7.107)

where

$$T'^p_{ijkl} = \delta_{ik}\delta_{jl} - T'_{ijkl}$$

Consequently, the two stress components $d\sigma_{ij}^p$ and $d\sigma_{ij}^f$ can be determined from the total stress relaxation $d\sigma_{ij}^{pf}$ by Eqs. (7.107) and (7.103), respectively provided that the stiffness degradation rate, $C'_{ijkl}$, is known.

### 7.4.3.4. CONSTITUTIVE RELATION

As soon as the relationships between the stress increments $d\sigma_{ij}^p$, $d\sigma_{ij}^f$, and the total increment $d\sigma_{ij}^{pf}$ have been established, the scalar $d\lambda$ in Eq. (7.95) can be derived from the consistency condition in the usual manner. Here, as in Eq. (7.81), $d\lambda$ has the form

$$d\lambda = \frac{1}{h}\frac{\partial F}{\partial \epsilon_{ij}} \, d\epsilon_{ij}$$
(7.108)

but the scalar function $h$ has a different form

$$h = -\left[\frac{\partial F}{\partial \epsilon_{ij}^p} D_{ijmn} T'^p_{mnkl} \frac{\partial F}{\partial \epsilon_{kl}} + \frac{\partial F}{\partial W^{pf}}\epsilon_{mn}^e\left(T'^p_{mnkl} + \frac{1}{2}T'_{mnkl}\right)\frac{\partial F}{\partial \epsilon_{kl}}\right]$$
(7.109)

Substituting Eq. (7.108) into Eq. (7.95), recalling that

$$d\sigma_{ij} = d\sigma^e_{ij} - d\sigma^{pf}_{ij}$$

and noting that

$$d\sigma^e_{ij} = C_{ijkl} \, d\epsilon_{kl}$$

we obtain the constitutive equation for a *plastic-fracturing solid*

$$d\sigma_{ij} = \left[ C_{ijkl} - \frac{1}{h} \frac{\partial F}{\partial \epsilon_{ij}} \frac{\partial F}{\partial \epsilon_{kl}} \right] d\epsilon_{kl} \qquad (7.110)$$

which has the same form as Eq. (7.74).

The general formulation given above is valid for the whole range of loading conditions (hardening or softening) and is suitable for modeling the stress–strain behavior of materials with elastoplastic coupling.

## 7.4.4. Remarks on Softening Modeling of Concrete Materials

The behavior of plastic (irreversible) deformation coupled with an elastic degradation is usually observed for concrete materials in the postfailure range. The approach of combining the theory of plasticity with the theory of damage (fracturing) in modeling such behavior is logical and reasonable. The strain-space formulation has provided a means for combining these two theories consistently.

To establish an analytic model for an actual material such as concrete, two functions must be defined: (1) the loading function (relaxation function), $F$, in strain space; (2) the rate of the stiffness degradation tensor $C'_{ijkl}$ as a function of the energy dissipation $W^{pf}$. The initial relaxation function represents all possible states of strain at which the plastic deformation and elastic degradation start to occur. This function varies with the increase of plastic deformation and the evolution of damage. Since experimental data for concrete in the softening range are lacking, a clear definition of the relaxation function for concrete is difficult at the present time. The stiffness degradation rate $C'_{ijkl}$ with 21 components is, in general, even more difficult to define. However, since the stiffness degradation behavior is generally induced by some kind of material damage, the continuous damage theory attempts to approach this problem based on the principles of continuum mechanics. This is still under active development for concrete materials. With the success of this theory and the availability of more experimental data later, the macroscopic strain-softening behavior of concrete materials will be better described.

The plastic-fracturing models for the strain-softening behavior have been criticized in that the stress–strain relationship in the softening range is merely a nominal property. It is true that in the postfailure range, strain localization usually occurs. The descending branch of the load–deformation curve may not be interpreted as the strain softening of the material. However, if the structural changes in the material are considered by some means (e.g., the model of Frantziskonis and Desai, 1987), the continuous description of the softening stress–strain relation is reasonable.

## References

Balmer, G.G., 1949. "Shearing Strength of Concrete Under High Triaxial Stress— Computation of Mohr's Envelope as a Curve," Structural Research Laboratory Report SP-23, Denver, Colorado, October, 1949.

Bazant, Z.P., and S.S. Kim, 1979. "Plastic-Fracturing Theory for Concrete," *Journal of Engineering Mechanics Division, ASCE* 105(EM3):407–428, with Errata in Vol. 106.

Bresler, B., and K.S. Pister, 1958. "Strength of Concrete Under Combined Stresses," *ACI Journal* 55:321–345.

Casey, J., and P.M. Naghdi, 1983. "On the Nonequivalence of the Stress Space and Strain Space Formulations of Plasticity Theory," *Journal of Applied Mechanics, ASME* 50:350–354.

Chen, W.F., 1982. *Plasticity in Reinforced Concrete*, McGraw-Hill, New York, 474 pp.

Chen, A.C.T., and W.F. Chen, 1975. "Constitutive Relations for Concrete," *Journal of Engineering Mechanics Division, ASCE* 101:465–481.

Chen, W.F., and H. Zhang, 1988. *Theory, Problems, and CAE Softwares of Structural Plasticity*, Springer-Verlag, New York.

Chen, E.Y.T., and W.C. Schnobrich, 1981. "Material Modeling of Plain Concrete," *Advanced Mechanics of Reinforced Concrete*, IABSE Colloquium Delft, pp. 33–51.

Chen, E.S., and O. Buyukozturk, 1985. "Constitutive Model for Concrete in Cyclic Compression," *Journal of the Engineering Mechanics Division, ASCE* 111(EM6):797–814.

Chinn, J., and R.M. Zimmerman, 1965. "Behavior of Plain Concrete Under Various High Triaxial Compression Loading Conditions," Air Force Weapons Laboratory, Technical Report WL TR 64-163, Kirtland Air Force Base, Albuquerque, New Mexico.

Dafalias, Y.F., 1977a. "Elasto-Plastic Coupling Within a Thermodynamic Strain Space Formulation of Plasticity," *International Journal of Non-linear Mechanics* 12:327–337.

Dafalias, Y.F., 1977b. "Il'yushin's Postulate and Resulting Thermodynamic Conditions on Elasto-Plastic Coupling," *International Journal of Solids and Structures* 13:239–251.

Dougill, J.W., 1975. "Some Remarks on Path Independence in the Small in Plasticity," *Quarterly of Applied Mathematics* 32:233–243.

Dougill, J.W., 1976. "On Stable Progressively Fracturing Solids," *Zeitschrift für Angewandte Mathematik und Physik* 27:423–437.

Fardis, M.N., B. Alibe, and J. Tassoulas, 1983. "Monotonic and Cyclic Constitutive Law for Concrete," *Journal of the Engineering Mechanics Division, ASCE* 109(EM2):516–536.

Frantziskonis, G., and C.S. Desai, 1987. "Constitutive Model with Strain Softening," *International Journal of Solids and Structures* 26(6):733–750.

Gerstle, K.H., D.H. Linse, et al., 1978. "Strength of Concrete Under Multi-Axial Stress States," Proc. McHenry Symposium on Concrete and Concrete Structures, Mexico City, 1976, Special Publication, SP55, ACI, pp. 103-131.

Han, D.J., and W.F. Chen, 1985. "A Nonuniform Hardening Plasticity Model for Concrete Materials," Journal of Mechanics of Materials 4(4):283-302.

Han, D.J., and W.F. Chen, 1986. "On Strain-Space Plasticity Formulation for Hardening-Softening Materials with Elasto-Plastic Coupling," International Journal of Solids and Structures 22(8):935-950.

Han, D.J., and W.F. Chen, 1987. "Constitutive Modeling in Analysis of Concrete Structures," Journal of Engineering Mechanics Division, ASCE 113:577-593.

Hsieh, S.S., E.C. Ting, and W.F. Chen, 1982. "A Plasticity-Fracture Model for Concrete," International Journal of Solids and Structures 18(3):181-197.

Il'yushin, A.A., 1961. "On the Postulate of Plasticity," Prikladnaya Matematika i Mekhanika AN SSSR, Moskva 25(3):503-507.

Kotsovos, M.D., and J.B. Newman, 1977. "Behavior of Concrete Under Multiaxial Stress," ACI Journal 74(9):443-446.

Kupfer, H., H.K. Hilsdorf, and H. Rusch, 1969. "Behavior of Concrete Under Biaxial Stresses," ACI Journal 66(8):656-666.

Launay, P., and H. Gachon, 1972. "Strain and Ultimate Strength of Concrete Under Triaxial Stresses," Special Publication, SP-34, ACI, 1, pp. 269-282.

Mills, L.L., and R.M. Zimmerman, 1970. "Compressive Strength of Plain Concrete Under Multiaxial Loading Conditions," ACI Journal 67(10):802-807.

Naghdi, P.M., and J.A. Trapp, 1975. "The Significance of Formulating Plasticity Theory with Reference to Loading Surface in Strain Space," International Journal of Engineering Science 13:785-797.

Ngo, D., and A.C. Scordelis, 1967. "Finite Element Analysis of Reinforced Concrete Beam," ACI Journal 64(3):152-163.

Ottosen, N.S., 1977. "A Failure Criterion for Concrete," Journal of Engineering Mechanics Division, ASCE 103(EM4):527-535.

Palaniswamy, R., and S.P. Shah, 1974. "Fracture and Stress-Strain Relationship of Concrete under Triaxial Compression," Journal of Structural Division, ASCE 100(ST5):901-916.

Peterson, P.E., 1981. "Crack Growth and Development of Fracture Zones in Plain Concrete and Similar Materials," Report No. TVBM-1106, Division of Building Materials, University of Lund, Lund, Sweden.

Qu, S.N., and Y.Q. Yin, 1981. "Drucker's and Il'yushin's Postulate of Plasticity," Acta Mechanica Sinica, No. 5, September, 1981, 465-473 (in Chinese).

Richart, F.E., A. Brandtzaeg, and R.L. Brown, 1928. "A Study of the Failure of Concrete Under Combined Compressive Stresses," University of Illinois Engineering Experimental Station Bulletin 185.

Scavuzzo, R., T. Stankowski, K.H. Gerstle, and H.Y. Ko, 1983. "Stress-Strain Curves for Concrete Under Multiaxial Load Histories," Department of Civil, Environmental and Architectural Engineering, University of Colorado, Boulder.

Schickert, G., and H. Winkler, 1977. "Results of Test Concerning Strength and Strain of Concrete Subjected to Multiaxial Compressive Stress," Deutscher Ausschuss für Stahlbeton, Heft 277, Berlin.

Shah, S.P., and S. Chandra, 1968. "Critical Stress Volume Change, and Microcracking of Concrete," ACI Journal 65(9):770-781.

Sinha, B.P., K.H. Gerstle, and L.G. Tulin, 1964. "Stress–Strain Relations for Concrete Under Cyclic Loading," *ACI Journal* **61**(2):195–211.

Tasuji, M.E., F.O. Slate, and A.H. Nilson, 1978. "Stress–Strain Response and Fracture of Concrete in Biaxial Loading," *ACI Journal* **75**(7):306–312.

Traina, L.A., S.M. Babcock, and H.L. Schreyer, 1983. "Reduced Experimental Stress-Strain Results for a Low Strength Concrete under Multiaxial State of Stress," AFWL-TR-83-3, Air Force Weapons Laboratory, New Mexico.

van Mier, J.G.M., 1984. "Complete Stress–Strain Behavior and Damaging Status of Concrete Under Multiaxial Conditions," RILEM-CEB-CNRS, *International Conference on Concrete Under Multiaxial Conditions*, Vol. 1, Presses de l'Université Paul Sabatier, Toulouse, France, pp. 75–85.

Vermeer, P.A., and R. De Borst, 1984. "Non-Associated Plasticity for Soils, Concrete and Rock," *Heron* **29**(3).

Willam, K.J., and E.P. Warnke, 1975. "Constitutive Model for the Triaxial Behavior of Concrete," International Association of Bridge and Structural Engineers, Seminar on Concrete Structure Subjected to Triaxial Stresses, Paper III-1, Bergamo, Italy, May, 1974, *IABSE Proc.* **19**.

Wischers, G., 1978. "Application of Effects of Compressive Loads on Concrete," *Betontech.*, Berlin, Nos. 2 and 3, Düsseldorf.

Yin, Y.Q., and S.N. Qu, 1982. "Elasto-Plastic Coupling and Generalized Normality Rule," *Acta Mechanica Sinica*, No. 1, 63–70 (in Chinese).

Yoder, P.J., and W.D. Iwan, 1981. "On the Formulation of Strain-Space Plasticity with Multiple Loading Surfaces," *Journal of Applied Mechanics, ASME* **48**:773–778.

# Appendix: Parameters and Coefficients Relevant to the Plasticity Model of Concrete Materials

## A.1. Shape Factor and Modification Factor

### A.1.1. SHAPE FACTOR OF THE LOADING SURFACE

To meet the basic shape requirements for the yield surface on the meridian plane, we choose the following functions for the shape factor $k$:

$$k = \begin{cases} 1 & \sigma_m \geq \xi_t \\ k_1(\sigma_m) & \xi_t > \sigma_m \geq \xi_c \\ k_0 & \xi_c > \sigma_m \geq \xi_k \\ k_2(\sigma_m) & \xi_k > \sigma_m \end{cases} \tag{A.1}$$

Function $k_1(\sigma_m)$ is assumed to have a quadratic form of $\sigma_m$, satisfying the following three conditions: at $\sigma_m = \xi_t$, $k_1 = 1$ and at $\sigma_m = \xi_c$, $k_1 = k_0$ and $dk_1/d\sigma_m = 0$, from which we obtain

$$k_1(\sigma_m) = \frac{1 + (1 - k_0)[-\xi_t(-2\xi_c + \xi_t) - 2\xi_c\sigma_m + \sigma_m^2]}{(\xi_c - \xi_t)^2} \tag{A.2}$$

Similarly, $k_2(\sigma_m)$ is also assumed to have a quadratic form of $\sigma_m$, satisfying the conditions: at $\sigma_m = \xi_k$, $k_2 = k_0$ and $dk_2/d\sigma_m = 0$, and at $\sigma_m = \bar{\xi}$, $k_2 = 0$, and we obtain

$$k_2(\sigma_m) = \frac{k_0(\bar{\xi} - \sigma_m)(\bar{\xi} + \sigma_m - 2\xi_k)}{(\bar{\xi} - \xi_k)^2} \tag{A.3}$$

### A.1.2. Modification Factor $M(\sigma_m, \theta)$

$$M(\sigma_m, \theta) = \begin{cases} f(\sigma_m, \theta) & \text{if } 0 < f \le 1 \\ 1 & \text{otherwise} \end{cases} \tag{A.4}$$

in which

$$f(\sigma_m, \theta) = -\frac{0.15}{(1.4 - \cos\theta)(\sigma_m + \frac{1}{3})(\sigma_m + 2.5)}$$

## A.2. Derivatives of the Loading Functions

### A.2.1. General Expressions

The derivatives of a general loading function

$$f(\sigma_{ij}, \epsilon_{ij}^p, k) = 0 \tag{A.5}$$

for an isotropic material can be expressed by the chain rule as

$$\frac{\partial f}{\partial \sigma_{ij}} = \frac{\partial f}{\partial I_1}\delta_{ij} + \frac{\partial f}{\partial J_2}s_{ij} + \frac{\partial f}{\partial J_3}t_{ij} \tag{A.6}$$

in which

$$\delta_{ij} = \frac{\partial I_1}{\partial \sigma_{ij}}$$

is the Kronecker delta,

$$s_{ij} = \frac{\partial J_2}{\partial \sigma_{ij}}$$

is the stress deviator tensor, and

$$t_{ij} = \frac{\partial J_3}{\partial \sigma_{ij}} = s_{ik}s_{kj} - \frac{2}{3}J_2\delta_{ij}$$

is the deviation of the square of the stress deviation.
Denoting

$$B_0 = \frac{\partial f}{\partial I_1}, \qquad B_1 = \frac{\partial f}{\partial J_2}, \qquad B_2 = \frac{\partial f}{\partial J_3} \tag{A.7}$$

the derivatives are further expressed as

$$\frac{\partial f}{\partial \sigma_{ij}} = B_0 \delta_{ij} + B_1 s_{ij} + B_2 t_{ij} \tag{A.8}$$

Specifically, the loading surface for concrete is defined by Eq. (7.15) as

$$f(\sigma_{ij}, \epsilon_{ij}^p, k) = \rho - k\rho_f = 0 \tag{A.9}$$

in which $k$ is a function of $\sigma_m$ (or $I_1$) and also of the plastic strain $\epsilon_{ij}^p$ or the plastic work $W_p$ implicitly.

From Eq. (A.9), the derivatives $B_0$, $B_1$, and $B_2$ can be written as

$$B_0 = \frac{\partial f}{\partial I_1} = -\frac{\partial k}{\partial I_1} \rho_f - kA_0$$

$$B_1 = \frac{\partial f}{\partial J_2} = \frac{1}{\rho} - kA_1 \tag{A.10}$$

$$B_2 = \frac{\partial f}{\partial J_3} = -kA_2$$

where $A_0$, $A_1$, and $A_2$ are the derivatives of the failure function $\rho_f$:

$$A_0 = \frac{\partial \rho_f}{\partial I_1}, \qquad A_1 = \frac{\partial \rho_f}{\partial J_2}, \qquad A_2 = \frac{\partial \rho_f}{\partial J_3} \tag{A.11}$$

which depend on the specific forms of $\rho_f$.

### A.2.2. OTTOSEN FOUR-PARAMETER MODEL, EQ. (7.10)

The failure function of this model is

$$\rho_f = \frac{1}{2a}[-\sqrt{2}\lambda + \sqrt{2\lambda^2 - 8a(3b\sigma_m - 1)}] \tag{A.12}$$

in which

$$\lambda = \begin{cases} k_1 \cos\left[\dfrac{1}{3}\cos^{-1}(k_2 \cos 3\theta)\right] & \text{for } \cos 3\theta \geq 0 \\[2ex] k_1 \cos\left[\dfrac{\pi}{3} - \dfrac{1}{3}\cos^{-1}(-k_2 \cos 3\theta)\right] & \text{for } \cos 3\theta < 0 \end{cases}$$

According to Eq. (A.12), the derivatives of this failure function can be obtained as

$$A_0 = \frac{\partial \rho_f}{\partial I_1} = \frac{\partial \rho_f}{3\partial \sigma_m} = \frac{-2b}{h_1}$$

$$A_1 = \frac{\partial \rho_f}{\partial J_2} = \frac{1}{2a}\left(-\sqrt{2} + \frac{2\lambda}{h_1}\right)\frac{d\lambda}{d\theta}\frac{\partial \theta}{\partial J_2} \tag{A.13}$$

$$A_2 = \frac{\partial \rho_f}{\partial J_3} = \frac{1}{2a}\left(-\sqrt{2} + \frac{2\lambda}{h_1}\right)\frac{d\lambda}{d\theta}\frac{\partial \theta}{\partial J_3}$$

in which

$$h_1 = \sqrt{2\lambda^2 - 8a(3b\sigma_m - 1)} \tag{A.14}$$

$$\frac{d\lambda}{d\theta} = \begin{cases} \dfrac{-k_1 k_2 \sin 3\theta \sin[\frac{1}{3} \cos^{-1}(k_2 \cos 3\theta)]}{\sin[\cos^{-1}(k_2 \cos 3\theta)]} & \text{for } \cos 3\theta \geq 0 \\[3ex] \dfrac{-k_1 k_2 \sin 3\theta \sin\left[\dfrac{\pi}{3} - \dfrac{1}{3} \cos^{-1}(-k_2 \cos 3\theta)\right]}{\sin[\cos^{-1}(-k_2 \cos 3\theta)]} & \text{for } \cos 3\theta < 0 \end{cases} \tag{A.15}$$

$$\frac{\partial \theta}{\partial J_2} = \frac{3\sqrt{3}}{4 \sin 3\theta} \frac{J_3}{J_2^{5/2}} \tag{A.16}$$

$$\frac{\partial \theta}{\partial J_3} = -\frac{\sqrt{3}}{2 \sin 3\theta} \frac{1}{J_2^{3/2}} \tag{A.17}$$

### A.2.3. HSIEH–TING–CHEN FOUR-PARAMETER MODEL, EQ. (7.11)

The failure function of this model is

$$\rho_f = \frac{1}{2a} [-(b \cos \theta + c) + \sqrt{(b \cos \theta + c)^2 - 4a(\sqrt{3} \, d\sigma_m - 1)}] \tag{A.18}$$

Its derivatives are given by

$$A_0 = -\frac{\sqrt{3} \, d}{3h_2}$$

$$A_1 = -\frac{b \sin \theta}{2a} \left[ 1 - \frac{(b \cos \theta + c)}{h_2} \right] \frac{\partial \theta}{\partial J_2} \tag{A.19}$$

$$A_2 = -\frac{b \sin \theta}{2a} \left[ 1 - \frac{(b \cos \theta + c)}{h_2} \right] \frac{\partial \theta}{\partial J_3}$$

in which $\partial \theta / \partial J_2$ and $\partial \theta / \partial J_3$ are represented by Eqs. (A.16) and (A.17) and

$$h_2 = \sqrt{(b \cos \theta + c)^2 - 4a(\sqrt{3} \, d\sigma_m - 1)} \tag{A.20}$$

### A.2.4. WILLAM–WARNKE FIVE-PARAMETER MODEL, EQ. (7.12)

The failure function of this model is expressed by the following equations:

$$\rho_f = \frac{s + t}{v} \tag{A.21}$$

where

$$s = 2\rho_c(\rho_c^2 - \rho_t^2) \cos \theta \tag{A.22}$$

$$t = \rho_c(2\rho_t - \rho_c)u^{1/2} \tag{A.23}$$

$$u = 4(\rho_c^2 - \rho_t^2) \cos^2 \theta + 5\rho_t^2 - 4\rho_t\rho_c \tag{A.24}$$

$$v = 4(\rho_c^2 - \rho_t^2) \cos^2 \theta + (\rho_c - 2\rho_t)^2 \tag{A.25}$$

and

$$\rho_c = -\frac{1}{2b_2}[b_1 + \sqrt{b_1^2 - 4b_2(b_0 - \sigma_m)}] \qquad (A.26)$$

$$\rho_t = -\frac{1}{2a_2}[a_1 + \sqrt{a_1^2 - 4a_2(a_0 - \sigma_m)}] \qquad (A.27)$$

The derivatives of this failure function are represented as

$$A_0 = -\frac{\rho_t'}{3v}\{4\rho(\rho_c - 2\rho_t) + 8\rho\rho_t \cos^2\theta$$

$$-[4\rho_c\rho_t \cos\theta - 2\rho_c u^{1/2} + \rho_c(2\rho_t - \rho_c)u^{-1/2}(4\rho_t \cos^2\theta - 5\rho_t + 2\rho_c)]\}$$

$$+\frac{\rho_c'}{3v}\{2\rho(\rho_c - 2\rho_t) + 8\rho\rho_c \cos^2\theta$$

$$-[2(3\rho_c^2 - \rho_t^2)\cos\theta + 2(\rho_t - \rho_c)u^{1/2}$$

$$+2\rho_c(2\rho_t - \rho_c)u^{-1/2}(2\rho_c \cos^2\theta - \rho_t)]\} \qquad (A.28)$$

where

$$\rho_t' = \frac{d\rho_t}{d\sigma_m} = \frac{1}{a_1 + 2a_2\rho_t}$$

$$\rho_c' = \frac{d\rho_c}{d\sigma_m} = \frac{1}{b_1 + 2b_2\rho_c}$$

and

$$A_1 = -\frac{3J_3}{2J_2}A_2 \qquad (A.29)$$

$$A_2 = \frac{\sqrt{3}(\rho_c^2 - \rho_t^2)}{v(4\cos^2\theta - 1)J_2^{3/2}}\{4\rho \cos\theta - \rho_c[1 + 2u^{-1/2}(2\rho_t - \rho_c)\cos\theta]\} \qquad (A.30)$$

# Part V: Limit Analysis

# 8

# General Theorems of Limit Analysis and Their Applications

## 8.1. Introduction

It has been noted in the preceding discussions that a complete elastoplastic analysis is generally quite complicated. The complexities arise mainly from the necessity of carrying out an analysis in an iterative and incremental manner. The development of efficient alternative methods that can be used to obtain the collapse load of a structural problem in a simple and more direct manner without recourse to an iterative and incremental analysis is, therefore, of great value to practicing engineers, despite the fact that the information so obtained is just a part of the total solution. *Limit analysis* is concerned with the development and applications of such methods that can furnish the engineer with an estimate of the *collapse load* of a structure in a direct manner. The estimation of the collapse load is of great value, not only as a simple check for a more refined analysis, but also as a basis for engineering design.

Figure 8.1a shows a simply supported rectangular beam subjected to a uniformly distributed load $q$. The beam is made of an elastic–perfectly plastic material. As the applied load $q$ is gradually increased starting from zero, the beam is stressed first in a purely elastic manner until the *elastic limit load* $q = q_e$ is reached, at which the stresses at the upper and lower edges of the center section have just reached the yield stress $\sigma_0$ (Fig. 8.1b). Further increase of the load will cause the yield zone to spread (Fig. 8.1c). At this stage, however, plastic flow in the yield zones is still contained by the surrounding elastic regions. This type of elastic-plastic behavior is termed *contained plastic flow*. The beam undergoes a definite overall displacement (Fig. 8.1d). As we keep increasing the load $q$, eventually , at $q = q_c = 1.5q_e$, the upper and lower plastic regions meet to form a *plastic hinge*. At this instant, the yielding has spread to such an extent that the remaining elastic material plays an insignificant role in sustaining the load. This is termed *uncontained or unrestricted plastic flow*. Now the structure becomes a mechanism and reaches its limit or collapse state. The ultimate load, $q_c = 1.5q_e$, is referred to as the *plastic collapse load* or simply the *collapse load* or *limit load*.

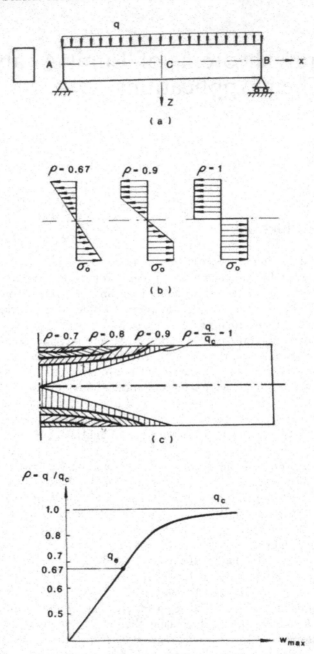

FIGURE 8.1. An illustration of a collapse process: (a) a simply supported rectangular beam; (b) stress distributions of the cross section at the midspan $C$; (c) spreading of the yield zone; (d) load–deflection curve.

Limit analysis deals directly with the estimation of the collapse load, bypassing the spreading process of the contained plastic flow. The collapse load so obtained can be used as a realistic basis for design. Therefore, this relatively simple method is of intense practical interest to engineers. On the other hand, the knowledge of the collapse load for an elastic–plastic problem is also of interest in assessing the stages of development of elastic–plastic solutions, which can only be solved numerically by methods such as the *finite-element method*. In this type of numerical analysis, the solution becomes more and more difficult to obtain as the plastic regions spread. Especially when the plastic regions first meet and merge together, the plastic-elastic boundaries start to spread rapidly and further development of the solution becomes extremely difficult. In many cases, the numerical analysis has to be stopped before an uncontained plastic flow occurs. Such a numerical analysis combined with the direct determination of the collapse load by limit analysis is therefore of great value in obtaining a better understanding of the development of the uncontained plastic flow through the contained plastic flow analysis.

It should be emphasized here that the collapse load as calculated in limit analysis is different from the actual plastic collapse load, as it occurs in a real structure or body. Herein, we shall calculate the plastic collapse load of an *ideal structure*, at which the deformation of the structure can increase without limit while the load is held constant. This, of course, rarely happens in a real structure, and hence, the limit analysis calculations to be presented in what follows apply strictly, not to the real structure, but to the idealized one, in which neither *work hardening* of the material nor significant *changes in geometry* of the structure occur. Nevertheless, a load computed on the basis of this definition or idealization may give a good approximation to the actual plastic collapse load of a real structure.

This chapter is concerned with the basic theorems of limit analysis for general three-dimensional bodies. The upper- and lower-bound techniques of limit analysis developed on the basis of the two basic limit theorems will provide a bound value of the collapse load, which, in general, is not exact but approximate with known limits of error. Applications of these theorems involve studies of the distributions of stresses, strain rates, and velocities. These will be given in some detail in this chapter. Limit analysis of engineering structures such as beams, frames, plates, and shells will be treated in the next chapter.

## 8.2. Theorems of Limit Analysis

### 8.2.1. Basic Assumptions

As mentioned previously, the term *limit load* or *collapse load* is defined as the plastic collapse load of an idealized body or structure for which the plastic deformation can increase without limit under a constant limit load. There are two basic assumptions made about the so-called idealized body or structure. These are:

(i) *Perfectly plastic assumption for the material*: the material exhibits perfect or ideal plasticity with the associated flow rule without strain hardening or softening (Fig. 4.1a).

(ii) *Small deformation assumption for the structure*: changes in geometry of the body or structure that occur at the limit load are negligible; hence, the geometric description of the body or structure remains unchanged during the deformation at the limit load.

The small deformation assumption for the structure allows the use of the *virtual work principle*, which is the key to proving the limit theorems. The *virtual work equation* has the form (see Chapter 3)

Equilibrium set

$$\int_A T_i u_i^* \, dA + \int_V F_i u_i^* \, dV = \int_V \sigma_{ij} \epsilon_{ij}^* \, dV \qquad (8.1)$$

Compatible set

in which the surface traction force $T_i$, the body force $F_i$, and the stresses $\sigma_{ij}$ form an *equilibrium set*, while the displacement $u_i^*$ and the strains $\epsilon_{ij}^*$ form a *compatible set*. The integration is over the whole surface area, $A$, and the whole volume, $V$, of the body.

Any equilibrium set and any compatible set may be substituted in Eq. (8.1). In particular, an increment or rate of change of external forces and interior stresses, $(dT_i, dF_i, d\sigma_{ij})$ or $(\dot{T}_i, \dot{F}_i, \dot{\sigma}_{ij})$, may be used as an equilibrium set, and an increment or rate of change of displacements and strains, $(du_i^*, d\epsilon_{ij}^*)$, may be used as a compatible set. Thus, the rate form given below is also valid as an alternative form of the *virtual work equation*:

$$\int_A \dot{T}_i u_i^* \, dA + \int_V \dot{F}_i u_i^* \, dV = \int_V \dot{\sigma}_{ij} \epsilon_{ij}^* \, dV \qquad (8.2)$$

The two forms of virtual work equations, (8.1) and (8.2), will be used later in proving the theorems of limit analysis.

## 8.2.2. Admissible Stress Field and Velocity Field

It is well known that there are three basic relations that must be satisfied for a valid solution of any problem in the mechanics of deformable solids, namely, the equilibrium equations, the constitutive relations, and the compatibility equations. In a limit state, plastic deformations of arbitrarily large magnitude can take place in a structure under constant load. In general, in a limit analysis problem, only the equilibrium equations and yield criterion need be satisfied for a lower-bound solution, and only the compatibility equations and the flow rule associated with a yield criterion need be satisfied for an upper-bound solution. However, an infinite number of stress states will generally satisfy the equilibrium equations and the yield criterion alone, and an infinite number of displacement modes will satisfy the kinematic conditions associated with the flow rule and the displacement boundary conditions. Like other dual principles in structural mechanics, the two theorems of limit analysis are obtained by comparing first only the conditions imposed on the solution by the equilibrium requirements and the constitutive relations, and second the conditions imposed only by the kinematic requirements and the constitutive relations, with the complete or exact solution to be satisfied by all three requirements: equilibrium, kinematics, and constitutive relations.

### 8.2.2.1. STATICALLY ADMISSIBLE STRESS FIELD

The stress field which (a) satisfies the equations of equilibrium, (b) satisfies the stress boundary conditions, and (c) nowhere violates the yield criterion is termed a *statically admissible stress field* for the problem under consideration. The external loads determined from a statically admissible stress field alone are not greater than the actual collapse load according to the lower-bound theorem of limit analysis. Hence, the *lower-bound theorem* may be stated simply as follows: *If a statically admissible stress distribution can be found, uncontained plastic flow will not occur at a lower load.* From these rules, it can be seen that the lower-bound technique considers only the equilibrium and yield conditions. It gives no consideration to kinematics.

### 8.2.2.2. KINEMATICALLY ADMISSIBLE VELOCITY FIELD

On the other hand, an assumed deformation mode (or velocity field) that satisfies (a) velocity boundary conditions, and (b) strain rate and velocity compatibility conditions is termed a *kinematically admissible velocity field*. The loads determined by equating the external rate of work to the internal rate of dissipation for this assumed deformation mode are not less than the actual collapse load according to the upper-bound theorem of limit analysis. Hence, the *upper bound theorem* states simply that: *If a kinematically admissible velocity field can be found, uncontained plastic flow must impend or have taken place previously.* The upper-bound technique considers only velocity or failure modes and energy dissipation. The stress distribution need not be in equilibrium, and is only defined in the deforming regions of the assumed failure mode.

By a suitable choice of stress and velocity fields, the above two theorems enable one to bracket the collapse load in a direct manner, and the bounds can be found as close as necessary for a particular problem under consideration.

## 8.2.3. Limit Theorems

### 8.2.3.1. PREPARATORY THEOREM

**Theorem 1.** *When the limit load is reached and the deformation proceeds under constant load, all stresses remain constant; only plastic (not elastic) increments of strain occur.*

PROOF. A proof of this theorem starts with the rate form of the virtual work equation (8.2):

$$\int_{A_T} \dot{T}_i^c \dot{u}_i^c \, dA + \int_{A_u} \dot{T}_i^c \dot{u}_i^c \, dA + \int_V \dot{F}_i^c \dot{u}_i^c \, dV = \int_V \dot{\sigma}_{ij}^c \dot{\epsilon}_{ij}^c \, dV \qquad (8.3)$$

In this equation, the equilibrium set consists of the body force rates, $\dot{F}_i^c$, the surface traction rates, $\dot{T}_i^c$, and the stress rates, $\dot{\sigma}_{ij}^c$, while the compatible set is formed by the strain rates, $\dot{\epsilon}_{ij}^c$, and the continuous displacement rates, $\dot{u}_i^c$. It is noted that the surface traction rates $\dot{T}_i^c$ are specified on the surface area $A_T$ while the displacement rates, $\dot{u}_i^c$, are prescribed to be zero on $A_u$. The superscript $c$ emphasizes the fact that all quantities used in Eq. (8.3) are the *actual state* at collapse.

Now, at the limit load, the left-hand side of Eq. (8.3) vanishes, by definition; $\dot{F}_i^c = 0$ everywhere in $V$, $\dot{T}_i^c = 0$ on $A_T$, and $\dot{u}_i^c = 0$ on $A_u$. Since the total strain rates $\dot{\epsilon}_{ij}^c$ consist of elastic and plastic parts, $\dot{\epsilon}_{ij}^c = \dot{\epsilon}_{ij}^{ec} + \dot{\epsilon}_{ij}^{pc}$, it follows from Eq. (8.3) that

$$\int_V \dot{\sigma}_{ij}^c \dot{\epsilon}_{ij}^c \, dV = \int_V \dot{\sigma}_{ij}^c (\dot{\epsilon}_{ij}^{ec} + \dot{\epsilon}_{ij}^{pc}) \, dV = 0 \qquad (8.4)$$

But from the associated flow rule (4.6), $\dot{\sigma}_{ij}^c \dot{\epsilon}_{ij}^{pc} = 0$ since for a perfectly plastic material, the vector $\dot{\sigma}_{ij}$ (or $d\sigma_{ij}$) is tangential to the yield surface wherever plastic strains occur. Therefore,

$$\int_V \dot{\sigma}_{ij}^c \dot{\epsilon}_{ij}^{ec} \, dV = 0 \qquad (8.5)$$

For linear elastic or stable nonlinear elastic materials, the work done by any system of stresses on the elastic deformation it produces is always *positive definite* (see Chapter 3). Hence, the vanishing of the integral (8.5) requires that $\dot{\sigma}_{ij}^c = 0$ throughout the body. Therefore, there is no change in stress, and correspondingly, there is no elastic change in strain during deformation at the limit load. All deformation is plastic.

This theorem states that *elastic characteristics play no part in the collapse state at the limit load.* Thus, the application of the elastic–perfectly plastic stress–strain rate relation becomes the same as the use of the rigid–perfectly plastic stress–strain rate relation for the material.

### 8.2.3.2. LOWER-BOUND THEOREM

**Theorem 2.** *If an equilibrium distribution of stress* $\sigma_{ij}^E$ *can be found which balances the body force* $F_i$ *in V and the applied loads* $T_i$ *on the stress boundary* $A_T$ *and is everywhere below yield,* $f(\sigma_{ij}^E) < 0$, *then the body at the loads* $T_i$, $F_i$ *will not collapse.*

PROOF. To prove the theorem, assume it false. We show that this leads to a contradiction. If the body at the loads $T_i$, $F_i$ collapses, a collapse pattern associated with the actual stresses, strain rates, and displacement rates, $\sigma_{ij}^c$, $\dot\epsilon_{ij}^c$, and $\dot u_i^c$, exists (see Fig. 8.2b). This collapse pattern corresponds to the collapse loads $T_i$ on $A_T$ and $F_i$ in $V$, with $\dot u_i^c = 0$ on $A_u$. Two equilibrium systems would exist—$T_i$, $F_i$, $\sigma_{ij}^c$ and $T_i$, $F_i$, $\sigma_{ij}^E$. From virtual work equation (8.1),

$$\int_{A_T} T_i^c \dot u_i^c \, dA + \int_V F_i^c \dot u_i^c \, dV = \int_V \sigma_{ij}^c \dot\epsilon_{ij}^c \, dV \tag{8.6a}$$

$$\int_{A_T} T_i^c \dot u_i^c \, dA + \int_V F_i^c \dot u_i^c \, dV = \int_V \sigma_{ij}^E \dot\epsilon_{ij}^c \, dV \tag{8.6b}$$

Hence,

$$\int_V (\sigma_{ij}^c - \sigma_{ij}^E) \dot\epsilon_{ij}^c \, dV = 0 \tag{8.7}$$

Since at collapse, all deformation is plastic (Theorem 1), it follows that

$$\int_V (\sigma_{ij}^c - \sigma_{ij}^E) \dot\epsilon_{ij}^{pc} \, dV = 0 \tag{8.8}$$

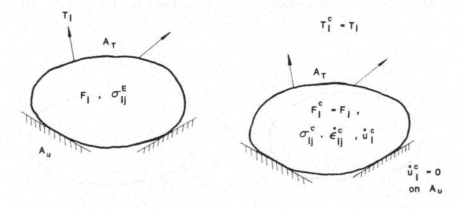

FIGURE 8.2. Proof of lower-bound theorem. (a) Any equilibrium, safe stress field, $f(\sigma_{ij}^E) < 0$; (b) assumed actual pattern at collapse.

In view of the fact that convexity and normality properties require $(\sigma_{ij}^c - \sigma_{ij}^E)\dot\epsilon_{ij}^{pc} > 0$ for $\sigma_{ij}^E$ below yield and since a sum of positive terms cannot vanish, Eq. (8.8) cannot be true, and the lower-bound theorem is proved. If $f(\sigma_{ij}^E) = 0$ is permitted, the body may be at the point of collapse.

The lower-bound theorem expresses the ability of the ideal body to adjust itself to carry the applied loads if at all possible.

### 8.2.3.3. UPPER-BOUND THEOREM

**Theorem 3.** *If a compatible mechanism of plastic deformation $\dot\epsilon_{ij}^{p*}$, $\dot u_i^{p*}$ is assumed which satisfies the condition $\dot u_i^{p*} = 0$ on the displacement boundary $A_u$, then the loads $T_i$, $F_i$ determined by equating the rate at which the external forces do work*

$$\int_{A_T} T_i \dot u_i^{p*} \, dA + \int_V F_i \dot u_i^{p*} \, dV \tag{8.9}$$

*to the rate of internal dissipation*

$$\int_V D(\dot\epsilon_{ij}^{p*}) \, dV = \int_V \sigma_{ij}^{p*} \dot\epsilon_{ij}^{p*} \, dV \tag{8.10}$$

*will be either higher than or equal to the actual limit load ($\sigma_{ij}^{p*}$ is the stress state associated with the strain rate $\dot\epsilon_{ij}^{p*}$).*

PROOF. Again, assume the theorem false; then we show that this leads to a contradiction. If the loads computed are less than the actual limit load, then the body will not collapse at this load. An equilibrium distribution of stress $\sigma_{ij}^E$ everywhere below yield, $f(\sigma_{ij}^E) < 0$, must therefore exist (converse of lower-bound theorem mentioned above; see Fig. 8.3b). From virtual work

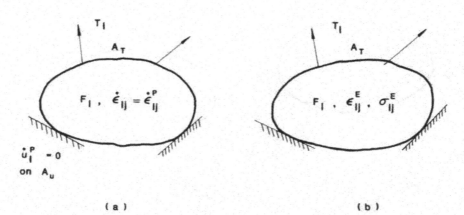

FIGURE 8.3. Proof of upper-bound theorem. (a) Any compatible plastic deformation mode: $\dot\epsilon_{ij}^P = \frac{1}{2}(\dot u_{i,j}^P + \dot u_{j,i}^P)$; (b) assumed equilibrium state: $F(\sigma_{ij}^E) < 0$.

equation (8.1):

$$\int_{A_T} T_i \dot{u}_i^{P^*} \, dA + \int_V F_i \dot{u}_i^{P^*} \, dV = \int_V \sigma_{ij}^E \dot{\epsilon}_{ij}^{P^*} \, dV \tag{8.11}$$

Since $T_i$ and $F_i$ are computed by equating Eq. (8.9) to Eq. (8.10), it follows that

$$\int_V (\sigma_{ij}^{P^*} - \sigma_{ij}^E) \dot{\epsilon}_{ij}^{P^*} \, dV = 0 \tag{8.12}$$

The convexity and normality properties require, however, that $(\sigma_{ij}^{P^*} - \sigma_{ij}^E) \dot{\epsilon}_{ij}^{P^*} > 0$ for $\sigma_{ij}^E$ below yield. This leads to a contradiction and thus proves Theorem 3.

The upper-bound theorem states that if *a path of failure exists, the ideal body will not stand up.*

## 8.2.4. Energy Dissipation Functions

In the application of the upper-bound theorem (Theorem 3), we are required to determine the rate of internal energy dissipation defined by

$$D = \sigma_{ij} \dot{\epsilon}_{ij}^P \tag{8.13}$$

where the stresses $\sigma_{ij}$ satisfy the yield condition

$$f(\sigma_{ij}) = F(\sigma_{ij}) - k^2 = 0 \tag{8.14}$$

We shall see that the specific dissipation rate is a unique function of the strain rate $\dot{\epsilon}_{ij}^P$.

We shall first assume that the yield function $f(\sigma_{ij})$ is strictly convex (i.e., there are no flats and corners). It is therefore continuously differentiable. In order to evaluate $D$ for a given $\dot{\epsilon}_{ij}^P$, we must solve the equations $\dot{\epsilon}_{ij}^P = \lambda \, \partial f / \partial \sigma_{ij}$ and $f(\sigma_{ij}) = 0$ for $\lambda$ and $\sigma_{ij}$. This inversion cannot be carried out in general form. An example of the inversion for the von Mises yield function will be given, but it is certainly true that $\sigma_{ij}$ can be uniquely determined (see Fig. 8.4; $\dot{\epsilon}_c^P$ determines $\sigma_c$ uniquely), as can the dissipation rate $D$.

We now examine the flat part of the yield surface. It can be seen from Fig. 8.4 that for a given $\dot{\epsilon}_{ij}^P$ ($\dot{\epsilon}_B^P$ in the figure) normal to the flat part on the yield surface $f(\sigma_{ij}) = 0$, the stress $\sigma_{ij}$ is not unique ($\sigma_{B1}, \sigma_{B2}$, for example), but $D = \sigma_{ij} \dot{\epsilon}_{ij}^P$ is uniquely determined.

The arguments can be extended to the corner point of the yield surface (point A in the figure), at which the same stress $\sigma_A$ is associated with a range of strain rates. Then the energy dissipation rate $D = \sigma_A \dot{\epsilon}_A^P$ is uniquely determined by a given strain rate $\dot{\epsilon}_A^P$.

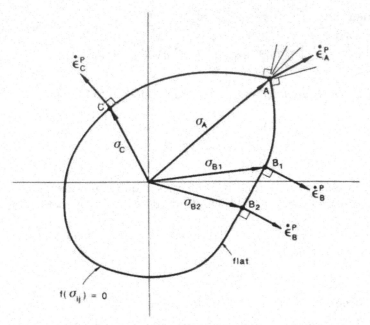

FIGURE 8.4. Energy dissipation as a function of strain rates $\dot{\epsilon}^P_{ij}$: (i) at convex part: $\dot{\epsilon}^P_C$ determines $\sigma_C$ and $D(\dot{\epsilon}^P_C)$ uniquely; (ii) at flat part: $\dot{\epsilon}^P_B$ determines $D(\dot{\epsilon}^P_B)$ uniquely, but not $\sigma_B$; (iii) at corner point: $\dot{\epsilon}^P_A$ determines $\sigma_A$ and $D(\dot{\epsilon}^P_A)$ uniquely.

### 8.2.4.1. ENERGY DISSIPATION FOR VON MISES, TRESCA, AND MOHR-COULOMB MATERIALS

Expressions for the energy dissipation function $D(\dot{\epsilon}^P_{ij})$ will now be given for the yield conditions of von Mises, of Tresca, and of Mohr-Coulomb, when the strain rates are continuous.

(i) If the von Mises yield condition is used, i.e.,

$$f(\sigma_{ij}) = F(\sigma_{ij}) - k^2 = \tfrac{1}{2}s_{ij}s_{ij} - k^2 = 0 \tag{8.15}$$

we have uniquely defined the plastic strain rates by the associated flow rule

$$\dot{\epsilon}^P_{ij} = \lambda \frac{\partial F}{\partial \sigma_{ij}} \tag{8.16}$$

so that, by Euler's theorem for homogeneous functions,

$$D = \sigma_{ij}\dot{\epsilon}^P_{ij} = \lambda \sigma_{ij} \frac{\partial F}{\partial \sigma_{ij}} = 2\lambda F = 2\lambda k^2 \tag{8.17}$$

since $F = \tfrac{1}{2}s_{ij}s_{ij}$ is homogeneous of second degree in the stress components. For this yield condition, $\dot{\epsilon}^P_{ij} = \lambda s_{ij}$, and $\dot{\epsilon}^P_{ij}\dot{\epsilon}^P_{ij} = \lambda^2 s_{ij}s_{ij}$; hence $2\lambda k = \sqrt{(2\dot{\epsilon}^P_{ij}\dot{\epsilon}^P_{ij})}$. Therefore, the required formula is

$$D(\dot{\epsilon}^P_{ij}) = k\sqrt{2\dot{\epsilon}^P_{ij}\dot{\epsilon}^P_{ij}} \tag{8.18}$$

For the plane strain case, $\dot{\epsilon}_3^P = 0$, Eq. (8.18) reduces to

$$D(\dot{\epsilon}_{ij}^P) = k\sqrt{2(\dot{\epsilon}_1^P)^2 + 2(\dot{\epsilon}_2^P)^2} = 2k|\dot{\epsilon}_1^P| \tag{8.19}$$

since the strain components satisfy the incompressibility condition $\dot{\epsilon}_1^P + \dot{\epsilon}_2^P = 0$. Noting that

$$|\dot{\gamma}_{max}^P| = |\dot{\epsilon}_1^P - \dot{\epsilon}_2^P| = 2|\dot{\epsilon}_1^P| \tag{8.20}$$

where $\gamma_{max}^P$ is the maximum rate of engineering plastic shear strain, then Eq. (8.19) further reduces to

$$D(\dot{\epsilon}_{ij}^P) = k\dot{\gamma}_{max}^P \tag{8.21}$$

(ii) If the Tresca yield condition is used, it has been shown, by considering typical points on the yield surface, together with the extended form of the flow rule and the incompressibility condition, that the plastic work rate, or the energy dissipation rate, is given by

$$D(\dot{\epsilon}_{ij}^P) = 2k \max|\dot{\epsilon}^P| \tag{8.22}$$

where $\max|\dot{\epsilon}^P|$ denotes the absolute value of the numerically largest component of the plastic strain rate [see Eq. (4.22) of Chapter 4 for the details].
For the particular case of plane strain, $\dot{\epsilon}_2 = -\dot{\epsilon}_1$ and $\dot{\epsilon}_3 = 0$, so that

$$\max|\dot{\epsilon}^P| = \frac{\dot{\epsilon}_1^P - \dot{\epsilon}_2^P}{2} = \frac{\dot{\gamma}_{max}^P}{2} \tag{8.23}$$

Thus, for the plane strain case, we have

$$D(\dot{\epsilon}_{ij}^P) = k\dot{\gamma}_{max}^P \tag{8.24}$$

(iii) If the Mohr–Coulomb yield condition is used, the plastic work rate $dW_p \equiv \dot{W}_p$ or the energy dissipation rate $D$, as given previously by Eqs. (4.36) and (4.37) (see Chapter 4), has the form

$$D = \dot{W}_p = f_c' \sum |\dot{\epsilon}_c^P| = \frac{f_c'}{m} \sum \dot{\epsilon}_t^P \tag{8.25}$$

where $\dot{\epsilon}_c^P$ and $\dot{\epsilon}_t^P$ denote the principal compressive and tensile components of the plastic strain rates, respectively; $f_c'$ is the uniaxial compression strength; and $m = f_c'/f_t'$, in which $f_t'$ is the uniaxial tension strength. $f_c', f_t',$ and $m$ are related to cohesion, $c$, and angle of internal friction, $\phi$, by the following equations:

$$f_c' = \frac{2c\cos\phi}{1 - \sin\phi} = 2c\tan\left(\frac{\pi}{4} + \frac{\phi}{2}\right) \tag{8.26a}$$

$$f_t' = \frac{2c\cos\phi}{1 + \sin\phi} = 2c\cot\left(\frac{\pi}{4} + \frac{\phi}{2}\right) \tag{8.26b}$$

$$m = \tan^2\left(\frac{\pi}{4} + \frac{\phi}{2}\right) \tag{8.27}$$

Using these relations, we rewrite Eq. (8.25) in the following form:

$$D = 2c \tan(\tfrac{1}{4}\pi + \tfrac{1}{2}\phi) \sum |\dot{\epsilon}_c^p| \qquad (8.28)$$

or

$$D = 2c \cot(\tfrac{1}{4}\pi + \tfrac{1}{2}\phi) \sum \dot{\epsilon}_t^p \qquad (8.29)$$

Equation (8.28) states that the rate of dissipation of energy per unit volume is $2c \tan(\tfrac{1}{4}\pi + \tfrac{1}{2}\phi)$ times the sum of the absolute values of the compressive plastic strain rate. As shown in Chapter 4, Eq. (8.25) is valid for all cases of stress state, and so are Eqs. (8.28) and (8.29). We can show that the rate of energy dissipation (8.25) has the alternative form

$$D = c \cot \phi (\dot{\epsilon}_1^p + \dot{\epsilon}_2^p + \dot{\epsilon}_3^p) \qquad (8.30)$$

In fact, Eq. (8.30) is seen most easily if one takes the end of a stress vector at the vertex of the Mohr–Coulomb pyramid, at which

$$\sigma_1 = \sigma_2 = \sigma_3 = c \cot \phi \qquad (8.31)$$

and using the definition of the energy dissipation

$$D = \sigma_1 \dot{\epsilon}_1^p + \sigma_2 \dot{\epsilon}_2^p + \sigma_3 \dot{\epsilon}_3^p \qquad (8.32)$$

we come up with Eq. (8.30).

For the particular case of plane strain, one of the principal components of the plastic strain rate is always zero, and the other two components have the relation

$$\dot{\epsilon}_t^p = m|\dot{\epsilon}_c^p| = |\dot{\epsilon}_c^p| \tan^2(\tfrac{1}{4}\pi + \tfrac{1}{2}\phi) \qquad (8.33)$$

on account of Eq. (4.33) (Chapter 4), so that the maximum rate of engineering plastic shear strain is given by

$$\dot{\gamma}_{max}^p = |\dot{\epsilon}_c^p| + \dot{\epsilon}_t^p = \frac{|\dot{\epsilon}_c^p|}{\cos^2(\tfrac{1}{4}\pi + \tfrac{1}{2}\phi)} \qquad (8.34)$$

Thus, for the plane strain case, Eq. (8.28) reduces to:

$$D = c \cos \phi \, \dot{\gamma}_{max}^p \qquad (8.35)$$

### 8.2.4.2. Energy Dissipation in a Surface of Discontinuity

In applications of the upper-bound theorem, it is often useful to consider a discontinuous velocity field. However, it should be kept in mind that in *plastic flow*, as distinct from *fracture*, actual discontinuity cannot occur across a fixed surface. The type of discontinuity to be considered in the following is simply an idealization of a continuous distribution in which the velocity changes very rapidly across a thin transition layer. The theorems are obviously valid in the presence of a transition layer. They will, therefore, remain valid in the limit as the thickness of the transition layer approaches zero. The rate of dissipation of energy in the transition layer approaches a finite value in the limit.

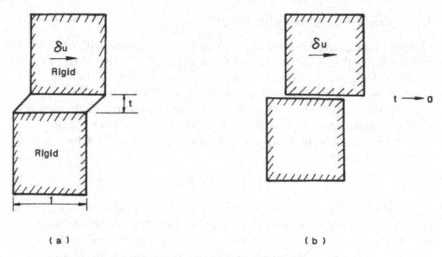

(a)                                          (b)

FIGURE 8.5. Narrow transition layer of Tresca material.

The discontinuity surface for Prandtl–Reuss and Tresca materials is the simplest one across which the tangential velocity changes, say by $\delta u$, as shown in Fig. 8.5. The top part of the block in Fig. 8.5 is moving to the right with the velocity $\delta u$ relative to the bottom part, and the two parts are separated by a narrow transition layer of plastic deformation of unit area and thickness $t$, in which the shearing strain rate is uniform. Since the mode of deformation is simple or pure shear with the large shear strain rate $\dot{\gamma}$ in the layer being equal to $\delta u / t$, the rate of energy dissipation in the narrow transition layer per unit area is computed from:

$$D = k\dot{\gamma}t = k\,\frac{\delta u}{t}\,t = k\,\delta u \qquad (8.36)$$

where $k$ is the yield stress in simple shear. The rate of energy dissipation in the narrow transition layer is seen to be independent of the thickness $t$, so $t$ may be taken as small as we please, including zero thickness as a matter of convenience. The zero-thickness idealization is certainly simple, but we should always keep in mind that it is the limiting case of a rapidly varying velocity layer. The rate of energy dissipation per unit area of this velocity discontinuity is simply the product of the yield stress in pure shear and the relative velocity of the two rigid blocks.

## 8.3. Applications of the General Theorems

It is the purpose of this section to give some simple examples, showing the simple techniques of applying the basic ideas of limit analysis aided only by a knowledge of ordinary mechanics of materials. Sometimes, remarkably good results can be obtained by the very elementary methods.

## 8.3.1. Tension of a Bar with Holes

Consider a long prismatic bar of rectangular cross section with one or more circular holes that is subjected to a force $P$ parallel to the edges (Figs. 8.6 and 8.7). Here, for simplicity, we adopt the Tresca yield condition, so the yield stress in simple tension has the value $\sigma_0 = 2k$.

If the bar has only one hole, a simple discontinuous stress field consisting of three longitudinal strips is assumed: two of the strips have simple tension $\sigma_0$, while the one containing the hole has no stress (Fig. 8.6a). A lower bound is obtained as

$$P^L = \sigma_0(b-d)t \tag{8.37}$$

To determine an upper-bound solution to the limit load, we need to assume a deformational failure mode or a velocity field which only needs to satisfy the compatibility condition. According to the upper-bound theorem, any compatible velocity field is permissible. In the following, we shall consider three different discontinuous failure modes as shown in Fig. 8.6b–d.

Mode 1 (Fig. 8.6b). In this mode, the upper and lower parts of the bar move as rigid bodies relative to each other by sliding along the planes $AB$ and $CD$, perpendicular to the face of the bar and making the angle $\alpha$ as shown. If the relative tangential velocity at the plane of sliding is $\dot{\delta}$, the velocity of separation is $\dot{\delta} \sin \alpha$. The rate of energy dissipation over the whole sliding surface is $k\dot{\delta}(b-d)t/\cos \alpha$. By setting this equal to the rate at which external work is done, $P_1^U \dot{\delta} \sin \alpha$, an upper bound is obtained as

$$P_1^U = \frac{k(b-d)t}{\sin \alpha \cos \alpha} = \frac{2k(b-d)t}{\sin 2\alpha} \tag{8.38}$$

This upper bound has a minimum value when $\alpha = 45°$:

$$P_1^U = 2k(b-d)t = \sigma_0(b-d)t \tag{8.39}$$

since the Tresca yield condition is adopted. It may be noted that the least upper bound of Eq. (8.39) is identical with the lower bound given by Eq. (8.37). Thus, it is the exact limit load

$$P_e = \sigma_0(b-d)t \tag{8.40}$$

Mode 2 (Fig. 8.6c). In this case, we assume that the bar is a thin plate, with thickness $t$ small compared to the hole dimensions, and use a velocity field that slides out of the plane of the plate, along a plane whose trace is $EF$ in the edge view. If the relative velocity along the plane $EF$ is $\dot{\Delta}$, the rate of separation of the parts above and below the sliding plane is $\dot{\Delta} \sin \alpha$. The rate of dissipation of energy at the sliding surface is $k\dot{\Delta}(b-d)t/\cos \alpha$, while the rate of external work is $P\dot{\Delta} \sin \alpha$. Hence, for a thin plate in this configuration, we again have $P_e = \sigma_0(b-d)t$.

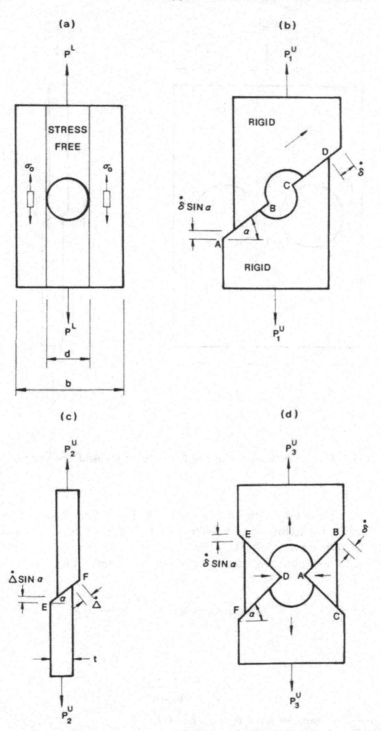

FIGURE 8.6. Tension of a bar with a hole: (a) statically admissible stress field; (b), (c), (d) kinematically admissible velocity fields.

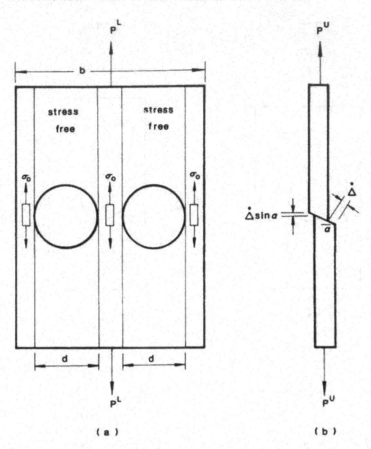

FIGURE 8.7. Tension of a sheet with two holes: (a) stress field, (b) velocity field.

*Mode 3* (Fig. 8.6d). In this mode, the bar is divided into four rigid parts, two of which move close to each other to form a neck. If the relative velocity along the slide planes $AB$, $AC$, $DE$, and $DF$ is $\dot\delta$, the rate of separation of upper and lower parts is $2\dot\delta \sin \alpha$. Applying the upper-bound theorem, one has

$$P_3^U(2\dot\delta \sin \alpha) = 4k\dot\delta \frac{(b-d)}{2} \frac{t}{\cos \alpha}$$

and

$$P_3^U = \frac{2k(b-d)t}{\sin 2\alpha}$$

For $\alpha = 45°$, again we have $P_c = \sigma_0(b-d)t$.

The limit load of Eq. (8.40) can be rewritten as

$$P_c = \sigma_0 A'$$

where $A'$ is the net area of the cross section, i.e., the cross-sectional area of the bar less the maximum area of the hole in a transverse plane. Clearly, this result can be extended to any shape of bar cross section; the hole could be rectangular, for example.

If the bar has several holes, no such general exact solution is immediately obtainable. However, if the bar is a thin plate, with $t \ll d$, a good solution may sometimes be quickly found. Consider a plate with two holes with centers on a transverse line as shown in Fig. 8.7a. Obviously, a lower bound is $\sigma_0 A'$. An upper bound can be found by assuming the velocity field shown in Fig. 8.7b. Finally, we come up with the same result as given by Eq. (8.40).

Obviously, this problem of the weakening effect of holes in tension sheets is of practical importance. Brady and Drucker (1955) have investigated plates with several rows of holes in various configurations and have compared the estimated limit loads with test results. They have also discussed the interpretation of the limit analysis results for practical design purposes.

### 8.3.2. Bending of a Notched Bar

Figure 8.8 shows a notched bar subjected to two equal and opposite pure couples $M$ at its ends. The minimum thickness of the bar is $a$ and the depth of the bar perpendicular to the plane of the paper is assumed to be very large so that the bar bends in plane strain condition.

#### 8.3.2.1. LOWER BOUND

The stress field shown in Fig. 8.8a consists of three regions. The regions to the left and right of the notch are assumed stress-free, while the regions below the notch are subjected to simple bending. At yield, there is a constant compressive stress $2k$ in the region above the neutral axis and a constant tensile stress $2k$ in the region below. The lower bound for the value of the bending couple is obtained as

$$M^L = 2k\left(\frac{a}{2}\right)\left(\frac{a}{2}\right) = 0.5ka^2 \tag{8.42}$$

#### 8.3.2.2. UPPER BOUND

The velocity field shown in Fig. 8.8b is considered to be a shearing action along two circular arcs of length $l$ and radius $r$. The rigid outer portions I and II of the bar rotate about the rigid hinge (the shaded area in Fig. 8.8b) with an angular velocity $\omega$, so that the constant discontinuous velocity along the arcs is $r\omega$. Thus, the rate of internal energy dissipation is, from Eq. (8.36),

$$D = 2k(r\omega)(2r\alpha) = 4kr^2\alpha\omega = 4ka\omega\left(\frac{a}{2\sin\alpha}\right)^2$$

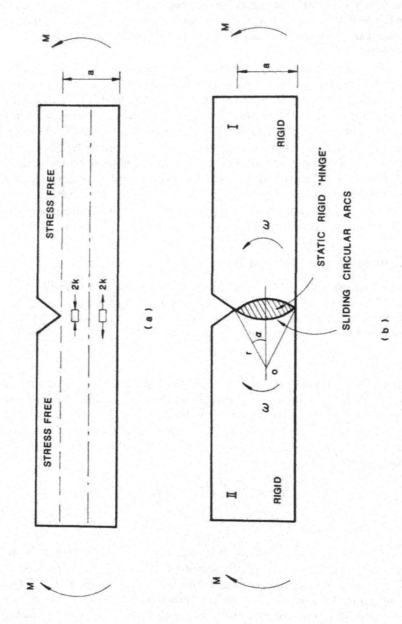

FIGURE 8.8. Bending of a notched bar: (a) stress field, (b) velocity field.

The external rate of work of the applied couples is $2M\omega$. The upper bound is

$$M^U = \frac{ka^2}{2} \frac{\alpha}{\sin^2 \alpha} \tag{8.43}$$

This upper bound has a minimum value when $\tan \alpha = 2\alpha$, or $\alpha = 67°$. Hence,

$$M^U = 0.69ka^2 \tag{8.44}$$

The limit load is therefore bound by

$$0.5 \le \frac{M}{ka^2} \le 0.69 \tag{8.45}$$

A slip line solution by Green (1954) gives a better value $0.63ka^2$ for an upper bound.

## 8.4. Discontinuous Stress Fields

In constructing an equilibrium distribution of the stress field which does not violate the yield condition, it may be advantageous to divide the body into several stress zones. In each zone, the stress field will satisfy the equation of equilibrium and will not violate the yield condition; further, the stress field will be continuous in each zone. However, the stress state on the boundary between two neighboring zones may have stress discontinuities. Such *discontinuities* of stress across a surface have been seen in the examples of Section 8.3 (see Figs. 8.6a, 8.7a, and 8.8a).

As an illustration, consider the development of the stress distribution of a bar bent by a couple as shown in Fig. 8.9. Figure 8.9a shows the usual linear elastic stress distribution. As the moment increases, the material near the upper and lower edges of the bar yields first, and the bar enters an elastic–plastic state as shown in Fig. 8.9b, with an elastic core in the central portion of the cross section. As the bending moment further increases toward its ultimate state corresponding to the limiting state configuration in Fig. 8.9c, the elastic core shrinks to a membrane, and we have the special case of a discontinuity with a jump of $4k$ in the stress component parallel to the line of stress discontinuity. Here, as in the previous case of velocity discontinuity, the line of stress discontinuity can be interpreted as the limiting case of a narrow transition zone between two distinct stress fields.

For a discontinuous stress field to be valid, the equilibrium conditions must be satisfied at any point on the discontinuous surfaces. First, we consider a boundary on two sides of which the stress systems are different. Figure 8.10 shows such a boundary plane between zone 1 and zone 2. Equilibrium requires only that the normal stress and shear stress be the same on both sides, namely,

$$\sigma_n^{(1)} = \sigma_n^{(2)} \quad \text{and} \quad \tau^{(1)} = \tau^{(2)} \tag{8.46}$$

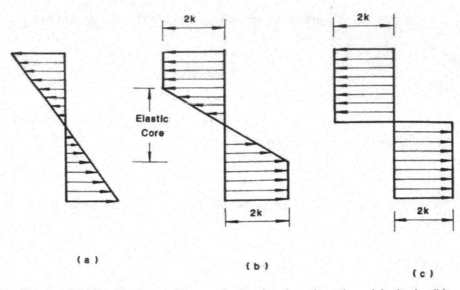

(a)          (b)          (c)

FIGURE 8.9. Development of stress distribution in a bent bar: (a) elastic, (b) elastic–plastic, (c) fully plastic.

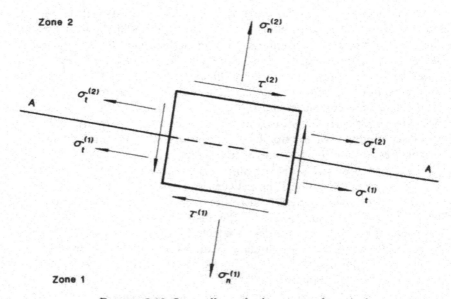

FIGURE 8.10. Stress discontinuity across plane A–A.

where the superscripts denote the zones. Equilibrium, however, places no restriction on the change of $\sigma_t$ across the boundary. Therefore, the components $\sigma_t^{(1)}$, $\sigma_t^{(2)}$ acting parallel to the boundary may be different.

In what follows, we shall discuss some discontinuous patterns of stress fields that are useful in estimating the lower bound of a limit load.

## 8.4.1. Triangular Stress Field

Figure 8.11a shows a right triangle $ABC$ subjected to a uniform pressure $p$ and $q$ on its two sides $AC$ and $BC$, respectively, so that the region $ABC$ forms a uniform stress field of biaxial compression type. Region $ABEF$ has a simple compression state, $\sigma_0 = 2k$, to keep the triangle on balance.

As shown in Fig. 8.11b, the resultant forces $P$, $Q$, and $F$ applied to the three sides of the triangle must be in equilibrium. This leads to

$$pb = \frac{fb \sin(\beta+\gamma) \cos \beta}{\sin \gamma} \quad \text{or} \quad p = f\frac{\sin(\beta+\gamma) \cos \beta}{\sin \gamma} \qquad (8.47)$$

and

$$qb \cot \gamma = \frac{fb \sin(\beta+\gamma) \sin \beta}{\sin \gamma} \quad \text{or} \quad q = f\frac{\sin(\beta+\gamma) \sin \beta}{\cos \gamma} \qquad (8.48)$$

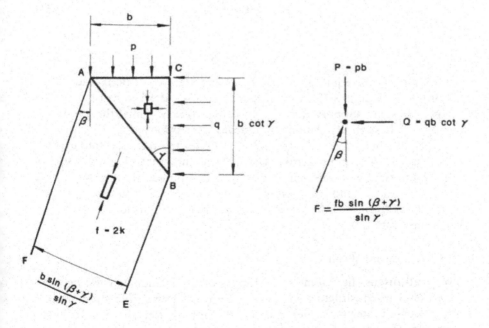

(a)                                    (b)

FIGURE 8.11. A rectangular-triangle stress field: (a) stress field, (b) equilibrium.

Assume that the depth of the triangle $ABC$ perpendicular to its plane is so large that the problem satisfies the condition of plane strain. If the Tresca yield condition is adopted, the yield condition for stresses $p$ and $q$ in the region $ABC$ is of the form:

$$p - q = 2k \tag{8.49}$$

Substituting Eqs. (8.47) and (8.48) into Eq. (8.49) and noting that $f = 2k$, one gets

$$\frac{\sin(\beta + \gamma) \cos(\beta + \gamma)}{\sin \gamma \cos \gamma} = 1$$

or

$$\sin 2(\beta + \gamma) = \sin 2\gamma = \sin(\pi - 2\gamma)$$

which implies

$$\gamma = \frac{\pi}{4} - \frac{\beta}{2} \tag{8.50}$$

Substituting Eq. (8.50) into Eqs. (8.47) and (8.48) yields

$$p = 2k(1 + \sin \beta) \tag{8.51}$$

and

$$q = 2k \sin \beta \tag{8.52}$$

It is seen that the equilibrium equations and yield conditions lead to Eqs. (8.50), (8.51), and (8.52), of which Eq. (8.50) is the geometric condition that relates the support angle $\beta$ with the angle $\gamma$. Thus, the stress field satisfying these three relations is *statically admissible*.

It should be noted that the side $AB$ of the triangle is a line of *stress discontinuity*. The jump across $AB$ in the component of normal stresses parallel to the line of discontinuity can be computed, but there is no need to do so. Normal and shear stresses across $AB$ are continuous because the overall equilibrium of the whole region has been established, which assures that the continuity requirements are met.

### 8.4.1.1. SPECIAL CASES

We shall discuss here some special cases of the triangular stress field that are useful in constructing some other forms of admissible stress fields to be discussed later in connection with the application of the lower-bound theorem of limit analysis of perfect plasticity.

(i) If $\gamma = 30°$, Eqs. (8.50), (8.51), and (8.52) lead to

$$\beta = 30°$$

$$p = 3k$$

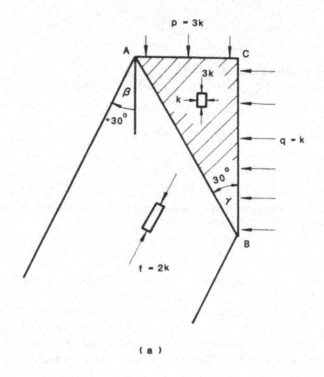

(a)

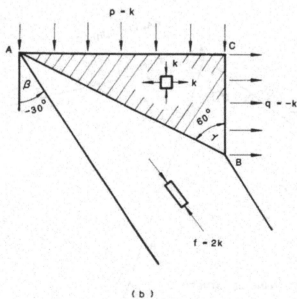

(b)

FIGURE 8.12. Special cases of triangular stress fields: (a) $\beta = 30°$, $\gamma = 30°$, $p = 3k$, $q = k$; (b) $\beta = -30°$, $\gamma = 60°$, $p = k$, $q = -k$.

and

$$q = k$$

This case is shown in Fig. 8.12a.

(ii) If $\gamma = 60°$, Eqs. (8.50), (8.51), and (8.52) lead to

$$\beta = -30°$$

$$p = k$$

and

$$q = -k$$

In this case, the region $ABC$ is subjected to a biaxial tension–compression state, as shown in Fig. 8.12b.

### 8.4.1.2. WEDGE UNDER UNILATERAL PRESSURE

The stress field of a wedge with a uniform pressure on one face can also be viewed as a special case of the triangular stress field as the two sides, $AB$ and $AC$, of the triangle element extend to infinity. Figure 8.13 shows an obtuse wedge $C'AE'$ loaded by a uniform pressure $p$ along $AC'$. The line $AB'$ is the line of stress discontinuity separating the two constant-stress regions $AC'B'$ and $AB'E'$ The state of stress for each region is represented

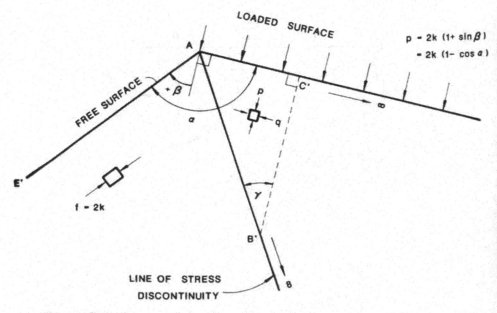

FIGURE 8.13. Obtuse wedge under unilateral pressure: wedge angle $\alpha = \pi/2 + \beta > \pi/2$.

by the small arrows marked in the figure. Since the pressure $p$ obtained previously is a function of $k$ and $\beta$ only, it can be concluded that here the value of $p$ is still given by Eq. (8.51). Figure 8.14 gives a stress field for an acute wedge and again the value of $p$ is still given by Eq. (8.51). By the lower-bound theorem of limit analysis, $p$ is therefore a lower bound for the uniform critical pressure of the wedges.

## 8.4.2. Discontinuous Fields of Stress Viewed as Pin-Connected Trusses

### 8.4.2.1. Triangular Stress Field Considered as a Truss Joint

Figure 8.15 shows that a triangular stress field $ABC$ can be considered to be a joint of three truss bars. In Fig. 8.15a, there are two compressive force fields, $P$ and $Q$, overlapping in region $ABC$. These are balanced by a third compressive force $F$. No internal equilibrium problems arise if $P$, $Q$, and $F$ are in equilibrium. The stress in region $ABC$ is just the sum of a uniaxial stress $Q/b$ and a uniaxial stress $P/a$ of the angles pictured. If $Q/b$, $P/a$, and $F/c$ are each at or below yield ($2k$), the only question which arises is whether $Q/b$ and $P/a$ sum to a state of stress below yield. Note that there are four lines of discontinuity meeting at each point (e.g., $AA'$, $AA''$, $AB$, $AC$). Normal and shear stress across each line of discontinuity will be

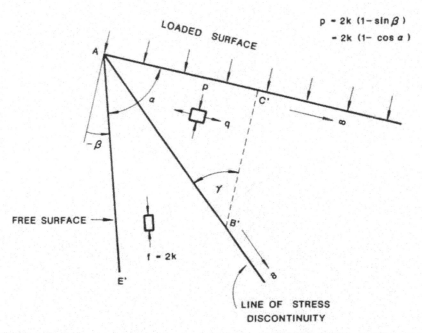

FIGURE 8.14. Acute wedge under unilateral pressure: wedge angle $\alpha = \pi/2 - \beta < \pi/2$.

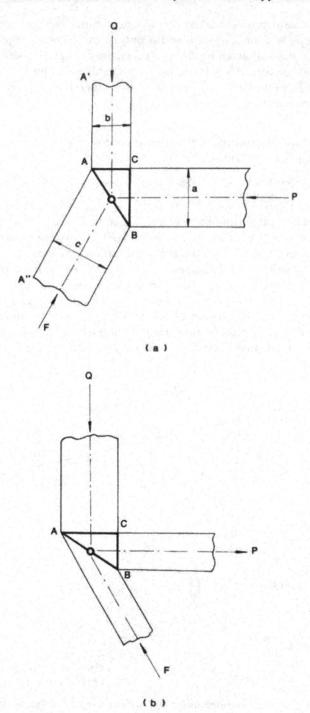

(a)

(b)

FIGURE 8.15. Triangular stress field considered as a truss joint.

continuous because equilibrium is satisfied in all regions. The different jumps across each side $AB$, $BC$, $CA$ can be computed, but there is no need to do so.

In Fig. 8.15b, the force $P$ is tensile instead of compressive. The combination of a tensile stress $P/a$ and a compressive stress $Q/b$ is likely to give a much higher maximum shear stress in $ABC$ of Fig. 8.15b than is produced by the combination of two compressive stresses in Fig. 8.15a.

### 8.4.2.2. DISCONTINUOUS STRESSES IN TRUNCATED WEDGE

Figure 8.16 shows an applied force $P$ carried by a rectangular block. Now, imagine a pin-connected truss to carry the load inside the body, as in Fig. 8.16b. The forces in the members of the truss are determined by equilibrium conditions at each joint. For a usual structural design, the cross-sectional area of each member must be taken large enough to give a safe or permissible axial stress. However, here, the stresses in each "member" and each "joint" must be chosen not to violate the yield condition so that the stress field obtained is statically admissible, and a lower bound on the limit load $P$ can be found from it.

Figure 8.17 gives the detail of such a stress field with a choice of $\beta = 30°$ and a minimum width of the inclined legs so as to give the yield value of $2k$. The state of stress in the overlap region $AAC$ is not above yield, and also the forces at the lower joints $A'B'C'$ can be carried without violating the yield condition of Tresca.

The stress pattern consists of four elementary stress fields as shown in Fig. 8.17, where the field I or II is the special case of the triangular field of Fig. 8.12a ($\beta = 30°$) while the field III or IV is the case of Fig. 8.12b ($\beta = -30°$).

The lower bound on the limit load $P$ determined from such a stress field is $P = 6kb$.

Figure 8.18 shows the stress field of a *truncated wedge* with the support angle $\beta$ taking a value between 0 and $\pi/2$. The lower bound on the limit load $P$ determined from this stress field is $P = 4kb(1 + \sin \beta)$, which turns out to be larger if $\beta$ takes a larger value.

### 8.4.2.3. COMPACT DISCONTINUOUS STRESS FIELD

Further example of discontinuous stress fields shown in Fig. 8.19 are the compact configurations of Figs. 8.17 and 8.18. The trapezoids $ABCD$ are in plastic states of stress due to the normal pressure $p$ and $p'$ on the parallel sides $AB$ and $CD$, the sides $AD$ and $BC$ being free from applied stress. The lines $AO$, $BO$, $CO$, and $DO$ are lines of stress discontinuity separating the regions of constant stress. The regions $ABO$ and $DCO$ are regions of constant biaxial compression and biaxial compression–tension, respectively. The regions $BCO$ and $ADO$ are regions of uniaxial compression.

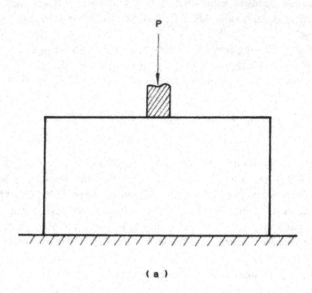

(a)

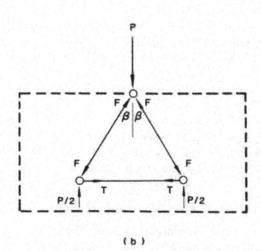

(b)

FIGURE 8.16. Truss action to carry load.

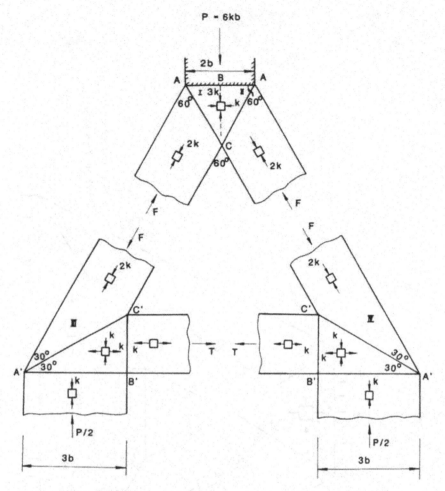

FIGURE 8.17. Discontinuous stresses in truncated wedge ($\beta = 30°$).

## 8.4.3. Examples of Applications

### EXAMPLE 8.1. TENSION OF A SPECIMEN WITH CIRCULAR NOTCHES

A notched tensile specimen is assumed to be stressed under the condition
of plane strain. A statically admissible stress field is shown in Fig. 8.20.
This field is symmetric with respect to the horizontal axis $OO$; the triangular
regions below this axis of symmetry form the trapezoidal pattern discussed
previously in connection with Fig. 8.19, except for the fact that the normal
stresses transmitted across the top and the base of the trapezoid are now
tensile rather than compressive. The rectangular region at the bottom of
the specimen is a region of pure axial stress, equal to the normal stress
transmitted across the base of the trapezoid. The largest lower bound on

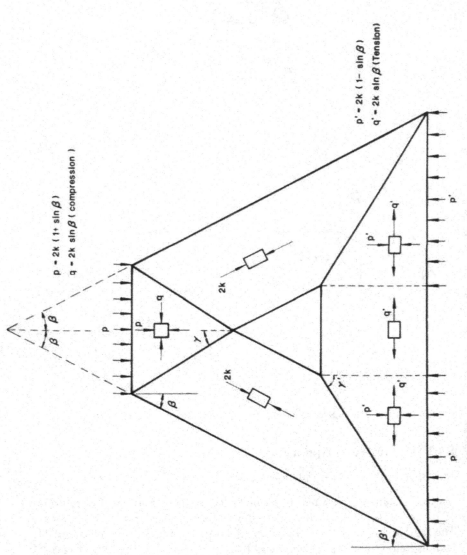

$p = 2k \, (1 + \sin\beta)$
$q = 2k \, \sin\beta \, (\text{compression})$

$p' = 2k \, (1 - \sin\beta)$
$q' = 2k \, \sin\beta \, (\text{Tension})$

FIGURE 8.18. Discontinuous stresses in truncated wedge: general case ($\beta \neq 30°$).

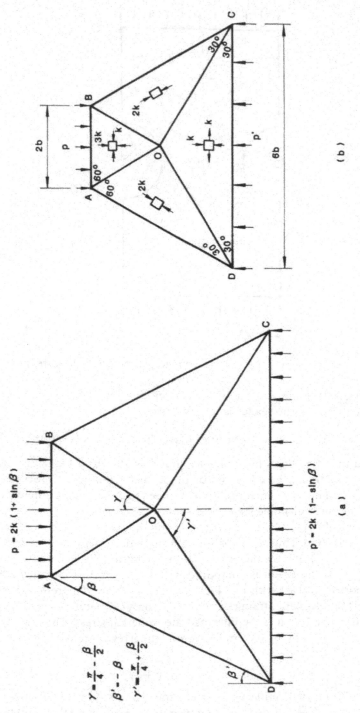

FIGURE 8.19. Compact discontinuous stress fields at yield.

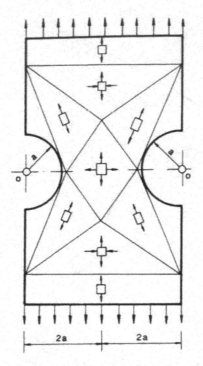

FIGURE 8.20. Tension of a specimen with circular notches (plane strain).

the axial stress determined from this stress field is $\sigma^L = 1.26k$. We leave the verification of this to the reader as an exercise.

## EXAMPLE 8.2. TENSION OF A PLATE WITH HOLES (PLANE STRESS)

Consider a thin plate with five holes arranged as shown in Fig. 8.21. This plate is stressed under the condition of plane stress. Before we proceed to the discussion of this particular problem, we need to point out the differences between a problem of plane stress and one of plane strains.

In the case of plane strain, $\dot\epsilon_z = \dot\gamma_{zx} = \dot\gamma_{zy} = 0$, it follows that $\dot\epsilon_z$ is always a principal direction. Thus, $\dot\epsilon_3 = \dot\epsilon_z = 0$. For a rigid–perfectly plastic material, we have $\dot\epsilon_z = \dot\epsilon_z^p = 0$, and the incompressibility condition leads to $\dot\epsilon_x^p + \dot\epsilon_y^p = 0$, or $\dot\epsilon_1^p + \dot\epsilon_2^p = 0$. If this result is compared with the strain rates predicted by the flow rule associated with the Tresca yield function, we are led to the conclusion that the plane strain restriction can only be met if $\sigma_3 = \sigma_z$ always takes a value such that it is the intermediate principal stress. Thus, we find that either $\sigma_1 > \sigma_z > \sigma_2$ or $\sigma_2 > \sigma_z > \sigma_1$, and the Tresca yield criterion has the form

$$|\sigma_1 - \sigma_2| = 2k \quad \text{or} \quad (\sigma_x - \sigma_y)^2 + 4\tau_{xy}^2 = 4k^2 \tag{8.53}$$

If the von Mises yield criterion is used, the plane strain condition $\dot\epsilon_z^p = d\lambda\,(\partial f/\partial\sigma_z) = s_z\,d\lambda = 0$ leads to $\sigma_z = (\sigma_x + \sigma_y)/2$ directly, and the von Mises

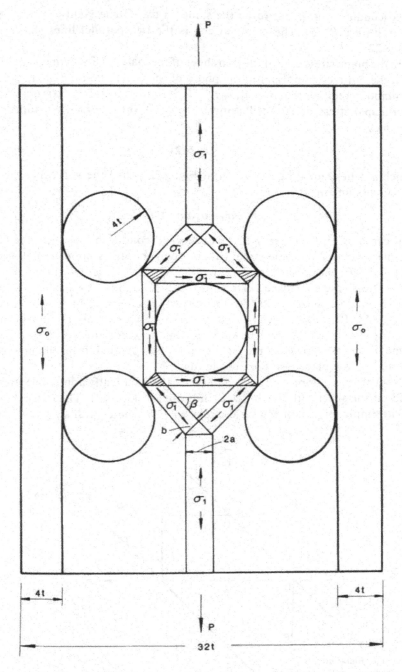

FIGURE 8.21. Stress field of a plate with five holes under tension ($t$ = thickness of the sheet).

yield condition turns out to be the same as the Tresca yield condition as given by Eq. (8.53). The yield surface is the two parallel lines shown in Fig. 8.22.

In the plane stress case, if the thin sheet of the material lies in the $xy$-plane, with the $z$-direction normal to the plane of the sheet, then the plane stress assumption requires that $\sigma_z = \tau_{zx} = \tau_{zy} = 0$. It follows that the $z$-axis is always a principal stress direction. Putting $\sigma_3 = \sigma_z = 0$, the Tresca yield function becomes

$$|\sigma_1| = 2k, \qquad |\sigma_1 - \sigma_2| = 2k, \qquad |\sigma_2| = 2k \tag{8.54}$$

which is a hexagonal figure shown in Fig. 8.22, while the von Mises yield function is an ellipse:

$$\sigma_1^2 + \sigma_2^2 - \sigma_1 \sigma_2 = 4k^2 \tag{8.55}$$

It can be seen from Fig. 8.22 that the yield conditions are the same for the plane stress and plane strain cases only if the stress point lies on lines $CD$ and $AF$; they are different otherwise. Since the yield surfaces in the plane stress case are closed curves, the stress is bounded by a value $M$ (for the Tresca criterion, $M = 2k$; for the von Mises criterion, $M = 4k/\sqrt{3}$), i.e. $|\sigma_{max}| \leq M$. If a plane sheet is subjected to an external traction on its boundary, say, $\sigma_n = p$, then the value $p$ cannot exceed the value $M$. In the plane strain case, however, the yield curve is two parallel lines, and no such restrictions are placed on the limit loads.

Now, the truss approach is used to construct a statically admissible stress field for the sheet with five holes as indicated in Fig. 8.21. The outer strips are in simple tension at the yield stress $\sigma_0 = 2k$. The central strip of width

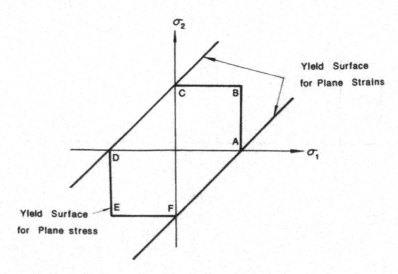

FIGURE 8.22. Differences in yield functions between plane stress and plane strains.

$2a$ has the tensile stress $\sigma_1$, which is carried around the central hole in a truss manner as indicated. The stresses in each "member" and each "joint" must not violate the yield condition of plane stress, as given by Eq. (8.54). In this case, the two triangles above and below the central hole have equal biaxial stress $\sigma_1$, while the stress in the four shaded triangles is obtained by superimposing the simple compression $\sigma_1$ in the horizontal strip on the tension $\sigma_1$ in the strip of width $b$ inclined at the angle $\beta$, as shown. The maximum shear stress in these shaded triangles is $\sigma_1 \sin \beta$. The Tresca yield condition is satisfied if the load carried by the central strip is $2a\sigma_1 = 2ak/\sin \beta = \sigma_0 b$. The maximum size of $b$, for the particular pattern of holes, may be found by trial and error. In this case, we find $b = 1.7t$, and hence the lower bound $P^L = 9.7\sigma_0 t$.

## EXAMPLE 8.3. PUNCH INDENTATION IN PLANE STRAIN

Consider the indentation of a rigid punch into a half-space of perfectly plastic Tresca material (Fig. 8.23). Assume that the width of the punch in the direction perpendicular to the plane of paper is so large that the problem is one of plane strain.

As a first attempt, consider the simple discontinuous stress field shown in Fig. 8.23a, which, yields a lower bound on the limit load $P$:

$$P_1^L = \sigma_0 b = 2kb \tag{8.56}$$

This is, of course, not a good lower bound, because the load is considered to be carried only by a single vertical strip of material directly beneath the punch. To improve the answer, we consider adding a horizontal pressure field as shown in Fig. 8.23b. In the overlapping region, the material is subjected to a biaxial compression so that the vertical stress can be increased to $2\sigma_0$ without violating the yield condition. The improved lower bound obtained is

$$P_2^L = 2\sigma_0 b = 4kb \tag{8.57}$$

Alternatively, we can use the concept of truss-action approach. Imagine the load $P$ to be carried by two inclined truss bars as shown in Fig. 8.24, and further, add a vertical "leg" directly below the punch area $AA$ of amount $2k$ to make a stress field as shown in Fig. 8.25. In this case, the stress discontinuities are admissible. We note, however, that the yield condition is violated in regions I and II. In region I, for example, the difference between the greatest and the least principal stress is $4k$. This violation can be accommodated by introducing at the free surface a horizontal strip in which there is a horizontal compressive stress $2k$. The width of this strip is as shown in Fig. 8.26a. Using this stress field, we obtain a better lower bound:

$$P_3^L = 5kb \tag{8.58}$$

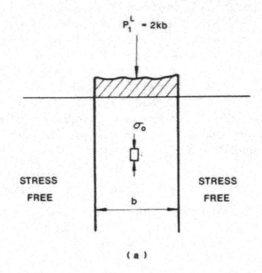

(a)

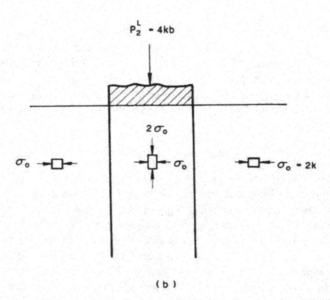

(b)

FIGURE 8.23. Stress fields for punch indentation in plane strain.

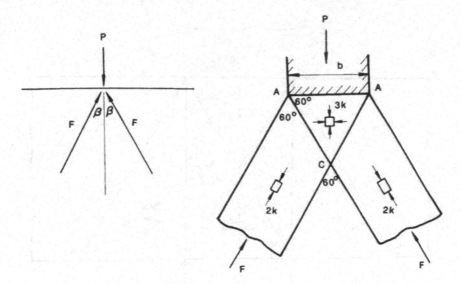

FIGURE 8.24. Load $P$ carried by two truss bars (two-leg stress field).

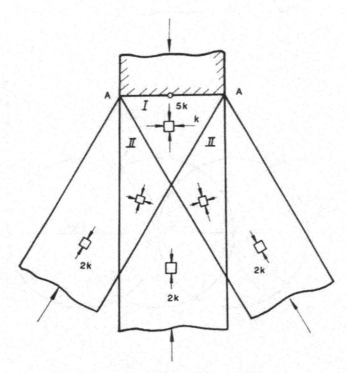

FIGURE 8.25. Three-leg stress field.

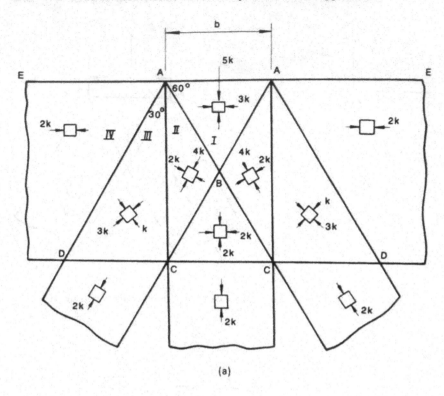

(a)

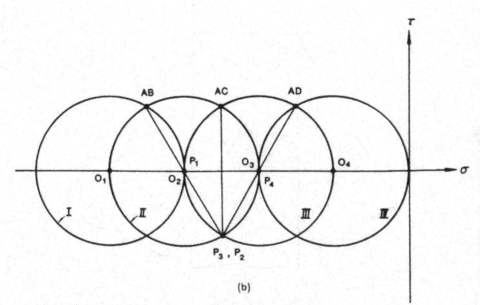

(b)

FIGURE 8.26. Combined stress field. (a) Discontinuous stress field gives a lower bound $P_3^L = 5kb$; (b) Mohr circles for regions I, II, III, and IV.

Figure 8.26b shows the Mohr circles with centers at $O_1$, $O_2$, $O_3$, and $O_4$ for the stress regions I, II, III, and IV, respectively. $P_1$, $P_2$, $P_3$, and $P_4$ are the poles of these Mohr circles. Points marked by $AD$, $AC$, and $AB$ represent the stresses at the lines of discontinuity $\overline{AD}$, $\overline{AC}$, and $\overline{AB}$, respectively. These four circles are built up in the following way. The Mohr circle $O_4$, representing a uniaxial compression of amount $\sigma_0 = -2k$ in region IV, is first drawn, and the pole $P_4$ is determined. Next, the line drawn through the pole $P_4$ of the Mohr circle parallel to the discontinuous line $\overline{AD}$ intersects a point labeled $AD$ on the circle $O_4$. When this point is known, the circle $O_3$, which corresponds to the state of stress in region III, is readily found by drawing a circle through the point $AD$ with radius $k$ and center $O_3$ on the $\sigma$-axis, and the pole $P_3$ of the circle $O_3$ is also determined. Again, circles $O_2$, $O_1$ and poles $P_2$, $P_1$ can be found in a similar manner. Readers can work out the details to check the validity of the stress field given by Fig. 8.26.

### 8.4.4. Remarks

Basic techniques of constructing a statically admissible stress field have been discussed in this section. Although the discussion has been restricted to Tresca or von Mises materials, these techniques can be applied to other types of materials. The differences lie in the yield conditions that must not be violated for an admissible stress field. The yield conditions of Tresca or von Mises material as given by Eq. (8.53) for the plane strain case or by Eqs. (8.54) and (8.55) for the plane stress case are simple and easy to use. However, other types of yield conditions, for example, the Mohr–Coulomb condition for soils, are not so straightforward to apply, special considerations are required in a lower-bound analysis (Chen, 1975).

## 8.5. Basic Techniques in Applications of the Upper-Bound Method

As we have seen from the preceding section, in applying the lower-bound theorem, we must assume a statically admissible stress field. Based on this assumed stress field, a lower-bound solution to the collapse load can be determined from the condition of equilibrium.

In contrast, in the application of the upper-bound theorem, the first step is to assume a kinematically admissible velocity field. An upper bound is then obtained by equating the rate of work done by the external forces to the internal rate of energy dissipation. Such a velocity field is a failure mechanism which must be continuous in the sense that no gaps or overlaps develop within the body, and the direction of the strains that are defined by the mechanism must in turn define the yield stresses required to calculate the dissipation.

In this section, we shall use some examples to illustrate the basic techniques in applying the upper-bound theorem.

### 8.5.1. Rigid-Block Sliding

The failure mechanism of rigid-block sliding is usually assumed in constructing a kinematically admissible velocity field. In this mechanism, the body is separated into several rigid blocks by some narrow transition layers.

EXAMPLE 8.4. UPPER-BOUND ANALYSIS OF THE PUNCH INDENTATION PROBLEM (PART 1: MECHANISMS OF RIGID-BLOCK SLIDING)

In this problem, the punch is assumed to be rigid; thus, the geometric boundary condition requires that the movement of the contact plane must always remain plane.

*Rotational Mechanisms (Figs. 8.27 and 8.28)*

Figure 8.27b shows a simple rigid-body rotational mechanism about O. This mechanism is geometrically admissible if there are no external constraints to hold the punch vertical. The block of material B rotates as a rigid body about O with an angular velocity $\dot{\alpha}$, and there is a semicircular transition layer between the rotating material and the remainder of the body. Since the angular velocity is $\dot{\alpha}$, the rate of work done by the external force P is the downward velocity at the center of the punch, $\dot{\alpha}b/2$, multiplied by the applied force P, while the total rate of energy dissipation along the semicircular discontinuity surface is found by multiplying the length of this discontinuity, $\pi b$, by the yield stress in pure shear, k, times the discontinuity in velocity across the surface $b\dot{\alpha}$. Equating the rate of external work to the rate of total internal energy dissipation gives:

$$P^U(\tfrac{1}{2}b\dot{\alpha}) = k(b\dot{\alpha})(\pi b)$$

or

$$P^U = 2\pi kb = 6.28 kb \qquad (8.59)$$

As we can see, the upper-bound load calculation is independent of the magnitude of the angular velocity $\dot{\alpha}$, so we can assume $\dot{\alpha}$ to be sufficiently small not to disturb the overall geometry. In other words, the proofs of the limit theorems can carry through using the initial geometry of the problem.

Now the rotational mechanism of Fig. 8.27b may be generalized by taking the radius and the position of the center of the circle as two independent variables; and we hope to find a lower, and therefore better, upper-bound solution by such a shifting of the center of rotation. If the center is shifted to O' as shown in Fig. 8.28a, the rate of external work is $P(r\cos\theta - \tfrac{1}{2}b)\dot{\alpha}$, where r is the radius of the surface of discontinuity and $\theta$ is the angle

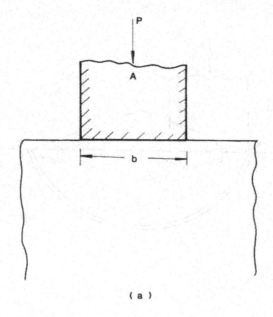

( a )

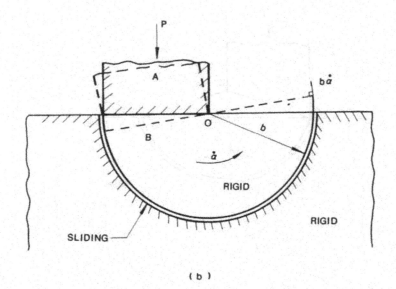

( b )

FIGURE 8.27. (a) Punch indentation problem; (b) rotational mechanism.

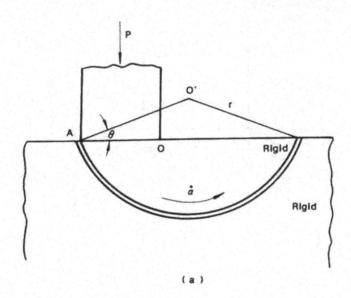

(a)

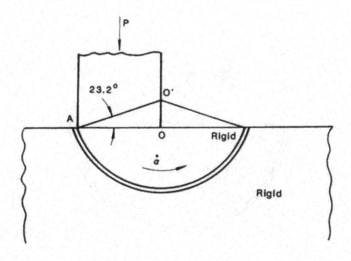

(b)

FIGURE 8.28. Rotating block with center at $O'$ for purpose of minimizing upper bound.

between the face of the half-space and the line $AO'$. The rate of internal work is given as $kr(\pi-2\theta)r\dot{\alpha}$, and the resulting upper-bound solution is:

$$P^U = \frac{k(\pi-2\theta)r^2}{r\cos\theta-\tfrac{1}{2}b} \qquad (8.60)$$

The upper-bound solution may be reduced by minimizing the solution with respect to the two variables $r$ and $\theta$:

$$\frac{\partial P^U}{\partial r}=0 \quad \text{or} \quad b=r\cos\theta \qquad (8.61a)$$

and

$$\frac{\partial P^U}{\partial \theta}=0 \quad \text{or} \quad b=r[2\cos\theta-(\pi-2\theta)\sin\theta] \qquad (8.61b)$$

Equation (8.61a) shows that the location of the center of rotation $O'$ is directly above $O$ as shown in Fig. 8.28b. Equating Eq. (8.61a) to Eq. (8.61b) yields:

$$\cos\theta = (\pi-2\theta)\sin\theta \qquad (8.62)$$

The critical value of the angle $\theta$ can be determined from Eq. (8.62) by trial and error. This is found to be $\theta_{cr}=23.2°$. The most critical layout of the mechanism is shown in Fig. 8.28b, and the corresponding least upper-bound value is:

$$P^U = \frac{4kb}{\sin 2\theta}=5.53kb \qquad (8.63)$$

*Translational Mechanisms (Figs. 8.29 and 8.30)*

Now, we consider two mechanisms involving only rigid-body translations. Figures 8.29 and 8.30 show two such examples of rigid-block sliding separated by plane velocity discontinuities. The mechanism of Fig. 8.29a represents a *rough* punch. It requires that the punch and the triangular block $ABC$ have the same velocity and therefore move together. However, Fig. 8.30 is referred to as a *smooth* punch, for it allows a relative sliding between the punch and the triangle $ABC$ along the surface $AB$.

Since plastic flow is confined to the lines of velocity discontinuity, we need to know all the relative sliding velocities between adjacent blocks in order to evaluate the dissipation of energy. The most direct method of determining these is by means of velocity diagrams (or hodographs). Before we proceed to construct these diagrams, we shall first briefly describe the kinematics of the mechanisms.

Since these mechanisms are symmetrical about the center line, it is only necessary to consider the movement on the right half of Figs. 8.29a and 8.30a. For a rough punch, the triangular region $ABC$ in Fig. 8.29a moves

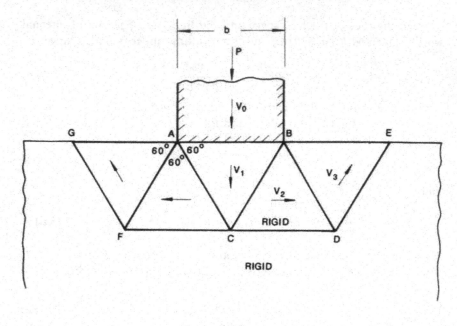

(a)

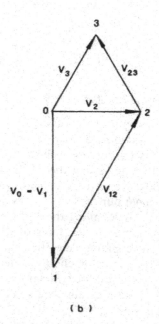

(b)

FIGURE 8.29. Simple rigid-block translation and associated velocity diagram for a rough punch.

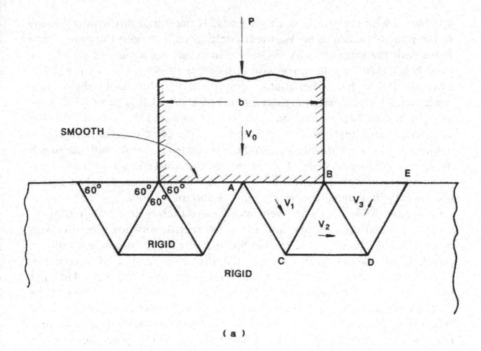

(a)

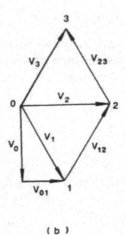

(b)

FIGURE 8.30. Rigid-block translation and associated velocity diagram for a smooth punch.

downward with the punch as a rigid body. If the initial downward velocity of the punch is taken to be $V_0$, the triangle in contact with the punch must move with the same velocity, $V_1 = V_0$. However, for a smooth punch, the rigid block $ABC$ would move with a different velocity $V_1$ (see Fig. 8.30). The velocity $V_1$ has a horizontal component in addition to the vertical component $V_0$. The two triangular regions of material $BCD$ and $BDE$ move as rigid bodies in the direction parallel to $CD$ and $DE$, respectively. The velocity of the triangle $BCD$ is determined by the condition that the relative velocity $V_{12}$ between this triangle and the triangle in contact with the punch must have the direction $BC$. The velocity of the third triangle is determined in a similar manner. The information regarding velocities is represented in the velocity diagrams as shown in Figs. 8.29b and 8.30b.

Referring to Fig. 8.29b, the velocity of the punch is first drawn as $V_0$ on a convenient arbitrary scale, and in the correct downward direction with respect to the layout of Fig. 8.29a. Since the triangle in contact with the punch has the same velocity as the punch, the velocities $V_1$ and $V_0$ coincide. The directions of the velocity of triangle $BCD$ relative to triangle $ABC$ and to the zone which remains stationary are the same as the directions of the corresponding interfaces, i.e., line $BC$ and line $CD$. The magnitude of the velocity $V_2$ and the relative velocity $V_{12}$ is therefore determined uniquely from points labeled 0 and 1, as shown. Similarly, the point labeled 3 is located from points 0 and 2. It can be seen that the velocity of each block is represented by a vector from the origin 0, and the vector joining two end points represents the relative velocity of the corresponding blocks.

Using the notation $l_{bc}$ for the length of the interface $BC$, etc., we write down the work equation for the mechanism of Fig. 8.29a, making use of symmetry:

$$P^U V_0 = 2k(l_{bc} V_{12} + l_{cd} V_2 + l_{bd} V_{23} + l_{de} V_3) \tag{8.64}$$

Since energy dissipation must always be positive, each of the terms on the right-hand side of the above equation presents no difficulty with regard to sign when the expression is evaluated. By inspection, we have

$$V_2 = V_3 = V_{23} = \frac{V_{12}}{2} = \frac{V_0}{\sqrt{3}} \tag{8.65}$$

and

$$l_{bc} = l_{cd} = l_{bd} = l_{de} = b \tag{8.66}$$

Substituting Eqs. (8.65) and (8.66) into (8.64), we have

$$P^U V_0 = 2kb \left( \frac{2}{\sqrt{3}} + \frac{1}{\sqrt{3}} + \frac{1}{\sqrt{3}} + \frac{1}{\sqrt{3}} \right) V_0$$

or

$$P^U = \frac{10kb}{\sqrt{3}} = 5.78kb \tag{8.67}$$

Such an upper bound for a rough punch is not as good as that of Eq. (8.63), corresponding to a rotational failure mechanism.

Now, referring to the mechanism and velocity diagram shown in Fig. 8.30 for a smooth punch, by inspection

$$P^U V_0 = 2k(l_{ac} V_1 + l_{bc} V_{12} + l_{cd} V_2 + l_{bd} V_{23} + l_{de} V_3) \tag{8.68}$$

in which

$$V_1 = V_2 = V_3 = V_{12} = V_{23} = \frac{2 V_0}{\sqrt{3}} \tag{8.69}$$

and

$$l_{ac} = l_{bc} = l_{cd} = l_{bd} = l_{de} = \frac{b}{2} \tag{8.70}$$

Hence, the upper-bound solution obtained from this mechanism is

$$P^U = \frac{10kb}{\sqrt{3}} = 5.78kb \tag{8.71}$$

which has the same value as the rough punch.

Different layouts of the mechanisms of Figs. 8.29 and 8.30 may be employed to reduce the upper bound. For example, a partial minimization may be accomplished analytically by keeping all the triangles as equal isosceles triangles. However, it can be shown that the limit load corresponding to these mechanisms is not sensitive to the particular layout of the mechanisms. In other words, the minimum is a rather flat curve, and a wide variety of geometric layouts within the same mechanism would be expected to give only slightly different upper bounds.

The exact answer for the limit load is $(2+\pi)kb$, which will be discussed later. Although the upper bounds $P^U = 6.28kb$, $5.78kb$, and $5.53kb$ are not extremely close to the exact answer, they are obtained quickly and easily. Were it not for the fact that this punch indentation problem is so well known, any one of these upper bounds would be considered to provide useful information.

## 8.5.2. Homogeneous Deforming Regions

The simplest homogeneous deforming fields are simple compression and simple shear flow. Shown in Figs. 8.31 and 8.32 are fields of homogeneous deformation denoted by the symbol $\dot{\varepsilon}$ for a field of simple vertical compression and lateral expansion and by $\dot{\gamma}$ for a simple shear. Since Tresca material is incompressible, volume is conserved in the plastic deformation. The lateral strain rate in the vertical compression mode must always be equal and opposite to the vertical strain rate (Fig. 8.31).

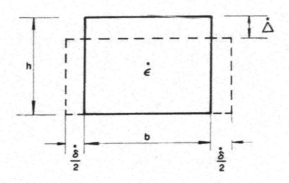

(a)

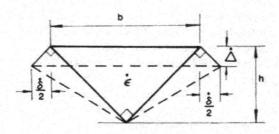

(b)

FIGURE 8.31. Simple vertical compression and lateral expansion deformation: incompressibility requires (a) $\dot{\Delta}/h = \dot{\delta}/b$; (b) $(\dot{\delta}/2)/\dot{\Delta} = (b/2)/h = 1$.

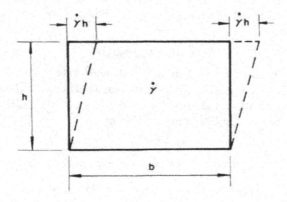

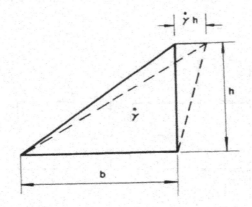

FIGURE 8.32. Simple shear deformation.

Let us evaluate first the rate of dissipation of energy per unit volume in the field of homogeneous compression. The mode of deformation is instantaneously one of pure vertical compression and horizontal extension; the vertical strain rate $\dot{\epsilon}_y$ is $-|\dot{\epsilon}|$ and the horizontal strain rate $\dot{\epsilon}_x$ is $|\dot{\epsilon}|$, so the rate of dissipation of energy is equal to $-\sigma_y|\dot{\epsilon}| + \sigma_x|\dot{\epsilon}|$ per unit volume, $\sigma_y$, $\sigma_x$ being the corresponding principal stresses in the plane of flow. Since the Tresca yield criterion must be satisfied in this plastically deformed region, the absolute value of the difference between $\sigma_y$ and $\sigma_x$ must be equal to $2k$, so:

$$D = |\dot{\epsilon}|(\sigma_x - \sigma_y) = 2k|\dot{\epsilon}| \qquad (8.72)$$

where $\dot{\epsilon}$ is the absolute value of the normal strain rate.

The plastic flow in Fig. 8.32 is of simple shear. Thus, the shear stress in the region must be the yield stress in pure shear, $k$. The rate of energy dissipation per unit volume is computed from

$$D = k\dot{\gamma}_{max} \qquad (8.73)$$

where $\dot{\gamma}_{max}$ is the maximum rate of engineering shear strain.

EXAMPLE 8.5. A STRIP SQUEEZED BETWEEN RIGID PUNCHES (PLANE STRAIN)

*Homogeneous Deformation Field (Fig. 8.33a)*

Figure 8.33a shows a failure mechanism with a strain field of homogeneous

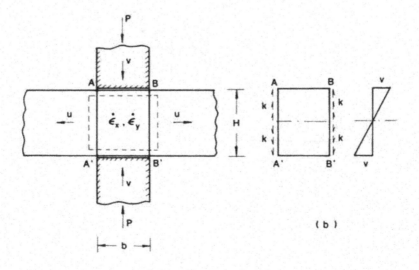

( a )

FIGURE 8.33. A strip squeezed between rigid punches: homogeneous deformation field.

compression accompanied by a lateral expansion in the region $ABB'A'$:

$$\dot{\epsilon}_y = \frac{2v}{H}, \qquad \dot{\epsilon}_x = \frac{2u}{b} \tag{8.74}$$

Volume is conserved in the plastic deformation, i.e., $\dot{\epsilon}_y + \dot{\epsilon}_x = 0$, which implies that the lateral strain is equal and opposite to the vertical strain. Therefore,

$$\frac{u}{b} = \frac{v}{H} \quad \text{or} \quad u = \frac{b}{H}v \tag{8.75}$$

The rate of energy dissipation throughout the volume is equal to the yield stress in compression $(2k)$ times the strain rate in compression $(2v/H)$ [see Eq. (8.72)] multiplied by the volume per unit dimension perpendicular to the paper, $bH$:

$$D_1 = (2k)\left(\frac{2v}{H}\right)(bH) = 4kbv \tag{8.76}$$

The rate of energy dissipation along the surfaces of discontinuity $AA'$ and $BB'$ is the yield stress $k$ multiplied by the relative velocity across the surfaces and integrated over the area of the surfaces:

$$D_2 = 2(k)(\tfrac{1}{2}v)(H) = kvH \tag{8.77}$$

in which the relative velocity $(\tfrac{1}{2}v)$ is the average between the maximum value of $v$ at the top and bottom of the lines of discontinuity, $AA'$ and $BB'$, and the value of zero at the mid-height (see Fig. 8.33b).

If the punches are smooth, there will be no dissipation at the interfaces, $AB$ and $A'B'$. If the punches are rough, the additional rate of dissipation of energy is given by the average value of the relative velocity $(\tfrac{1}{2}u)$ multiplied by the yield stress $k$ in shear and by the total length of contact $2b$:

$$D_3 = (k)(\tfrac{1}{2}u)(2b) = \frac{kvb^2}{H} \tag{8.78}$$

For a smooth punch, the upper bound is obtained by equating $2Pv$, the rate of work done by the external pair of forces $P$, to the total rate of internal dissipation $(D_1 + D_2)$:

$$P^U = 2kb + \frac{1}{2}kH = 2kb\left(1 + \frac{H}{4b}\right) \tag{8.79}$$

For a rough punch, the total rate of dissipation is $(D_1 + D_2 + D_3)$; the upper bound is obtained as

$$P^U = 2kb + \frac{1}{2}kH + \frac{1}{2}k\frac{b^2}{H} = 2kb\left(1 + \frac{H}{4b} + \frac{b}{4H}\right) \tag{8.80}$$

*Rigid-Block Sliding (Fig. 8.34)*

The deformational mode shown in Fig. 8.34 is constructed by a four-rigid-block sliding. There is no relative motion across $AB$ and $A'B'$, so that no distinction need be made between rough and smooth punches. Calculation of the energy dissipation requires the expression of the relative velocity across the line of discontinuity $AB'$ or $A'B$, which is equal to

$$\sqrt{u^2+v^2} = \frac{v}{H}\sqrt{b^2+H^2} \qquad (8.81)$$

The relation of Eq. (8.75) between $u$ and $v$ holds because there is no volume change. The upper bound is then obtained as

$$P^U = 2kb\left(\frac{H}{2b}+\frac{b}{2H}\right) \qquad (8.82)$$

Each of these simple fields of velocity with their simple discontinuities requires very little computational effort, but neither represents the actual field. As $H/b$ becomes large, the two punches will not interact, and a local solution, the indentation of a very large body loaded by a rigid punch, will take over. Some upper-bound solutions to this problem have been obtained in Example 8.4 using rigid-block sliding mechanisms. The intermixing of inhomogeneous deforming regions and rigid-block sliding, described in the next section, will give the "exact" indentation force, $(2+\pi)kb$, for the semi-infinite block.

These two results for the narrow punch, together with the local solution for the semi-infinite block, and with the well-known slip-line solution by Hill (1950), are plotted in Fig. 8.35. The upper-bound solutions are not

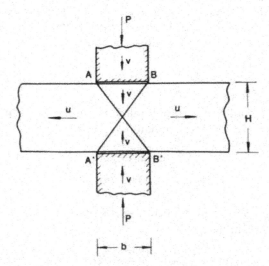

FIGURE 8.34. A strip squeezed between rigid punches: rigid-block sliding.

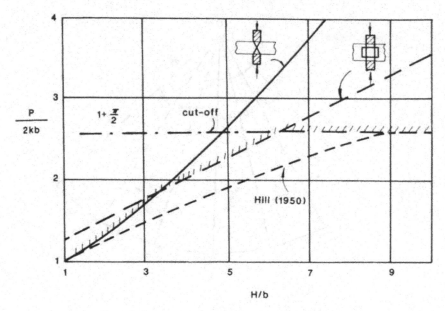

H/b

FIGURE 8.35. Upper bounds for narrow punch.

extremely close, in general, to Hill's slip-line solution, which is considered to be the "exact" answer, but the diagram gives a fairly clear impression of the part played by the thickness of the strip in the indentation behavior.

### 8.5.3. An Inhomogeneous Deforming Region: Radial Shear Velocity Field

The inhomogeneous deformation field of radial shear involving straight and curve failure lines is frequently encountered in applications. In this field, one family of failure lines consists of concurrent straight lines, and the other of concentric circles.

An approximation to this zone is given in Fig. 8.36, where a picture for six rigid triangles at an equal central angle $\Delta\theta$ to each other is shown. Energy dissipation takes place along the radial lines $OA$, $OB$, $OC$, etc., due to the discontinuity in velocity between the triangles. Energy also is dissipated on the discontinuous surface $DABCEFG$ since the material below this surface is considered at rest. Since the material must remain in contact with the surface $DABCEFG$, the triangles must move parallel to the arc surfaces. Also, the rigid triangles must remain in contact with each other; the compatible velocity diagram of Fig. 8.36b shows that each triangle of the mechanism must have the same speed $V_1 = V_2 = V_n = V$.

With Eq. (8.36), the rate of dissipation of energy can easily be calculated. The energy dissipation along the radial line $OB$, for example, is the yield stress in pure shear, $k$, multiplied by the relative velocity, $\delta u$, and the length

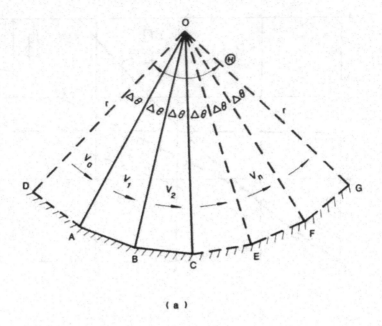

( a )

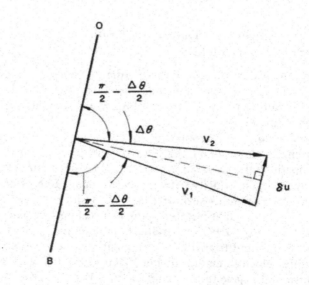

( b )

FIGURE 8.36. Radial shear zone: (a) rigid triangles; (b) velocity relation.

of the line discontinuity:

$$D|_{\text{line } OB} = kr\left(2V\sin\frac{\Delta\theta}{2}\right) \tag{8.83}$$

in which the relative velocity $\delta u$ appears as $2V\sin(\Delta\theta/2)$. Similarly, the energy dissipation along the discontinuous surface $AB$ is

$$D|_{\text{line } AB} = k\left(2r\sin\frac{\Delta\theta}{2}\right)V \tag{8.84}$$

where the length of $AB$ is $2r\sin(\Delta\theta/2)$ and $\delta u = V$. Since the energy dissipation along the radial line $OB$ is the same as that along line $AB$, it is natural to expect that the total energy dissipation in the zone of radial shear, $DOG$, with a central angle $\Theta$, will be identical with the energy dissipated along the arc $DG$. This is evident since Fig. 8.36a becomes closer and closer to the zone of radial shear as the number $n$ grows. In the limit when $n$ approaches infinity, the zone of radial shear is recovered. The total energy dissipated in the zone of radial shear is the sum of the energy dissipated along each radial line when the number $n$ approaches infinity:

$$D|_{\text{region } ODG} = \lim_{n\to\infty} n\left[2kr\, V\sin\frac{\Theta}{2n}\right] = kVr\Theta \tag{8.85}$$

Indicated in Fig. 8.37a is the displaced pattern of the radial shear zone $ODG$. This figure shows a small displacement of the field that would result if the initial velocity along $OD$ was maintained for a short period of time. In the zone, the velocity along each radial line is constant in the direction perpendicular to the radial line. The material below the line $DG$ is assumed to be at rest. The initial position of the zone is indicated by solid lines in the figure.

An alternative derivation for the dissipation of energy in the radial shear zone can also be obtained by visualizing the field of flow as a series of transition layers. Each of these layers is bounded by concentric circular arcs, and each of these arcs is rotating as a rigid body about the center of rotation $O$ with an angular velocity—for example, $V/\bar{r}$ for the arc surface $ab$ or $V/(\bar{r}+d\bar{r})$ for the arc surface $a'b'$ (see Fig. 8.37b). With Eq. (8.36), the differential rate of dissipation of energy along the layer surface is found by multiplying the difference in arc length between $ab$ and $a'b'$, i.e., $\Theta\, d\bar{r}$, of this layer by $k$ times the velocity, $V$, across the layer. The total internal dissipation of the energy is then found by integration over the entire radial line:

$$D|_{\text{region } ODG} = \int_0^r kV\Theta\, dr = kVr\Theta \tag{8.86}$$

agreeing with the value obtained previously in Eq. (8.85).

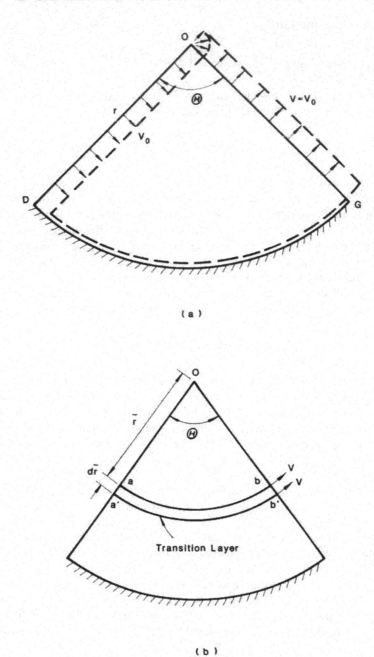

FIGURE 8.37. Radial shear zone: (a) displaced pattern; (b) alternative derivation.

EXAMPLE 8.6. UPPER BOUND ANALYSIS OF PUNCH INDENTATION
PROBLEM (PART 2: MECHANISMS INVOLVING RADIAL SHEAR ZONES)

The punch indentation problem has been studied previously by using mechanisms involving only rigid-block sliding. Herein, we shall try to improve the upper bound by introducing radial shear zones to the velocity fields. To reduce the dissipation rate along line $BD$ (see Figs. 8.29 and 8.30), we need to eliminate the velocity discontinuity across line $BD$. This is accomplished if a radial shear zone is used to replace the rigid triangle $BCD$ and $\angle ABC = 45°$ instead of $60°$ as shown in Figs. 8.38 and 8.39 for rough and smooth punches, respectively.

Consider first the solution represented by Fig. 8.38. We need to discuss only the right-hand plastic flow region because of symmetry. The mechanism consists of three rigid triangles, $ABC$, $BDE$, and $AFG$, and two radial shear zones, $BCD$ and $ACF$. As shown by the velocity diagram of Fig. 8.38b, triangle $ABC$ moves downward as a rigid body with velocity $V_1 = V_0$, which can decompose into a normal component $V_{1n}$ to line $BC$ and a shear component $V_{1t}$ along $BC$. $BC$ is a line of discontinuity, at which the energy dissipation is

$$D|_{BC} = kV_{1t}l_{bc} = k\left(\frac{\sqrt{2}}{2}V_0\right)\left(\frac{\sqrt{2}}{2}b\right) = \frac{1}{2}kbV_0$$

The tangential velocity at any point in the quadrant $BCD$ is $V_{1n} = (\sqrt{2}/2)V_0$. With Eq. (8.85), the energy dissipation in zone $BCD$ is calculated as

$$D|_{BCD} = k\left(\frac{\sqrt{2}}{2}V_0\right)\left(\frac{\sqrt{2}}{2}b\right)\left(\frac{\pi}{2}\right) = \frac{\pi}{4}kbV_0$$

Along arc $CD$, the energy dissipation is the same as in region $BCD$, i.e.,

$$D|_{CD} = \frac{\pi}{4}kbV_0$$

Since the translation velocity of the block $BDE$ is $V_3 = V_{1n} = (\sqrt{2}/2)V_0$, there is no discontinuity across the line $BD$ as mentioned above, and the energy dissipation along line $DE$ is obtained as

$$D|_{DE} = \tfrac{1}{2}kbV_0$$

Equating the total rate of internal dissipation of energy to the external rate of work done by load $P$ yields

$$P^U = (2 + \pi)kb \tag{8.87}$$

Referring to the mechanism and velocity diagram given in Fig. 8.39, we come up with the same answer of Eq. (8.87) for a smooth punch.

In fact, this result is the correct value of the limit load, and these two mechanisms correspond to the two slip-line solutions proposed by Prandtl in 1921 and by Hill in 1950, respectively.

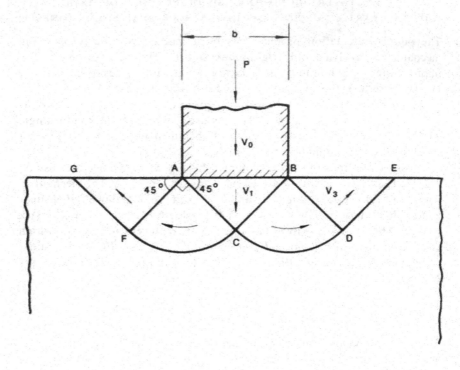

(a)

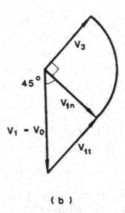

(b)

FIGURE 8.38. Velocity field with a radial shear zone for a rough punch. This mechanism was proposed by Prandtl in 1921.

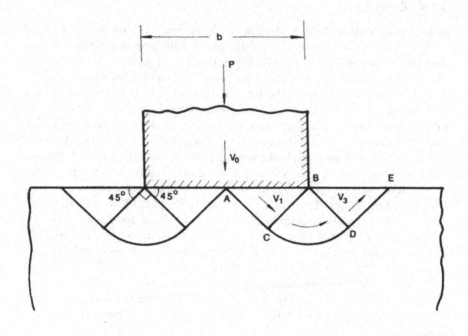

( a )

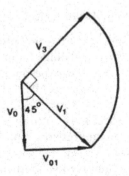

( b )

FIGURE 8.39. Velocity field with a radial shear zone for a smooth punch. The mechanism was proposed by Hill in 1950.

### 8.5.4. Remarks

Some typical velocity fields in the upper-bound analysis have been discussed in this section. The kinematic condition of plastic flow is governed by the flow rule associated with a yield function. For Tresca or von Mises materials, no volumetric deformation occurs in plastic flow so that the velocity across a narrow transition zone or the line of discontinuity is tangential to that line. However, for materials whose yield stress in shear depends upon normal stress, such as soils, volume expansion is seen to be a necessary accompaniment to shear deformation. Consequently, the velocity across a surface of discontinuity will make an angle $\phi$, which is the angle of internal friction, to the surface. The circular sliding surfaces for metals must now be replaced by logarithmic spirals and the circular radial shear zones by logspiral shear zones. The book by Chen (1975) should be referred to in order to study the construction of these velocity fields for soils.

In general, an infinite variety of such fields can be drawn for this problem or for any other problems in accord with the intuitive feeling of the designer or analyst for the appropriate mode of failure.

Discontinuous velocity fields not only prove convenient but often are contained in the actual collapse mode. This is in marked contrast to the stress situation, where discontinuity is useful and permissible but rarely the actual state.

## 8.6. Example Problems in Plane Stress, Plane Strain, and 3-D

### 8.6.1. Analysis of Ice Sheet Indentation (Ralston, 1978)

The indentation of an ice sheet by a flat indenter is of engineering interest because of its similarity to the crushing failure of ice moving against a vertical pier. Several analytical and experimental studies have been discussed in the literature (Croasdale, et al., 1976; Frederking and Gold, 1975; Michel and Toussaint, 1976; Ralston, 1977, 1978, among others). The strength of columnar-grained ice is anisotropic, sensitive to confining stress, and differs in tension and in compression. An appropriate failure criterion that describes these effects has been discussed in Chapter 2. Herein, the method of limit analysis is employed to predict the ice indentation forces.

The present discussion is limited to two-dimensional analyses of the initial indentation of an ice sheet by a flat indenter that is in perfect contact with a flat edge of the sheet. The special cases of plane strain and plane stress in the plane of the ice sheet are of particular significance. These represent limiting cases corresponding to very narrow indenters in thick ice sheets $(b/t \to 0)$ and very wide indenters in thin ice sheets $(b/t \to \infty)$, respectively. In principle, the analyses presented here could be extended to three dimensions. A 3-D analysis would provide a curve of ice indentation pressure as a function of $b/t$. Such an analysis would also provide definite values for how small $b/t$ must be to justify the plane strain analysis and how large $b/t$ must be for the plane stress approximation to be valid.

### 8.6.1.1. Yield Criterion

For the ice indentation problem, ice is considered as an elastic–perfectly plastic material. Plastic flow occurring in ice crystals is primarily due to the movement of dislocations along the basal planes where sliding resistance to shear stresses is low. The yield strength of ice in a ductile mode of deformation corresponds to the maximum stress level attained either before complete failure occurs or before higher strains are reached with no increase in stresses.

An anisotropic yield function proposed by Reinicke and Ralston (1977) implies a parabolic increase in strength as a function of hydrostatic pressure (see Section 2.4.2, Chapter 2). Nine material parameters are necessary to define adequately the strengths with respect to three mutually orthogonal planes of symmetry. The function is given by

$$f(\sigma_{ij}) = a_1(\sigma_y - \sigma_z)^2 + a_2(\sigma_z - \sigma_x)^2 + a_3(\sigma_x - \sigma_y)^2 + a_4\tau_{yz}^2$$

$$+ a_5\tau_{zx}^2 + a_6\tau_{xy}^2 + a_7\sigma_x + a_8\sigma_y + a_9\sigma_z - 1 = 0 \qquad (8.88)$$

Symmetry in the material strength properties imposes restrictions on the possible values of the coefficients. Since the strength of this ice is isotropic within the horizontal plane, Eq. (8.88) can be reduced to:

$$f(\sigma_{ij}) = a_1[(\sigma_y - \sigma_z)^2 + (\sigma_z - \sigma_x)^2] + a_3(\sigma_x - \sigma_y)^2$$

$$+ a_4(\tau_{yz}^2 + \tau_{zx}^2) + a_6\tau_{xy}^2 + a_7(\sigma_x + \sigma_y) + a_9\sigma_z - 1 = 0 \qquad (8.89)$$

where $a_6 = 2(a_1 + 2a_3)$, and the $x$, $y$, $z$ coordinate system is now fixed in the ice sheet with the $xy$-plane in the plane of the ice sheet.

Plastic deformation is related to stress by means of a flow rule. Herein, we shall use the associated flow rule, which implies that the yield function is a potential function for the plastic strain rate, i.e.,

$$\dot{\epsilon}_{ij}^p = \lambda \frac{\partial f}{\partial \sigma_{ij}} \qquad (8.90)$$

Plane strain in the plane of the ice sheet requires that

$$\dot{\epsilon}_z^p = \dot{\epsilon}_{xz}^p = \dot{\epsilon}_{yz}^p = 0$$

When this restriction is imposed on the yield function (8.89) via the flow rule, the resulting plane strain yield function is

$$f(\sigma_{ij}) = a_6[(\sigma_x - \sigma_y)^2/4 + \tau_{xy}^2] + (2a_7 + a_9)(\sigma_x + \sigma_y)/2 - (1 + a_9^2/8a_1) = 0$$

$$(8.91a)$$

Plane stress in the plane of the ice sheet implies that

$$\sigma_z = \tau_{xz} = \tau_{yz} = 0$$

When this restriction is imposed on the yield function (8.89), the resulting plane stress yield function is

$$f(\sigma_{ij}) = a_1(\sigma_x^2 + \sigma_y^2) + a_3(\sigma_x - \sigma_y)^2 + a_6\tau_{xy}^2 + a_7(\sigma_x + \sigma_y) - 1 = 0 \quad (8.91b)$$

*Determination of the Material Constants*

The values of the coefficients in Eq. (8.89) are selected to match the yield function to the available data for plane stress and plane strain loading of the ice sheet. Four coefficients, $a_1$, $a_3$, $a_7$, and $a_9$, are required for these representations of the ice yield function.

In Eq. (2.205), Chapter 2, the coefficients $a_1$, $a_3$, $a_7$, and $a_9$, were determined from $T_x$, $C_x$, $T_z$, and $C_z$, the tensile and compressive strengths in the x- and z-directions, respectively (see Fig. 2.31).

Herein, we shall not use the strengths $T_z$ and $C_z$ in the z-direction since we are studying the plane stress and plane strain problems. The four types of ice strength tests used to determine the constants $a_1$, $a_3$, $a_7$, and $a_9$ are the in-plane unconfined compressive strength $C_x$, the in-plane uniaxial tensile strength $T_x$, Frederking's type A plane strain strength, $\sigma_{ps}^A$, and Frederking's type B plane strain strength, $\sigma_{ps}^B$ (see Fig. 8.40).

Now, we get four yield points in stress space:

(i) The uniaxial compressive yield strength $C_x$:

$$\sigma_x = -C_x, \qquad \sigma_y = \sigma_z = \tau_{yz} = \tau_{zx} = \tau_{xy} = 0$$

(ii) The uniaxial tensile yield strength $T_x$:

$$\sigma_x = T_x, \qquad \sigma_y = \sigma_z = \tau_{yz} = \tau_{zx} = \tau_{xy} = 0$$

(iii) Frederking's type A strength $\sigma_{ps}^A$:

$$\sigma_x = -\sigma_{ps}^A, \qquad \sigma_z = \tau_{yz} = \tau_{zx} = \tau_{xy} = 0$$

And from $\dot{\varepsilon}_y^p = 0$, we have

$$\sigma_y = \frac{-a_3\sigma_{ps}^A}{a_1 + a_3} - \frac{a_7}{2(a_1 + a_3)}$$

(iv) Frederking's type B strength $\sigma_{ps}^B$:

$$\sigma_x = -\sigma_{ps}^B, \qquad \sigma_y = \tau_{yz} = \tau_{zx} = \tau_{xy} = 0$$

And from $\dot{\varepsilon}_z^p = 0$, we have

$$\sigma_z = -\frac{\sigma_{ps}^B}{2} - \frac{a_9}{4a_1}$$

Substituting these four stress points into the yield function (8.89) leads to four equations to determine $a_1$, $a_3$, $a_7$, and $a_9$.

CARTER  AND  MICHEL'S

UNCONFINED  STRENGTH  TESTS

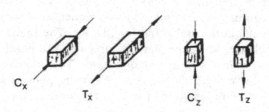

FREDERKING'S  PLANE  STRAIN

STRENGTH  TESTS

TYPE A                    TYPE B

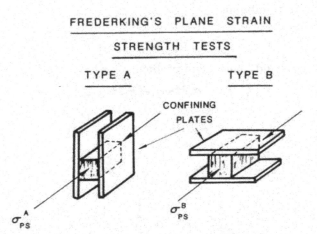

FIGURE 8.40. Unconfined and plane strain ice strength tests for columnar-grained ice.

The ice indentation analysis conducted by Ralston (1978) uses the following ice strength ratios:

$$T_x/C_x = 0.142, \qquad \sigma^A_{ps}/C_x = 3.6, \qquad \sigma^B_{ps}/C_x = 1.01$$

The corresponding values of $a_1$, $a_3$, $a_7$, and $a_9$ can be obtained as

$$a_1 = 1.77 C_r^{-2}, \qquad a_3 = 5.27 C_x^{-2}, \qquad a_7 = 6.04 C_x^{-1}, \qquad a_9 = -3.54 C_x^{-1}$$

in which $C_x$ is the in-plane unconfined compressive strength.

The plane stress and plane strain yield functions corresponding to these coefficients are plotted in Fig. 8.41. The symmetry of these curves with respect to the coordinate directions is a consequence of the isotropy of ice strength within the horizontal plane for this ice. The plane stress curve is a long narrow ellipse, while the plane strain curve is a parabola that opens into the compression–compression quadrant. These curves provide the description of ice strength that is needed for the plane stress and plane strain analysis of ice indentation. These functions are quite different from those that are usually used in applications of metal plasticity as shown in Fig. 8.22.

### 8.6.1.2. ENERGY DISSIPATION FOR CONTINUOUS STRAIN RATES

As discussed in Section 8.2.4, the rate of internal energy dissipation per unit volume $D_V$ is a unique function of the plastic strain rates $\dot\epsilon^p_{ij}$. The functional form of the energy dissipation will now be given for the yield functions of plane strain and plane stress, Eqs. (8.91a) and (8.91b), when strain rates are continuous.

By definition,

$$D_V = \sigma_{ij}\dot\epsilon^p_{ij}$$

For plane stress and plane strain problems, we have

$$D_V = \sigma_x\dot\epsilon^p_x + \sigma_y\dot\epsilon^p_y + 2\tau_{xy}\dot\epsilon^p_{xy} \tag{8.92}$$

where

$$\dot\epsilon^p_x = \lambda\frac{\partial f}{\partial\sigma_x} \tag{8.92a}$$

$$\dot\epsilon^p_y = \lambda\frac{\partial f}{\partial\sigma_y} \tag{8.92b}$$

$$\dot\epsilon^p_{xy} = \lambda\frac{\partial f}{\partial\tau_{xy}} \tag{8.92c}$$

and the stresses $\sigma_x$, $\sigma_y$, and $\tau_{xy}$ satisfy the yield condition (8.91a) or (8.91b), namely,

$$f(\sigma_x, \sigma_y, \tau_{xy}) = 0 \tag{8.92d}$$

Now, we have four relations, (8.92a) through (8.92d), for four unknowns, $\sigma_x$, $\sigma_y$, $\tau_{xy}$, and $\lambda$. Therefore, $\sigma_x$, $\sigma_y$, $\tau_{xy}$, and $\lambda$ can be determined from

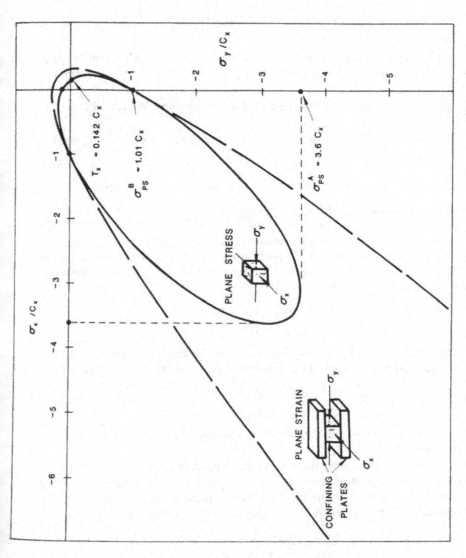

FIGURE 8.41. Plane stress and plane strain ice yield surfaces.

given strain rates, $\dot{\epsilon}_x^p$, $\dot{\epsilon}_y^p$, and $\dot{\epsilon}_{xy}^p$, and so can the rate of internal energy dissipation, $D$, given by Eq. (8.92).

(i) For the plane strain case, Eq. (8.91a) is relevant. Derivation of the functional form of $D_V$ can be carried out to give

$$D_V = \frac{3a_7}{4a_6} \frac{(\dot{\epsilon}_x - \dot{\epsilon}_y)^2 + 4\dot{\epsilon}_{xy}^2}{\dot{\epsilon}_x + \dot{\epsilon}_y} + \frac{\dot{\epsilon}_x + \dot{\epsilon}_y}{3a_7}\left(1 + \frac{a_7^2}{8a_1}\right) \tag{8.93}$$

in which the superscript "$p$" of the plastic strain rates has been dropped since in the limit state, the strain rates are all plastic (Theorem 1 of Section 8.2.3).

(ii) For the plane stress case, Eq. (8.91b) is relevant. We can obtain

$$D_V = \frac{1}{a_1}\sqrt{\frac{2a_1 + a_7^2}{a_6}}\sqrt{(a_1 + a_3)(\dot{\epsilon}_x + \dot{\epsilon}_y)^2 + 2a_1(\dot{\epsilon}_{xy}^2 - \dot{\epsilon}_x\dot{\epsilon}_y)} - \frac{a_7}{2a_1}(\dot{\epsilon}_x + \dot{\epsilon}_y)$$

$$\tag{8.94}$$

### 8.6.1.3. Energy Dissipation in a Surface of Discontinuity

In the application of the upper-bound theorem of limit analysis, the construction of discontinuous velocity fields is usually convenient. Figure 8.42a shows a failure mechanism of ice crushing consisting of rigid blocks separated by planes of velocity discontinuities. Figure 8.42b shows a typical boundary between zone 1 and zone 2 across which the velocity components $v_x$ and $v_y$ change by the amounts

$$\delta v_x = v_x^{(2)} - v_x^{(1)}$$
$$\delta v_y = v_y^{(2)} - v_y^{(1)}$$

The velocity jumps $\delta v_x$ and $\delta v_y$ may also be expressed in terms of the normal component $\delta v$ and the tangential components $\delta u$ (see Fig. 8.42b):

$$\delta v = \delta v_x n_x + \delta v_y n_y$$
$$\delta u = -\delta v_x n_y + \delta v_y n_x$$

in which $n_x$ and $n_y$ are the direction cosines of the unit normal $\mathbf{n}$ of the discontinuity surface.

The surface of discontinuity is an idealization of a continuous distribution of a narrow transition zone of some thickness $t$ (Fig. 8.42b). The strain rates in the transition zone can be assumed to be homogeneous and can be computed from a given value of the velocity jumps $\delta v_x$ and $\delta v_y$:

$$\dot{\epsilon}_x = \frac{\partial v_x}{\partial x} = \frac{v_x^{(2)} - v_x^{(1)}}{dx} = \frac{1}{t}\delta v_x n_x$$

$$\dot{\epsilon}_y = \frac{\partial v_y}{\partial y} = \frac{v_y^{(2)} - v_y^{(1)}}{dy} = \frac{1}{t}\delta v_y n_y$$

$$\dot{\epsilon}_{xy} = \frac{1}{2}\left(\frac{\partial v_x}{\partial y} + \frac{\partial v_y}{\partial x}\right) = \frac{1}{2t}(\delta v_x n_y + \delta v_y n_x)$$

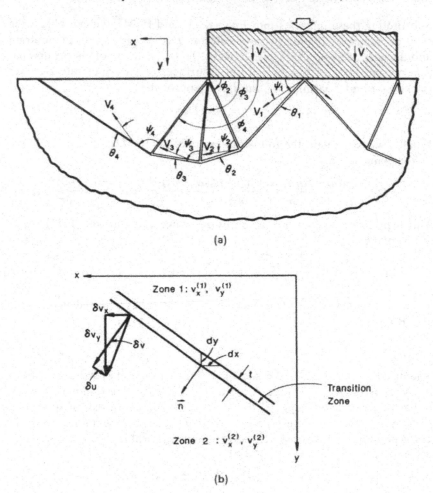

(a)

(b)

FIGURE 8.42. (a) Upper-bound velocity field (Reinicke and Ralston, 1977). (b) Surface of discontinuity viewed as a limit of a transition zone.

Using the relationships given above, we can find that

$$(\dot{\epsilon}_x + \dot{\epsilon}_y)^2 = \frac{1}{t^2}(\delta v_x n_x + \delta v_y n_y)^2 = \frac{1}{t^2}\delta v^2$$

$$\dot{\epsilon}_{xy}^2 - \dot{\epsilon}_x \dot{\epsilon}_y = \frac{1}{4t^2}\delta u^2$$

and

$$(\dot{\epsilon}_x - \dot{\epsilon}_y)^2 + 4\dot{\epsilon}_{xy}^2 = \frac{1}{t^2}(\delta u^2 + \delta v^2)$$

Substituting these relations into Eqs. (8.93) and (8.94), we can obtain the rates of energy dissipation per unit volume, $D_V$, for the case of plane strain and the case of plane stress, respectively. To find the rate of energy dissipation per unit area of the discontinuity surface, $D_A$, we integrate $D_V$ over thickness $t$ and determine the limit as $t$ approaches zero

$$D_A = \lim_{t \to 0} t D_V \qquad (8.95)$$

Finally, we can obtain the following expressions for $D_A$.
    For plane strain:

$$D_A = \frac{2a_1 a_6 + 2a_1 a_7(a_7 + a_9) + a_9^2(a_1 + a_3)}{2a_1 a_6(2a_7 + a_9)} \delta v + \frac{2a_7 + a_9}{4a_6} \frac{(\delta u)^2}{\delta v} \qquad (8.96)$$

This expression was first presented by Reinicke and Ralston (1977).
    For plane stress:

$$D_A = \frac{1}{2a_1} \left[ \sqrt{\frac{(a_7^2 + 2a_1)}{a_6}} \sqrt{4(a_1 + a_3)(\delta v)^2 + 2a_1(\delta u)^2} - a_7 \, \delta v \right] \qquad (8.97)$$

This expression was first given by Prodanovic (1976) as reported by Reinicke and Ralston (1977).

### 8.6.1.4. LOWER BOUND

The discontinuous stress field shown in Fig. 8.43 can be used to construct a lower bound for the failure load. This field is described by the six independent parameters $p_1$, $p_2$, $\sigma_1$, $\sigma_2$, $\theta_1$, and $\theta_2$ and consists of 10 regions of constant stress. The best lower bound $\sigma^L$ for the indentation pressure, corresponding to this stress field, is the maximum value of the stress on the indenter, i.e.,

$$\sigma^L = p_2 + 2\sigma_1 \cos^2 \theta_1 + 2\sigma_2 \cos^2 \theta_2 \qquad (8.98)$$

subject to eleven constraints. In each of the ten regions of constant stress, we require that

$$f(\sigma_x, \sigma_y, \tau_{xy}) \leq 0$$

where $f$ is either the plane strain yield function given by Eq. (8.91a) or the plane stress yield function given by Eq. (8.91b), and $\sigma_x, \sigma_y$, and $\tau_{xy}$ are given in terms of the stress field parameters in Table 8.1. An additional constraint $\theta_1 \geq \theta_2$ is also added to avoid ambiguity in the definition of the ten regions. Any set of parameters that satisfies all of the constraints determines a lower bound for the indentation pressure. The best lower bound can be determined by numerically maximizing the expression for $\sigma^L$ subject to the eleven constraints. The best lower-bound solutions for the plane strain and plane stress cases are listed in Table 8.2 and plotted in Fig. 8.44. They will be compared later with the best upper-bound solutions in the same table and figure.

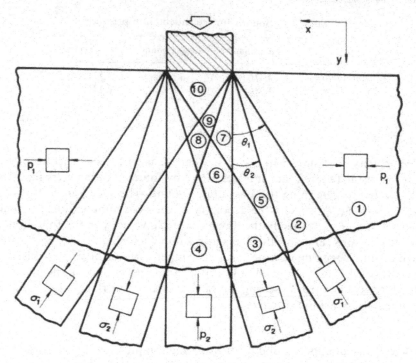

FIGURE 8.43. Lower-bound stress field (Reinicke and Ralston, 1977).

TABLE 8.1. Regional stress components in the lower bound stress field.[a]

| Region | $-\sigma_x$ | $-\sigma_y$ | $\tau_{xy}$ |
|---|---|---|---|
| 1 | $p_1$ | $0$ | $0$ |
| 2 | $p_1 + \sigma_1 \sin^2 \theta_1$ | $\sigma_1 \cos^2 \theta_1$ | $\sigma_1 \sin \theta_1 \cos \theta_1$ |
| 3 | $p_1 + \sigma_2 \sin^2 \theta_2$ | $\sigma_2 \cos^2 \theta_2$ | $\sigma_2 \sin \theta_2 \cos \theta_2$ |
| 4 | $p_1$ | $p_2$ | $0$ |
| 5 | $p_1 + \sigma_1 \sin^2 \theta_1 + \sigma_2 \sin^2 \theta_2$ | $\sigma_1 \cos^2 \theta_1 + \sigma_2 \cos^2 \theta_2$ | $\sigma_1 \sin \theta_1 \cos \theta_1 + \sigma_2 \sin \theta_2 \cos \theta_2$ |
| 6 | $p_1 + \sigma_2 \sin^2 \theta_2$ | $p_2 + \sigma_2 \cos^2 \theta_2$ | $\sigma_2 \sin \theta_2 \cos \theta_2$ |
| 7 | $p_1 + \sigma_1 \sin^2 \theta_1 + \sigma_2 \sin^2 \theta_2$ | $p_2 + \sigma_1 \cos^2 \theta_1 + \sigma_2 \cos^2 \theta_2$ | $\sigma_1 \sin \theta_1 \cos \theta_1 + \sigma_2 \sin \theta_2 \cos \theta_2$ |
| 8 | $p_1 + 2\sigma_2 \sin^2 \theta_2$ | $p_2 + 2\sigma_2 \cos^2 \theta_2$ | $0$ |
| 9 | $p_1 + \sigma_1 \sin^2 \theta_1 + 2\sigma_2 \sin^2 \theta_2$ | $p_2 + \sigma_1 \cos^2 \theta_1 + 2\sigma_2 \cos^2 \theta_2$ | $\sigma_1 \sin \theta_1 \cos \theta_1$ |
| 10 | $p_1 + 2\sigma_1 \sin^2 \theta_1 + 2\sigma_2 \sin^2 \theta_2$ | $p_2 + 2\sigma_1 \cos^2 \theta_1 + 2\sigma_2 \cos^2 \theta_2$ | $0$ |

[a] Reinicke and Ralston, 1977.

TABLE 8.2. Bounds for ice indentation pressure.

|  | Upper bound | Lower bound | Average |
|---|---|---|---|
| Plane strain | $4.47C_x$ | $3.77C_x$ | $4.12C_x \pm 8\%$ |
| Plane stress | $3.38C_x$ | $2.98C_x$ | $3.13C_x \pm 5\%$ |

### 8.6.1.5. UPPER BOUND

The velocity field shown in Fig. 8.42a can be used to construct an upper bound for the failure load. It is described by fourteen parameters that can be selected to determine the least upper bound. The seven angles $\psi_1$, $\psi_2$, $\psi_3$, $\psi_4$, $\phi_2$, $\phi_3$, $\phi_4$ specify the geometry of the deforming region, the four angles $\theta_1$, $\theta_2$, $\theta_3$, $\theta_4$ specify the direction of the velocity in the four rigidly moving regions, and the three magnification factors $\alpha$, $\beta$, $\gamma$ specify the changes in velocity magnitude across the inner surfaces of velocity discontinuity, i.e., $V_2 = \alpha V_1$, $V_3 = \beta V_2$, $V_4 = \gamma V_3$. An upper bound for the indentation force is obtained by setting the rate of external work equal to the rate of internal dissipation of energy, i.e.,

$$P^U V = P^U V_1 \sin(\psi_1 - \theta_1) = D_T \qquad (8.99)$$

where $P^U$ is an upper bound for the indentation pressure, $V$ is the relative velocity between the indenter and the ice sheet, and $D_T$ is the total rate of energy dissipation from the seven surfaces of velocity discontinuity of each side of the indenter. An upper bound for the indentation of an ice sheet of thickness $t$ by an indenter of width $b$ is thus

$$\sigma^U = \frac{P^U}{bt} = \frac{D_T}{bt \, V_1 \sin(\psi_1 - \theta_1)} \qquad (8.100)$$

The total dissipation $D_T$ is the sum of seven terms of the form $D_A A$, where $D_A$ is the rate of energy dissipation per unit area of the discontinuity surface and $A$ is the area of the surface. For plane strain (Reinicke and Ralston, 1977), $D_A$ is given by Eq. (8.96):

$$D_A = \frac{2a_1 a_6 + 2a_1 a_7(a_7 + a_9) + a_9^2(a_1 + a_3)}{2a_1 a_6(2a_7 + a_9)} \delta v + \frac{2a_7 + a_9}{4a_6} \frac{(\delta u)^2}{\delta v}$$

where $\delta v$ and $\delta u$ are the normal and tangential velocity jumps at the discontinuity, respectively. The constraint $\delta v > 0$ must also be satisfied in the plane strain bound. In the plane stress case (Reinicke and Ralston, 1977), $D_A$ is given by Eq. (8.97):

$$D_A = \frac{1}{2a_1}\left[ \sqrt{\frac{(a_7^2 + 2a_1)}{a_6}} \sqrt{4(a_1 + a_3)(\delta v)^2 + 2a_1(\delta u)^2} - a_7 \delta v \right]$$

The values of the velocity field and stress field parameters that produced the bounds are listed in Table 8.3. Both the upper and lower bounds for

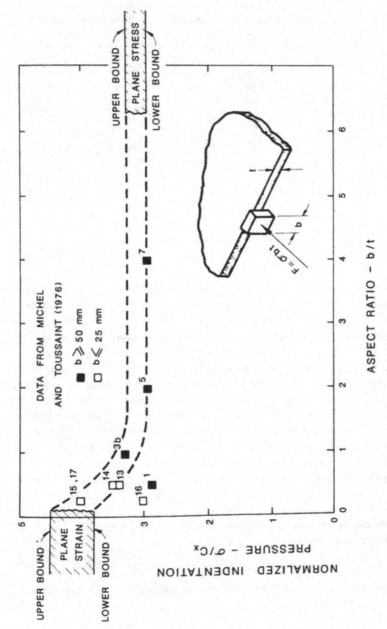

FIGURE 8.44. Comparison of computed bounds for indentation pressure with test data.

TABLE 8.3. Stress and velocity field parameters[a]

| | Upper bound | | | Lower bound | |
| Parameter | Plane stress | Plane strain | Parameter | Plane stress | Plane strain |
| --- | --- | --- | --- | --- | --- |
| $\psi_1$ | 0.750 | 0.802 | $p_1/C_x$ | 0.976 | 1.00 |
| $\psi_2$ | 1.209 | 1.802 | $p_2/C_x$ | 0.685 | 0.759 |
| $\psi_3$ | 1.466 | 1.701 | $\sigma_1/C_x$ | 0.876 | 1.092 |
| $\psi_4$ | 1.629 | 1.505 | $\sigma_2/C_x$ | 0.670 | 0.917 |
| $\phi_2$ | 1.032 | 1.385 | $\theta_1$ | 0.369 | 0.300 |
| $\phi_3$ | 1.697 | 1.679 | $\theta_2$ | 0.707 | 0.729 |
| $\phi_4$ | 2.233 | 2.089 | | | |
| $\theta_1$ | 0.000 | 0.182 | | | |
| $\theta_2$ | 0.034 | 0.369 | | | |
| $\theta_3$ | 0.179 | 0.419 | | | |
| $\theta_4$ | 0.313 | 0.353 | | | |
| $\alpha$ | 1.060 | 1.019 | | | |
| $\beta$ | 1.045 | 1.142 | | | |
| $\gamma$ | 1.127 | 1.302 | | | |

[a] All angles are listed in radians.

the plane stress case are less than $\sigma_{ps}^A$, the laterally confined plane stress (i.e., $3.6C_x$). This implies that ice may fail against very wide structures at less than the type A plane strain strength.

The upper and lower bounds for indentation pressure are listed in Table 8.2 and are also compared with Michel and Toussaint's (1976) constant effective strain rate indentation data in Fig. 8.44. The average value for the pressure is $4.12C_x$ for the plane strain case with a maximum possible error of $\pm 8\%$. For the plane stress case, the average value is 3.13 with a possible error of $\pm 5\%$. The plane stress bounds agree very well with the test data at the higher aspect ratios. At low aspect ratio, the plane strain bounds are only about 30% higher than the plane stress bounds and show good agreement with most of the data obtained with the smaller indenters. This suggests that the data from tests 13 to 17 may be representative of the proper material response, and not a consequence of the ice crystal size. Although the size of these indenters was comparable to the crystal diameter, the size of the plastic deforming region may have been sufficiently large to obscure individual crystal effects.

## 8.6.2. Indentation of a Semi-Infinite Medium by a Square or Rectangular Punch

Three-dimensional problems of rigid-punch indentation are considered in this section. The material is assumed to obey Tresca's yield criterion of constant maximum shearing stress during plastic deformation.

### 8.6.2.1. Lower Bound

The lower bound is obtained here for the value of the average pressure $q_0$ over the rectangular area of contact.

A three-dimensional stress field for the rectangular punch is shown in Fig. 8.45. Figure 8.45b is a direct extension of the two-dimensional stress field in Fig. 8.24, while Fig. 8.45a is a hydrostatic stress field beneath the punch. The stress field of Fig. 8.45b establishes $3k$ as a lower bound for the ultimate pressure. It is easily verified that superposition of the stress field of Fig. 8.45a on that of Fig. 8.45b does not lead to stresses in excess of the yield limit. Thus, $5k$ is a lower bound for the ultimate pressure in the three-dimensional rectangular punch or the square punch.

### 8.6.2.2. Upper Bound by Hill Mechanism

A simple failure mechanism for the rectangular punch is shown diagrammatically in Fig. 8.46. The area of the punch is *lmno*, and the downward movement of the punch is accommodated by movement of the material as indicated by the small arrows in Fig. 8.46a. In Fig. 8.46b and c are shown the plan view and vertical section of Fig. 8.46a, and it can be seen that this mechanism is an extension into three dimensions of a simple modification of the two-dimensional Hill's mechanism in Fig. 8.39a. Since the movement is symmetric about *xy*, it is only necessary to consider the right-hand side of Fig. 8.46a for the following upper-bound computations.

The rates of internal dissipation of energy on the discontinuous surface between the material at rest and the material in motion and in the radial

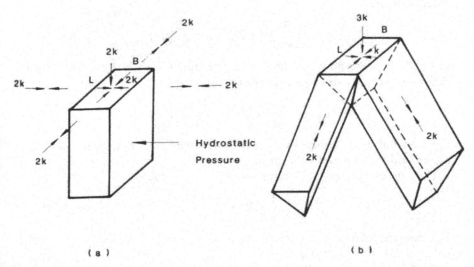

(a)                                              (b)

FIGURE 8.45. Three-dimensional stress field: (a) hydrostatic stress field; (b) extension of Fig. 8.24.

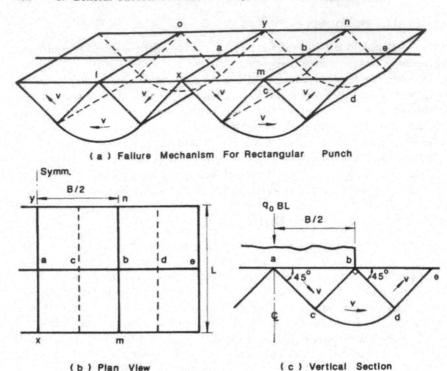

**(a) Failure Mechanism For Rectangular Punch**

**(b) Plan View**

**(c) Vertical Section**

FIGURE 8.46. Three-dimensional Hill mechanism.

shear zone are:

$$2kv\left(\frac{B}{2\sqrt{2}}L\right)+2kv\left(\frac{1}{2}\pi\frac{B}{2\sqrt{2}}L\right) \qquad (8.101)$$

in which $B$ is the punch width and $L$ the punch length. Rates of energy dissipated on the two end surfaces are:

$$2kv\left(\frac{B}{2\sqrt{2}}\right)\left(\frac{B}{2\sqrt{2}}\right)+2kv\left(\frac{1}{4}\pi\right)\left(\frac{1}{8}B^2\right) \qquad (8.102)$$

For a smooth punch, there is no energy dissipated by friction between the punch and the material beneath it.

The rate of external work done by the pressure $q_0$ is:

$$\frac{1}{2}q_0BL\left(\frac{v}{\sqrt{2}}\right) \qquad (8.103)$$

Equating the total rates of internal and external work yields an upper-bound solution for the pressure $q_0$:

$$q_0^U=k\left(5.14+1.26\frac{B}{L}\right) \qquad (8.104)$$

Thus, for a square, smooth punch, for which $B = L$, Eq. (8.104) gives the value $6.4k$, and a value of $5.46k$ is found for a ratio $L/B = 4$. This equation tends to the value of $5.14k$ for rectangles whose length is great compared with their width, i.e., $L \gg B$.

### 8.6.2.3. UPPER BOUND BY MODIFIED HILL MECHANISM FOR SQUARE PUNCH

A better upper bound for a square punch is obtained as follows: In Fig. 8.47a, *lmno* is the square area of the punch which moves downward with

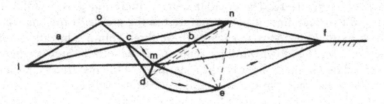

( a )  Failure  Mechanism  for  Square  Punch

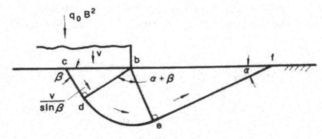

( b )  Vertical  Section

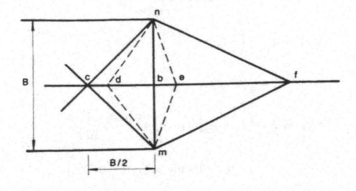

( c )  Plan  View

FIGURE 8.47. Modified Hill mechanism for a square punch (Shield and Drucker, 1953).

initial velocity $v$. The square is divided into four equal triangles by the diagonals $ln$ and $mo$. Taking a typical triangle $cmn$, the downward movement of the triangle is accommodated by lateral movement in the volume $cdefmn$. The volumes $dcmn$ and $efmn$ are tetrahedra, the points $d$ and $e$ being vertically below the line $cf$. The volumes $mbde$ and $nbde$ are two similar sections of right circular cones, the axes of which lie on $mn$. Figures 8.47b and c show the vertical section and the plan through $cf$. Vertical sections by planes parallel to $cf$ through the volume $cdefmn$ are similar in shape to the section shown in Fig. 8.47b but are of varying size. The tetrahedral volumes $dcmn$ and $efmn$ move as rigid bodies in the directions parallel to $cd$ and $ef$, respectively. The circular cone volume $mned$ is a radial shear zone, and its streamline of flow is parallel to the arc $de$. If the angle $bcd$ is denoted by $\beta$, the velocity in each of these three volumes has the constant value $v \csc \beta$. The downward movement of the other three triangles is accommodated in the same way, the remainder of the material being at rest.

Energy is dissipated in the discontinuity surface between the material at rest and the material moving in the volume $cdefmn$. The rate of dissipation of energy due to this discontinuity surface is equal to the area of the surface multiplied by $kv \csc \beta$, since the change in velocity across the surface has the constant value $v \csc \beta$. It is a simple matter to calculate the three parts of the area of the surface of discontinuity. Referring now to Fig. 8.47a, we have:

$$\text{Area } mef = (\text{Area } bef) \frac{(1+\sin^2 \beta)^{1/2}}{\sin \beta}$$

$$= \frac{1}{8} B^2 \sin \beta \cot \alpha (1+\sin^2 \beta)^{1/2} \tag{8.105a}$$

where $\alpha$ is the angle $bfe$ (Fig. 8.47b) and:

$$\text{Area } cmd = \frac{1}{2}\left(\frac{1}{2} B \cos \beta\right)\left(\frac{1}{2} B \sin \beta\right) \frac{(1+\sin^2 \beta)^{1/2}}{\sin \beta}$$

$$= \frac{1}{8} B^2 \cos \beta (1+\sin^2 \beta)^{1/2} \tag{8.105b}$$

$$\text{Area } mde = \frac{1}{2}(\alpha + \beta)\left(\frac{1}{2} B \sin \beta\right)^2 \frac{(1+\sin^2 \beta)^{1/2}}{\sin \beta}$$

$$= \frac{1}{8} B^2(\alpha + \beta) \sin \beta (1+\sin^2 \beta)^{1/2} \tag{8.105c}$$

Energy is also dissipated in the circular cone volume $mned$ where the material is in a state of plane strain motion (Fig. 8.37) so that expression (8.85) can be used to calculate the rate of dissipation of energy per unit thickness in the cone axis direction *mn*. Since only the radius of the cone changes with respect to the axis of the cone, it follows that the rate of dissipation of energy in the cone volume *mned* is equal to the area *mne* = $\frac{1}{4}B^2 \sin \beta$, multiplied by $k(v \csc \beta)(\alpha + \beta)$. When the total rate of internal dissipation of energy is equated to the external rate of work due to the punch load, $q_0 B^2 v$, we obtain an upper-bound solution as:

$$q_0^U(\alpha, \beta) = k[\alpha + \beta + \sqrt{1 + \sin^2 \beta}(\alpha + \beta + \cot \alpha + \cot \beta)] \quad (8.106)$$

The function $q_0^U(\alpha, \beta)$ has the minimum value $5.80k$ when $\alpha$ and $\beta$ are approximately 47° and 34°, respectively.

### 8.6.2.4. MECHANISM FOR RECTANGULAR PUNCHES

Further improvement in the upper bound for a square punch would require the more elaborate failure mechanism discussed by Shield and Drucker (1953). The failure mechanism which was used by Shield and Drucker provides an upper bound for the indentation pressure in the rectangular punch problem and also gives a better upper bound for the square punch than those obtained in the foregoing. Their more elaborate failure mechanism is essentially an extension to a rectangular punch of a simple modification of the failure mechanism in Fig. 8.47 for the case of a square punch. In this improved mechanism, the lateral movement of the material accompanied by the downward movement of the two triangles *mnc* and *col* (Fig. 8.47) is defined by two unknown angles $\alpha$ and $\beta$ (Fig. 8.47b) while the material accompanied by the downward movement of the other two triangles *mcl* and *cno* is now defined by another two unknown angles, $\alpha_1$ and $\beta_1$, which moves in the directions perpendicular to the plane through *cf*. The indentation pressure $q_0$ for a rectangular punch can then be expressed as a function of the four unknown angles, $\alpha$, $\beta$, $\alpha_1$, and $\beta_1$. For rectangles for which $B/L \geq 0.53$, the indentation pressure function $q_0(\alpha, \beta, \alpha_1, \beta_1)$ has the minimum value:

$$q_0^U = k\left(5.24 + 0.47\frac{B}{L}\right) \quad (8.107)$$

when $\alpha = 47°4'$, $\beta = 34°$, $\alpha_1 = 46°17'$, and $\beta_1 = 39°$. Thus, for a smooth square punch, for which $B = L$, Eq. (8.107) gives the value $5.71k$, which is a better upper bound for the smooth, square punch than those obtained previously. For rectangles for which $B/L < 0.53$, a better upper bound than that given by Eq. (8.107) is obtained by putting $\alpha = \beta = 45°$, $\alpha_1 = 46°17'$, and $\beta_1 = 39°$ in the function $q_0^U(\alpha, \beta, \alpha_1, \beta_1)$ to give:

$$q_0^U = k\left(5.14 + 0.66\frac{B}{L}\right) \quad (8.108)$$

This equation tends to the value $5.14k$ for rectangles whose length is much greater compared than their width ($L \gg B$), in agreement with the upper bound for the two-dimensional flat punch.

## References

Brady, W.G., and D.C. Drucker, 1955. "Investigation and Limit Analysis of Net Area in Tension," *Transaction, ASCE* **120:** 1133–1164.

Chen, W.F., 1975. *Limit Analysis and Soil Plasticity*, Elsevier, Amsterdam.

Croasdale, K.R., N.R. Morgenstern, and J.B. Nuttall, 1976. "Indentation Tests to Investigate Ice Pressure on Vertical Piers" (Preprint), Symposium on Applied Glaciology, International Glaciological Society, Cambridge, England.

Drucker, D.C., and W.F. Chen, 1968. "On the Use of Simple Discontinuous Fields to Bound Limit Loads," in *Engineering Plasticity* (J. Heyman and F.A. Leckie, eds.), Cambridge University Press, Cambridge, p. 129.

Drucker, D.C., W. Prager, and H.J. Greenberg, 1952. "Extended Limit Design Theorems for Continuous Media," *Quarterly of Applied Mathematics* **9:** 381–389.

Frederking, R.M.W., and L.W. Gold, 1975. "Experimental Study of Edge Loading of Ice Plates," *Canadian Geotechnical Journal* **12**(4): 456–463.

Green, A.P., 1954. "The Plastic Yielding of Notched Bars Due to Bending," *Quarterly Journal of Mechanics and Applied Mathematics* **6:** 223.

Hill, R., 1950. *The Mathematical Theory of Plasticity*, Oxford University Press, New York.

Johnson, W., and P.B. Mellor, 1973. *Engineering Plasticity*, Van Nostrand, London.

Michel, B., and N. Toussaint, 1976. "Mechanics and Theory of Indentation of Ice Plates" (Preprint), Symposium on Applied Glaciology, International Glaciological Society, Cambridge, England.

Prager, W., and P.G. Hodge, Jr., 1951. *Theory of Perfectly Plastic Solids*, Wiley, New York.

Ralston, T.D., 1977. "Yield and Plastic Deformation in Ice Crushing Failure," ICSI/AIDJEX Symposium on Sea Ice-Processes and Models, Seattle, Washington.

Ralston, T.D., 1978. "An Analysis of Ice Sheet Indentation," Fourth International Symposium on Ice Problems, International Association for Hydraulic Research, Luleå, Sweden, August 7–9, 1978, pp. 13–31.

Ralston, T.D., 1979. "Plastic Limit Analysis of Sheet Ice Loads on Conical Structures," in *Physics and Mechanics of Ice*, IUTAM Symposium, Copenhagen, August 6–10, 1979, pp. 289–308.

Reinicke, K.M., 1979. "Analytical Approach for the Determination of Ice Forces Using Plasticity Theory," in *Physics and Mechanics of Ice*, IUTAM Symposium, Copenhagen, August 6–10, 1979, pp. 325–331.

Reinicke, K.M., and T.D. Ralston, 1977. "Plastic Limit Analysis with an Anisotropic, Parabolic Yield Function," *International Journal of Rock Mechanics and Mining Sciences* **14:** 147–154.

Shield, R.T., 1955. "The Plastic Indentation of a Layer by a Flat Punch," *Quarterly of Applied Mathematics* **13:** 27–46.

Shield, R.T., and D.C. Drucker, 1953. "The Application of Analysis to Punch-Indentation Problems," *Journal of Applied Mechanics* **20:** 453–461.

PROBLEMS

8.1. In Section 8.3.1, we consider, for simplicity, the rectangular bar with a central circular hole. If the Tresca yield condition is applied, the limit load is found to be $P_c = \sigma_0(b-d)t = \sigma_0 A'$, where $A'$ is the net area of the cross section. Clearly, the result $P_c = \sigma_0 A'$ can be applied to a hole of any shape located anywhere in the bar (the line of action of $P$ not being specified) and even holds for any cross section of the bar, not merely a rectangular one. Show that this general conclusion is true.

8.2. Consider a plate with two holes with centers on a transverse line as shown in Fig. P8.2. Find the limit load for the following cases if the Tresca yield criterion is used:

(a) the bar is a thin plate ($t$ small compared to $d$);
(b) the bar is a very thick plate (plane strain).

8.3. Draw the four Mohr's circles in one diagram for the regions $A - A - B$, $A - B - C$, $A - C - D$, and $A - D - E$ of the discontinuous stress field developed in Example 8.3, Section 8.4.3 (see Fig. 8.26). Show clearly the discontinuous lines in the diagram. Hence, verify that the discontinuous stress field so developed is a statically admissible stress field.

8.4. Find the upper and lower bounds for a 90° notched bar as shown in Fig. P8.4, assuming perfect plasticity and employing the Tresca yield criterion, for the following cases:

(a) the bar is a very thin plate (plane stress);
(b) the bar is a very thick plate (plane strain).

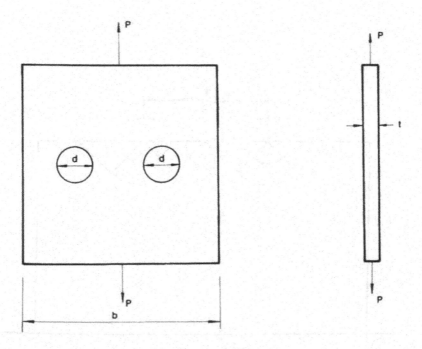

FIGURE P8.2

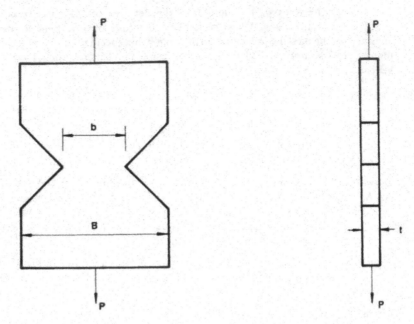

FIGURE P8.4

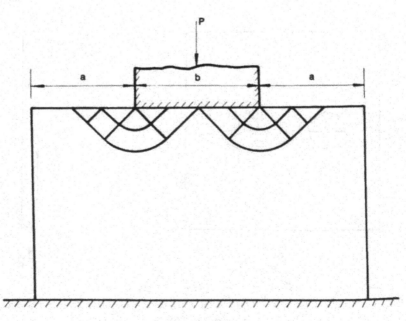

FIGURE P8.5

8.5. A slip-line field solution for the plane strain problem of a rigid punch of width $b$ on a perfectly plastic column of width $2a+b$ is given to you as drawn in Fig. P8.5.

(a) Explain the meaning of such fields in general.

(b) Is the solution valid if $a=b$? Explain. If not, find reasonable upper and lower bounds on the limit load $P$ for this particular case.

8.6. Verify the upper bounds to the limit load for the deformation modes shown in Fig. P8.6.

8.7. Nonhomogeneous media are very often encountered in practical applications. The limit theorems are apt in fact to be of greatest use in such cases because of the enormous difficulty in obtaining an exact solution. A simple plane strain example of soil mechanics is shown in Fig. P8.7. To avoid obscuring the basic points, the friction angle $\phi$ is taken as zero. Hence, the Mohr-Coulomb yield criterion of soil plasticity is reduced to the familiar Tresca yield criterion of metal plasticity. The problem is to find the maximum uniform surcharge $p_{max}$ which can be carried for various ratios of the cohesion $k_2$ of the weak soil to the cohesion $k_1$ of the stronger soil which lies under the load:

(a) if the weak soil has no strength, $k_2=0$;

(b) if $k_2=k_1/2$.

8.8. Impose the plane strain condition on Eq. (8.89) and show that Eq. (8.91a) is true.

8.9. Derive the energy dissipation function $D_e$ [Eqs. (8.93) and (8.94)] for the plane strain and plane stress cases in ice crushing problems.

ANSWERS TO SELECTED PROBLEMS

8.2. (a) $P^C = \sigma_0 t(b-2d)$.

(b) $c=$ clear distance between two holes

If $c \approx 0$, $P^C = \sigma_0 bt\left(1-2\dfrac{d}{b}\right)$

If $c>d$,

$$\sigma_0 bt\left(1-\dfrac{c+2d}{b}\sin\beta\right) < P^C < \sigma_0 bt\left(1-\dfrac{d}{b}\right)$$

8.4. (a) $P^C = 2ktb$.

(b) $P_1^L = 2ktB\left(\dfrac{b}{B}\right)$

$$P_2^L = 2kBt\dfrac{2\dfrac{b}{B}}{1+\dfrac{b}{B}}$$

$$P_1^U = 2kBt\left(2.414\dfrac{b}{B}\right)$$

$$P_2^U = 2kBt\left[\dfrac{1}{2}\left(1+\dfrac{b}{B}\right)\right]$$

8.5. (b) For the case $a=b$,

$3kb < P^C < (2+\pi)kb$

8.7. (a) $P^C = 2kb$.

(b) $3k_1b < P^C < 3.464k_1b$.

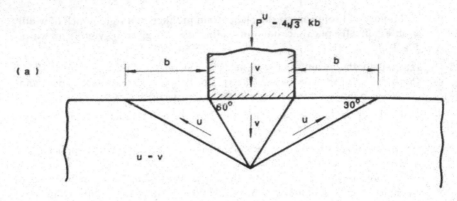

(a)

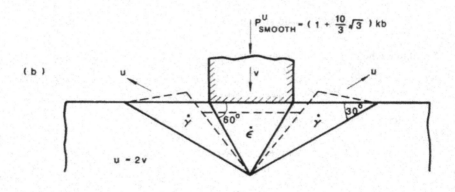

(b)

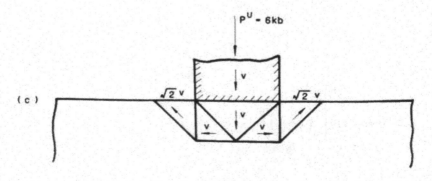

(c)

FIGURE P8.6

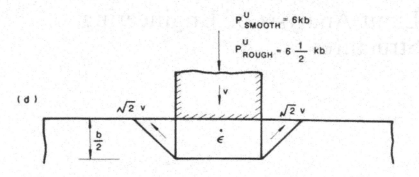

(d)

$$P^U_{SMOOTH} = 6kb$$

$$P^U_{ROUGH} = 6\frac{1}{2}\ kb$$

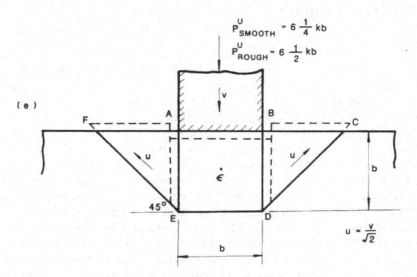

(e)

$$P^U_{SMOOTH} = 6\frac{1}{4}\ kb$$

$$P^U_{ROUGH} = 6\frac{1}{2}\ kb$$

$$u = \frac{v}{\sqrt{2}}$$

FIGURE P8.6 *Continued*

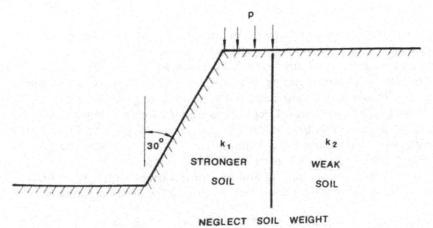

p

30°

$k_1$
STRONGER
SOIL

$k_2$
WEAK
SOIL

NEGLECT SOIL WEIGHT

FIGURE P8.7

# 9
# Limit Analysis of Engineering Structures

## 9.1. Introduction

### 9.1.1. Generalized Variables

In most engineering structures such as beams, frames, plates, and shells, it is more convenient to work with stress and strain resultants, rather than directly with stress and strain. The stress resultants are obtained in a structural element by suitably integrating the physical components of stress. Thus, if the yield stress of a beam element is assumed, the yield moment can be computed as can the yield curve for combined bending and compression for an arch element. Similarly, the strain resultants are obtained in the same element by a suitable kinematic assumption relating the physical components of strain in the element to the rotation or displacement of the element. For example, in a simple bending of beams, the *generalized stress variable* is the bending moment $M$ and the corresponding *generalized strain variable* is the bending curvature $\phi$. The bending strain $\epsilon$ and the bending curvature $\phi$ are related through the powerful kinematic assumption that plane sections remain plane after bending for the beam element under consideration.

#### 9.1.1.1. GENERALIZED STRESSES

Generalized stresses are stress resultants in a structural element. For the general case of a structural element, $n$ stress-type variables, $Q_1, Q_2, \ldots, Q_n$, known as *generalized stresses*, will be required to describe its state. For example, in a simple bending of beams, $Q_1 = M$ is the only stress variable, but for beams under combined tension and bending, we need $Q_1 = N$ and $Q_2 = M$ to describe them. Deep beams under combined tension, bending, and shear are described by $Q_1 = N$, $Q_2 = M$, and $Q_3 = V$ (Fig. 9.1a). A rotationally symmetric plate would have $Q_1 = M_r$, $Q_2 = M_\theta$, and $Q_3 = M_{r\theta}$ (Fig. 9.1b). A rectangular plate would be described by $Q_1 = M_x$, $Q_2 = M_y$, and $Q_3 = M_{xy}$, but we have to add two more variables $Q_4 = V_x$ and $Q_5 = V_y$ if we must also consider the effect of shear force on its deformation (Fig.

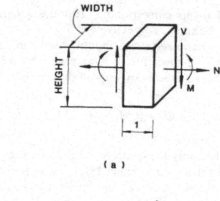

(a)

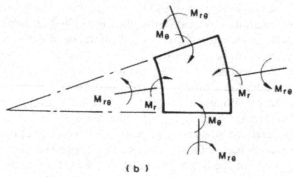

(b)

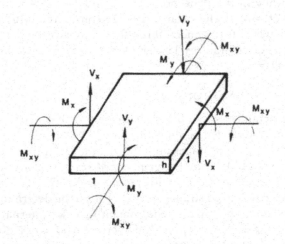

(c)

FIGURE 9.1. Examples of generalized stresses: (a) unit length of the beam taken as a basic unit length. (b), (c) unit element with plate thickness $h$ taken as a basic unit element.

9.1c). The structural elements corresponding to these generalized stresses are shown in Fig. 9.1a, b, and c, respectively.

The choice of which generalized stress should be used in a given problem is *not unique*. For example, for a rectangular plate one may choose, for convenience, the generalized stresses in terms of the dimensionless variables as

$$Q_1 = \tfrac{1}{2}(M_x + M_y)/M_0, \qquad Q_2 = \tfrac{1}{2}(M_x - M_y)/M_0, \qquad Q_3 = M_{xy}/M_0$$

in which $M_0$ is the *limit moment*.

It should be noted that only those stress quantities which contribute to internal work should be taken as generalized stresses. In problems of beams and plates, the transverse shear force may be treated as the generalized stress only for an elaborate analysis. However, it has been shown by tests that shear deformation in a beam or a plate is negligible; hence, the shear force does not do any work, within the limits of the simple theory. Therefore, for simplicity, the shear force is not treated as a generalized stress but rather as a *reaction*. The reactions are the stress resultants that are needed for equilibrium consideration, but their corresponding deformations are neglected. The shear force belongs to this type of force and is therefore considered to be a reaction, rather than a generalized stress.

Once the generalized stresses for a problem have been chosen, the state of stress of a unit element of a structure can be expressed by a stress vector $Q_i$ $(i = 1, 2, \ldots, n)$. In the bar problem of Fig. 9.1a, for example, there are two generalized stress components and hence $n = 2$, and we can put $Q_1 = M$ and $Q_2 = N$, while in the plate problem of Fig. 9.1c, $n = 3$, and $Q_1 = M_x$, $Q_2 = M_y$, and $Q_3 = M_{xy}$. The shear forces are considered to be the reaction forces. As a matter of fact, general three-dimensional stress analysis can be included in our discussion, by letting $Q_1 = \sigma_x$, $Q_2 = \sigma_y$, $Q_3 = \sigma_z$, $Q_4 = \tau_{yz}$, $Q_5 = \tau_{zx}$, and $Q_6 = \tau_{xy}$.

### 9.1.1.2. GENERALIZED STRAINS

The state of generalized strain of a unit element of a structure is described by a number of generalized strain components. We denote these strains as the components of an $n$-dimensional generalized strain vector $q_i$ with $i = 1, 2, \ldots, n$. We shall define the *generalized strain components* in such a way that for a chosen set of generalized stresses $Q_1, Q_2, \ldots, Q_n$, the corresponding set of generalized strains $q_1, q_2, \ldots, q_n$ to be determined to within a multiplicative constant by the requirement that the internal work be of the form $Q_1 q_1 + Q_2 q_2 + \cdots + Q_n q_n$. The increment of work done per unit volume of the structural element when the strains are increased from $q_i$ to $q_i + dq_i$ is therefore given by

$$dW = Q_i \dot{q}_i = Q_i \, dq_i = Q_1 \, dq_1 + Q_2 \, dq_2 + \cdots + Q_n \, dq_n \qquad (9.1)$$

The work increment $dW$ is thus the scalar product of the vectors $Q_i$ and $dq_i$. In the bar problem, therefore, with the choice of $Q_1 = M$, $Q_2 = N$, the strain variables corresponding to $M$ and $N$ are the curvature $\phi$ and the axial strain $\epsilon$. Thus, we put $q_1 = \phi$ and $q_2 = \epsilon$. In this problem, the work increment given by Eq. (9.1) becomes

$$dW = Q_i \, dq_i = M \, d\phi + N \, d\epsilon \tag{9.2}$$

## 9.1.2. Relations Between Generalized Stress and Generalized Strain

### 9.1.2.1. Yield Criterion and Flow Rule

The yield criterion can now be written as

$$f(Q_1, Q_2, \ldots, Q_n) = f(Q_i) = 0 \tag{9.3}$$

which defines the elastic limit for any possible combination of generalized stresses.

As in the plastic theory of a continuum, the generalized strain rates are assumed to be the sum of the elastic strain and plastic strain rates:

$$dq_i = dq_i^e + dq_i^p \tag{9.4}$$

The *generalized plastic strain rates*, $dq_i^p$, are again assumed to be derived from a *plastic potential*, taken as identical with the yield function (9.3). Regarding the stress space simultaneously as a strain space, each coordinate axis corresponds to both the relevant generalized stresses and the corresponding generalized plastic strain rate. The *flow rule* requires the generalized plastic strain rate vector to have the direction of the normal to the yield surface at all points where this is uniquely defined, namely,

$$dq_i = \mu \frac{\partial f}{\partial Q_i} \tag{9.5}$$

where $\mu$ is a positive factor of proportionality. At corners or vertices where the yield surface does not have a continuously turning tangent plane, the generalized strain rate vector may have any direction within the fan or cone defined by the normals of the contiguous surfaces.

### 9.1.2.2. Virtual Work Principle

Consider an arbitrary structure which occupies a volume $V$ bounded by a surface $A$. Let $Q_i$ be any statically admissible generalized stress vector which is balanced with the external traction $T_i$, and let $\dot{u}_i^*$ be a kinematically admissible velocity vector which is associated with the generalized strain rate vector $\dot{q}_i^*$. We assume small deformation so that the changes in geometry of the structure are negligible and the virtual work principle is valid. Hence, the total rate of internal work done by $Q_i$ on $\dot{q}_i^*$ is equal to the total rate of external work done by the traction $T_i$ on the velocities $\dot{u}_i^*$:

$$\int_A T_i \dot{u}_i^* \, dA = \int_V Q_i \dot{q}_i^* \, dV$$

A similar statement can also be made about the stress rate:

$$\int_A \dot{T}_i \dot{u}_i^* \, dA = \int_V \dot{Q}_i \dot{q}_i^* \, dV$$

The proof is precisely analogous to that of the *principle of virtual work* (see Section 3.4, Chapter 3).

### 9.1.2.3. DRUCKER'S STABILITY POSTULATE AND ITS INFERENCES

If the generalized stress and strain vectors $Q_i$ and $q_i$ are used to replace the ordinary stress and strain tensors $\sigma_{ij}$ and $\epsilon_{ij}$, the statements of Drucker's *stability postulate* for a structural problem are identical to those for a continuum of work-hardening type (see Section 5.4.2, Chapter 5). The inferences of Drucker's stability postulate for a work-hardening material can be carried over for a general structure. These include:

 (i)  *Convexity.* The initial yield surface $f(Q_i) = 0$ and all subsequent loading surfaces established by the path of loading must be *convex.*
 (ii) *Normality.* The generalized plastic strain increment vector $dq_i^p$ must be *normal* to each yield or loading surface at a smooth point and lie between adjacent normals at a corner.
(iii) *Linearity.* The generalized strain increment components are *linear* in the components of the generalized stress increments for a smooth loading surface.

A direct consequence of Drucker's stability postulate was given by Eq. (5.61) in Chapter 5, which can be rewritten in terms of the generalized stress and strain as

$$(Q_i - Q_i^*) \, dq_i^p \geq 0 \quad \text{or} \quad Q_i \, dq_i^p \geq Q_i^* \, dq_i^p$$

If $Q_i^*$ is inside the yield surface, then the inequality sign holds. Referring to Eq. (4.41) of Chapter 4, we find that the above relation is also valid for a structural element with a generalized stress–strain relationship of the elastic–perfectly plastic type.

This relation states that *the rate at which plastic work is done on a given plastic strain rate system has a maximum value for the actual stress state.* This is known as the *principle of maximum plastic work.*

Further, the rate of plastic work or the rate of energy dissipation $D$ can also be shown to be uniquely defined by a given generalized strain rate $\dot{q}_i$, namely,

$$D = Q_i \dot{q}_i^p = D(\dot{q}_i^p)$$

The proof is analogous to that for a general body (see Section 8.2 and Fig. 8.4).

### 9.1.3. Limit Analysis Theorems

The three theorems of limit analysis, as stated for general bodies, furnish the corresponding theorems for any special class of engineering structures, if the appropriate generalized stresses are used to replace the stress components, and the corresponding generalized strain rates are used to replace the strain rate components. Proofs are analogous to those given in Chapter 8.

Thus, a lower bound will be furnished by showing that a set of generalized stresses $Q_i$'s nowhere violates the yield condition (9.3) and satisfies the equation of equilibrium throughout the structure with a load $P^L$; these may be algebraic or differential equations. An upper bound will be obtained by considering any compatible velocity field agreeing with the constraints. The assumed velocities in the field determine the generalized plastic strain rates throughout the structure, which, in turn, determine the rate of energy dissipation by the flow rule (9.5). An upper bound $P^U$ is found by equating the rate of work of the external loads to the rate of total energy dissipation.

## 9.2. Bending of Beams and Frames

### 9.2.1. Assumptions

The cross section of a beam or a member of a frame is assumed to have an axis of symmetry (Fig. 9.3b) which is in the plane of loading. In a frame, the members are assumed to lie in one plane and to be subjected to loads which also lie in the same plane. The plastic deformations due to shear and normal forces are generally neglected. Consequently, the *simple plastic theory* is concerned with the development of the relationship between the bending moment $M$ and the curvature $\phi$ for the segment of beams or frames.

Figure 9.2b shows a typical relationship between the bending moment $M$ and the curvature $\phi$ for a beam segment whose material can be idealized as elastic–perfectly plastic (Fig. 9.2a). If the bending moment is applied to the previously unloaded and unstrained beam segment, as at $O$ in the figure, the curvature at first increases linearly with the bending moment along $OA$. This linear elastic range is terminated when the bending moment $M_y$ is attained at $A$. The moment at which the yield stress at the most stressed outer fibers of the beam is just reached is referred to as the *yield moment* and is denoted by $M_y$. When the bending moment is further increased, the curvature begins to increase more rapidly along $AB$. This corresponds to the spread of the yield from the outmost fibers inwards towards the neutral axis of the beam. This is known as the *plastification* or the *contained plastic flow*. Finally, the curvature tends to infinity as the limiting value of the bending moment is approached. This limiting bending moment is called the *fully plastic moment* and is denoted by $M_p$. $M_p$ is often used as the *limit moment*, $M_0$, of the cross section.

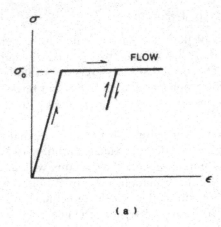

(a)

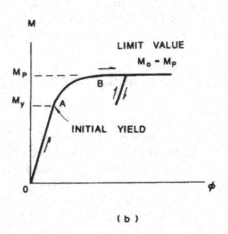

(b)

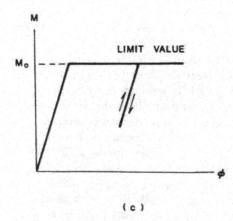

(c)

FIGURE 9.2. Generalized stress–strain relation: (a) material stress–strain relation; (b) $M$-$\phi$ relation; (c) further idealization.

For brevity, we have not distinguished between the limit moments for bending in positive and negative senses. The two magnitudes are identical if the tension and compression yield stresses are equal or if the cross section has an axis of symmetry perpendicular to the plane of loading.

In the limit analysis of beams and frames, the *generalized stress-strain relation* or the *moment-curvature relation* is further idealized by neglecting the contained plastic flow effect, thus leading to the elastic-perfectly plastic type of relation of the kind shown in Fig. 9.2c.

Figure 9.2c introduces the fundamental concept of *plastic hinge* in a simple plastic theory. It shows that as the bending moment reaches the value $M_p$, the curvature becomes infinitely large so that a finite change of slope can occur over an infinitely small length of the member at this cross section. Thus, the behavior of the section where $M_p$ is attained can be described by imagining a hinge to be inserted in the member at this section, the hinge being capable of completely resisting the relative rotation until the fully plastic moment $M_p$ is attained, and then permitting a positive relative rotation of any magnitude, while the bending moment remains constant at the value $M_p$. If the bending moment is reduced below $M_p$, elastic unloading occurs and the relative hinge rotation remains constant. Precisely corresponding statements can be made concerning the hinge action which is supposed to occur when the negative fully plastic moment $-M_p$ is attained.

To sum up, the basic assumption used in the limit analysis of beams and rigid frames is that at any section of a member, the bending moment must lie between the positive and negative fully plastic moments, $\pm M_p$, i.e.,

$$|M| \le M_p \qquad (9.6)$$

## 9.2.2. Evaluation of the Fully Plastic Moment $M_p$

The plastic moment $M_p$ is the maximum moment capacity, $M_0$, a beam segment can carry. This moment can be determined by the method of limit analysis. The quantity $M_0$ depends on the yield stress values in simple tension and compression, which are assumed to be equal here and denoted by $\sigma_0$ (Fig. 9.2a), and on the geometry of a given cross section. Consider a beam element (Fig. 9.3a) subjected to the limit moment $M_0$. A suitable stress field is shown in Fig. 9.3c. Above the plane N.A., the material is in simple compression, and, below this plane, in simple tension, in both cases at the yield stress $\sigma_0$. Equilibrium requires

$$\Sigma H = 0 \quad \text{or} \quad A_1 = A_2 \qquad (9.7)$$

$$\Sigma M = 0 \quad \text{or} \quad M_0^L = \int_A \sigma_x y \, dA = \sigma_0 \int_{A_1} y \, dA - \sigma_0 \int_{A_2} y \, dA \qquad (9.8)$$

Equation (9.8) gives a lower bound.

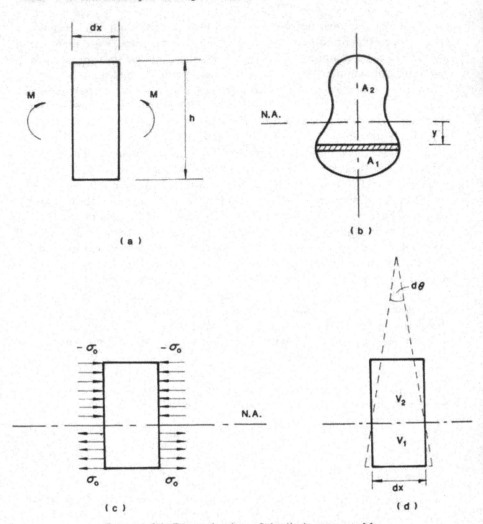

FIGURE 9.3. Determination of the limit moment $M_0$.

To find an upper bound, we consider a field of linear displacement and strain increment as shown in Fig. 9.3d, the fibers at the plane N.A. being unstrained. Integration over the slice gives the total plastic energy dissipation:

$$W_I = \int_V \sigma \epsilon \, dV$$

$$= \int_{V_1} \sigma_0 \left( \frac{y \, d\theta}{dx} \right) dV + \int_{V_2} (-\sigma_0) \left( \frac{y \, d\theta}{dx} \right) dV$$

$$= \left[ \sigma_0 \int_{A_1} y \, dA - \sigma_0 \int_{A_2} y \, dA \right] d\theta$$

and equating this to the external work

$$W_E = M^U \, d\theta$$

gives the upper bound $M^U$, which is the same as the lower bound $M^L$. Equation (9.8) gives, therefore, the correct limit moment $M_0$ or the fully plastic moment $M_p$:

$$M_p = M_0 = \sigma_0 \int_{A_1} y \, dA - \sigma_0 \int_{A_2} y \, dA \tag{9.9}$$

For a rectangular section with depth $h$ and width $b$, we have $M_p = \sigma_0 bh^2/4$. Recall that the yield moment $M_y$ for a rectangular section is $M_y = \sigma_0 bh^2/6$. The ratio $M_p/M_y$ for the rectangular section is 1.5. This ratio is called the *shape factor*. As can be seen, the value of the shape factor depends only on the geometry of the cross section, and is about 1.10 to 1.25 for rolled I-section beams. The shape factor for a solid circular section is $16/3\pi$; and that for a very thin-walled, hollow circular tube is $4/\pi$.

### 9.2.3. Limit Analysis of Beams and Frames

Limit load is defined as the load at which plastic collapse occurs for the ideal structure which has the same dimensions as the actual structure, but with no work-hardening for the material. The limit theorems stated in forms appropriate to beams and frames are summarized as follows:

*Lower-bound theorem*: If an equilibrium distribution of moment can be found, which balances the applied load, and is everywhere below yield or at yield, the structure will not collapse or will just be at the point of collapse.

*Upper-bound theorem*: The structure will collapse if there is any compatible pattern of plastic deformation for which the rate of work done by the external forces exceeds the rate of internal dissipation.

In applying the upper-bound theorem, any compatible pattern of plastic deformation can be used, but, in practice, only those which convert the structure into a mechanism need to considered. The term *mechanism* implies that a system of rigid bars linked by hinges is developed in the framed structure such that motion of the system takes place through rotations at the hinges. In a continuous beam or frame, such hinges are plastic hinges where discontinuities in slope occur in a member or at a joint. This is the distinctive type of discontinuity in the limit analysis for framed structures. At the plastic hinge location, the *plastic energy dissipation* is given by $M_0\theta$, where $\theta$ is the change in slope across the discontinuity and $M_0$ is the plastic limit moment of the cross section. An upper bound is furnished by any set of plastic hinges which converts the structure into a mechanism. The virtual work equation for the mechanism is

$$W_E(P) = \sum M_{0i}\theta_i$$

where $W_E(P)$ is the work rate of external forces, $\theta_i$ is the rate of rotation at the $i$th hinge, $M_{0i}$ is the limit moment at the hinge section, and the sum includes all plastic hinges. Each term in the sum is always positive for the plastic work or energy dissipation. There always exists a set of plastic hinges in a framed structure to convert the structure (or part of it) into a mechanism. It is usually a simple matter to check that the plastic moment conditions, $|M| \leq M_0$, and the equilibrium equations are satisfied for the critical mechanism that gives the least upper bound to the limit load. Since conditions of both the upper-bound and lower-bound theorems are satisfied, the load so determined is the correct limit load.

Based on the two theorems, techniques have been developed that have enabled the limit load to be computed for complex frames with a remarkable rapidity. Such techniques include the method of combining mechanisms and the method of inequalities and moment distributions.

In what follows, we shall consider a simple example of the application of limit theorems to a framed structure to illustrate the basic concept of limit analysis. We shall not discuss here various special techniques of limit analysis as applied to framed structures. Interested readers may refer to the books by Baker and Heyman (1969), Hodge (1959), and Neal (1963) for such applications.

## 9.2.4. A Convincing Demonstration

Discussed here is a simple example by which some main points are indicated. Consider a rectangular portal of uniform section with one foot fixed and the other pinned as shown in Fig. 9.4a.

By an equilibrium approach, we may assume that the beam $BD$ carries the vertical load $P$ as a simply supported beam and that the left column carries the horizontal load $Q$ as a cantilever (Fig. 9.4b). The corresponding moment distribution satisfies equilibrium and will be everywhere at or below yield if

$$\frac{PL}{4} \leq 3M_0 \quad \text{and} \quad Qh \leq M_0$$

Then, the lower-bound theorem assures that the structure will not collapse. Hence, the actual collapse loads will be bounded by

$$\frac{12M_0}{L} \leq P^c \quad \text{and} \quad \frac{M_0}{h} \leq Q^c \tag{9.10}$$

By the mechanism approach, we may assume a sidesway mechanism as shown in Fig. 9.4c. Equating the external rate of work done, $\dot{W}_E = Qh\theta$, to

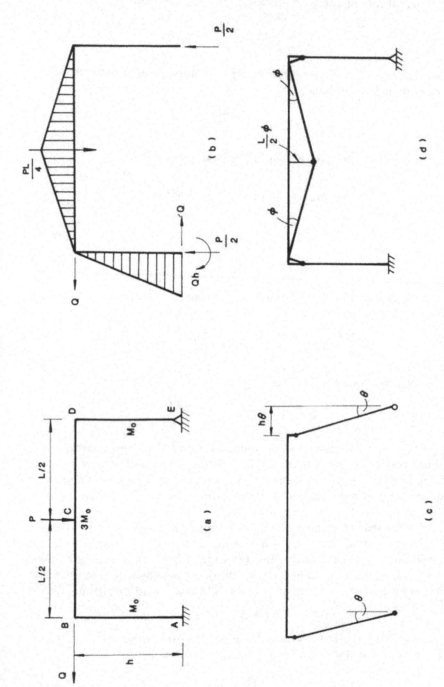

FIGURE 9.4. (a) A rectangular portal frame. (b) An equilibrium stress field. (c), (d) Independent mechanisms.

the internal rate of energy dissipation, $\dot{W}_I = 3M_0\theta$, leads to

$$Q = \frac{3M_0}{h}$$

The structure will collapse, according to the upper-bound theorem, if $Q$ is greater than $3M_0/h$; hence,

$$Q^c \leq \frac{3M_0}{h} \qquad (9.11)$$

Similarly, the mechanism shown in Fig. 9.4d gives

$$M_0\phi + 3M_0(2\phi) + M_0\phi = P\phi\frac{L}{2}$$

or

$$P^c \leq 16\frac{M_0}{L} \qquad (9.12)$$

Now, we can see that from Eqs. (9.10) through (9.12), a very crude picture gives rough bounds on the collapse load, namely,

$$\frac{12M_0}{L} \leq P^c \leq \frac{16M_0}{L} \quad \text{and} \quad \frac{M_0}{h} \leq Q^c \leq \frac{3M_0}{h}$$

If an average value is taken, say,

$$P^c = 14\frac{M_0}{L}, \qquad Q^c = 2\frac{M_0}{h}$$

then the maximum errors will be about 14% and 50%, respectively.

Such results can be obtained with very little effort and often will suffice for an immediate practical answer. A correct answer can be obtained by taking a slightly more elaborate equilibrium moment distribution and also mechanism.

Different answers can be obtained for different ratios of $P$ to $Q$ and $L$ to $h$. Now, choose $P = Q$, $L = 4h$, assume that the left column carries $\frac{2}{3}Q$ and $\frac{2}{3}P$ and the right column carries $\frac{1}{3}Q$ and $\frac{1}{3}P$ (Fig. 9.5a). The corresponding moment diagram is obtained by equilibrium as shown in Fig. 9.5b and represents a statically admissible stress field if the load $Q$ satisfies

$$Qh \leq 3M_0$$

and, hence, by the lower-bound theorem, the structure will not collapse. Therefore, the collapse load is bounded below by

$$\frac{3M_0}{h} \leq Q^c \qquad (9.13a)$$

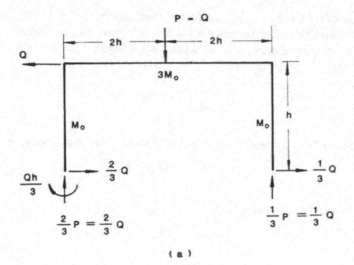

(a)

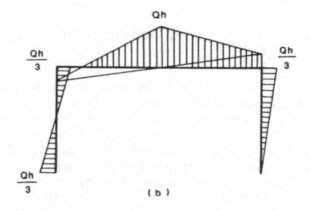

(b)

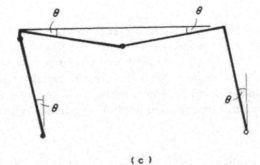

(c)

FIGURE 9.5. A more elaborate analysis: (a) sketch of the frame; (b) the equilibrium field; (c) the failure mechanism.

On the other hand, we can assume a mechanism that is a combination of the previous two mechanisms with the right corner hinge eliminated (Fig. 9.5c). Equating the external rate of work, $\dot{W}_E = Qh\theta + PL\theta/2 = 3Qh\theta$, to the internal rate of energy dissipation, $\dot{W}_I = M_0\theta + M_0(2\theta) + 3M_0(2\theta) = 9M_0\theta$, gives

$$Q = \frac{3M_0}{h}$$

By the upper-bound theorem, the collapse load is bounded above by

$$Q^c \leq 3\frac{M_0}{h} \tag{9.13b}$$

Hence, from Eqs. (9.13a) and (9.13b),

$$Q^c = 3\frac{M_0}{h} \tag{9.14}$$

is the correct answer.

# 9.3. Combined Axial and Bending Forces in Frames and Arches

In the general cases of frames and arches, the resulting forces are described by the bending moment, axial force, and shear force at each cross section. In the preceding discussion, we have assumed that the yield strength of a unit element of beams or frames depends only on the bending moment, while the shear and axial forces were considered to be the reactions related to the bending moment by equilibrium. Thus, in calculating the fully plastic moment $M_p$, the member has been assumed to be subjected only to pure bending, so that the axial and shear forces are zero. However, as will be discussed in this section, the value of the fully plastic moment is affected by the presence of axial and shear forces. Since in the applications of the plastic methods, fully plastic moments and plastic hinges usually occur at positions where one or both of these influences is present, it is of considerable importance to be able to predict the changes in the values of the fully plastic moment due to these causes. Although in many practical cases, the effects of axial and shear forces are very small, there are certain types of structures in which it is important to account for these effects. This is the case, for example, in columns carrying axial load and in arches subjected to normal pressure.

## 9.3.1. Yield Condition—Axial and Bending Interaction Curves

Consider a bar element subjected to combined bending and tension. For simplicity, the section is assumed to be rectangular with two axes of

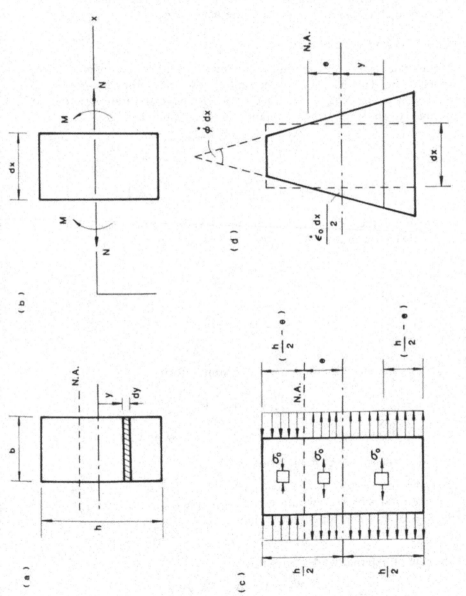

FIGURE 9.6. Axial force–bending moment interaction.

symmetry and with equal yield stresses in tension and compression (Fig. 9.6). The generalized stresses now include the moment $M$ and the axial force $N$. The corresponding generalized strain rates are the curvature rate $\dot{\phi}$ and the axial strain rate at the neutral plane $\dot{\epsilon}_0$. The general theorems of limit analysis are now applied to the bar element.

### 9.3.1.1. LOWER-BOUND APPROACH

We assume a stress distribution satisfying the yield condition as indicated in Fig. 9.6c. This has a simple tension below and a simple compression above the plane N.A. located a distance $e$ above the transverse axis of symmetry. Then, the stress resultants can be computed:

$$N = \sigma_0 b(2e) = 2be\sigma_0 \tag{9.15a}$$

and

$$M = \sigma_0 b\left(\frac{h}{2} - e\right)\left(\frac{h}{2} + e\right) = \sigma_0 b\left(\frac{h^2}{4} - e^2\right) \tag{9.15b}$$

in which $\sigma_0$ is the yield stress in simple tension and compression. Eliminating the parameter $e$ in the above two equations leads to

$$M = \frac{1}{4}\sigma_0 bh^2\left(1 - \frac{N^2}{\sigma_0^2 b^2 h^2}\right) \tag{9.15c}$$

If we define

$$M_0 = M_p = \tfrac{1}{4}\sigma_0 bh^2$$
$$N_0 = \sigma_0 bh \tag{9.16}$$

then the relation (9.15c) reduces to the simple form

$$\frac{M}{M_0} = 1 - \left(\frac{N}{N_0}\right)^2 \tag{9.17}$$

This is a lower-bound interaction condition for $M$ and $N$.

### 9.3.1.2. UPPER-BOUND APPROACH

Consider the velocity field shown in Fig. 9.6d, in which one face of the element rotates by $\dot{\phi}\,dx$ with respect to the other, and the strain is zero at the axis N.A. The compatibility condition is

$$\dot{\phi}e = \dot{\epsilon}_0 \tag{9.18}$$

The rate of external work is given by

$$W_E = M\dot{\phi}\,dx + N\dot{\epsilon}_0\,dx \tag{9.19}$$

and the rate of internal energy dissipation is

$$\dot{W}_I = \int_V \dot{D}\,dV = \int_{-h/2}^{-e}(-\sigma_0)\dot{\epsilon}b\,dx\,dy + \int_{-e}^{h/2}\sigma_0\dot{\epsilon}b\,dx\,dy \tag{9.20}$$

Equating Eq. (9.19) to Eq. (9.20) and applying the compatibility condition (9.18) leads to

$$M\dot{\phi} + N\dot{\epsilon}_0 = b\sigma_0\left(\frac{h^2}{4}\dot{\phi} + \frac{\dot{\epsilon}_0^2}{\dot{\phi}}\right) \tag{9.21}$$

Minimizing $M$ and $N$ with respect to $\dot{\phi}$ and $\dot{\epsilon}_0$, we have

$$N = 2b\sigma_0 \frac{\dot{\epsilon}_0}{\dot{\phi}} = 2b\sigma_0 e$$

$$M = \sigma_0 b\left(\frac{h^2}{4} - \frac{\dot{\epsilon}_0^2}{\dot{\phi}^2}\right) = \sigma_0 b\left(\frac{h^2}{4} - e^2\right)$$

which are the same as Eqs. (9.15a) and (9.15b). Hence, the upper-bound solution coincides with the lower-bound solution. It follows that the correct axial-bending interaction equation or the yield condition takes the simple form

$$\frac{M}{M_0} + \left(\frac{N}{N_0}\right)^2 = 1 \tag{9.22}$$

where $M_0$ and $N_0$ defined by Eq. (9.16) are the limit moment in pure bending and the limit force in simple tension or compression, respectively.

It can easily be checked that the flow rule is satisfied by differentiating Eq. (9.22) and using relations (9.15), (9.16), and (9.18):

$$-\frac{dN}{dM} = \frac{1}{e} = \frac{\dot{\phi}}{\dot{\epsilon}_0} \tag{9.23}$$

which is the *normality condition*. Or, more precisely, denote $f = M/M_0 + (N/N_0)^2 - 1$, and recalling the flow rule (9.5), we have

$$\left.\begin{array}{l} \dot{\phi} = \mu \dfrac{\partial f}{\partial M} = \dfrac{\mu}{M_0} \\[3mm] \dot{\epsilon}_0 = \mu \dfrac{\partial f}{\partial N} = 2\mu \dfrac{N}{N_0^2} \end{array}\right\} \tag{9.24}$$

Elimination of the proportionality $\mu$ yields

$$\frac{\dot{\phi}}{\dot{\epsilon}_0} = \frac{N_0^2}{2M_0 N} \quad \text{or} \quad \frac{M_0\dot{\phi}}{N_0\dot{\epsilon}_0} = \frac{N_0}{2N} \tag{9.25}$$

It is convenient to use $M/M_0$ and $N/N_0$ as nondimensionalized stress resultants, to which the corresponding generalized plastic strain rates are $M_0\dot{\phi}$ and $N_0\dot{\epsilon}_0$, respectively. The interaction curve of Eq. (9.22) is drawn in terms of these quantities in Fig. 9.7a. Any point on this curve corresponds to a fully plastic section of the bar. The generalized plastic strain rate vector $(M_0\dot{\phi}, N_0\dot{\epsilon}_0)$ is also shown normal to this curve. The *flow rule* (9.25) applies at all points except the vertices $M/M_0 = 0$, $N/N_0 = \pm1$. At these points, the plastic strain rate vector may have any direction between normals defined by the adjacent elements of the curves.

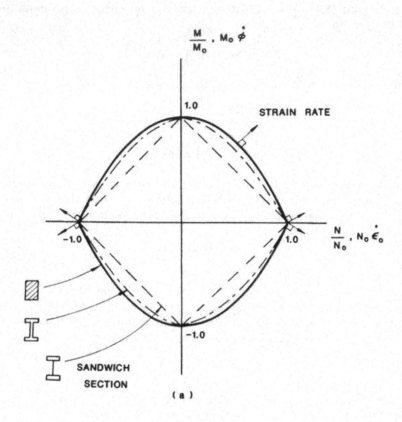

(a)

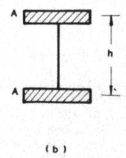

(b)

FIGURE 9.7. (a) Axial and bending interaction curves. (b) A sandwich section.

For a typical I section, the yield curve is like the dashed-dotted curve shown in Fig. 9.7. The extreme case of an I section, i.e., the sandwich section with an infinitely thin web and all sectional area contained in the two flanges, is represented by the square yield curve with straight lines joining points $\pm 1$ on the axes. The curve for a circular cross section lies outside but very close to that for the rectangle, the $M/M_0$ value differing at most by 3% (not shown in the figure).

## 9.3.2. Tension of a Circular Ring

Consider a circular ring subjected to tension by two opposite forces $Q$ (Fig. 9.9a). The material is assumed to be rigid–perfectly plastic, and an idealized square yield curve (Fig. 9.8) for a sandwich section is used as the yield condition for the axial force $N$ and the moment $M$, because it represents a safe approximation and is simple to use.

### 9.3.2.1. RATE OF ENERGY DISSIPATION

Denote $D$ as the energy dissipation per unit length, defined by

$$D = M\dot{\phi} + N\dot{\epsilon}_0$$

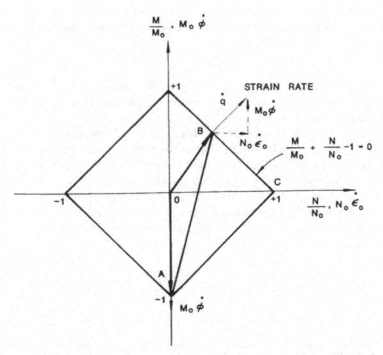

FIGURE 9.8. An approximate yield curve. $OA$ and $OB$ are the loading paths for plastic hinges $A$ and $B$, respectively. $AB$ is the "stress profile" for arc $AB$ at collapse load.

At a stress point $B$ (Fig. 9.8), for instance, the generalized stresses $M$ and $N$ satisfy the relation

$$f = \frac{M}{M_0} + \frac{N}{N_0} - 1 = 0 \tag{9.26}$$

and due to the flow rule, with the derivatives $\partial f / \partial M$ and $\partial f / \partial N$ equal, the strain rates satisfy

$$M_0 \dot{\phi} = N_0 \dot{\epsilon}_0 \geq 0$$

Hence, the energy dissipation $D$ reduces to

$$D = M_0 \dot{\phi} \left( \frac{M}{M_0} + \frac{N}{N_0} \right) = M_0 \dot{\phi} \tag{9.27}$$

Alternatively, we may derive this from its geometric interpretation, namely,

$$D = \mathbf{OB} \cdot \mathbf{q} = \mathbf{OC} \cdot \mathbf{q} = (1, 0) \cdot (N_0 \dot{\epsilon}_0, M_0 \dot{\phi}) = N_0 \dot{\epsilon}_0 = M_0 \dot{\phi}$$

or

$$D = M_0 \dot{\phi} \tag{9.28}$$

It can easily be shown that this expression remains valid when the stress point lies on another side of the square yield curve or coincides with a vertex.

### 9.3.2.2. Preliminary Analysis

To obtain a preliminary answer, we neglect first the effect of the axial force on the plastic-moment-carrying capacity of the section; then the yield condition is simply $M = M_0$.

Assume a four-hinge mechanism (Fig. 9.9a), and denote $\dot{\phi}\, dx$ as the rate of relative rotation of the element $dx$ at the plastic hinge location. The rate of work done by the externally applied load is $\dot{W}_E = 2Q(\frac{1}{2}\dot{\phi}\, dx\, R)$. The rate of internal dissipation of energy is $\dot{W}_I = 4M_0 \dot{\phi}\, dx$. Thus, $\dot{W}_E = \dot{W}_I$ gives the upper bound as $Q^U = 4M_0 / R$. On the other hand, referring to the stress field shown in Fig. 9.9b, we obtain a lower-bound solution as $Q^L = 4M_0 / R$, which is the same as the upper bound. Therefore, the collapse load is obtained as

$$Q_1^c = 4 \frac{M_0}{R} \tag{9.29}$$

### 9.3.2.3. Analysis Considering the Effect of Axial Force

In this case, we also assume the same four-hinge mechanism but the corresponding velocity field (Fig. 9.10a) is somewhat different from that of Fig. 9.9a, because the bending curvature will now be accompanied by the axial deformation.

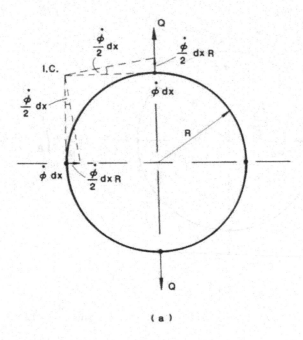

(a)

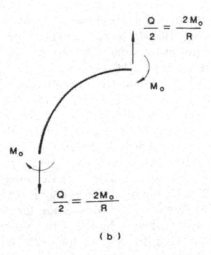

(b)

FIGURE 9.9. Velocity and stress fields for a preliminary analysis of a circular ring.

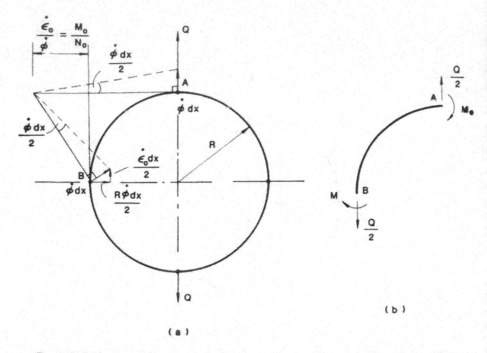

FIGURE 9.10. An elaborate analysis of a circular ring: (a) the velocity field; (b) the stress field.

Due to symmetry, the axial force at hinge $A$, denoted by $N_A$, is zero, while the moment $M_A$ takes a negative value, because it produces a compression inside the ring. This stress state is represented by the stress point $A$ that lies on the vertex of the yield curve of Fig. 9.8. We may choose the associated strain rate vector $(0, -M_0\dot{\phi})$ that gives only the rotation without the component of axial extension. The rate of energy dissipation at hinge $A$ is therefore $D_A = M_0\dot{\phi}$, the same as Eq. (9.28). At the plastic hinge $B$, both the axial force and the bending moment are positive. The stress state can be represented by a point such as the point $B$ on the yield curve $M/M_0 + N/N_0 - 1 = 0$ (Fig. 9.8), and the corresponding strain rate vector is $(N_0\dot{\epsilon}_0, M_0\dot{\phi})$ with the equal rates of rotation and expansion, $M_0\dot{\phi} = N_0\dot{\epsilon}_0$. The rate of energy dissipation at hinge $B$ is calculated as $D_B = M_0\dot{\phi}$, same as given by Eq. (9.28). The total rate of energy dissipation is obtained as the sum of the dissipations at the four hinge locations $\dot{W}_I = 4M_0\dot{\phi}\,dx$. Referring to the velocity field given by Fig. 9.10a, the rate of work done by the external load is $\dot{W}_E = 2Q[R + (M_0/N_0)](\dot{\phi}\,dx/2)$. Hence, $\dot{W}_E = \dot{W}_I$ gives the upper bound

$$Q^U = \frac{4M_0}{R + \dfrac{M_0}{N_0}}$$

Consider the stress field shown in Fig. 9.10b. Equilibrium requires that

$$M + M_0 = \frac{Q}{2} R$$

The yield condition at section $B$ is represented by

$$\frac{M}{M_0} + \frac{Q}{2N_0} - 1 = 0$$

Solving these two equations for $Q$, we obtain a lower bound

$$Q^L = \frac{4M_0}{R + \dfrac{M_0}{N_0}}$$

which is the same as the upper bound. Hence, the collapse load is obtained as

$$Q_2^c = \frac{4M_0}{R + \dfrac{M_0}{N_0}} \tag{9.30}$$

Comparison of Eq. (9.29) with Eq. (9.30) reveals that the preliminary analysis, which neglects the influence of the axial force, overestimates the load-carrying capacity of the ring. For example, if the ring has a sandwich section as shown in Fig. 9.7b, then the limit moment and limit axial force are given by

$$M_0 = \sigma_0 h A \quad \text{and} \quad N_0 = 2\sigma_0 A$$

respectively. Substituting these into Eqs. (9.29) and (9.30), we can calculate the overestimation, in the form $(Q_1^c - Q_2^c)/Q_1^c$:

$$\text{Overestimation} = 50 \left(\frac{h}{R}\right) \% \tag{9.31}$$

The state of stresses at different sections along the arc $AB$ is plotted as a "stress profile" in the stress space. At different stages of loading, different stress profiles can be plotted. See, for example, the line segment $AB$ plotted inside the square yield curve (Fig. 9.8), representing the stress profile of arc $AB$ at the instant of collapse.

## 9.3.3. Compression of a Curved Cantilever Beam

Figure 9.11a shows a cantilever beam with a circular shape subjected to a compression load $P$. When the effect of the axial force is neglected, the collapse load is easy to find as

$$P = \frac{M_0}{l} \tag{9.32}$$

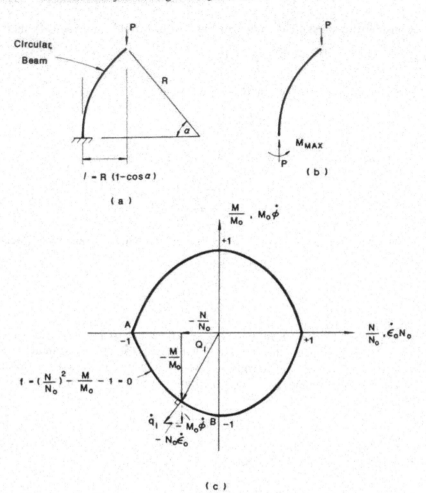

FIGURE 9.11. Compression of a curved cantilever beam. (a) Cantilever beam; (b) free body diagrams; (c) yield criterion.

To account for the effect of the axial force $P$ on the moment $M$, the yield condition

$$f = \left(\frac{N}{N_0}\right)^2 + \left|\frac{M}{M_0}\right| - 1 = 0 \tag{9.33}$$

as given by Eq. (9.22) and shown in Fig. 9.11c should be used in the analysis.

We shall derive in the following the rate of energy dissipation for a state of yield stress lying on the curve $AB$ in Fig. 9.11c. In this case, the yield condition is

$$f = \left(\frac{N}{N_0}\right)^2 - \frac{M}{M_0} - 1 = 0 \tag{9.34}$$

Recalling the general expression for the energy dissipation, $D = M\dot{\phi} + N\dot{\epsilon}_0$, and the flow rule, $\dot{\phi} = \mu \, \partial f/\partial M = -\mu/M_0$ and $\dot{\epsilon}_0 = \mu \, \partial f/\partial N = 2\mu N/N_0^2$, we have

$$D = M\dot{\phi} + N\dot{\epsilon}_0 = \mu\left[-\frac{M}{M_0} + 2\left(\frac{N}{N_0}\right)^2\right]$$

$$= \mu\left[1 + \left(\frac{N}{N_0}\right)^2\right]$$

in which the yield condition (9.34) has been used. Since $\mu = -M_0\dot{\phi}$ and $N/N_0 = \frac{1}{2}\dot{\epsilon}_0 N_0/\mu$, the energy dissipation expression $D$ becomes, after substitution,

$$D = -M_0\dot{\phi}\left[1 + \frac{1}{4}\left(\frac{N_0\dot{\epsilon}_0}{M_0\dot{\phi}}\right)^2\right] \tag{9.35}$$

### 9.3.3.1. UPPER-BOUND APPROACH

Assume a *rotation and contraction plastic hinge* at the fixed end. Then, the rate of energy dissipation $D$ is given by Eq. (9.35), while the rate of external work done is $\dot{W}_E = -P(\dot{\epsilon}_0 + \dot{\phi}l)$. Thus, letting $\dot{W}_E = D$, one gets

$$P^U = \frac{\left[1 + \frac{1}{4}\left(\frac{N_0}{M_0}\right)^2\left(\frac{\dot{\epsilon}_0}{\dot{\phi}}\right)^2\right]M_0}{\frac{\dot{\epsilon}_0}{\dot{\phi}} + l} \tag{9.36}$$

The ratio $\dot{\epsilon}_0/\dot{\phi}$ determines the direction of the plastic strain rate vector $\dot{\vec{q}}$, which varies along the yield curve $AB$. To minimize $P^U$, we use the condition

$$\frac{\partial P^U}{\partial(\dot{\epsilon}/\dot{\phi})} = 0$$

and obtain the relation

$$\left(\frac{\dot{\epsilon}_0}{\dot{\phi}}\right)^2 + 2l\frac{\dot{\epsilon}_0}{\dot{\phi}} - 4\left(\frac{M_0}{N_0}\right)^2 = 0$$

Solving the equation for $\epsilon_0/\dot{\phi}$, one gets

$$\frac{\dot{\epsilon}_0}{\dot{\phi}} = l\left[-1 + \sqrt{1 + \frac{4}{l^2}\left(\frac{M_0}{N_0}\right)^2}\right] \tag{9.37}$$

Substitution of Eq. (9.37) in Eq. (9.36) leads to

$$P^U = \frac{1}{2}\left(\frac{N_0}{M_0}\right)^2\left(\frac{\dot{\epsilon}_0}{\dot{\phi}}\right)M_0$$

or

$$P^U = \frac{1}{2}M_0\left(\frac{N_0}{M_0}\right)^2 l\left[-1 + \sqrt{1 + \frac{4}{l^2}\left(\frac{M_0}{N_0}\right)^2}\right] \tag{9.38}$$

### 9.3.3.2. LOWER-BOUND APPROACH

Referring to the stress field in Fig. 9.11b, one can write the equations of equilibrium:

$$N = -P$$
$$M - -M_{max} = -Pl \tag{9.39}$$

The yield condition at the fixed end requires

$$\left(\frac{N}{N_0}\right)^2 - \frac{M}{M_0} - 1 = 0 \tag{9.40}$$

Solving Eqs. (9.39) and (9.40) for $P$, we obtain the same load as the upper-bound solution (9.38). Hence, the collapse load is

$$P^c = \frac{1}{2} M_0 \left(\frac{N_0}{M_0}\right)^2 l \left[-1 + \sqrt{1 + \frac{4}{l^2}\left(\frac{M_0}{N_0}\right)^2}\right] \tag{9.41}$$

For a rectangular section, $M_0 = (\sigma_0/4)bh^2$, $N_0 = \sigma_0 bh$, and $M_0/N_0 = h/4$, the collapse load becomes

$$\frac{P^c l}{M_0} = 8 \left(\frac{l}{h}\right)^2 \left[-1 + \sqrt{1 + \frac{1}{4}\left(\frac{h}{l}\right)^2}\right] \tag{9.42}$$

If the ratio $h/l$ is reasonably small, then by using binomial series, Eq. (9.42) can be reduced to

$$\frac{P^c l}{M_0} = 1 - \frac{1}{16}\left(\frac{h}{l}\right)^2 + \cdots = 1 - \left(\frac{h}{4l}\right)^2 \tag{9.43}$$

Comparison of this with the answer of Eq. (9.32) based on the simple plastic theory reveals that the correction for the limit load due to the presence of the axial force is about $(h/4l)^2$. For example, if $l = h/2$, then, by Eq. (9.43), the limit load can be overestimated by as much as 25% by the simple plastic theory.

It may be seen from the above two examples that bending usually predominates so that the concept of simple plastic hinges is sufficient in most cases. Should there be an axial force in addition, the *extending* or *contracting hinge* would take care of the situation.

## 9.4. Effect of Shear Force

### 9.4.1. Combined Shear and Bending of Beams

Consider a beam under combined shear and bending. If the x-axis is taken as the beam axis, the nonzero stress components at a point are $\sigma_x$ and $\tau_{xy}$

(assume plane stress). The generalized stresses involve both the bending moment $M$ and the shear force $V$, which are the stress resultants:

$$M = \int_A y\sigma_x \, dA \tag{9.44}$$

and

$$V = \int_A \tau_{xy} \, dA \tag{9.45}$$

If the Tresca yield condition is used, then the stresses must satisfy

$$\sigma_x^2 + 4\tau_{xy}^2 \leq \sigma_0^2 \tag{9.46}$$

At first, we need to find the shear and bending interaction curve, i.e., the yield curve in terms of $M$ and $V$. We shall do this by finding the coincident upper and lower bounds for the yield curve. The problem may be stated as follows: Determine the distributions of the stresses $\sigma_x$ and $\tau_{xy}$ which satisfy Eq. (9.46), give a prescribed value of $M$ according to Eq. (9.44), and maximize (or minimize) $V$ as given by Eq. (9.45). We shall discuss the interaction curve for a rectangular section.

### 9.4.1.1. LOWER BOUNDS

(i) Assume a stress distribution given by the elastic solution (Fig. 9.12b):

$$\sigma_x = \frac{2\sigma_0}{h} y \quad \text{and} \quad \tau_{xy} = \frac{\sigma_0}{2}\left[1 - \frac{y^2}{(h/2)^2}\right] \tag{9.47}$$

It is easy to check that the yield condition (9.46) is satisfied. Hence, this assumed stress distribution leads to a lower-bound solution for $M$ and $V$

$$M^L = \tfrac{1}{6}\sigma_0 bh^2 \quad \text{and} \quad V^L = \tfrac{1}{3}\sigma_0 bh \tag{9.48}$$

Denote $M_0$ as the limiting moment and $V_0$ as the limiting shear force:

$$M_0 = \tfrac{1}{4}\sigma_0 bh^2 \quad \text{and} \quad V_0 = \tfrac{1}{2}\sigma_0 bh \tag{9.49}$$

Equation (9.48) can be rewritten as

$$\frac{M^L}{M_0} = \frac{2}{3} \quad \text{and} \quad \frac{V^L}{V_0} = \frac{2}{3} \tag{9.50}$$

This is shown as a dashed curve marked (i) in Fig. 9.13.

(ii) A yield or interaction curve must be convex, and we already know that the points $(0, 1)$ and $(1, 0)$ in Fig. 9.13 represent the yield states of stresses. Thus, lines connecting the points $(\tfrac{2}{3}, \tfrac{2}{3})$ with points $(0, 1)$ and $(1, 0)$ will give a better lower bound. This is shown as the second dashed curve marked (ii) in Fig. 9.13.

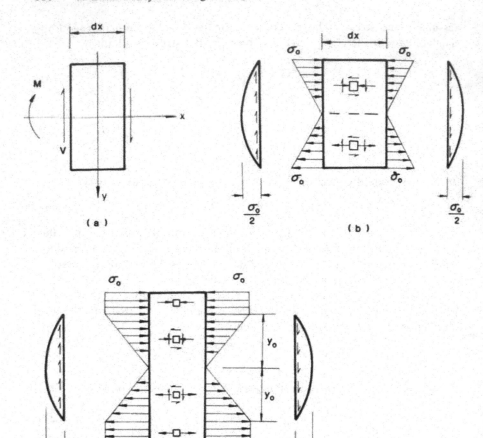

FIGURE 9.12. Stress fields for lower-bound solutions. (a) Beam element; (b) elastic stress distribution; (c) assumed elastic–plastic stress distribution.

(iii) To improve the lower bound, we can assume a distribution of stress as shown in Fig. 9.12c which also satisfies the yield condition (9.46). Hence, we have

$$M^L = M_0 \left[ 1 - \frac{1}{3}\left(\frac{2y_0}{h}\right)^2 \right] \tag{9.51}$$

$$V^L = \tfrac{2}{3}\sigma_0 b y_0 \tag{9.52}$$

or

$$\frac{M}{M_0} = 1 - \frac{3}{4}\left(\frac{V}{V_0}\right)^2 \tag{9.53}$$

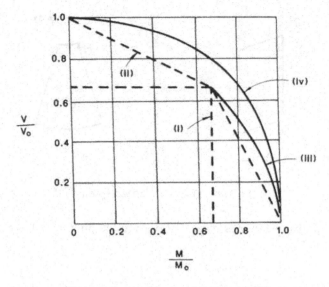

FIGURE 9.13. Lower bounds of the interaction curve.

which is valid for $y_0$ up to $h/2$, i.e., $V/V_0 \leq \frac{2}{3}$. This is shown as the solid curve marked (iii) in Fig. 9.13.

(iv) A more elaborate effort was made by Hodge (1956), who carried out the maximization process by introducing a Lagrange multiplier $\lambda$.

The interaction curve is expressed in terms of the positive parameter $\lambda$

$$M = 2\lambda\sigma_0 b \int_{-h/2}^{h/2} y^2 (1 + 4\lambda^2 y^2)^{-1/2} \, dy \qquad (9.54)$$

and

$$V = \frac{\sigma_0}{2} b \int_{-h/2}^{h/2} (1 + 4\lambda^2 y^2)^{-1/2} \, dy \qquad (9.55)$$

The curve is shown as the solid curve marked (iv) in Fig. 9.13.

### 9.4.1.2. UPPER BOUNDS

Consider now the same problem from the viewpoint of the upper-bound theorem. To this end, we first define the rates of generalized strains $\dot{\phi}$ and $\dot{\gamma}$. The former represents the curvature rate and the latter represents the rate of shearing of the section, being consistent with the stress resultants $M$ and $V$; then the relevant strain rate components at any point are

$$\dot{\epsilon}_x = y\dot{\phi} \quad \text{and} \quad \dot{\gamma}_{xy} = \dot{\gamma} \qquad (9.56)$$

Assume the velocity field in which the two neighboring cross sections are rotated and transversely moved with respect to each other as shown in Fig.

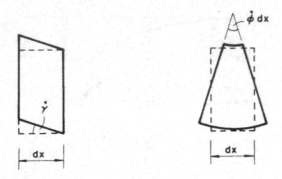

FIGURE 9.14. Generalized strains.

9.14. Recall the energy dissipation $D$ per unit volume for a Tresca material [see Eq. (8.22)]:

$$D(\dot{\epsilon}_{ij}^{p}) = 2k|\dot{\epsilon}_{max}| = \sigma_0|\dot{\epsilon}_{max}| \qquad (9.57)$$

According to the Tresca criterion (9.46), it follows that all deformations take place in the $xy$-plane, so that $\epsilon_z = 0$. Incompressibility thus requires $\dot{\epsilon}_y = -\dot{\epsilon}_x$. Since $\dot{\epsilon}_x = y\dot{\phi}$ and $\dot{\gamma}_{xy} = \dot{\gamma} = 2\dot{\epsilon}_{xy}$, the maximum strain $|\dot{\epsilon}_{max}|$ can be found from Mohr's circle as

$$|\dot{\epsilon}_{max}|^2 = \dot{\epsilon}_x^2 + \left(\frac{\dot{\gamma}_{xy}}{2}\right)^2 \qquad (9.58)$$

Substituting Eq. (9.58) into (9.57), noting Eq. (9.56), and integrating over the section lead to the rate of internal energy dissipation:

$$\dot{W}_I = b \int_{-h/2}^{h/2} D\,dy = \frac{b\sigma_0}{2} \int_{-h/2}^{h/2} \dot{\gamma} \left[4\left(\frac{\dot{\phi}}{\dot{\gamma}}\right)^2 y^2 + 1\right]^{1/2} dy \qquad (9.59a)$$

The rate of external work done is

$$\dot{W}_E = M\dot{\phi} + V\dot{\gamma} = \dot{\gamma}\left(M\frac{\dot{\phi}}{\dot{\gamma}} + V\right) \qquad (9.59b)$$

Letting $\dot{W}_E = \dot{W}_I$ and taking the derivative with respect to the ratio $\dot{\phi}/\dot{\gamma}$, one gets

$$M = 2\left(\frac{\dot{\phi}}{\dot{\gamma}}\right) b\sigma_0 \int_{-h/2}^{h/2} y^2 \left[1 + 4\left(\frac{\dot{\phi}}{\dot{\gamma}}\right)^2 y^2\right]^{-1/2} dy \qquad (9.60)$$

Finally, substituting Eq. (9.60) into the equation $\dot{W}_E = \dot{W}_I$ for $M$ leads to an expression for the shear force $V$:

$$V = \left(\frac{b\sigma_0}{2}\right) \int_{-h/2}^{h/2} \left[1 + 4\left(\frac{\dot{\phi}}{\dot{\gamma}}\right)^2 y^2\right]^{-1/2} dy \qquad (9.61)$$

It can be seen that if we choose $\dot{\phi}/\dot{\gamma} = \lambda$, the upper-bound interaction equations (9.60) and (9.61) will have the same forms as the lower-bound equations (9.54) and (9.55).

### 9.4.1.3. Remarks on the Bending and Shear Interaction Curve

The deformation pattern used in Fig. 9.14 for the beam element will not be permissible for the beam problem as a whole because of the remaining portions of the beam. For example, the shear strain will be restrained by the neighboring elastic (or rigid) regions, and furthermore, the bending strain will produce mismatching which requires a large energy dissipation (see Fig. 9.15). Hence, all the previous discussions are limited to a "local" yield criterion (Drucker, 1956).

Since at the limit load, the variation of $\epsilon_y$ is significant and the deformation is entirely plastic and is strongly localized, the transition between the deforming and undeforming materials is abrupt, causing a mismatching problem. Mismatch could be lessened with a gradual transition from the section of maximum moment to the section of zero moment. For example, if the length of the beam is several times greater than the depth, then $\epsilon_y$ is not significant. For a pure bending problem, the mismatch trouble is avoided by the plastic hinge as shown in Fig. P9.5(i) (Problem 9.5).

It is therefore evident that the interaction curve is not a local affair but depends on the loading and geometry of the entire beam. There is no unique interaction curve for a given section.

Fortunately, the effect of shear force is generally small in practice, and so it is not wholly unreasonable to propose a unique interaction relation for the effect of shear force on the fully plastic moment. In general, when the length of the beam is more than 10 times the beam height ($h/L < 0.1$), the effect of shear is negligible and hence shear can reasonably be neglected. For a length down to the beam height for rectangular sections ($0.1 < h/L < 1$), the interaction curve given by Eqs. (9.60) and (9.61) would be reasonable.

A simplified interaction relation for rectangular sections was proposed by Drucker as an approximation to the actual interaction curve:

$$\frac{M}{M_0} = 1 - \left(\frac{V}{V_0}\right)^4 \tag{9.62}$$

This curve is shown in Fig. 9.16.

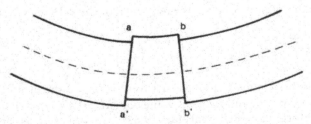

FIGURE 9.15. Bending produces mismatching at $aa'$ and $bb'$.

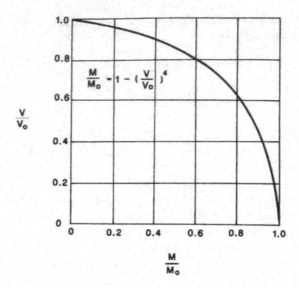

FIGURE 9.16. Approximated bending–shear interaction curve.

## 9.4.2. Combined Axial, Shear, and Bending of Beams

The approach taken to investigate the effect of shear and axial forces acting simultaneously with the moment on the fully plastic section is very similar to the case of the shear and bending combination, and will therefore not be presented in detail here. Based on a study of the rectangular section of a cantilever beam, Neal (1961a) suggested an interaction relation:

$$\frac{M}{M_0} + \left(\frac{N}{N_0}\right)^2 + \frac{\left(\frac{V}{V_0}\right)^4}{1 - \left(\frac{N}{N_0}\right)^2} = 1 \tag{9.63}$$

which represents a good approximation to the lower-bound interaction relation. For $V = 0$, the relation reduces to Eq. (9.17):

$$\frac{M}{M_0} = 1 - \left(\frac{N}{N_0}\right)^2 \tag{9.17}$$

which is exact, while for $N = 0$, the relation reduces to Drucker's approximation (9.62):

$$\frac{M}{M_0} = 1 - \left(\frac{V}{V_0}\right)^4 \tag{9.62}$$

In general, the relation (9.63) is exact for $V = 0$, and the discrepancy never exceeds 5% over the full range of values of $(M/M_0)$, $(N/N_0)$, and $(V/V_0)$. It is therefore suggested that Eq. (9.63) can be used as an empirical interaction relation between $M$, $N$, and $V$ for rectangular cross sections.

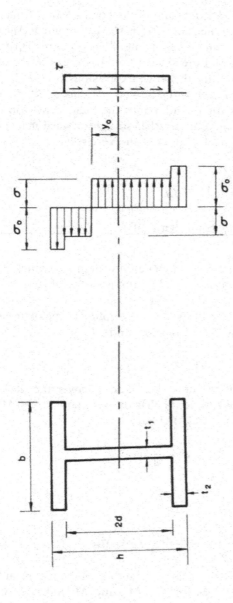

FIGURE 9.17. A stress distribution for an I-section.

For an I-section, Horne (1958) assumed a stress distribution as shown in Fig. 9.17, and also assumed the von Mises yield criterion:

$$\sigma^2 + 3\tau^2 = \sigma_0^2 \qquad (9.64)$$

The moment of the section, $M$, is the sum of the contributions from $M_w$ of the web and $M_f$ of the flange, i.e., $M = M_w + M_f$. The moment $M_f$ can be approximated by $\sigma_0 b t_2 (2d)$, while the moment $M_w$ is given by $\sigma t_1 (d^2 - y_0^2)$. Similarly, the axial force $N$ and the shear force $V$ are easily obtained as $N = \sigma t_1 (2y_0)$ and $V = 2 d t_1 \tau$. With the three expressions for $M$, $N$, and $V$ given, $y_0$ can be eliminated and $\sigma$ and $\tau$ solved in terms of $M$, $N$, and $V$. Substituting the expressions for $\sigma$ and $\tau$ into the yield condition (9.64), one obtains the interaction curve in a rather straightforward manner. The details are left to the reader to work out as an exercise.

## 9.5. Limit Analysis of Plates

### 9.5.1. Yield Condition and Flow Rule of a Rigid–Perfectly Plastic Plate

Figure 9.18b shows a unit element of plate with depth $h$, which has the generalized stresses $M_x$, $M_y$, and $M_{xy} = M_{yx}$ and the generalized strains $\dot{\phi}_x$, $\dot{\phi}_y$, and $\dot{\phi}_{xy}$. The shear and normal forces are considered as the "reactions." The general theorems of limit analysis are now applied to the plate element to obtain the interaction relation of $M_x$, $M_y$, and $M_{xy}$.

#### 9.5.1.1. Lower-Bound Approach

The assumed stress distribution for $\sigma_x$, $\sigma_y$, and $\tau_{xy}$ over the depth $h$ is shown in Fig. 9.18c. It is easy to verify that the resultant forces $M_x$, $M_y$, and $M_{xy}$ corresponding to this stress field are

$$M_x = \int \sigma_x z \, dz = \frac{h^2}{4} \sigma_x$$

$$M_y = \int \sigma_y z \, dz = \frac{h^2}{4} \sigma_y \qquad (9.65)$$

$$M_{xy} = \int \tau_{xy} z \, dz = \frac{h^2}{4} \tau_{xy} = M_{yx}$$

which are simply the stresses $\sigma_x$, $\sigma_y$, and $\tau_{xy}$ multiplied by a constant $h^2/4$. It follows that the yield condition for $M_x$, $M_y$, and $M_{xy}$ takes exactly the same form as it does for the stresses $\sigma_x$, $\sigma_y$, and $\tau_{xy}$.

For example, the von Mises yield condition

$$\sigma_x^2 - \sigma_x \sigma_y + \sigma_y^2 + 3\tau_{xy}^2 = \sigma_0^2$$

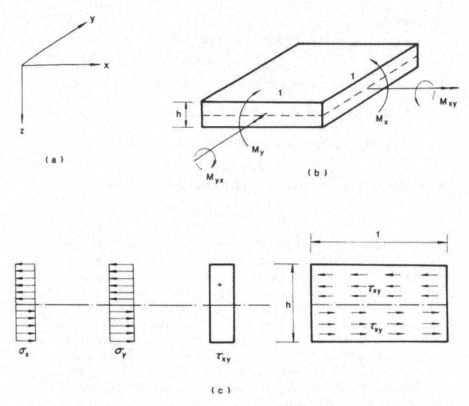

FIGURE 9.18. (a) Coordinate system. (b) Generalized stresses of a plate element. (c) Assumed stress field.

multiplied by $(h^2/4)^2$ on both sides leads to the yield condition for the moments:

$$M_x^2 - M_x M_y + M_y^2 + 3M_{xy}^2 = M_0^2 \qquad (9.66)$$

Similarly, the Tresca yield condition

$$\max(|\sigma_1|, |\sigma_2|, |\sigma_1 - \sigma_2|) = \sigma_0$$

will lead to the corresponding condition for the moments:

$$\max(|M_x|, |M_y|, |M_x - M_y|) = M_0 \qquad (9.67)$$

### 9.5.1.2. UPPER-BOUND APPROACH

The assumed velocity field involves the pure curvature rates $\dot{\phi}_x$ and $\dot{\phi}_y$ about the $y$-axis and the $x$-axis, respectively, and the pure rate of twist per unit length $\frac{1}{2}\dot{\phi}_{xy} = \frac{1}{2}\dot{\phi}_{yx}$ about the $x$- and $y$-axes. Then, the strain rates at any point are

$$\dot{\epsilon}_x = z\dot{\phi}_x, \qquad \dot{\epsilon}_y = z\dot{\phi}_y$$
$$\dot{\epsilon}_{xy} = \tfrac{1}{2}z\dot{\phi}_{xy}, \qquad \dot{\epsilon}_{yx} = \tfrac{1}{2}z\dot{\phi}_{yx} \tag{9.68}$$

The total energy dissipation per unit area of the plate is

$$\dot{W}_I = \dot{\phi}_x \int_{-h/2}^{h/2} \sigma_x z \, dz + \dot{\phi}_y \int_{-h/2}^{h/2} \sigma_y z \, dz + \dot{\phi}_{xy} \int_{-h/2}^{h/2} \tau_{xy} z \, dz \tag{9.69}$$

which is equal to the rate of external work done

$$\dot{W}_E = M_x \dot{\phi}_x + M_y \dot{\phi}_y + M_{xy} \dot{\phi}_{xy} \tag{9.70}$$

We regard the plate material as rigid-plastic and satisfying the flow rule,

$$d\epsilon_{ij}^p = d\epsilon_{ij} = d\lambda \frac{\partial f}{\partial \sigma_{ij}}$$

It follows from Eq. (9.68) that the state of strain is proportional to the distance $z$ from the central plane. Therefore, the strain rate vector with components $(\dot{\epsilon}_x, \dot{\epsilon}_y, \dot{\gamma}_{xy})$ must have the same direction at all points on one side of the central plane and must have the opposite direction on the other. It follows that the stress state must be constant on each side of the central plane, and, from the assumed symmetry, these values must be negative of each other. Therefore, if $(\sigma_x, \sigma_y, \tau_{xy})$ represents the state of stress for positive $z$ at a fully plastic moment, the state of stress for negative $z$ is $(-\sigma_x, -\sigma_y, -\tau_{xy})$. Substituting these into Eq. (9.69) for $\sigma_x, \sigma_y, \tau_{xy}$ and noting that they are constant, one can carry out the necessary integration. Comparing this with Eq. (9.70) yields

$$M_x = \tfrac{1}{4}h^2 \sigma_x, \qquad M_y = \tfrac{1}{4}h^2 \sigma_y, \qquad M_{xy} = \tfrac{1}{4}h^2 \tau_{xy}$$

This upper bound is the same as that given by Eq. (9.65). It follows that the yield condition assumes exactly the same form for the moments at it does for the stresses.

## 9.5.2. Load-Carrying Capacity of Square and Rectangular Plates

Exact solution is difficult for square and rectangular plates but close upper and lower bounds on the answer can easily be obtained.

Consider a small element of the plate, $dx \, dy$, subjected to the internal forces $M_x, M_y, M_{xy}, V_x, V_y$, as shown in Fig. 9.19. The internal forces must satisfy the equilibrium equations:

$$\frac{\partial M_x}{\partial x} + \frac{\partial M_{xy}}{\partial y} = V_x$$

$$\frac{\partial M_{xy}}{\partial x} + \frac{\partial M_y}{\partial y} = V_y \tag{9.71}$$

$$\frac{\partial V_x}{\partial x} + \frac{\partial V_y}{\partial y} = -q$$

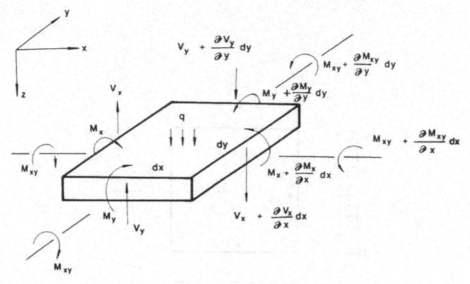

FIGURE 9.19. Equilibrium of a rectangular plate element.

in which the shear forces $V_x$ and $V_y$ are needed for equilibrium but their corresponding deformations are neglected. They have been considered as reactions. $V_x$ and $V_y$ may be eliminated by substituting the first two of equations (9.71) into the third:

$$\frac{\partial^2 M_x}{\partial x^2} + \frac{\partial^2 M_y}{\partial y^2} + 2\frac{\partial^2 M_{xy}}{\partial x \, \partial y} = -q \qquad (9.72)$$

EXAMPLE 9.1. SIMPLY SUPPORTED SQUARE PLATE UNDER UNIFORM LOAD

*Lower Bounds—Stress Fields*

To find the lower bounds, several stress fields are assumed.

(i) Intuitively, we assume that the load $q$ is carried by two independent beams as shown in Fig. 9.20b. Obviously, the critical situation will occur in the center of the plate. Hence, according to the Tresca yield criterion,

$$M_x = M_y = \frac{1}{8}\left(\frac{q}{2}\right) a^2 \le M_0$$

which yields

$$q^L = \frac{16 M_0}{a^2} \qquad (9.73)$$

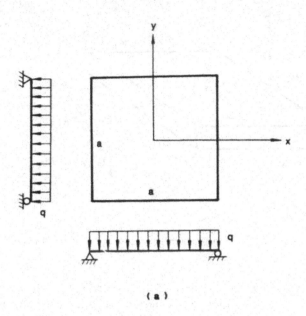

(a)

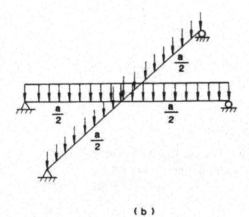

(b)

FIGURE 9.20. (a) Simply supported square plate under uniform load. (b) An intuitive assumption of stress field.

or, formally, we assume a stress field which is a modification of a stress distribution in a uniformly loaded and simply supported beam, i.e.,

$$M_x = M_0\left[1-\left(\frac{2x}{a}\right)^2\right]$$

$$M_y = M_0\left[1-\left(\frac{2y}{a}\right)^2\right] \tag{9.74}$$

$$M_{xy} = 0$$

It is easy to check that such a stress field does not violate the yield condition. Substituting into the equilibrium equation (9.72) will lead to the same lower bound as given by Eq. (9.73).

(ii) To improve the lower bound, we may modify the stress field of Eq. (9.74) by assuming

$$M_x = M_0\left[1-\left(\frac{2x}{a}\right)^2\right]$$

$$M_y = M_0\left[1-\left(\frac{2y}{a}\right)^2\right] \tag{9.75}$$

$$M_{xy} = -\frac{M_0}{2}\left(\frac{2x}{a}\right)\left(\frac{2y}{a}\right)$$

we can show that this stress distribution nowhere violates the yield condition and is therefore statically admissible. Substituting this into Eq. (9.72), one gets

$$q^L = 20\frac{M_0}{a^2} \tag{9.76}$$

which is a better lower bound.

(iii) We may directly use the elastic solution for a square plate because it always provides a permissible stress field. The maximum moments of the plastic solution occur at the center of the plate:

$$M_x = M_y = M_{max} = 0.0479qa^2 \quad \text{and} \quad M_{xy} = 0$$

Thus, from the Tresca yield condition, we have

$$q^L = 20.8\frac{M_0}{a^2} \tag{9.77}$$

*Upper Bounds—Velocity Fields*

To find an upper bound, assume a pyramidal mode of deformation in which the center of the plate descends with the speed $\dot{\Delta}$ (Fig. 9.21a). The plate is divided into four triangles, each of which acts as a rigid body rotating with an angular velocity with respect to its edge, e.g., $\triangle ABE$ rotates with an

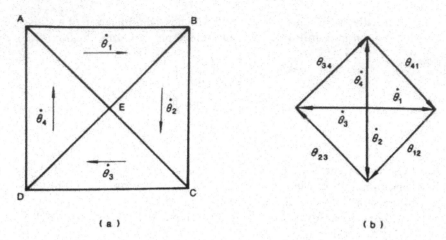

FIGURE 9.21. Failure mechanism of a square plate.

angular velocity $\dot{\theta}_1$ with respect to edge $AB$ (see Fig. 9.21a). Thus, $EA$, $EB$, $EC$, and $ED$ are the lines of angular *velocity discontinuity*, which are also referred to as the *plastic hinge lines* or *yield lines*. The rate of energy dissipation per unit length at the yield line, $D$, is calculated by multiplication of the limit moment with the relative angular velocity, e.g., at line $EB$, $D = M_0 \dot{\theta}_{12}$, in which $\dot{\theta}_{12}$ is the angular velocity of $\Delta EBC$ with respect to $\Delta EAB$. To determine the relative angular velocities, we can use the velocity diagram (hodograph) similarly to what we have done for the mechanisms of rigid-block sliding in Section 8.5.1 (see Fig. 8.29). In this case, the angular velocity diagram is shown in Fig. 9.21b, which shows that $\dot{\theta}_1 = \dot{\theta}_2 = \dot{\theta}_3 = \dot{\theta}_4 = 2\dot{\Delta}/a$ and $\dot{\theta}_{12} = \dot{\theta}_{23} = \dot{\theta}_{34} = \dot{\theta}_{41} = \sqrt{2}\dot{\theta}_1 = 2\sqrt{2}\dot{\Delta}/a$. Now, it is easy to verify that the rate of external work is

$$\dot{W}_E = \tfrac{1}{3}qa^2\dot{\Delta}$$

Numerically, this is equal to the load intensity $q$ multiplied by the volume of the pyramid. The total rate of internal energy dissipation along the yield lines $AE$, $BE$, $CE$, and $DE$ is

$$\dot{W}_I = 4M_0(2\sqrt{2}\dot{\Delta}/a)\left(\frac{\sqrt{2}}{2}a\right) = 8M_0\dot{\Delta}$$

Hence, the upper bound of the collapse load is obtained:

$$q^U = 24\frac{M_0}{a^2} \tag{9.78}$$

EXAMPLE 9.2. SIMPLY SUPPORTED RECTANGULAR PLATE UNDER UNIFORM LOAD

*Lower Bounds—Stress Fields*

For a simply supported rectangular plate subjected to uniform loading (see

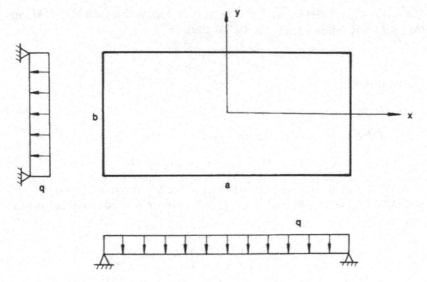

FIGURE 9.22. Simply supported rectangular plate under uniform load.

Fig. 9.22), we may assume a stress field as follows

$$M_x = M_0\left[1 - \left(\frac{2x}{a}\right)^2\right]$$

$$M_y = M_0\left[1 - \left(\frac{2y}{b}\right)^2\right] \qquad (9.79)$$

$$M_{xy} = -\gamma M_0\left(\frac{2x}{a}\right)\left(\frac{2y}{b}\right)$$

in which $\gamma$ is a constant, taking different values in order to satisfy different yield criteria. The principal moments $M_1$ and $M_2$ can be found from Mohr's circle and can be expressed as

$$M_{1,2} = \frac{M_x + M_y}{2} \pm \sqrt{\left(\frac{M_x - M_y}{2}\right)^2 + M_{xy}^2}$$

$$= M_0\left\{\left(1 - \frac{\xi^2 + \eta^2}{2}\right) \pm \sqrt{\left(\frac{\xi^2 + \eta^2}{2}\right)^2 - (1 - \gamma^2)\xi^2\eta^2}\right\} \qquad (9.80)$$

in which

$$\xi = \frac{2x}{a} \quad \text{and} \quad \eta = \frac{2y}{b}$$

(i) If a square yield criterion is used (see Fig. 9.23), then $|M_1| \le M_0$ and $|M_2| \le M_0$. It follows from Eq. (9.80) that

$$\gamma \le 1$$

We choose

$$\gamma = 1 \tag{9.81}$$

(ii) If the Tresca yield condition is used, it requires that

$$\max(|M_1|, |M_2|, |M_1 - M_2|) \le M_0$$

We know from the results of (i) that the first two conditions—$|M_1| \le M_0$ and $|M_2| \le M_0$—are satisfied if $\gamma \le 1$. Now, we only need to check the third condition

$$|M_1 - M_2| = 2M_0 \sqrt{\left(\frac{\xi^2 + \eta^2}{2}\right)^2 - (1 - \gamma^2)\xi^2\eta^2} \le M_0$$

To this end, we first find the stationary value of $|M_1 - M_2|$ by letting $\partial(M_1 - M_2)/\partial\xi^2 = 0$ and $\partial(M_1 - M_2)/\partial\eta^2 = 0$, which yields

$$\eta^2 = \xi^2$$

This implies that the critical point lies on the diagonal of the plate. Hence, the maximum value of $|M_1 - M_2|$ is

$$|M_1 - M_2| = 2\gamma\xi^2 M_0$$

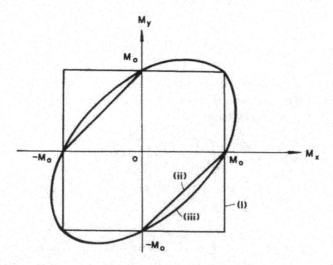

FIGURE 9.23. Yield conditions: (i) square yield condition; (ii) Tresca yield condition; (iii) von Mises yield condition.

and the condition $|M_1 - M_2| \le M_0$ leads to

$$\gamma \le \tfrac{1}{2}$$

We choose

$$\gamma = \tfrac{1}{2} \tag{9.82}$$

(iii) For the von Mises yield condition, $\gamma = \tfrac{1}{2}$ will be a safe value (see Fig. 9.23).

With the stress field (9.79) given, the lower bound of the collapse load, $q^L$, can be obtained by substituting Eq. (9.79) into equilibrium equation (9.72):

$$q^L = \frac{8M_0}{b^2}\left(1 + \gamma\frac{b}{a} + \frac{b^2}{a^2}\right) \tag{9.83}$$

Alternatively, we may use the stress field given by the elastic solution (see Timoshenko and Woinowsky-Krieger, 1959). The maximum value of $M_x$ and of $M_y$ occurs at the center of the plate. Since $a > b$, $(M_x)_{max} > (M_y)_{max}$, the lower bounds will be obtained by equating $(M_x)_{max}$ to $M_0$. Results for different values of the $a/b$ ratio are listed in Table 9.1.

*Upper Bounds—Velocity Fields*

To obtain upper bounds, three different velocity fields are assumed.

1. Similarly to the case of the square plate, we assume two yield lines occurring along the diagonals of the plate such that the plate is divided into four rigid triangles as shown in Fig. 9.24a. The center of the plate is assumed to descend with the velocity $\dot{\Delta}$, while the triangles rotate with respect to their edges to form a pyramid shape. It is easy to see by inspection that the angular velocities of the four rigid triangles, $\dot{\theta}_1$, $\dot{\theta}_2$, $\dot{\theta}_3$, and $\dot{\theta}_4$, can be expressed in terms of $\dot{\Delta}$ as follows:

$$\dot{\theta}_1 = \dot{\theta}_3 = \frac{2\dot{\Delta}}{b}$$

$$\dot{\theta}_2 = \dot{\theta}_4 = \frac{2\dot{\Delta}}{a} \tag{9.84}$$

TABLE 9.1. Lower bounds given by the elastic stress field.

| $a/b$ | $(M_x)_{max}/(qb^2)$ | $q^L/(M_0/b^2)$ |
|---|---|---|
| 1 | 0.0479 | 20.8 |
| 2 | 0.1017 | 9.8 |
| 4 | 0.1235 | 8.1 |
| $\infty$ | 0.125 | 8 |

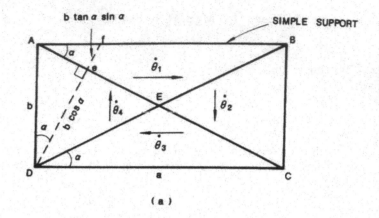

(a)

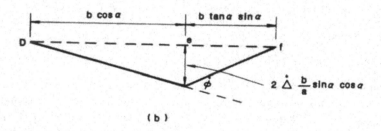

(b)

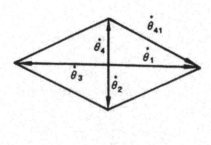

(c)

FIGURE 9.24. (a) Failure mechanism of a rectangular plate (mode 1). (b) Determination of $\dot{\phi}$ by geometry. (c) Velocity diagram.

During deformation, the rigid triangles remain flat. The relative rotation rate, $\dot{\phi}$, between any two adjacent triangles along a yield line, say, $AE$, can be found by consideration of the geometry as shown in Fig. 9.24b:

$$\dot{\phi} = \frac{2\dot{\Delta}}{a}(\sin\alpha + \cos\alpha \cot\alpha) = \frac{2\dot{\Delta}}{a\sin\alpha} \tag{9.85}$$

or more concisely by the use of the velocity diagram as shown in Fig. 9.24c, in which the relative rotation rate along $AE$ is represented by $\dot{\theta}_{41}$, i.e.,

$$\dot{\theta}_{41} = \dot{\phi} = \sqrt{\dot{\theta}_1^2 + \dot{\theta}_4^2} = \frac{2\dot{\Delta}}{a}\sqrt{\frac{a^2+b^2}{b^2}} \tag{9.86}$$

which is the same as that given by the geometry, as it should be.

The internal rate of energy dissipation of the plate is the energy dissipated along the two yield lines:

$$\dot{W}_I = 2M_0\dot{\phi}\sqrt{a^2+b^2} = 4M_0\dot{\Delta}\frac{a^2+b^2}{ab} \tag{9.87}$$

The rate of external work is equal to $q$ times the volume of the pyramid:

$$\dot{W}_E = \tfrac{1}{3}qab\dot{\Delta} \tag{9.88}$$

Equating $\dot{W}_I$ to $\dot{W}_E$ gives

$$q^U = 12M_0\left(\frac{1}{a^2} + \frac{1}{b^2}\right) = \frac{12M_0}{b^2}\left(1 + \frac{b^2}{a^2}\right) \tag{9.89}$$

2. The deformation mode shown in Fig. 9.25a has five hinge lines dividing the plate into four rigid regions, each of which rotates with an angular velocity with respect to the edge of the plate. The velocity diagram is shown in Fig. 9.25b. If the rate of displacement of the yield line $EF$ is $\dot{\Delta}$, the angular velocity rates $\dot{\theta}_1$, $\dot{\theta}_2$, $\dot{\theta}_3$, and $\dot{\theta}_4$ are given by

$$\dot{\theta}_1 = \dot{\theta}_2 = \dot{\theta}_3 = \dot{\theta}_4 = \frac{2\dot{\Delta}}{b} \tag{9.90}$$

The velocity discontinuities across the yield lines $AE$, $ED$, $BF$, and $FC$ are easy to find using the velocity diagram:

$$\dot{\theta}_{41} = \dot{\theta}_{12} = \dot{\theta}_{23} = \dot{\theta}_{34} = (\sqrt{2})\left(\frac{2\dot{\Delta}}{b}\right) = \frac{2\sqrt{2}\dot{\Delta}}{b} \tag{9.91}$$

These are represented by lines 41, 12, 23, and 34 in the velocity diagram, respectively. The velocity discontinuity across the yield line $EF$ is represented by line 31 in the velocity diagram and equals

$$\dot{\theta}_{31} = 2\dot{\theta}_1 = \frac{4\dot{\Delta}}{b} \tag{9.92}$$

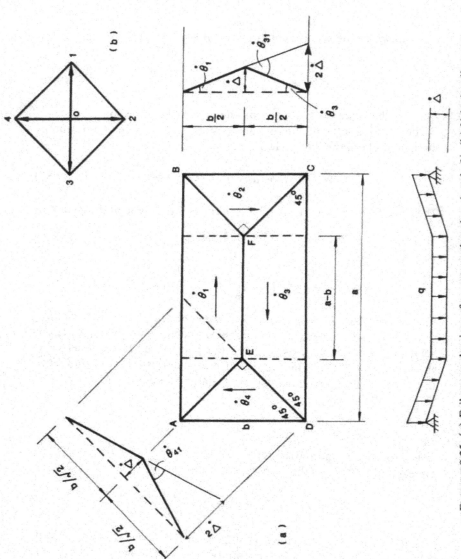

FIGURE 9.25. (a) Failure mechanism of a rectangular plate (mode 2). (b) Velocity diagram.

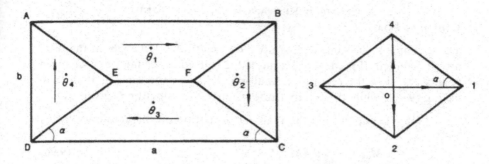

FIGURE 9.26. Failure mechanism of a rectangular plate (mode 3) and its velocity diagram.

This quantity has been shown geometrically in Fig. 9.25a. It is easy to verify that the work rates $\dot{W}_E$ and $\dot{W}_I$ are expressed by

$$\dot{W}_E = q[\tfrac{1}{2}(a-b)b\dot{\Delta} + \tfrac{1}{3}b^2\dot{\Delta}] \tag{9.93}$$

and

$$\dot{W}_I = M_0\left[\frac{4\dot{\Delta}}{b}(a-b) + \frac{2\sqrt{2}\dot{\Delta}}{b}2\sqrt{2}b\right] = 4M_0\left(1+\frac{a}{b}\right)\dot{\Delta} \tag{9.94}$$

Hence, $\dot{W}_E = \dot{W}_I$ gives

$$q^U = \frac{8M_0}{b^2}\frac{1+b/a}{1-\tfrac{1}{3}(b/a)} \tag{9.95}$$

3. The hinge line pattern of mode 3 as shown in Fig. 9.26 is similar to that of mode 2 except that the angle $\alpha$ now is not 45° but is to be determined by minimizing the upper bound. The solution is obtained by a procedure similar to that for mode 2 and, therefore, will not be repeated here. The reader may work out the details of this as an exercise.

All the numerical results are summarized in Table 9.2.

TABLE 9.2. Collapse load of a simply supported rectangular plate under uniform load.

| Ratio $a/b$ | Upper bounds in $(M_0/b^2)$ | | | Lower bounds in $(M_0/b^2)$ | | Elastic solution |
|---|---|---|---|---|---|---|
| | Mode 1 | Mode 2 | Mode 3 | $\gamma = 1$ | $\gamma = \tfrac{1}{2}$ | |
| 2 | 15.00 | 14.40 | 14.14 | 14.00 | 12.00 | 9.80 |
| 4 | 12.75 | 10.90 | 10.67 | 10.50 | 9.50 | 8.10 |
| $\infty$ | 12.00 | 8.00 | 8.00 | 8.00 | 8.00 | 8.00 |

## EXAMPLE 9.3. A CLAMPED RECTANGULAR PLATE UNDER UNIFORM LOAD

This example is the same as Example 9.2 except that the edges of the plate are all clamped. If mode 2 of Example 9.2 is used as the failure mechanism (Fig. 9.27), then the rate of external work done is the same as that given by Eq. (9.93), while the internal rate of energy dissipation is given by

$$\dot{W}_I = \text{Dissipation in the plate} + \text{Dissipation on the boundary}$$

$$= 4M_0\left(1+\frac{a}{b}\right) + 4M_0\left(1+\frac{a}{b}\right) \tag{9.96}$$

in which the two dissipations are the same. This is true, not by chance, but rather as a consequence of the following geometric property: the sum of the angle changes on the boundary is equal to the sum of the angle changes in the plate.

Hence, in general, the upper bound of the collapse load for a clamped plate is twice as large as that predicted by the same failure mechanism for the simply supported plate.

### 9.5.3. Load-Carrying Capacity of Circular Plates

Thanks to the axisymmetry of the structure and the loading condition, exact solutions for the limit load for circular plates can be obtained in most cases.

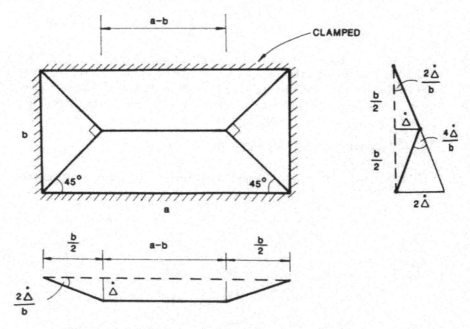

FIGURE 9.27. Failure mechanism of a clamped rectangular plate.

### 9.5.3.1. Generalized Stresses in a Circular Plate Under Symmetric Load

Consider a small element of a circular plate, as shown in Fig. 9.28a. Because of the axisymmetry of the structure and the loading, the shearing stresses $\tau_{r\theta} = \tau_{z\theta} = 0$. Since for a thin plate, the ratio of the plate thickness to the radius $R$, $h/R$, is assumed to be small, $\sigma_z$ and $\tau_{rz}$ will be small in comparison with $\sigma_r$ and $\sigma_\theta$. Hence, the state of stress of a typical plate element is plane stress, and $\sigma_\theta$ and $\sigma_r$ are the relevant principal stresses.

All the stress resultants over the depth $h$ of the plate element are shown in Fig. 9.28b, among which $M_r$ is the radial bending moment per unit length and $M_\theta$ is the circumferential bending moment per unit length:

$$M_r = \int_{-h/2}^{h/2} \sigma_r z\, dz, \qquad M_\theta = \int_{-h/2}^{h/2} \sigma_\theta z\, dz$$

$M_r$ and $M_\theta$ are the generalized stresses in circular plate problems.

As a result of the symmetry, only one shear force, $V$, is present:

$$V = V_r = \int_{-h/2}^{h/2} \tau_{rz}\, dz$$

The shear force $V_r$ is not considered to be a generalized stress because its corresponding shear deformation is neglected. The shear $V_r$ is a reaction force, needed in equilibrium considerations.

Consideration of the equilibrium of the plate element (Fig. 9.28b) in the vertical direction and in rotation provides the following differential equations:

$$\frac{d}{dr}(rV) = -qr \tag{9.97}$$

and

$$\frac{d}{dr}(rM_r) = M_\theta + rV \tag{9.98}$$

Equations (9.97) and (9.98) are valid for a thin plate within the theory of small deflection. On the one hand, the plate must be thin enough, i.e., the thickness $h$ is small in comparison with the other dimensions of the plate, so that $\sigma_z$ can be neglected in comparison with the bending stresses. On the other hand, the plate must be not very thin so that the deflection $w$ of such a plate is small in comparison with its thickness $h$; hence, it produces no membrane stresses, and therefore, the linear character of the theory of bending is preserved.

### 9.5.3.2. Yield Condition and Energy Dissipation

Since the direction of the principal axes and the relative magnitudes of the principal moments, namely, $M_r$ and $M_\theta$, are known due to the axisymmetry

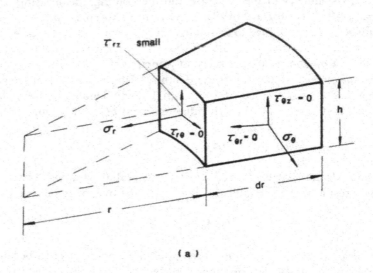

(a)

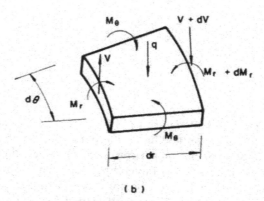

(b)

FIGURE 9.28. Stresses and generalized stresses in a circular plate element. (a) Stresses; (b) generalized stresses.

and the thin-plate idealization, the Tresca criterion provides a simpler mathematical form than the other criteria, which states

$$\max(|M_r|, |M_\theta|, |M_r - M_\theta|) \le M_0 \tag{9.99}$$

Since the shear deformations have been neglected, displacement of the plate is then characterized by the deflection $w = w(r)$. Let $\dot{w}$ be the rate of deflection; then the curvature rates are given by

$$\dot{\phi}_r = -\frac{d^2\dot{w}}{dr^2} \tag{9.100}$$

$$\dot{\phi}_\theta = -\frac{1}{r}\frac{d\dot{w}}{dr} \tag{9.101}$$

where $\dot{\phi}_r$ and $\dot{\phi}_\theta$ are the principal curvatures due to the axisymmetry. The internal energy dissipation per unit area is given by

$$D = M_r\dot{\phi}_r + M_\theta\dot{\phi}_\theta$$

According to the Tresca criterion and its associated flow rule, the above equation can be further expressed as

$$D = \tfrac{1}{2}M_0(|\dot{\phi}_r| + |\dot{\phi}_\theta| + |\dot{\phi}_r + \dot{\phi}_\theta|) \tag{9.102}$$

### 9.5.3.3. DISCONTINUITIES

As shown in Fig. 9.29, the arc $C$ represents a circular line of *stress discontinuity*. Equilibrium requires that $M_r$ and $V$ be continuous across the common boundary $C$ of two different regions of stresses, but $M_\theta$ may be discontinuous. From Eq. (9.98), since $M_\theta$ could be discontinuous, so is $dM_r/dr$.

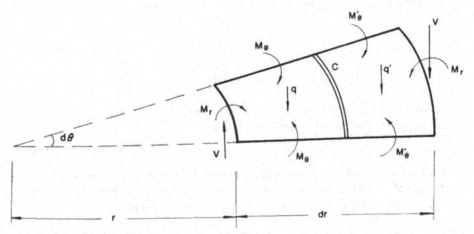

FIGURE 9.29. A line of discontinuity separating the plastic stress fields.

The continuity of the plate requires that the displacement rate $\dot{w}(r)$ be continuous, but the curvatures $\dot{\phi}_r$ and $\dot{\phi}_\theta$ could be discontinuous, i.e., the slope of $\dot{w}$, $d\dot{w}/dr$, could be discontinuous or jumped.

By analogy to the plastic hinge in beams, circular plates could have a *plastic hinge circle*. It should be noted that the hinge circle is considered to be the limit of a narrow annulus across which the rate of slope $d\dot{w}/dr$ varies rapidly but continuously, as the width of this annulus tends towards zero. At the hinge circle, the second derivative $d^2\dot{w}/dr^2$, i.e., the curvature $\dot{\phi}_r$, would become infinite in the limit.

### EXAMPLE 9.4. SIMPLY SUPPORTED PLATE UNDER CONCENTRATED LOAD

*Upper-Bound Solution*

Intuitively, we choose a mechanism as shown in Fig. 9.30 in which $\dot{\phi}_\theta > 0$ and $\dot{\phi}_r = 0$. Since $M_r > 0$ and $M_\theta > 0$, the stress field associated with this mechanism will be on the regime $AB$ of the Tresca hexagon (Fig. 9.30b). In particular, the center of the plate will correspond to the point $A$, where $M_r = M_\theta = M_0$. The associated flow rule will lead to $\dot{\phi}_r = 0$ and $\dot{\phi}_\theta \neq 0$. The energy dissipation per unit area will be $M_0\dot{\phi}_\theta$, according to Eq. (9.102).

Since we assume $\dot{\phi}_r = 0$, we have, using Eq. (9.100),

$$\dot{\phi}_r = -\frac{d^2\dot{w}}{dr^2} = 0 \tag{9.103}$$

The boundary conditions for $\dot{w}$ are

$$\dot{w} = 0 \qquad \text{at } r = a$$
$$\dot{w} = \dot{w}_0 \qquad \text{at } r = 0 \tag{9.104}$$

Integrating Eq. (9.103) and using Eq. (9.104) yields

$$\frac{d\dot{w}}{dr} = -\frac{\dot{w}_0}{a}$$

$$\dot{w} = \dot{w}_0\left(1 - \frac{r}{a}\right) \tag{9.105}$$

Equation (9.105) represents the conical shape of the failure mode.

Obviously, the external work rate is

$$\dot{W}_E = P\dot{w}_0 \tag{9.106}$$

With the deflection (9.105) given, $\dot{\phi}_\theta$ is easy to obtain as $\dot{\phi}_\theta = \dot{w}_0/(ra)$, and the rate of internal energy dissipation per unit area is $M_0\dot{w}_0/(ra)$. Integrating over the plate area, we get the total rate of energy dissipation:

$$\dot{W}_I = \int_A M_0\dot{\phi}_\theta \, dA = \int_0^a M_0\left(\frac{\dot{w}_0}{ra}\right)(2\pi r) \, dr = 2\pi M_0\dot{w}_0 \tag{9.107}$$

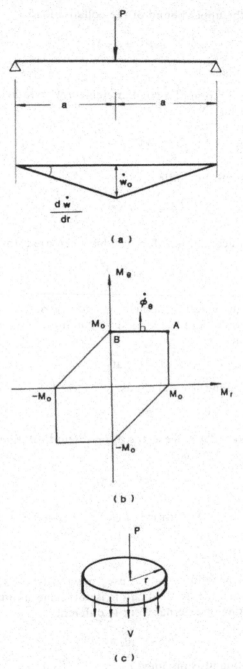

FIGURE 9.30. A simply supported circular plate under concentrated load $P$: (a) failure mechanism; (b) Tresca yield curve and flow rule; (c) equilibrium of a circular area.

$\dot{W}_E = \dot{W}_I$ gives the upper bound of the collapse load $P$:

$$P^U = 2\pi M_0 \tag{9.108}$$

*Lower-Bound Solution*

Consider the equilibrium of a circular element with radius $r$ in the vertical direction (Fig. 9.30c). We have

$$rV = -\frac{P}{2\pi} \tag{9.109}$$

Substituting this into Eq. (9.98) yields

$$\frac{d(rM_r)}{dr} = rV + M_\theta = -\frac{P}{2\pi} + M_0 \tag{9.110}$$

in which the yield condition, $M_\theta = M_0$, has been used. Integrating leads to

$$M_r = M_0 - \frac{P}{2\pi} + \frac{C_1}{r} \tag{9.111}$$

The condition that the radial moment $M_r$ must be bounded at $r = 0$ implies $C_1 = 0$. On the other hand, the boundary condition at a simply supported edge requires that

$$M_r = 0 \quad \text{at} \quad r = a$$

Thus, the lower bound is obtained as

$$P^L = 2\pi M_0$$

which happens to be the same as the upper-bound solution (9.108). Hence, the correct collapse load is

$$P^c = 2\pi M_0 \tag{9.112}$$

EXAMPLE 9.5. SIMPLY SUPPORTED PLATE UNDER UNIFORM LOAD (FIG. 9.31)

*Upper-Bound Solution*

We can use the same failure mode as used previously in Example 9.4. Then, the rate of internal energy dissipation is the same as that given by Eq. (9.107), but the rate of external work is different:

$$\dot{W}_E = \tfrac{1}{3}q(\pi a^2)\dot{w}_0$$

Thus, the upper bound is obtained as

$$q^U = 6\frac{M_0}{a^2} \tag{9.113}$$

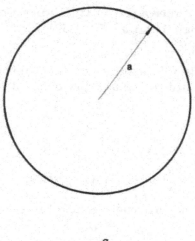

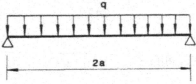

FIGURE 9.31. A simply supported plate under uniform load.

*Lower-Bound Solution*

Since the load $q$ is constant, the equilibrium equation (9.97) can be integrated to give

$$rV = -\frac{q}{2}r^2 \tag{9.114}$$

in which the integration constant has been set to zero because the shear force $V$ must be bounded at $r = 0$. Substituting the above relation into the equilibrium equation (9.98) leads to

$$\frac{d}{dr}(rM_r) = rV + M_\theta = -\frac{q}{2}r^2 + M_0 \tag{9.115}$$

Integrating and applying the boundary condition at $r = 0$ yields

$$M_r = -\tfrac{1}{6}qr^2 + M_0 \tag{9.116}$$

The boundary condition at $r = a$ requires that $M_r = 0$; thus, we obtain a lower-bound solution $q^L = 6M_0/a^2$, which is the same as the upper-bound solution and is therefore the correct answer

$$q^c = 6\frac{M_0}{a^2} \tag{9.117}$$

EXAMPLE 9.6. SIMPLY SUPPORTED PLATE UNDER CONCENTRATED
LOAD $P$ AND UNIFORM LOAD $q$ (FIG. 9.32)

*Upper-Bound Solution*

The same failure mode of Example 9.4 is also used here. The rate of work
done by the external loads $P$ and $q$ is now obtained as

$$\dot{W}_E = \left[\frac{q}{3}(\pi a^2) + P\right]\dot{w}_0$$

Equating this with $\dot{W}_I$ as given by Eq. (9.107)

$$\left[\frac{q}{3}(\pi a^2) + P\right]\dot{w}_0 = 2\pi M_0 \dot{w}_0$$

leads to an upper-bound interaction relation between the loads $P$ and $q$:

$$\frac{qa^2}{6M_0} + \frac{P}{2\pi M_0} = 1 \tag{9.118}$$

*Lower-Bound Solution*

Resulting from the equilibrium equation (9.97) in the vertical direction,

$$rV = -\frac{P}{2\pi} - \frac{qr^2}{2} \tag{9.119}$$

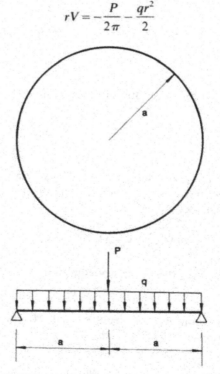

FIGURE 9.32. A simply supported plate under concentrated load $P$ and uniform
load $q$.

and from (9.98), and using the yield condition $M_\theta = M_0$, we have

$$\frac{d}{dr}(rM_r) = -\frac{P}{2\pi} - \frac{qr^2}{2} + M_0 \qquad (9.120)$$

Similarly, integrating and using the boundary conditions, we obtain a lower-bound interaction relation between the loads $P$ and $q$, which turns out to be the same as the upper-bound curve, as given by Eq. (9.118). Thus, Eq. (9.118) is the correct interaction curve.

EXAMPLE 9.7. SIMPLY SUPPORTED PLATE LOADED BY A UNIFORM LOAD OVER THE CIRCULAR AREA $0 \le r \le c$ (FIG. 9.33)

*Lower-Bound Solution*

In the region $r \le c$, the equilibrium equations for the uniform load case of Example 9.5 can be applied, while in region $r > c$, some equations for the concentrated load case of Example 9.4 can also be applied, if the concentrated load $P$ is visualized as a distributed load over this circular region. Now, we denote

$$q = \frac{P}{\pi c^2} \qquad (9.121)$$

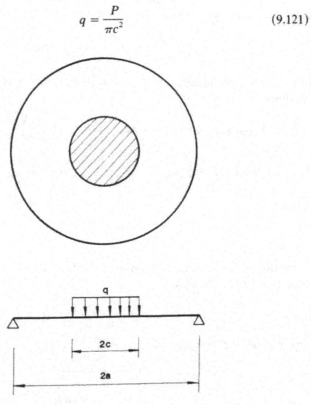

FIGURE 9.33. A simply supported plate loaded by a uniform load over the circular area $0 < r < c$.

For $r \leq c$, Eqs. (9.114) through (9.116) apply and we have

$$rV = -\frac{q}{2}r^2 = -\frac{1}{2}\left(\frac{P}{\pi c^2}\right)r^2 \tag{9.122}$$

$$\frac{d}{dr}(rM_r) = M_0 - \left(\frac{P}{\pi c^2}\right)\frac{r^2}{2}$$

and

$$M_r = M_0 - \frac{1}{6}\left(\frac{P}{\pi c^2}\right)r^2 \tag{9.123}$$

For $r > c$, Eqs. (9.109) and (9.110) apply; then we have

$$rV = -\frac{P}{2\pi} \tag{9.124}$$

and

$$\frac{d}{dr}(rM_r) = M_0 - \frac{P}{2\pi}$$

Integrating, we obtain

$$rM_r = M_0 r - \frac{P}{2\pi}r + b \tag{9.125}$$

in which $b$ is an integration constant to be determined by the continuity condition of $M_r$ at $r = c$:

from Eq. (9.123),     $M|_{r=c} = M_0 - \left(\dfrac{P}{\pi c^2}\right)\dfrac{c^2}{6} = M_0 - \dfrac{P}{6\pi}$

from Eq. (9.125),     $M|_{r=c} = M_0 - \dfrac{P}{2\pi} + \dfrac{b}{c}$

Hence,

$$b = \frac{Pc}{3\pi} \tag{9.126}$$

On the other hand, the boundary condition at $r = a$ requires that $M_r = 0$. Equation (9.125) becomes

$$M_0 - \frac{P}{2\pi} + \frac{Pc}{3\pi a} = 0$$

which gives the lower bound

$$P^L = \frac{2\pi M_0}{1 - \dfrac{2}{3}\left(\dfrac{c}{a}\right)} \tag{9.127}$$

*Upper-Bound Solution*

Again, assume the same failure mechanism as that of the previous examples

(9.4 and 9.5). Then, the rate of external work done by the load $q$ is given by

$$\dot{W}_E = q(\pi c^2)\left(\frac{a-c}{a}\,\dot{w}_0\right) + \frac{q}{3}(\pi c^2)\left(\frac{c}{a}\,\dot{w}_0\right) = P\left(1 - \frac{2}{3}\frac{c}{a}\right)\dot{w}_0$$

The rate of energy dissipation is the same as given by Eq. (9.107), i.e., $\dot{W}_I = 2\pi M_0 \dot{w}_0$. Thus, $\dot{W}_E = \dot{W}_I$ gives an upper bound on the collapse load, which is the same as the lower bound (9.127). Hence, the correct answer is given by

$$P^c = \frac{2\pi M_0}{1 - \frac{2}{3}\left(\frac{c}{a}\right)} \tag{9.128}$$

### 9.5.3.4. Remarks on the Stress Profiles

*Stress Profile at the Instant of Collapse*

In Example 9.4, $M_\theta = M_0$ and $M_r = 0$ at all points except the center of the plate, where $M_r = M_\theta = M_0$. The stress profile jumps from plastic regime $B$ to plastic regime $A$ around the neighborhood of the plate center. This is a peculiar case as shown in Fig. 9.34a.

In Example 9.5, $M_\theta = M_0$ and $M_r$ is given by (9.116). Since at the instant of collapse, $q = 6M_0/a^2$, we have

$$M_r = M_0\left[1 - \left(\frac{r}{a}\right)^2\right]$$

which gives $M_r = 0$ at $r = a$ and $M_r \to M_0$ when $r \to 0$. The stress profile moves continuously from plastic regime $A$ to regime $B$ (see Fig. 9.34b).

In Example 9.7, we have

$$M_\theta = M_0$$

and for $r \geq c$

$$M_r = M_0\left[1 - \frac{1 - \frac{2}{3}\left(\frac{c}{r}\right)}{1 - \frac{2}{3}\left(\frac{c}{a}\right)}\right]$$

which gives $M_r = 0$ at $r = a$ and $M_r = [(a-c)/(1.5a-c)]M_0$ at $r = c$; and for $r < c$

$$M_r = M_0\left[1 - \frac{a}{(3a-2c)}\left(\frac{r}{c}\right)^2\right]$$

which gives $M_r = M_0$ at $r = 0$ and $M_r = [(a-c)/(1.5a-c)]M_0$ at $r = c$. The stress profile moves continuously along $AB$ as shown in Fig. 9.34c.

*Successive Stages of Stress Profile*

As the load is slowly increased from zero, the successive stages of stress profile can be seen clearly by a stress plot, which is shown for Example 9.5

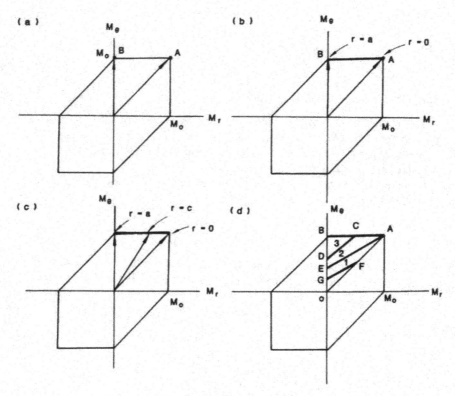

FIGURE 9.34. Stress profiles.

in Fig. 9.34d. The stress at $r = 0$ is increased along the line $OA$ while the stress at $r = a$ is increased along the line $OB$. The stress profiles for the following four different loading stages are plotted in Fig. 9.34d:

Stage 1: Profile $G - F$ represents an elastic stage.
Stage 2: Profile $A - E$ represents a critical stage at which the plate center has just reached yielding.
Stage 3: Profile $A - C - D$ represents an elastic–plastic stage at which the center part of the plate has been yielded.
Stage 4: Profile $A - C - B$ represents the stage of plastic collapse.

# 9.6. Limit Analysis of Plates on Elastic Foundations

The interaction between foundations and the supporting soil media has been a subject of interest to both geotechnical and structural engineers for many years. In approaching this problem, it is generally accepted that the plastic or the limit analysis methods are preferable to the elastic method. It has been shown that plastic theories not only lead to simpler methods of analysis but also give results that are more reliable and are less sensitive to the exact distribution of the contact stress between the foundation and the supporting soil than those of the elastic theories.

## 9.6.1. Load-Carrying Capacity of Concrete Pavements

In what follows, the load-carrying capacity of concrete slabs on an elastic foundation is studied. As a matter of fact, a considerable literature on finding the upper bounds of the collapse load of a concrete slab is available under the heading of *rupture-line method* or *yield-line theory*. The concept was first developed independently of the limit analysis theorems by Johansen in 1932. A comprehensive presentation of the method is contained in Johansen's book (1943), among many others.

Consider now the load-carrying capacity of plain concrete pavement under a central uniform load, as shown in Fig. 9.35. The pavement slab is assumed to be infinitely large and resting on a linear–elastic (Winkler) subgrade.

### 9.6.1.1. UPPER-BOUND SOLUTION

As the load $p$ is slowly applied to the pavement over a small circle of radius $a$, the slab will be gradually driven into the subgrade until fully plastic radial moments are realized and a *plastic mechanism* develops in the slab, as shown in Fig. 9.35. The collapse mechanism consists of an infinite number of radial yield lines and a circular yield line of radius $c$. With further increase of the load, the slab deforms into a conical surface but no sudden failure is observed; i.e., although the external forces continue to increase, equilibrium

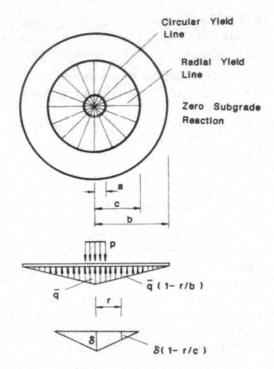

FIGURE 9.35. Failure mechanism and pressure distribution.

is maintained because subgrade reactions also increase. However, the rate of settlement under the applied external load grows rapidly at the formation of the failure mechanism until, eventually, the load and the collapsed concrete sink into the ground. The interest of limit analysis is to determine the load at which a plastic collapse mechanism develops in the slab. In the present calculation, the subgrade reaction $\bar{q}$ is represented by a conical distribution with radius $b$.

The work done by the applied load $W_p$, the upward subgrade reaction $W_{\bar{q}}$, and the internal energy dissipation by the radial yield lines $W_r$ and circular yield line $W_c$ have the following values:

$$W_p = \int_0^{2\pi} \int_0^a \frac{P_0}{\pi a^2} r\, d\theta\, dr\, \delta\left(1 - \frac{r}{c}\right) = P_0\delta\left(1 - \frac{2a}{3c}\right) \tag{9.129}$$

$$W_{\bar{q}} = -\int_0^{2\pi} \int_0^c \bar{q}\left(1 - \frac{r}{b}\right) r\, d\theta\, dr\, \delta\left(1 - \frac{r}{c}\right) = -\bar{q}\delta\frac{\pi c^2}{3}\left(1 - \frac{c}{2b}\right)$$

$$= -P_0\delta\left(\frac{c}{b}\right)^2\left(1 - \frac{c}{2b}\right) \tag{9.130}$$

$$W_r = 2\pi M_0\delta \tag{9.131}$$

$$W_c = 2\pi M_0\delta \tag{9.132}$$

From the work equation of this mechanism

$$W_p + W_{\bar{q}} = W_r + W_c \tag{9.133}$$

we obtain

$$P_0 = \frac{4\pi M_0}{1 - \frac{2}{3}\frac{a}{b}\frac{b}{c} - \left(\frac{c}{b}\right)^2 + \frac{1}{2}\left(\frac{c}{b}\right)^3} \tag{9.134}$$

The position of the circular yield line can be determined from

$$\frac{dP_0}{dc} = 0 \quad \text{or} \quad 4\left(\frac{a}{b}\right) - 12\left(\frac{c}{b}\right)^3 + 9\left(\frac{c}{b}\right)^4 = 0 \tag{9.135}$$

Given the ratio $a/b$, the relative radius of the circular yield line $c/b$ can be found from Eq. (9.135). By substituting the value of $c/b$ so obtained into Eq. (9.134), an upper-bound collapse load $P_0$ is obtained.

### 9.6.1.2. LOWER-BOUND SOLUTION

The vertical and the moment equilibrium of a circular plate element with an axisymmetric loading have two differential equations, (9.97) and (9.98), relating the principal moments $M_r$ and $M_\theta$, shear $V$, and distributed load $q(r)$ in the usual polar coordinates $r$ and $\theta$

$$\frac{d}{dr}(rV) + rq = 0, \qquad \frac{d}{dr}(rM_r) - M_\theta - rV = 0 \tag{9.136}$$

Since $q$ depends only on $r$, the first equation can be integrated and substituted into the second to eliminate $V$

$$\frac{d}{dr}(rM_r) - M_\theta = -\int_0^r rq(r)\,dr \tag{9.137}$$

Equation (9.137) can be integrated for any given plastic collapse mechanism. For the mechanism shown in Fig. 9.35, we have

$$M_\theta = M_0 \qquad M_r = \frac{1}{r}\int\left[M_0 - \int_0^r rq(r)\,dr\right]dr \tag{9.138}$$

Using the lateral load distribution

$$q(r) = \begin{cases} p - \bar{q}\left(1 - \dfrac{r}{b}\right) & \text{for } 0 \le r \le a \\[2mm] -\bar{q}\left(1 - \dfrac{r}{b}\right) & \text{for } a \le r \le b \end{cases} \tag{9.139}$$

and the overall vertical equilibrium condition

$$\tfrac{1}{3}\pi b^2 \bar{q} = p\pi a^2 = P_0 \tag{9.140}$$

together with the appropriate boundary conditions at $r=0$ and $r=c$, with $M_r = M_0$, and the continuity condition of $M_r$ at $r=a$, the radial moment $M_r$ along the radius can be calculated from Eq. (9.138)

$$M_r = \begin{cases} M_0 + \dfrac{P_0 r^2}{2\pi b^2}\left(1 - \dfrac{r}{2b}\right) - \dfrac{P_0 r^2}{6\pi a^2} & \text{for } 0 < r \leq a \quad\quad (9.141a) \\[3mm] M_0 + \dfrac{P_0 r^2}{2\pi b^2}\left(1 - \dfrac{r}{2b}\right) - \dfrac{P_0}{2\pi}\left(1 - \dfrac{2a}{3r}\right) & \text{for } a \leq r \leq b \quad\quad (9.141b) \end{cases}$$

The radial moment for $a/b = 0.3$ obtained using Eq. (9.134) is shown as the solid curve in Fig. 9.36. Since the radial moment in the slab is either smaller than or equal to the yield moment, the yield criterion is fulfilled everywhere. Thus, the load $P_0$ as given by Eq. (9.134) is also a lower bound on the limit load. Therefore, $P_0$ is the true collapse load, according to the two limit theorems.

### 9.6.1.3. MEYERHOF'S SOLUTION

Meyerhof (1962) obtained the following collapse load for a concentrated load over a small circular area on a large slab in full contact with the base:

$$P_0 = \frac{8\pi M_0}{1 - \dfrac{4a}{3b}} \quad\quad (9.142)$$

In deriving this expression, a failure mechanism similar to that of Fig. 9.35 is used. However, the circular yield line is assumed to coincide with the circle of zero upward subgrade reaction. Meyerhof's solution is therefore an upper-bound solution. This can be demonstrated by substituting the upper-bound solution (9.142) into Eqs. (9.141); the result for $a/b = 0.3$ is plotted in Fig. 9.36 as the dashed curve. It can be seen that except in the central part of the slab and at the circular yield line, the bending moments are equal to or less than the yield moment of the slab. The moments in the region of $0.3 < r/b < 1.0$ all exceed the yield moment $M_0$, and the yield

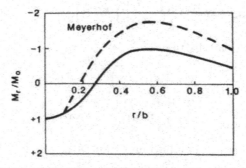

FIGURE 9.36. Moment distribution with $a/b = 0.3$.

criterion is not satisfied. Therefore, the system of yield lines for Meyerhof's mechanism does not form the correct collapse mechanism. The correct position of the circular yield line must lie somewhere inside the circle of zero subgrade reaction, as shown in Fig. 9.35.

For dual, triple, and quadruple circular loads and a strip load, similar expressions can be derived. They are listed in Table 9.3 as derived originally by Jiang (1979).

## 9.6.2. Progressive Collapse of Rigid–Plastic Circular Foundations

### 9.6.2.1. DESCRIPTION OF THE PROBLEM

Consider a rigid-perfectly plastic circular foundation plate resting on a uniform soil layer of thickness $H$ underlain by a rigid block as shown in

TABLE 9.3. Different loading cases (Jiang, 1979).

| Loading case | $P_a/M_0$ |
|---|---|
| | $2\dfrac{c}{b} - \left(\dfrac{2}{3}+\dfrac{4}{3\pi}\right)\dfrac{a}{b} - \dfrac{4[(\pi c/b)+s/b]}{\dfrac{2\pi}{3}\left(\dfrac{c}{b}\right)^3\left(1-\dfrac{1}{2}\dfrac{c}{b}\right)+2\dfrac{s}{b}\left(\dfrac{c}{b}\right)^2\left(1-\dfrac{1}{3}\dfrac{c}{b}\right)}\Big/\big[(\pi/3)+s/b\big]$ |
| | $3\dfrac{c}{b} - \left(\dfrac{2}{3}+\dfrac{2}{\pi}\right)\dfrac{a}{b} - \dfrac{4[(\pi c/b)+1.5s/b]}{\pi\left(\dfrac{c}{b}\right)^3\left(1-\dfrac{1}{2}\dfrac{c}{b}\right)+4.5\dfrac{s}{b}\left(\dfrac{c}{b}\right)^2\left(1-\dfrac{1}{3}\dfrac{c}{b}\right)+\dfrac{3\sqrt{3}}{4}\left(\dfrac{s}{b}\right)^2\dfrac{c}{b}}\Big/\big[(\pi/3)+(1.5s/b)+(\sqrt{3}/4)(s/b)^2\big]$ |
| | $4\dfrac{c}{b} - \left(\dfrac{2}{3}+\dfrac{8}{3\pi}\right)\dfrac{a}{b} - \dfrac{4[(\pi c/b)+2s/b]}{\dfrac{4\pi}{3}\left(\dfrac{c}{b}\right)^3\left(1-\dfrac{1}{2}\dfrac{c}{b}\right)+8\dfrac{s}{b}\left(\dfrac{c}{b}\right)^2\left(1-\dfrac{1}{3}\dfrac{c}{b}\right)+4\left(\dfrac{s}{b}\right)^2\dfrac{c}{b}}\Big/\big[(\pi/3)+(2s/b)+(s/b)^2\big]$ |
| | $\dfrac{c}{b} - \dfrac{\dfrac{\pi}{3}\dfrac{a}{b}+\dfrac{1}{2}\dfrac{s}{b}}{(\pi/2)+s/b} - \dfrac{4[(\pi c/b)+s/b]}{\dfrac{\pi}{3}\left(\dfrac{c}{b}\right)^3\left(1-\dfrac{1}{2}\dfrac{c}{b}\right)+\dfrac{s}{b}\left(\dfrac{c}{b}\right)^2\left(1-\dfrac{1}{3}\dfrac{c}{b}\right)}\Big/\big[(\pi/3)+s/b\big]$ |

Fig. 9.37. The interface between the foundation and the soil is assumed to be smooth or frictionless, and the plate material obeys the square yield criterion of Fig. 9.38 with the associated flow rule. The stress–strain relations of the soil will be described later.

A uniformly distributed load is slowly applied over a small circle of radius $c$ concentric with the foundation plate so that the axisymmetry is preserved. Thereafter, the load is gradually incremented and the plate is driven into the soil layer as a rigid flat indenter until fully plastic radial moments are developed and a plastic *mechanism* forms in the plate. With further increase of the load, the plate deforms into a conical surface, but no sudden failure is observed. Although the external forces continue to increase, equilibrium is maintained because soil reactions also increase. At a certain level of the load intensity, a change in the geometry of the soil–foundation interface takes place as the foundation *lifts off* the base. Therefore, the rate of settlement under the applied external load grows rapidly until, eventually, the foundation "sinks" into the ground. Details of the analysis are given elsewhere (Gazetas and Tassios, 1978).

In general, there are three distinct stages of foundation deformation identified as shown in Fig. 9.39.

(a) For small values of the external load, $P < P_0$, the bending moments that develop in the foundation plate are smaller than the corresponding

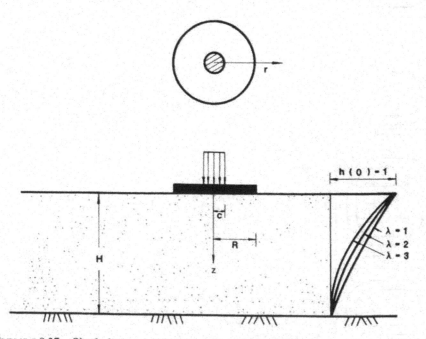

FIGURE 9.37. Single-layer two-parameter soil model and the $h(r)$ vertical displacement distribution function.

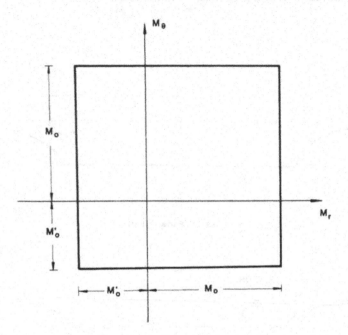

FIGURE 9.38. Yield criterion of foundation plate.

plastic moment $M_0$ or $M_0'$. Therefore, no yielding takes place and the foundation penetrates the soil layer elastically as a rigid punch.

(b) Beyond a critical *threshold* value of the load, $P_0$, an infinite number of radial yield lines develop and the plate deforms into a conical shape. Full contact is still maintained between the foundation and the soil during this phase.

(c) Beyond a certain value of the load, $P_s$, called the *separation load*, the foundation lifts off the soil as the center of the plate is driven further into the ground while the edges move upward.

### 9.6.2.2. Two-Parameter Soil Model

Consider a uniform soil layer of thickness $H$ as shown in Fig. 9.37. If the vertical and horizontal displacements, $w(r, z)$ and $u(r, z)$, can be found at all points in the layer, then the stresses and strains can be obtained from the established linear stress-strain and strain-displacement relations of the theory of elasticity. In the analysis of soil-foundation interaction problems, the horizontal displacements $u(r, z)$ may be considered to be negligibly small in comparison with the vertical displacements $w(r, z)$, and the unknown function $w(r, z)$ is assumed to be separable and can be expressed as

$$w(r, z) = W(r) \, h(z) \qquad (9.143)$$

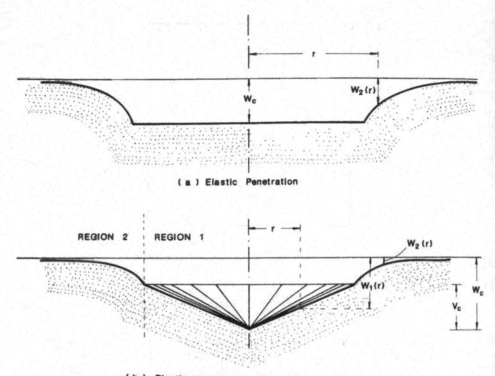

(a) Elastic Penetration

(b) Plastic penetration With Full Contact

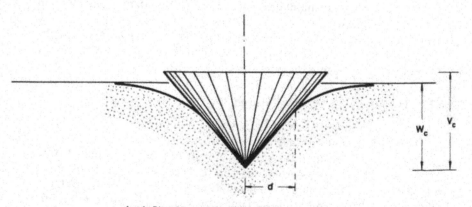

(c) Plastic penetration With Separation

FIGURE 9.39. Three stages of deformation of a rigid-plastic circular foundation.

where $h(z)$ may be assumed to be a linear function of depth $z$ or a nonlinear decreasing function of $z$ as shown in Fig. 9.37 with $\lambda = 1, 2$, and 3:

$$h(z) = 1 - \frac{z}{H} \quad \text{or} \quad h(z) = \frac{\sinh \lambda\left(\dfrac{H-z}{R}\right)}{\sinh \lambda\left(\dfrac{H}{R}\right)}$$

With the displacements of all points expressed by Eq. (9.143), the second-order equilibrium differential equation of the soil layer subjected to an axisymmetrically distributed load $q(r)$ on the surface of the soil can be derived by using the principle of virtual displacements, the elastic stress–strain relations, and the linear strain–displacement relations. It has the form (see Vlasov and Leont'ev, 1966)

$$-2t \, \nabla^2 W(r) + k \, W(r) = q(r) \tag{9.144}$$

in which

$$\nabla^2 = \frac{d^2}{dr^2} + \frac{1}{r}\frac{d}{dr}$$

is the Laplace operator in cylindrical coordinates, and

$$k = D \int_0^H \left[\frac{dh(z)}{dz}\right]^2 dz, \qquad t = \frac{G}{2} \int_0^H [h(z)]^2 \, dz \tag{9.145}$$

are the two elastic parameters of the foundation, where $D = E(1-v)/(1+v)(1-2v)$ is the *constrained modulus* of the soil, while $G = E/2(1+v)$ is the *shear modulus*. The two parameters $k$ and $t$, called *dilatational* and *shear* parameters, depend on the selected vertical displacement function $h(z)$.

### 9.6.2.3. PLASTIC PENETRATION WITH FULL CONTACT

In the case of plastic penetration with full contact (Fig. 9.39b), there are two regions of soil: one under and one beyond the foundation. Let $W_1$ and $W_2$ be the (unknown) settlements in the two regions (measured from the original ground surface) and $W_c$ the settlement of the foundation center. By Eq. (9.144), the two differential equations governing the spatial variation of $W_1$ and $W_2$ are:

$$\text{for } 0 \le r \le R, \qquad \frac{d^2 W_1}{dr^2} + \frac{1}{r}\frac{dW_1}{dr} - \alpha^2 W_1 = -\frac{\sigma_0(r)}{2t} \tag{9.146}$$

$$\text{for } R \le r < \infty, \qquad \frac{d^2 W_2}{dr^2} + \frac{1}{r}\frac{dW_2}{dr} - \alpha^2 W_2 = 0 \tag{9.147}$$

where $\sigma_0(r)$ is the as yet unknown contact stress at the foundation–soil interface and $\alpha^2 = k/2t$.

The governing equation for the foundation plate can be derived using the principle of virtual work. Note that for stage (b), the plate has been yielded to form a mechanism of a conical shape. Allowing the slab mechanism to deflect a small displacement such that the center moves an arbitrary distance, say, unity, the internal and the external work can be calculated and equated. This leads to

$$2\pi M_0 = P\left(1 - \frac{2}{3}\beta\right) - 2\pi \int_0^R \sigma_0(r)\, r\left(1 - \frac{r}{R}\right) dr \qquad (9.148)$$

in which $\beta = c/R$ and $P = p\pi c^2$ is the total applied load.

Equations (9.146) through (9.148) constitute a system of three differential-integral equations with three unknowns ($W_1$, $W_2$, and $\sigma_0$). Equation (9.147) can be directly solved for $W_2$

$$W_2(r) = \bar{c}\, K_0(\alpha r) + \bar{c}'\, I_0(\alpha r) \qquad (9.149a)$$

where $K_0$ and $I_0$ are the modified Bessel functions of order zero of the second and first kind, respectively. Since $W_2$ must vanish at infinity, $\bar{c}' = 0$ ($I_0 \to \infty$ as $r \to \infty$). Thus,

$$W_2 = \bar{c}\, K_0(\alpha r) \qquad (9.149b)$$

To integrate Eq. (9.146), we observe that due to the rigid–plastic behavior of the plate, $W_1$ can be expressed in terms of $W_c$ and the slope $\theta$ of the deformed foundation:

$$W_1 = W_c - \theta r \qquad (9.150)$$

Introducing Eq. (9.150) in Eq. (9.146) yields the explicit form of the contact stress distribution:

$$\sigma_0(r) = kW_c + \frac{2t\theta}{r} - k\theta r \qquad (9.151)$$

which, upon substitution in Eq. (9.148), results in

$$\left(1 - \frac{2}{3}\beta\right) P = 2\pi M_0 + \frac{\pi k W_c R^2}{3} + \pi R\left(2t - \frac{kR^2}{6}\right)\theta \qquad (9.152)$$

which, for a given plastic moment of the foundation, $M_0$, relates the applied load $P$ to the resulting deformations $W_c$ and $\theta$.

In addition to the distributed soil reactions against the foundation that are described by Eq. (9.151), fictitious reactions $Q_f$ per unit length act along the contour of the circular plate. These are due to the deformations of the soil beyond the plate region and correspond to the infinitely large stresses beneath the edges of rigid foundations predicted by the theory of elasticity for a semi-infinite continuum (e.g., see Poulos and Davis, 1974). To see how they are created, consider an infinitesimally thin hollow cylinder of soil with height $H$, internal radius $(R-f)$, and external radius $(R+f)$, where $f \to 0$. The condition of equilibrium can be written by equating to zero the total work done by all forces acting on the cylinder for a virtual displacement $\delta w(r, z) = h(z)$:

$$2\pi RQ_f h(0) + \int_0^H \int_0^{2\pi} h(z)\, \tau_{rz}^{(2)}(R+f)\, d\phi\, dz$$

$$- \int_0^H \int_0^{2\pi} h(z)\, \tau_{rz}^{(1)}(R-f)\, d\phi\, dz = 0 \qquad (9.153)$$

where the shear stresses $\tau_{rz}^{(1)}$ and $\tau_{rz}^{(2)}$, corresponding to the two regions of Fig. 9.39b, are obtained from the elastic relation $\tau_{rz} = G[\partial w/\partial r]$ and Eq. (9.143). Substitution in Eq. (9.153) and integration yields the fictitious force:

$$Q_f = 2t \left[ \frac{dW_1}{dr} - \frac{dW_2}{dr} \right]_{r=R} \qquad (9.154)$$

Thus, $Q_f$ is caused by the different slopes of the settling surface at the edge of the foundation and the ability of the soil to take up the shearing stresses.

The boundary conditions of the problem can now be stated as follows:

$$W_1(R) = W_2(R) \qquad (9.155)$$

$$P = 2\pi \int_0^R \sigma_0(r)\, r\, dr + 2\pi RQ_f \qquad (9.156)$$

and the system of six equations (9.149b), (9.151), (9.152), and (9.154)–(9.156) can be analytically solved for the six unknown quantities $W_2$, $\bar{c}$, $W_c$, $\theta$, $Q_f$, and $\sigma_0$.

After some lengthy but straightforward algebraic operations, the following closed-form relations are derived:

$$W_c = \left[ \frac{P\xi}{k\pi R^2} - \frac{2M_0}{kR^2} \right] \omega \qquad (9.157a)$$

with

$$\xi = 1 - \frac{2}{3}\beta + \frac{3 - \dfrac{(\alpha R)^2}{2}}{g(\alpha R)^2} \qquad (9.157b)$$

$$\omega = \frac{g}{\dfrac{(g-1)}{6} + \dfrac{(g+1)}{(\alpha R)^2}} \qquad (9.157c)$$

and

$$g = 2\left[ 1 + 3\frac{K_1(\alpha R)}{\alpha R\, K_0(\alpha R)} \right] \qquad (9.157d)$$

$$\theta = \frac{(1+g)}{g} \frac{W_c}{R} - \frac{3}{g} \frac{P}{k\pi R^3} \qquad (9.158)$$

$$W_2 = \frac{(W_c - \theta R)}{K_0(\alpha R)} K_0(\alpha r) \qquad (R < r < \infty) \qquad (9.159)$$

Equations (9.157) to (9.159) can be used to compute the deformation of the soil surface for any applied load $P$ with the known moment capacity of the plate, $M_0$, provided that

$$P_0 \le P \le P_s \tag{9.160}$$

The contact stress distribution $\sigma_0(r)$ is then obtained from Eq. (9.151) after substituting the computed values of $W_c$ and $\theta$.

### 9.6.2.4. THRESHOLD LOAD

The load $P_0$ required to transform the foundation plate into a mechanism and, thus, bring it into the second phase of deformation is obtained by setting $\theta = 0$ in the previous relations. Denoting by $W_0$ the *threshold settlement* of the plate at this particular load ($W_0 = W_1 = W_c$), Eqs. (9.157) and (9.158) give

$$P_0 = \frac{3\pi}{(1.5 - \beta) - \dfrac{0.5}{\left[1 + 2\dfrac{K_1(\alpha R)}{\alpha R K_0(\alpha R)}\right]}} M_0 \tag{9.161}$$

for the threshold load, and

$$\tilde{W}_0 = \frac{W_0}{\dfrac{P_0}{k\pi R^2}} = \frac{1}{\left[1 + 2\dfrac{K_1(\alpha R)}{\alpha R K_0(\alpha R)}\right]} \tag{9.162}$$

for the threshold settlement. It is interesting to compare the above expression for $P_0$ with those resulting from other soil models such as the *Winkler model* and the *continuum model*:

$$(P_0)_{\text{Winkler}} = \frac{3\pi}{(1 - \beta)} M_0 \tag{9.163}$$

$$(P_0)_{\text{continuum}} = \frac{3\pi}{(1.18 - \beta)} M_0 \tag{9.164}$$

Figure 9.40 shows the reduced threshold load, $P_0/M_0$, predicted from Eqs. (9.161), (9.163), and (9.164) as a function of the reduced radius of loading, $\beta$. Since the value $\alpha$ in Eq. (9.161) is a function of the relative thickness, $H/R$, of the soil layer, the empirical parameter $\lambda$, and Poisson's ratio $\nu$ of the soil, a family of curves is shown in the figure for the soil model that has been presented here (the *Vlasov model*). The ratio of $H/R$ ranges from 1 to 4, and $\lambda$ ranges from 1 to 2 while $\nu = 0.30$. Only a single curve is obtained for the *Winkler soil* or the *half-space continuum*. The agreement of the three models ranges, in general, from satisfactory at low

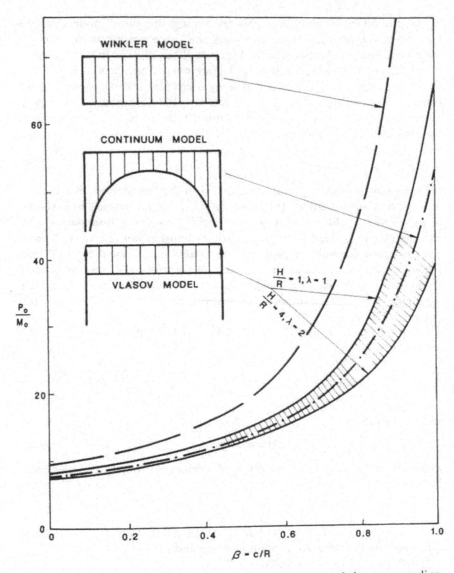

FIGURE 9.40. Comparison of the reduced threshold loads and the corresponding contact stress distributions from three soil models.

$\beta$ ratios (i.e., for a nearly concentrated load) to rather poor at very high $\beta$ ratios (i.e., for a uniformly applied pressure over almost the whole foundation area). Notice, however, that the Vlasov model leads to threshold loads that are much closer to those predicted by the continuum model throughout the range of $\beta$. In fact, the curve plotted from Eq. (9.164) falls almost in the center of the band comprising the curves of Eq. (9.161).

These similarities and discrepancies among the three models can be qualitatively explained if one considers the three corresponding contact stress distributions, shown also in Fig. 9.40. It is obvious that the concentrated forces $(Q_f)$ at the contour of a plate pushed into a Vlasov soil and the infinite stresses at the edges of a rigid circular slab supported by an elastic homogeneous half-space produce higher bending moments and, thus, transform the foundation into a mechanism faster (i.e., at a smaller load) than the uniform reactions of a Winkler soil.

### 9.6.2.5. PLASTIC PENETRATION WITH SEPARATION

Since no tensile stresses can develop between the foundation and the soil, beyond the critical load $P_s$, the foundation lifts off the ground as is shown in Fig. 9.39c. Problems of a similar nature, involving boundaries with variable geometry, lead to nonlinear force-deformation relations. A complete solution for such a problem must yield not only the foundation settlement and soil reactions but also the *transversality conditions* for locating the position of the variable contact surface. In our case, due to the axial symmetry, the latter requirement is translated into the determination of the radius $d$ (Fig. 9.39c).

Equations (9.146), (9.147), and (9.150) still hold, while Eq. (9.148) changes to

$$2\pi M_0 = P\left(1 - \frac{2}{3}\beta\right) - 2\pi \int_0^d \sigma_0(r)\, r\left(1 - \frac{r}{R}\right) dr \qquad (9.165)$$

and Eq. (9.155) to

$$W_1(d) = W_2(d) \qquad (9.166)$$

where $W_2$ in this case refers to soil displacements beyond the contact area. The boundary condition at $r = d$ is

$$Q_f = 0 \qquad (9.167)$$

which is equivalent to stating that the slope of the deformed surface is continuous at the separation point of plate and soil.

Finally, Eq. (9.156) changes to

$$P = 2\pi \int_0^d \sigma_0(r)\, r\, dr \qquad (9.168)$$

and the new set of equations can be solved for $W_c$, $\theta$, and $\sigma_0$, in terms of the *contact radius, d*. The *transversality* condition [Eq. (9.167)] is then used to determine $d$ for a particular intensity of the applied force. After some lengthy but straightforward algebraic operations, one arrives at

$$W_c = \frac{P}{\pi k R^2}\left(\frac{R}{d}\right)^2 (1 + ab)^{-1} \qquad (9.169)$$

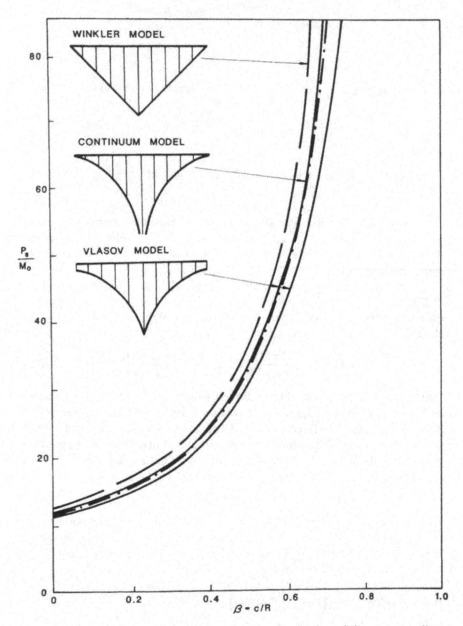

FIGURE 9.41. Comparison of the reduced separation loads and the corresponding contact stress distributions from three soil models.

with

$$a = \frac{2\alpha d\, K_1(\alpha d)/K_0(\alpha d)}{1 + \alpha d\, K_1(\alpha d)/K_0(\alpha d)} \quad \text{and} \quad b = \frac{1}{(\alpha d)^2} - \frac{1}{3} \tag{9.169a}$$

$$\theta = \frac{a}{2d}\, W_c \tag{9.170}$$

and $\sigma_0(r)$ is given by Eq. (9.151). The *transversality condition* is

$$\frac{P}{M_0} = \frac{12\pi}{6 - 4\beta - F(\alpha, d)} \tag{9.171}$$

where

$$F(\alpha, d) = \left\{ 6 - 2a + 6a(\alpha d)^{-2} - [4 - 1.5a + 3a(\alpha d)^{-2}]\frac{d}{R} \right\}(1 + ab)^{-1} \tag{9.171a}$$

For any applied load $P \geq P_s$ and knowing the plastic moment of the plate, $M_0$, Eq. (9.171) gives by trial and error the *contact radius d*. Equations (9.169), (9.170), and (9.151) can then be used to specify the surface settlement, the foundation distortion, and the soil reactions.

To determine the *separation load*, $P_s$, it is sufficient to set $d = R$ in Eqs. (9.171) and (9.169a). It is interesting to compare this load with the corresponding separation loads of a foundation on a Winkler base (Gazetas, et al., 1978) or on an elastic continuum (Krajcinovic, 1976):

$$P_{s_{\text{Winkler}}} = \frac{12\pi M_0}{3 - 4\beta}, \quad P_{s_{\text{continuum}}} = \frac{12\pi M_0}{\pi - 4\beta} \tag{9.172}$$

Figure 9.41 compares the three reduced separation loads, $P_s/M_0$, shown as functions of $\beta$. The agreement between them is excellent, especially in the low-$\beta$ range (nearly concentrated loads). This is hardly surprising in view of the very similar contact stress distributions predicted by the theories and shown also in Fig. 9.41. Note also that, as can be confirmed with Eqs. (9.172) (or Fig. 9.41), no separation can occur if

$$\beta \geq 0.75 \text{ to } 0.80$$

That is, with such a large loading area, the slab will never lose contact with the soil. On the other hand, for a concentrated load, i.e., $\beta = 0$, the separation load is only about 50% larger than the threshold load, irrespective of the particular type of soil model used.

## 9.7. Limit Analysis of Shells

### 9.7.1. Cylindrical Shells Under Axially Symmetric Loading

#### 9.7.1.1. BASIC EQUATIONS

Consider a circular cylindrical shell subjected to an axially symmetric radial pressure distribution and no end load (Fig. 9.42a). We choose a cylindrical

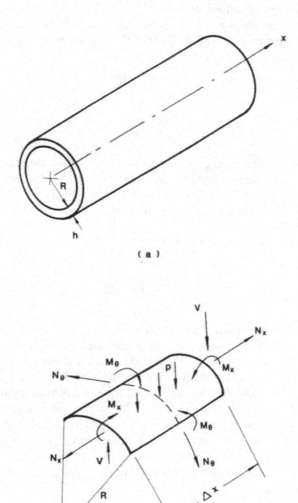

(a)

(b)

FIGURE 9.42. (a) A thin circular cylindrical shell. (b) Equilibrium of a shell element.

coordinate system with the x-axis along the axis of the cylinder. Owing to axisymmetry, the only displacements of the median surface of the shell are $u$ in the axial direction and $w$ in the outward radial direction; these displacements are functions of $x$ only. Here, as in the elastic theory, we assume that elements normal to the median surface remain straight and normal after deformation. It follows that the displacements at any point in the shell are

$$u_x = u - z \frac{dw}{dx}$$

$$u_\theta = 0 \tag{9.173}$$

$$u_r = w$$

where $z$ is the distance through the shell thickness, measured positive outward from the median surface. The shear strains $\gamma_{x\theta}$, $\gamma_{r\theta}$ are zero from the consideration of axisymmetry, $\gamma_{rx}$ is neglected by assumption, and the radial strain $\epsilon_r$ will not appear in the present analysis. Therefore, the only strains of interest are

$$\epsilon_\theta = \frac{w}{R}$$

$$\epsilon_x = \frac{du}{dx} - z \frac{d^2 w}{dx^2} = \frac{du}{dx} - z\phi_x \tag{9.174}$$

in which $\phi_x$ is the curvature in the x-direction.

The state of stress in a shell element is specified by its normal resultants $N_x$ and $N_\theta$, moment resultants $M_x$ and $M_\theta$, and a shear resultant $V$ (Fig. 9.42b). Since the transverse shear deformation is neglected and, because of axisymmetry, the circumferential curvature change is infinitesimal, it follows therefore that the resultant forces $V$ and $M_\theta$ corresponding to these shear deformation and bending curvature are not considered to be the generalized stresses. Further, for the case under consideration, the axial equilibrium requires that $N_x = 0$. Hence, only the normal force $N_\theta$ and moment $M_r$ are considered to be the generalized stresses, and the yield condition is generally expressed as

$$f(N_\theta, M_x) = 0 \tag{9.175}$$

In addition to satisfy the yield criterion, the stress resultants must be in equilibrium. Axial equilibrium requires $N_x$ to be constant and hence zero since there are no applied end loads. Vertical and bending equilibrium lead to the equations

$$\frac{dV}{dx} + \frac{N_\theta}{R} + p = 0 \tag{9.176a}$$

$$\frac{dM_x}{dx} - V = 0 \tag{9.176b}$$

The shear force $V$ can be eliminated from these two equations to yield

$$\frac{d^2 M_x}{dx^2} + \frac{N_\theta}{R} + p = 0 \tag{9.177}$$

### 9.7.1.2. YIELD CRITERION

The material of the shell is assumed to be perfectly plastic and obey Tresca's yield condition. Since the principal directions are axial and circumferential, this condition may be written as

$$\max(|\sigma_x|, |\sigma_\theta|, |\sigma_x - \sigma_\theta|) \le \sigma_0 \tag{9.178}$$

In the plate theory, the yield condition assumed the same form for the generalized stresses as for the stress components [see Eq. (9.67)]. However, this is not the case for shells. In finding the yield condition in terms of $N_\theta$ and $M_x$, we shall use the limit analysis concept. A material element can carry as much normal force and moment as may be obtained by any assumed distribution of stress as long as it does not violate the yield criterion (lower-bound theorem). On the other hand, if a deformation or velocity pattern is chosen that is compatible with the flow rule associated with the yield criterion, then collapse must occur if the rate at which $N_\theta$ and $M_x$ do work equals or exceeds the rate of internal energy dissipation (upper-bound theorem).

The lower bound of the $N_\theta$-$M_x$ interaction relation may be obtained by assuming a permissible stress distribution as shown in Fig. 9.43b. The normal stresses are $\sigma_0$ or zero, and we have

$$\begin{aligned} N_\theta &= \sigma_0(h - \delta) \\ M_x &= \sigma_0\delta(h - \delta) \end{aligned} \tag{9.179}$$

Denote

$$n_\theta = \frac{N_\theta}{\sigma_0 h} \quad \text{and} \quad m_x = \frac{M_x}{\frac{1}{4}\sigma_0 h^2} \tag{9.180}$$

Then, Eq. (9.179) reduces to

$$m_x = 4n_\theta(1 - n_\theta) \tag{9.181}$$

Absolute value signs placed around $m_x$ and $n_\theta$ will make the yield limit valid for all combinations of positive and negative values of $M_x$ and $N_\theta$. This yield condition is shown as the bold line in Fig. 9.44.

As a matter of fact, Eq. (9.179) or (9.181) is the best lower-bound solution that gives the maximum $N_\theta$ for a given $M_x$, or conversely.

To show that Eq. (9.179) or (9.181) is the true limit yield condition, the upper-bound theorem may be used by assuming a rate of deformation pattern of permissible type as shown in Fig. 9.43c. The faces of the unit element on which the moment acts rotate with respect to each other about

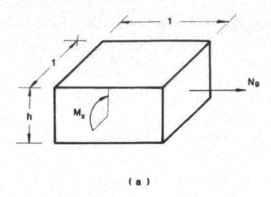

(a)

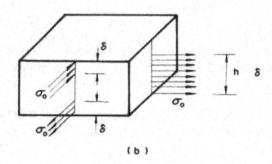

(b)

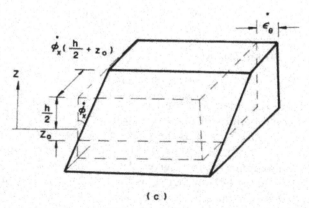

(c)

FIGURE 9.43. (a) Bending moment and normal force on perpendicular faces of a cylindrical shell. (b) Critical stress distribution. (c) Permissible deformation pattern.

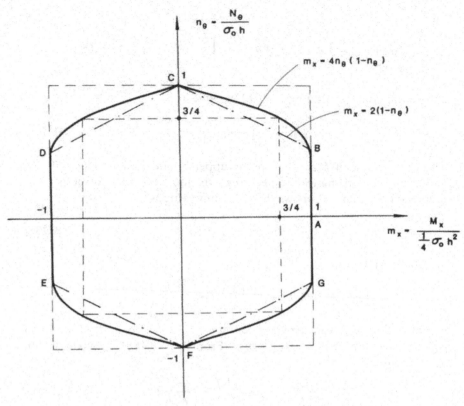

FIGURE 9.44. Yield curves for circular cylindrical shell.

the axis $z = z_0$ at a rate $\dot{\phi}_x$, and the two surfaces on which the normal force acts separate at a rate $\dot{\epsilon}_\theta$.

Upper bounds for the yield condition or collapse set $M_x, N_\theta$ are obtained by equating the rate of external work to the rate of internal energy dissipation per unit area

$$M_x \dot{\phi}_x + N_\theta \dot{\epsilon}_\theta = D_A \tag{9.182}$$

where $D_A$ is found by integrating the rate of dissipation per unit volume $D_V$ through the thickness, $h$:

$$D_V = \sigma_{ij} \dot{\epsilon}_{ij} = \sigma_0 |\dot{\epsilon}|_{max} \tag{9.183}$$

where $|\dot{\epsilon}|_{max}$ is the numerically largest principal strain rate [see Eq. (4.22) or Eq. (8.19), and note that $\sigma_0 = 2k$]. In Fig. 9.43c, which applies to positive $M_x$ and $N_\theta$, $|\dot{\epsilon}|_{max}$ is $\dot{\epsilon}_\theta + \dot{\phi}_x(z - z_0)$ for $z \geq z_0$ and is the larger of $\dot{\epsilon}_\theta$ and $\dot{\phi}_x(z_0 - z)$ for $z \leq z_0$. Calling $z_1$ the transition value of $z$,

$$z_0 - z_1 = \frac{\dot{\epsilon}_\theta}{\dot{\phi}_x} \tag{9.184}$$

and

$$D_A = \sigma_0 \int_{z_0}^{h/2} [\dot{\epsilon}_\theta + \dot{\phi}_x(z - z_0)] \, dz + \sigma_0 \int_{z_1}^{z_0} \dot{\epsilon}_\theta \, dz + \sigma_0 \int_{-h/2}^{z_1} \dot{\phi}_x(z_0 - z) \, dz \tag{9.185}$$

Integrating and substituting Eq. (9.184) gives

$$\frac{D_A}{\sigma_0} = \dot{\epsilon}_\theta \left(\frac{h}{2} - z_0\right) + \dot{\phi}_x \left(\frac{h^2}{4} + z_0^2\right) + \frac{1}{2} \frac{\dot{\epsilon}_\theta^2}{\dot{\phi}_x} \tag{9.186}$$

The largest value of $D_A$ will make the upper-bound collapse loads $M_x$, $N_\theta$ minimum. Minimizing with respect to $z_0$, the location of the as yet undetermined axis of rotation, shows the optimum value is

$$z_0 = \frac{\dot{\epsilon}_\theta}{2\dot{\phi}_x} = -z_1 \tag{9.187}$$

and the least $D_A$ is

$$D_A = \frac{\sigma_0 h}{4} \left(2\dot{\epsilon}_\theta + \dot{\phi}_x h + \frac{\dot{\epsilon}_\theta^2}{\dot{\phi}_x h}\right) = \dot{\phi}_x \frac{\sigma_0 h^2}{4} \left(1 + \frac{\dot{\epsilon}_\theta}{\dot{\phi}_x h}\right)^2 \tag{9.188}$$

except at the corners of the yield domain $n = \pm 1$, when $|\dot{\epsilon}_\theta/(\dot{\phi}_x h)| > 0$ and $|z_1| = h/2$. At the corners, $D_A = \sigma_0 h \dot{\epsilon}_\theta$. From Eq. (9.188), in analogy with $\sigma_{ij} = \partial D_A/\partial \dot{\epsilon}_{ij}$,

$$\frac{\partial D_A}{\partial \dot{\phi}_x} = M_x = \frac{\sigma_0 h}{4} \left(h - \frac{\dot{\epsilon}_\theta^2}{\dot{\phi}_x^2 h}\right)$$

or

$$\frac{4M_x}{\sigma_0 h^2} = m_x = 1 - \frac{\dot{\epsilon}_\theta^2}{(\dot{\phi}_x h)^2} \tag{9.189}$$

$$\frac{\partial D_A}{\partial \dot{\epsilon}_\theta} = N_\theta = \frac{\sigma_0 h}{4} \left(2 + 2\frac{\dot{\epsilon}_\theta}{\dot{\phi}_x h}\right)$$

or

$$\frac{2N_\theta}{\sigma_0 h} = 2n_\theta = 1 + \frac{\dot{\epsilon}_\theta}{\dot{\phi}_x h} \tag{9.190}$$

Eliminating $\dot{\epsilon}_\theta/(\dot{\phi}_x h)$ between Eqs. (9.189) and (9.190) results in the yield condition (9.181). Upper and lower bounds thus coincide and the solid line in Fig. 9.44 is proved to be the exact yield condition for a perfectly plastic material.

### 9.7.1.3. Approximate Yield Condition

Safe approximate yield conditions of a square or hexagonal shape may be chosen lying entirely inside the yield domain as shown in Fig. 9.44. A square is also shown circumscribing the domain. It is interesting to note that the lower-bound inscribed hexagon may be derived directly by considering an idealized sandwich section as in the case of arches (Hodge, 1959).

Dissipation functions for the squares are

$$D_A = N'_\theta |\dot{\epsilon}_\theta| \quad \text{or} \quad M'_x |\dot{\phi}_x| \quad \text{or} \quad N'_\theta |\dot{\epsilon}_\theta| + M'_x |\dot{\phi}_x| \qquad (9.191)$$

depending upon whether the stress point is on the horizontal sides (the rate of curvature $\dot{\phi}_x = 0$), on the vertical sides (the rate of strain $\dot{\epsilon}_\theta = 0$), or at the corners. In Eq. (9.191) $N'_\theta = \sigma_0 h$ and $M'_x = \sigma_0 h^2/4$ and for the circumscribed square yield surface and $\frac{3}{4}$ these values for the inscribed square yield surface (Fig. 9.44). The dissipation function for the hexagon is

$$D_A = \sigma_0 h |\dot{\epsilon}_\theta| \quad \text{or} \quad \frac{\sigma_0 h^2}{4} |\dot{\phi}_x| \quad \text{or} \quad \frac{\sigma_0 h}{2} |\dot{\epsilon}_\theta| + \frac{\sigma_0 h^2}{4} |\dot{\phi}_x| \qquad (9.192)$$

for the stress point on the horizontal sides, on the vertical sides, or on the sloping sides including the corners where $|m_x| = 1$ and $|n_\theta| = \frac{1}{2}$, respectively.

### 9.7.1.4. DISCONTINUITIES

Here, as in plates, certain discontinuities are expected in the fully plastic solutions of circular cylindrical shell problems. Since the shell is axially symmetric, the only conceivable lines of discontinuity are circles $x = \text{const.}$ If we consider the equilibrium of an element which straddles a proposed discontinuity circle (Fig. 9.42b), it is evident that $M_x$, $V$, and $N_x$ must be continuous but that $N_\theta$ and $M_\theta$ need not be. Therefore, in view of Eq. (9.176b), the continuity of $M_x$ and its first derivative $dM_x/dx$ is the only such requirement on the stress resultants.

With regard to velocities, the same reasoning as in the circular plate case shows that $\dot{w}$ must be continuous but $d\dot{w}/dx$ may be discontinuous at a hinge circle. This condition implies that the ratio $\dot{\phi}_x/\dot{\epsilon}_\theta$ tends to infinity. Therefore, a hinge circle must correspond to the vertical sides or their end points. Further, the sign of the discontinuity in slope will determine whether $m_x = +1$ or $-1$ at the hinge circle.

### 9.7.1.5. EXAMPLE: RING OF PRESSURE (DRUCKER, 1954)

Consider a cylindrical shell subjected to a ring of force $P$ per unit length (Fig. 9.45a).

*Upper Bound*

Assume the kinematically admissible plastic pattern of Fig. 9.45b and choose the circumscribing rectangle of Fig. 9.44 for the yield condition. The rate of dissipation per unit circumferential length at each of the "hinges" is the maximum limit moment, $\sigma_0 h^2/4$, times the rate of change of angle. In the

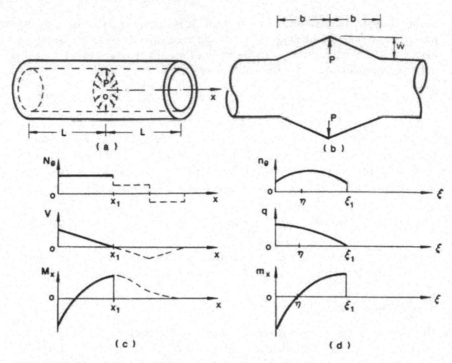

FIGURE 9.45. (a) A cylindrical shell subjected to a ring of pressure. (b) Kinematic pattern. (c) Admissible stress field based on square yield condition. (d) Admissible stress field based on hexagonal yield condition.

conical regions, the rate of dissipation is the integral of the maximum circumferential force, $\sigma_0 h$, times the rate of extension. The extensional strain rate varies linearly from zero at the outer hinges to $\dot{w}/R$ at the center hinge and so averages $\frac{1}{2}\dot{w}/R$. Equating the external rate of doing work to the internal rate of dissipation leads to

$$P^U \dot{w} = \frac{1}{4}\sigma_0 h^2 \left(\frac{\dot{w}}{b} + \frac{2\dot{w}}{b} + \frac{\dot{w}}{b}\right) + \sigma_0 h \left(\frac{1}{2}\frac{\dot{w}}{R}\right) \quad (2b)$$

in which the multiplier $2\pi R$ has been omitted. The upper bound is obtained as

$$P^U = \frac{\sigma_0 h^2}{b} + \frac{\sigma_0 h b}{R}$$

The minimum value is

$$P^U = 2\sigma_0 h \left(\frac{h}{R}\right)^{1/2} \quad (9.193)$$

for

$$b = (Rh)^{1/2}$$

0129

*Lower Bound*

Any equilibrium system of moments and normal forces which does not violate the yield criteria will provide a lower bound. There is no distributed lateral pressure on the shell except the ring pressure, so we set $p = 0$ in Eq. (9.176a). If we assume normal force $N_\theta$, $N'$ taking as the constant, then the equilibrium equations (9.176) can be integrated:

$$V = -N'\frac{x}{R} + A$$

$$M_x = -\frac{N'x^2}{2R} + Ax + B \tag{9.194}$$

The constants $A$ and $B$ are determined from the four boundary conditions:

$$V = \tfrac{1}{2}P \quad \text{and} \quad M_x = -M' \qquad \text{at } x = 0$$

$$V = 0 \quad \text{and} \quad M_x = M' \qquad \text{at } x = x_1$$

Equations (9.194) yield

$$P = 4\left(\frac{M'N'}{R}\right)^{1/2} \tag{9.195a}$$

and

$$x_1 = 2\left(\frac{M'R}{N'}\right)^{1/2} \tag{9.195b}$$

If the circumscribing yield rectangle were the true yield condition, it would be permissible to take $M' = \sigma_0 h^2/4$, $N' = \sigma_0 h$, and the answer (9.193) would be obtained as a lower bound. The coincidence of upper and lower bounds would mean the exact answer had been found. However, the maximum permissible inscribed rectangle is only three-fourths as large. Thus, we may write

$$P^L = 1.5\sigma_0 h\left(\frac{h}{R}\right)^{1/2} \tag{9.196}$$

It remains to be shown that the equilibrium field shown in Fig. 9.45c can be extended beyond the region $x = x_1$ without violating the yield condition, so that the load so obtained is a lower bound. Extension of the field to a point of zero $M_x$, $N_\theta$, and $V$ can be accomplished easily by taking $N_\theta$ as a permissible possible constant until $M_x$ is reduced to half of $M'$, then taking $N_\theta$ of the same numerical value but negative for an equal distance (Fig. 9.45c).

The simple calculation given above leads to

$$1.5 \le \frac{P}{\sigma_0 h\left(\dfrac{h}{R}\right)^{1/2}} \le 2.0 \tag{9.197}$$

where the upper and the lower bounds are in this special case the correct answers for the circumscribed and inscribed rectangular yield conditions, respectively. This problem is sufficiently simple. Such results are obtained with little effort and the average value of 1.75 will suffice for analysis or design.

### Solution for the Case of a Hexagonal Yield Condition

When the hexagonal yield condition in dimensionless form is used, it is convenient to write the equilibrium equations (9.176) and (9.177) in the form

$$\frac{1}{2}\frac{dq}{d\xi} + n_\theta + \psi = 0 \tag{9.198}$$

$$\frac{dm_x}{d\xi} - q = 0 \tag{9.199}$$

$$\frac{1}{2}\frac{d^2 m_x}{d\xi^2} + n_\theta + \psi = 0 \tag{9.200}$$

in which

$$\xi = \frac{x}{\sqrt{Rh/2}}, \qquad \psi = \frac{pR}{\sigma_0 h}$$

$$m_x = \frac{4M_x}{\sigma_0 h^2}, \qquad n_\theta = \frac{N_\theta}{\sigma_0 h}$$

$$q = \frac{4V\sqrt{Rh/2}}{\sigma_0 h^2}$$

To solve the problem, we first assume a stress profile. At $\xi = 0$, the state of stress corresponds to point $D$, and at $\xi = \xi_1$, to point $B$ in Fig. 9.44. We assume the stress profile extends from $D$ to $B$ along $DC$ and $CB$. The value $\xi = \eta$ at point $C$ is to be determined. Therefore, the yield conditions to be used are

$$m_x = 2(n_\theta - 1) \qquad \text{for } 0 \le \xi \le \eta \tag{9.201a}$$

and

$$m_x = 2(1 - n_\theta) \qquad \text{for } \eta \le \xi \le \xi_1 \tag{9.201b}$$

Noting that there is no laterally distributed load, i.e., $\psi = 0$, and substituting the linear yield conditions (9.201) in equilibrium equation (9.200), we can solve Eq. (9.201) to obtain the general solutions:

for $0 \le \xi \le \eta$

$$n_\theta = C_1 \sin \xi + C_2 \cos \xi$$

$$m_x = 2(n_\theta - 1) = 2(C_1 \sin \xi + C_2 \cos \xi - 1) \tag{9.202}$$

$$q = 2(C_1 \cos \xi - C_2 \sin \xi)$$

and for $\eta \le \xi \le \xi_1$

$$n_\theta = \bar{C}_1 \sinh \xi + \bar{C}_2 \cosh \xi$$

$$m_x = 2(1 - n_\theta) = 2(1 - \bar{C}_1 \sinh \xi - \bar{C}_2 \cosh \xi) \qquad (9.203)$$

$$q = -2(\bar{C}_1 \cosh \xi + \bar{C}_2 \sinh \xi)$$

Constants $C_1$ and $C_2$, $\bar{C}_1$ and $\bar{C}_2$ are determined from the boundary conditions

$$q = \frac{1}{2}\bar{P} = \frac{1}{2}P\left[\frac{4\sqrt{Rh/2}}{\sigma_0 h^2}\right] \quad \text{and} \quad m_x = -1 \qquad \text{at } \xi = 0 \qquad (9.204a)$$

and

$$q = 0 \quad \text{and} \quad m_x = 1 \qquad \text{at } \xi = \xi_1 \qquad (9.204b)$$

respectively, and are given by

$$C_1 = \tfrac{1}{4}\bar{P}, \qquad C_2 = \tfrac{1}{2} \qquad (9.205a)$$

and

$$\bar{C}_1 = -\tfrac{1}{2}\sinh \xi_1, \qquad \bar{C}_2 = \tfrac{1}{2}\cosh \xi_1 \qquad (9.205b)$$

Three additional conditions are available at $\xi = \eta$, namely, $m_x(\eta^+) = m_x(\eta^-) = 0$ and the continuity of $q(\eta)$. Thus, we are provided with the following three equations to determine the boundaries $\eta$ and $\xi_1$ and the collapse load $P$:

$$\tfrac{1}{4}\bar{P} \sin \eta + \tfrac{1}{2}\cos \eta = 1 \qquad (9.206a)$$

$$\tfrac{1}{2}\cosh (\xi_1 - \eta) = 1 \qquad (9.206b)$$

and

$$\frac{1}{2}\sinh(\xi_1 - \eta) = \left(\frac{\bar{P}}{4}\cos \eta - \frac{1}{2}\sin \eta\right) \qquad (9.206c)$$

After some computations, we obtain

$$\eta = 0.47, \qquad \xi_1 = 1.79, \qquad \tfrac{1}{4}\bar{P} = 1.225 \qquad (9.207)$$

or

$$x_\eta = 0.47\sqrt{\frac{Rh}{2}} = 0.33\sqrt{Rh}$$

$$x_1 = 1.79\sqrt{\frac{Rh}{2}} = 1.27\sqrt{Rh} \qquad (9.208)$$

$$P = \frac{\bar{P}\left(\dfrac{\sigma_0 h^2}{4}\right)}{\sqrt{\dfrac{Rh}{2}}} = 1.73\,\sigma_0 h \left(\frac{h}{R}\right)^{1/2}$$

Extension of the static field beyond the point $x = x_1$ may be achieved as before (Fig. 9.45c). Since the stress distribution is statically admissible, the third of Eqs. (9.208) provides a lower bound on the collapse load.

The kinematic pattern of Fig. 9.45b can be employed to obtain an upper bound for the hexagonal condition. However, normality of the strain rate vector with respect to the slope sides of the hexagon leads to $\dot{\phi}_x h/\dot{\epsilon}_\theta = \pm 2$. Recalling that $\dot{\phi}_x = d^2\dot{w}/dx^2$ and $\dot{\epsilon}_\theta = \dot{w}/R$, we have

$$Rh \frac{d^2\dot{w}}{dx^2} = \pm 2\dot{w} \tag{9.209}$$

Therefore, the actual variation of $\dot{w}$ is trigonometric in the neighborhood of $P$ and hyperbolic further away. It can be shown that the deformation pattern derived from Eq. (9.209) corresponding to the stress profile $D - C - B$ is kinematically admissible, and that the solution (9.208) gives the exact answer for the case of a hexagonal yield surface. However, the assumption of a linear variation of $\dot{w}$ in Fig. 9.45b is equivalent to imposing restraints and so gives an upper bound which is too high.

The problem can be solved for different approximate yield conditions. For the exact yield condition, the answer can be obtained by numerical integration. Defining $\beta$ by

$$P = \beta \sigma_0 h \left( \frac{h}{R} \right)^{1/2} \tag{9.210}$$

answers for the limit load are given in Table 9.4.

*Solution for Short Shells*

In the previous discussion, we have assumed that the shell is sufficiently long so that the end effects can be neglected. We shall determine the minimum shell length for which the above solution remains valid and then extend the solution to shorter shells (Hodge, 1959).

Recalling the solutions in (ii) and (iii), we see that the statically admissible stress fields have been found in the shell region $0 \le x \le x_1$, where $x_1 = \sqrt{(Rh)}$ for the square yield conditions [see Eq. (9.195b)] and $x_1 = 1.27\sqrt{(Rh)}$ for the hexagon (Eq. 9.208). Evidently, one restriction on the shell length is that $2L$ must be greater than $2x_1$.

To find additional conditions necessary for the solution to be statically admissible in the range $x_1 \le x \le L$, we must construct some equilibrium stress distribution which satisfies the yield condition, the boundary conditions

$$M_x = M_0 = \frac{\sigma_0 h^2}{4} \quad \text{and} \quad V = \frac{dM_x}{dx} = 0 \quad \text{at } x = x_1 \tag{9.211}$$

and the prescribed conditions at $x = L$. For simplicity, we shall use the circumscribed square as the yield condition in the following discussion.

In the particular case of a shell whose ends are restrained against rotation, it is evident that $M_x = M_0$ in $x_1 \le x \le L$ is one such distribution, so that the above solutions (Table 9.4) furnish the true collapse load for this case. There is no further restriction on the length $L$, but $L \ge x_1$.

TABLE 9.4. Limit load of a circular cylinder subjected to a ring pressure.

| Yield condition | Limit load factor $\beta = P / \left[ \sigma_0 h \left( \dfrac{h}{R} \right)^{1/2} \right]$ |
|---|---|
| Inscribed square | 1.50 |
| Hexagon | 1.73 |
| Exact | 1.82 |
| Circumscribed square | 2.00 |

If the shell is simply supported at both ends, we consider the stress distribution

$$M_x = - \frac{M_0(L-x)(2x_1 - x - L)}{(L-x_1)^2}$$

$$N_\theta = \frac{2RM_0}{(L-x_1)^2}$$

$$(x_1 < x \le L) \qquad (9.212)$$

Evidently, this satisfies the equilibrium equation, the boundary condition Eq. (9.211), and the condition at $x = L$, $M_x = 0$. Further, $M_x$ is everywhere less than $M_0$ in magnitude. Therefore, the distribution will be statically admissible provided only that $N_\theta$ is less than $N_0 = \sigma_0 h$ in magnitude. We see that this condition is satisfied if

$$L \ge \sqrt{\frac{Rh}{2}} + x_1 \qquad (9.213)$$

Therefore, for such values of $L$, Table 9.4 provides the answers for the collapse load.

For a simply supported shell with length $2L$ less than $2[\sqrt{(Rh/2)} + x_1]$, i.e.,

$$L \le \sqrt{\frac{Rh}{2}} + x_1$$

We may solve the equilibrium equation (9.177) with $N_\theta = N_0$ for all $x$ to give [see Eq. (9.194)].

$$M_x = - \frac{N_0 x^2}{2R} + Ax + B \qquad (9.214)$$

and use the boundary conditions

$$V = \tfrac{1}{2}P \quad \text{and} \quad M_x = -M_0 \qquad \text{at } x = 0$$

$$V = 0 \quad \text{and} \quad M_x = M_0 \qquad \text{at } x = L \qquad (9.215)$$

From Eqs. (9.214) and (9.215) we obtain the answer for load $P$:

$$P = \frac{N_0 L}{R} + \frac{2M_0}{L} = \frac{2M_0}{L}\left(1 + \frac{2L^2}{Rh}\right) \qquad (9.216)$$

For $L = \sqrt{(Rh/2)} + x_1 = \sqrt{(Rh/2)} + \sqrt{(Rh)} = (1 + \sqrt{2}/2)\sqrt{(Rh)}$, Eq. (9.216) gives the same answer as Eq. (9.195a) with $N' = N_0$ and $M' = M_0$. For all $L$ less than $(1 + \sqrt{2}/2)\sqrt{(Rh)}$, the collapse load is given by Eq. (9.216), which depends upon the length of the shell, $L$.

### 9.7.1.6. CYLINDRICAL SHELLS WITH END LOAD

In the preceding discussion, we have assumed that the axial normal force $N_x$ is equal to zero so that only the resultant forces $N_\theta$ and $M_x$ appear in the yield condition (9.175). If the shell is subjected to both radial and axial forces, there will be three generalized stresses, $M_x$, $N_\theta$, and $N_x$, where

$$N_x = \int_{-h/2}^{h/2} \sigma_x \, dz \tag{9.217}$$

Now, we must retain this term rather than setting it equal to zero because the shell is subjected to end loads. In this case, the yield condition is the same as that for a symmetrically loaded shell of revolution, which will be discussed in the next section.

## 9.7.2. Thin Shells of Revolution Under Symmetric Loading

### 9.7.2.1. GENERALIZED STRESSES IN A SHELL OF REVOLUTION

An element of a shell of revolution is shown in Fig. 9.46, where $r_0$ is the distance from the axis of the shell, $r_1$ is the meridional radius of curvature, $\phi$ is the inclination of the meridional normal to the vertical, $V$ is the transverse shear force per unit length, $T$ is the force per unit area in the meridional direction, and $p$ is the interior pressure. The stress resultants $N_\phi$, $N_\theta$, $M_\phi$, $M_\theta$, and $V$ must satisfy the equilibrium equations:

$$\frac{d}{d\phi}(r_0 N_\phi) - r_1 N_\theta \cos\phi - r_0 V + r_0 r_1 T$$

$$\equiv r_0 \frac{dN_\phi}{d\phi} + (N_\phi - N_\theta)r_1 \cos\phi - r_0 V + r_0 r_1 T = 0 \tag{9.218}$$

$$r_0 N_\phi + r_1 N_\theta \sin\phi + \frac{d}{d\phi}(r_0 V) - r_0 r_1 p = 0 \tag{9.219}$$

$$\frac{d}{d\phi}(r_0 M_\phi) - r_1 M_\theta \cos\phi - r_1 r_0 V$$

$$\equiv r_0 \frac{dM_\phi}{d\phi} + (M_\phi - M_\theta)r_1 \cos\phi - r_0 r_1 V = 0 \tag{9.220}$$

If the bending action of an element is to contribute at all significantly to the load-carrying capacity of the shell, the shear force $V$ must be of the order of magnitude of thickness $h$ times the yield stress $\sigma_0$. As the magnitudes

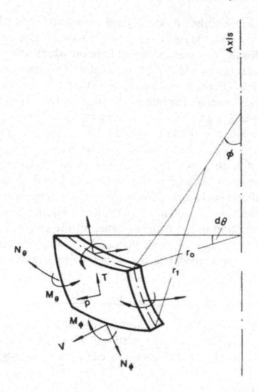

FIGURE 9.46. An element of a shell of revolution.

of the bending moments and of their algebraic difference cannot exceed $\frac{1}{4}\sigma_0 h^2$, $V r_0$ will far exceed $M_\phi$ or $M_\theta$ wherever $V$ is important. Therefore, the important terms in the third equation of equilibrium (9.220) are $V$ and the derivative of $M_\phi$. The term containing the moments $M_\phi$ and $M_\theta$ themselves can be dropped, and the equation which is left is essentially the same as for the cylindrical case:

$$\frac{dM_\phi}{d\phi} - V r_1 = 0 \tag{9.221}$$

Equation (9.221) can also be thought of as the result of taking $M_\theta = M_\phi$ in Eq. (9.220). For some configurations, the equilibrium equations with $M_\theta$ alone omitted may be more convenient to apply than Eq. (9.221):

$$\frac{d}{d\phi}(r_0 M_\phi) - r_0 r_1 V = 0 \tag{9.222}$$

Thus, with the exception of portions of the shell near the axis of revolution, for which $h/r_0$ may not be small, $M_\theta$ may be ignored in the equations of equilibrium. The use of Eq. (9.221) or (9.222) instead of (9.220) will have little effect on the lower bounds of the load-carrying capacity of the shell obtained from the equilibrium stress and moment fields which are at or below yield. In this sense, $M_\theta$ may again be considered as induced or passive.

As for the cylinder, the meridional planes on which the circumferential moment $M_\theta$ acts remain meridional during small deformations of the middle surface of the shell. On the other hand, deformations of the middle surface change the circumferential curvature in the general case (but not in the cylinder). The moment $M_\theta$ therefore in general does work during deformation. However, when the thickness of the shell is small compared to the local radius $r_0$, the rate at which work is done by the stress system on the change in circumferential curvature is small compared with the rate of work done on the change in circumferential and meridional strains, and on the change in meridional curvature. This might be inferred from the discussion of the unimportance of $M_\theta$ in the equations of equilibrium and can also be seen by examination of the orders of magnitude of the various terms in the expression for the rate of work done. Thus, only the stress resultants $N_\phi$, $N_\theta$, and $M_\phi$ are considered the generalized stresses, and the yield surface is generally expressed as

$$F(N_\phi, N_\theta, M_\phi) = k \qquad (9.223)$$

It appears, therefore, that the calculation of the limit load by the use of either Eq. (9.221) or (9.222) together with the yield condition (9.223) should lead to answers close to the true carrying capacity of the shell.

### 9.7.2.2. YIELD SURFACE (HODGE, 1954)

The yield condition for a shell element may be obtained by limit analysis concepts as used in the last section. Assume the material obeys the Tresca yield condition. Since the state of stress at a point is specified by the values of $\sigma_\phi$ and $\sigma_\theta$, the Tresca criterion requires that

$$\max\{|\sigma_\phi|, |\sigma_\theta|, |\sigma_\phi - \sigma_\theta|\} \le \sigma_0 \qquad (9.224)$$

Let the thickness of the shell be $h$ and let $z$ be measured from the middle surface. At a fully plastic section of the shell, the relation (9.224) will be satisfied as an equality for all $-h/2 \le z \le h/2$. In the first octant of the generalized stress space, the problem of finding the yield condition may be formulated as "given values of any two of the stress resultants, distribute the stresses so as to maximize the third." If positive bending moments are defined as those producing compression in the surface $z = h/2$, it follows that for a given $N_\phi$, the maximum value of $N_\theta$ will be independent of the particular arrangement of the positive and negative $\sigma_\phi$ stresses, while $M_\phi$ will be maximized by concentrating the negative $\sigma_\phi$ stresses near the top surface $z = h/2$, and positive stresses near the bottom surface. The critical $\sigma_\phi$ stress distribution is shown in Fig. 9.47e. The maximum value of $N_\theta$ associated with this $\sigma_\phi$ distribution is obtained for the $\sigma_\theta$ distribution shown in Fig. 9.47f, and the minimum value of $N_\theta$ is obtained when $\sigma_\theta$ is as given

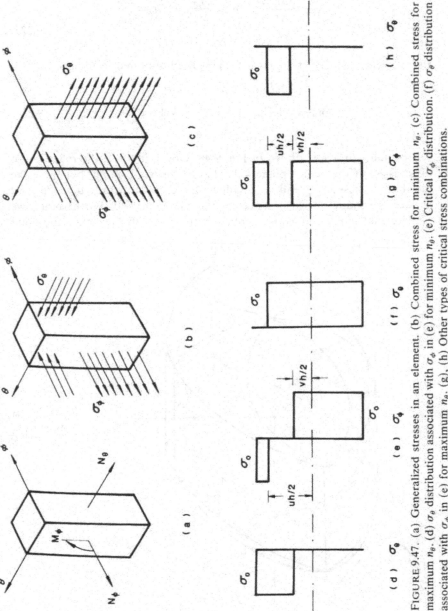

FIGURE 9.47. (a) Generalized stresses in an element. (b) Combined stress for maximum $n_\theta$. (c) Combined stress for minimum $n_\theta$. (d) $\sigma_\theta$ distribution associated with $\sigma_\phi$ in (e) for minimum $n_\theta$. (e) Critical $\sigma_\phi$ distribution. (f) $\sigma_\theta$ distribution associated with $\sigma_\phi$ in (e) for maximum $n_\theta$. (g), (h) Other types of critical stress combinations.

in Fig. 9.47d. Figure 9.47b,c show the perspective pictures of these two stress combinations. If reduced stress resultants are defined by

$$m_\phi = \frac{M_\phi}{M_0}, \qquad n_\phi = \frac{N_\phi}{N_0}, \qquad n_\theta = \frac{N_\theta}{N_0}$$

$$M_0 = \tfrac{1}{4}\sigma_0 h^2, \qquad N_0 = \sigma_0 h \tag{9.225}$$

the following reduced stress resultants are then obtained from Figs. 9.47d–f:

$$m_\phi^+ = 1 - \tfrac{1}{2}(u^2 + v^2), \qquad n_\phi^+ = \tfrac{1}{2}(u + v)$$

$$n_\theta^+ = \tfrac{1}{2}(1 + u), \qquad n_\theta^- = -\tfrac{1}{2}(1 - v) \tag{9.226}$$

$$-1 \le v \le u \le 1$$

The yield surface is drawn in Fig. 9.48. Observing that the reflection of all stresses in the center plane will change the sign of $m_\phi$ but leave the membrane stresses, $n_\phi$ and $n_\theta$, unaffected, we can conclude that the yield surface is symmetric with respect to the plane $m_\phi = 0$. Furthermore, we observe that if $m_\phi$, $n_\phi$, $n_\theta$ is a point on the yield surface, then the point

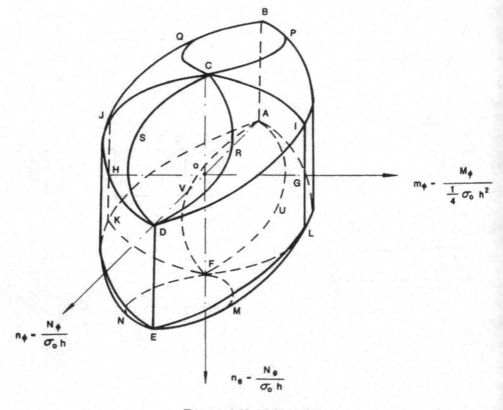

FIGURE 9.48. Yield surface.

$m_\phi, -n_\phi, -n_\theta$ is also. Thus, the yield surface is skew-symmetric with respect to the $m_\phi$ axis as shown. The yield surface consists of the following parts of surfaces:

(i) *EMFL, DRCI, ENFK,* and *DSCJ* are parts of four quadric surfaces

$$m_\phi = \pm m_\phi^+, \qquad n_\phi = n_\phi^+, \qquad n_\theta = n_\theta^\pm \qquad (9.227)$$

(ii) Noting that when $v = u$, the section is fully plastic due to the $\sigma_\phi$ stresses alone, and $n_\theta$ may have any value between $n_\theta^+$ and $n_\theta^-$. Thus, the parabolic cylinder

$$m_\phi = \pm(1 - u^2), \qquad n_\phi = u$$
$$-\tfrac{1}{2}(1 - u) \le n_\theta \le \tfrac{1}{2}(1 + u), \qquad -1 \le u \le 1 \qquad (9.228)$$

represented by *DELABI* and *DEKABJ* are parts of the yield surface.

(iii) The bottom end surface *EMFN* is obtained by setting $u = 1$ so that the section is fully plastic due to the $\sigma_\theta$ stresses. The resulting plane section can be defined by

$$n_\theta = 1, \qquad -2n_\phi(1 - n_\phi) \le m_\phi \le 2n_\phi(1 - n_\phi) \qquad (9.229)$$

(iv) The remaining section *DSCR* can be defined by considering stress distributions of the type shown in Fig. 9.47g,h, leading to

$$n_\phi = 1 + n_\theta, \qquad -2(1 + n_\theta)^2 \le m_\phi \le -2n_\theta(1 + n_\theta) \qquad (9.230)$$

*Approximation to the Yield Surface*

The yield surface (Fig. 9.48) is quite difficult to use for either equilibrium or energy dissipation computations. Circumscribed and inscribed surfaces can be drawn that are far more convenient. Use of the lower-bound or equilibrium theorem with an inscribed yield surface clearly gives a lower bound. Use of the upper-bound or kinematic theorem with a circumscribed convex yield surface clearly gives an upper bound on the limit load. On the other hand, merely satisfying equilibrium with a circumscribed figure or computing dissipation rates from an inscribed figure gives an approximate result which cannot be identified as either an upper or a lower bound. However, an answer which is exact for an inscribed surface provides a lower bound, while that for a circumscribed surface provides an upper bound on the limit loading. This result is especially helpful in problems which are essentially statically determinate.

In general, the complete approximating surface must be convex for the upper-bound theorem to give a valid answer. For many problems, however, the range of the variables is restricted to a small region of the total yield surface. Within this range, a plane or other surface lying inside the yield surface will give a lower bound and the one outside an upper bound. It is of no importance that the approximating surface cuts through the actual yield surface outside of the region of interest if care is taken to restrict its use to within the region of validity.

As in the cylinder case, assuming an ideal sandwich section, the exact yield surface can be linearized and an inscribing polyhedron yield surface is obtained (Hodge, 1954). Other approximations proposed by Drucker and Shield (1959) are shown in Fig. 9.49a,b. A convex circumscribing surface of (a) consists of a parabolic cylinder and four cutoff planes. The hexagonal prism of (b) is simpler but is by no means as good a fit at some points. Reducing all of its dimensions by the factor $\frac{1}{2}(\sqrt{5}-1)$ or 0.618 produces an inscribed surface. Upper bounds based upon the circumscribed hexagonal prism and lower bounds computed with the inscribed prism will be too far apart for many engineering purposes. If, however, a $\frac{3}{4}$-size prism is drawn, it lies inside the actual yield surface of Fig. 9.48 over an extended range. This may be seen from the $\frac{3}{4}$-prism sections shown by dashed lines in Fig. 9.49c,d.

Some problems of shells of revolution have been treated by Onat and Prager (1954) for a spherical cap and by Drucker and Shield (1959) for a

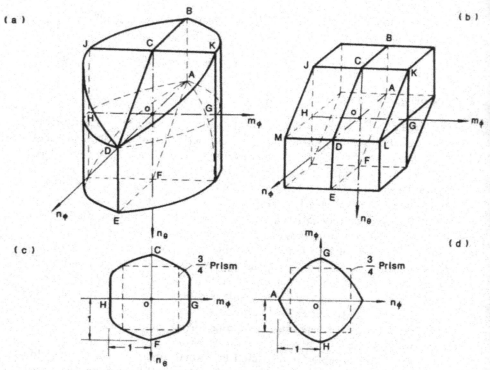

FIGURE 9.49. (a) Circumscribed yield surface. (b) Simpler circumscribed yield surface. (c) Yield curve for $n_\phi = 0$. (d) Yield curve for $n_\theta = 0$.

toroidal knuckle. These solutions will not be studied here because of space limitations.

# References

Baker, J.F., and J. Heyman, 1969. *Plastic Analysis of Frames*, Cambridge University Press, Cambridge.

Baker, J.F., M.R. Horne, and J. Heyman, 1956. *The Steel Skeleton*, Cambridge University Press, Cambridge.

Chen, W.F., 1982. *Plasticity in Reinforced Concrete*, McGraw-Hill, New York.

Drucker, D.C., 1954. "Limit Analysis of Cylindrical Shells Under Axially Symmetric Loading," Proc. 1st Midwestern Conference Solid Mechanics, Urbana, Illinois, pp. 158-163.

Drucker, D.C., 1956. "The Effect of Shear on the Plastic Bending of Beams," *Journal of Applied Mechanics* 23: 509-514.

Drucker, D.C., and H.G. Hopkins, 1954. "Combined Concentrated and Distributed Load on Ideally-Plastic Circular Plates," Proc. 2nd U.S. National Congress on Applied Mechanics, Ann Arbor, Michigan, pp. 517-520.

Drucker, D.C., and R.T. Shield, 1959. "Limit Analysis of Symmetrically Loaded Thin Shells of Revolution," *Journal of Applied Mechanics* 26: 61-68.

Gazetas, G., 1982. "Progressive Collapse of Rigid-Plastic Circular Foundations," *Journal of the Engineering Mechanics Division, ASCE* 108(EM3): 493-508.

Gazetas, G., and T.P. Tassios, 1978. "Elastic-Plastic Slabs on Elastic Foundation," *Journal of the Structural Division, ASCE* 104(ST4): 621-636.

Green, A.P., 1954. "A Theory of the Plastic Yielding Due to Bending of Cantilevers and Fixed-Ended Beams," *Journal of Mechanics and Physics of Solids* 3(1): 143.

Heyman, J., 1968. "Bending Moment Distributions in Collapsing Frames," in *Engineering Plasticity* (J. Heyman and F.A. Leckie, eds.), Cambridge University Press, Cambridge, 219.

Hodge, P.G., Jr., 1954. "The Rigid-Plastic Analysis of Symmetrically Loaded Cylindrical Shells," *Journal of Applied Mechanics* 21: 336-342.

Hodge, P.G., Jr., 1956. "Displacements in an Elastic-Plastic Cylindrical Shell," *Journal of Applied Mechanics* 23: 73-79.

Hodge, P.G., Jr., 1957. "Interaction Curves for Shear and Bending of Plastic Beams," *Journal of Applied Mechanics* 24: 453.

Hodge, P.G., Jr., 1959. *Plastic Analysis of Structures*, McGraw-Hill, New York.

Hopkins, H.G., and W. Prager, 1953. "The Load Carrying Capacities of Circular Plates," *Journal of Mechanics and Physics of Solids* 2(1): 1-15.

Horne, M.R., 1958. "The Full Plastic Moments of Sections Subjected to Shear Force and Axial Load," *British Welding Journal*, April, 170-178.

Jiang, D.H., 1979. "On the Load-Carrying Capacity of Concrete Pavements," Tong Ji University Report, Shanghai, China.

Johansen, K.W., 1943. *Brudlinieteorier*, Gjellerup (English translation: *Yield Line Theory*, Cement and Concrete Association, London, 1962).

Krajcinovic, D., 1976. "Rigid-Plastic Circular Plates on Elastic Foundation," *Journal of the Engineering Mechanics Division, ASCE* 102(EM2): 213-224.

Meyerhof, G.G., 1962. "Load-Carrying Capacity of Concrete Pavements," *Journal of the Soil Mechanics and Foundation Division, ASCE* 88: 89-116.

Neal, B.G., 1961a. "The Effect of Shear and Normal Forces on the Fully Plastic Moment of a Beam of Rectangular Cross-Section," *Journal of Applied Mechanics* **28**: 269–274.

Neal, B.G., 1961b. "Effect of Shear Force on the Fully Plastic Moment of an I-Beam," *Journal of Mechanical Engineering Science* **3**: 258.

Neal, B.G., 1963. *The Plastic Methods of Structural Analysis*, 2nd Ed., Chapman and Hall, London.

Neal, B.G., 1967. "Effect of Shear and Normal Forces on the Fully Plastic Moment of an I-Beam," *Journal of Mechanical Engineering Science* **3**: 279.

Onat, E.T., and W. Prager, 1953. "Limit Analysis of Arches," *Journal of Mechanics and Physics of Solids* **1**: 77.

Onat, E.T., and W. Prager, 1954. "Limit Analysis of Shells of Revolution," *Proc. Roy. Netherlands Acad. Sci.* **B57**: 534–548.

Poulos, H.G., and E.H. Davis, 1974. *Elastic Solutions for Soil and Rock Mechanics*, Wiley, New York.

Prager, W., 1952. "General Theory of Limit Design," Proceedings, Eighth International Congress of Applied Mechanics, Istanbul, Vol. II, pp. 65–72.

Timoshenko, S.P. and S. Woinowsky-Krieger, 1959. *Theory of Plates and Shells*, McGraw-Hill, New York.

Vlasov, V.Z., and N.N. Leont'ev, 1966. *Beams, Plates and Shells on Elastic Foundation* (translated from Russian), National Science Foundation, Washington, D.C. and Department of Commerce, U.S.A., by the Israel Program for Scientific Translations, Jerusalem.

## PROBLEMS

9.1. A continuous beam has uniform cross section and is subjected to the loads $P_1$ and $P_2$, as shown in Fig. P9.1, which may act jointly or separately. Show that within the framework of limit analysis, joint action of the loads constitutes the most dangerous state of loading. Determine the yield moment $M_0$ such that the load-carrying capacity is just exhausted under this state of loading.

9.2. The circular sandwich arch shown in Fig. P9.2 has a uniform cross section with the yield moment $M_0$ and the yield force $N_0$. Determine its load-carrying capacity $P$ if

   (a) the influence of the axial force is neglected;
   (b) the influence of the axial force is taken into account.

9.3. Derive the axial and bending forces interaction equation for the circular cross section. Plot your result and compare it with the curve for the rectangular section.

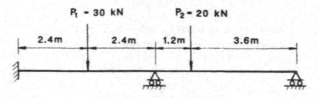

FIGURE P9.1

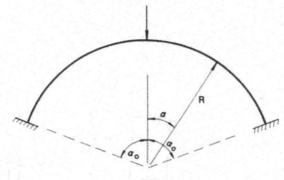

FIGURE P9.2

9.4. The moment $M$ and the axial force $P$ are applied at the ends of the bar shown in Fig. P9.4. Assuming plane strain conditions, determine the maximum load-carrying capacity of the bar under the combined action of $M$ and $P$, using the suggested upper-bound picture. Verify that the result so obtained is correct by constructing a simple lower-bound stress field. Note: $e$ is as yet undefined.

9.5. (a) A simply supported rectangular beam with a concentrated load $2P$ applied at its midpoint is shown in Fig. P9.5. Using the suggested picture of failure modes, derive the shear and bending force interaction relations. The material is assumed to satisfy the Tresca yield criterion. Note that picture (i) is a simple plastic hinge. It will be seen that pictures (ii) and (iii) are a much better answer than (i) for a very short beam (or for small $M/M_0$).

  (b) Assume that the interaction relation Eq. (9.62) proposed by Drucker (1956) is a lower-bound solution. Plot and compare with the results given in part (a).

  (c) Evaluate the integrals (due to Hodge, 1957) for $M$ and $V$ [see Eqs. (9.54) and (9.55)]. Again, plot and compare with part (a).

  (d) If your upper-bound solutions in part (a) are found below Hodge's solution of part (c), what is your explanation?

  (e) What conclusions as to the effect of shear can be drawn from the example of a simply supported beam?

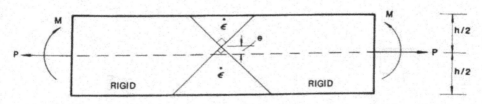

FIGURE P9.4

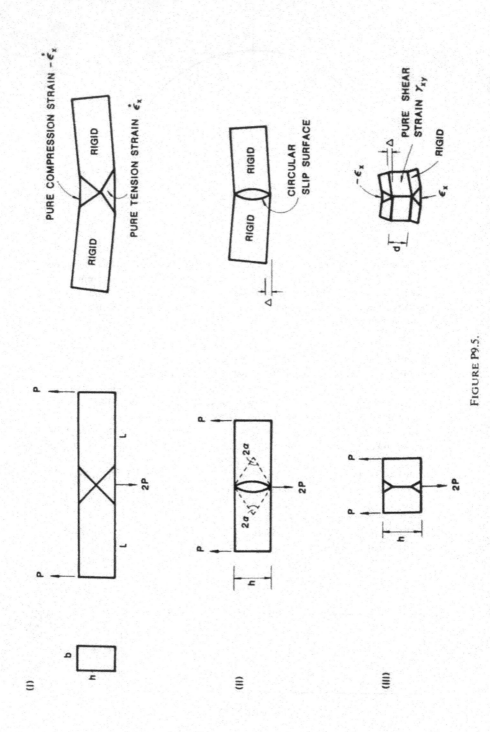

FIGURE P9.5.

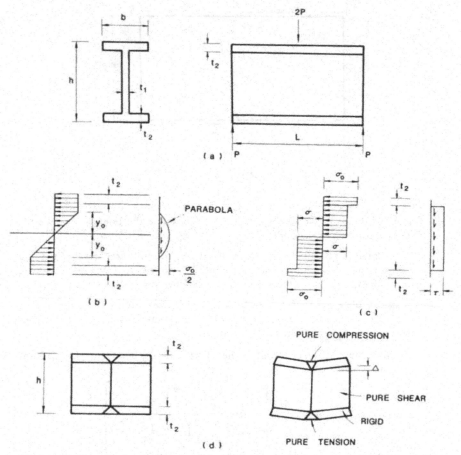

FIGURE P9.6. (a) A simply supported I-beam. (b) Neal's distribution. (c) Heyman and Dutton's distribution. (d) Leth's deformation field.

9.6. Derive the shear and bending force interaction relations for the simply supported I-beam, using the suggested stress and deformation fields (Fig. P9.6):

  (a) Neal's distribution, Fig. P9.6(b).
  (b) Heyman and Dutton's distribution, Fig. P9.6(c).
  (c) Leth's deformation field, Fig. P9.6(d).

Assume that the material obeys the Tresca yield criterion and that the shear forces are carried by web only. Denote $A = A_f + A_w$, $A_f = 2bt_2$, $A_w = ht_1$ and assume $t_2 \ll h$.

  Compare the results by plotting the interaction relations of parts (a), (b), and (c). Take $A_f = \frac{2}{3}A$.

9.7. A rectangular plate is simply supported on its two opposite sides and fixed along one of the other two sides (see Fig. P9.7). The entire plate is subjected to a uniform load $q$. Compute one upper and one lower bound on the limit

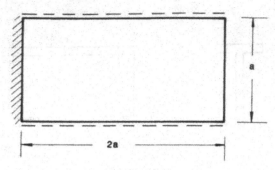

FIGURE P9.7

load. If they are not within a factor 2 to 1, refine the bounds. Assume the Tresca criterion of yielding.

9.8. A rectangular plate with three sides fixed and one side free is subjected to a uniform load $q$. Determine the least upper bound for the assumed mechanism as shown in Fig. P9.8. Assume the plate satisfies the square yield criterion.

9.9. A square plate has four clamped edges and carries a uniform load $q$. Figure P9.9 depicts a failure mechanism, containing four fans, each extending $\phi$. Assume the Tresca criterion of yielding.

   (a) Show that the upper bound on the collapse load $q$ derived by this mechanism is

$$q = \frac{48 M_0}{L^2} \left[ \frac{\dfrac{\phi}{2} + \tan\left(\dfrac{\pi}{4} - \dfrac{\phi}{2}\right)}{\dfrac{\phi}{2} \sec^2\left(\dfrac{\pi}{4} - \dfrac{\phi}{2}\right) + \tan\left(\dfrac{\pi}{4} - \dfrac{\phi}{2}\right)} \right]$$

   (b) Find a critical value of $\phi$ and show that the corresponding upper-bound solution is

$$q = \frac{43.5 M_0}{L^2}$$

*Hint*: Total energy dissipation = dissipation in the plate + dissipation in the boundaries, i.e., $D_t = D_p + D_b$, and $D_p = D_b$; therefore, $D_t = 2D_b$. Referring to the hodograph (b), $Oa$ is the angular velocity $\dot\theta_1$ of the rigid triangle $OAB$ and $Ob$ is the angular velocity $\dot\theta_f$ of the fan $OBC$. Thus, we have $D_b = 4[M_0 \dot\theta_1 \overline{AB} + M_0 \dot\theta_1 \cos(\pi/4 - \phi/2) \overline{BC}]$.

9.10. Find the limit loads for the circular plates shown in Fig. P9.10. Assume the Tresca yield criterion.

9.11. Find the limit loads for the annular plates shown in Fig. P9.11. Assume the Tresca yield criterion.

9.12. Find the collapse load for a circular unsupported cylindrical shell subjected to an internal band pressure with bandwidth $2c$. Use the circumscribing square yield curve.

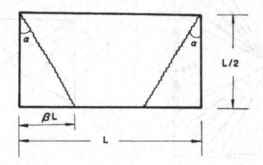

( a )

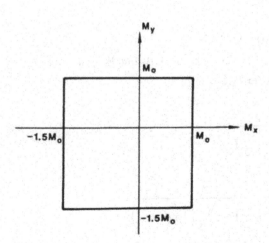

( b )

FIGURE P9.8.

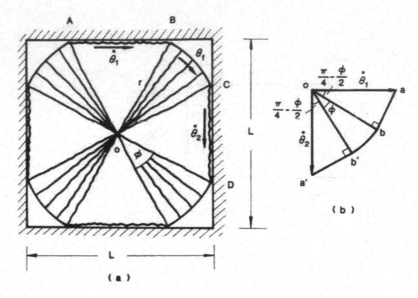

FIGURE P9.9

ANSWERS TO SELECTED PROBLEMS

9.1. $M_0 = 18$ kN-m if load $P_1$ acts only, or if $P_1$ and $P_2$ act jointly. $M_0 = 10.3$ kN-m if load $P_2$ acts only.

9.2. (a) $P = \dfrac{4M_0}{R} \dfrac{\sin \dfrac{\alpha_0}{2}}{\left(1 - \cos \dfrac{\alpha_0}{2}\right)}$

(b) $P = \dfrac{4M_0}{R} \dfrac{\sin \dfrac{\alpha_0}{2}}{\left(1 - \cos \dfrac{\alpha_0}{2}\right) + \dfrac{M_0}{N_0 R}\left(1 + \cos \dfrac{\alpha_0}{2}\right)}$

9.3. $M/M_0 = \cos^3 \theta_0$, $N/N_0 = (2/\pi)(\theta_0 + \tfrac{1}{2}\sin 2\theta_0)$, with $M_0 = \tfrac{4}{3}\sigma_0 R^3$, $N_0 = \pi R^2 \sigma_0$, and $\sin \theta_0 = e/R$.

9.4. $M/M_0 + (P/P_0)^2 = 1$.

9.5. (a) (i) $\begin{cases} \dfrac{M}{M_0} = \dfrac{1}{1 - \dfrac{h}{2L}} \\[3ex] \dfrac{V}{V_0} = \dfrac{h}{2L\left(1 - \dfrac{h}{2L}\right)} \end{cases}$

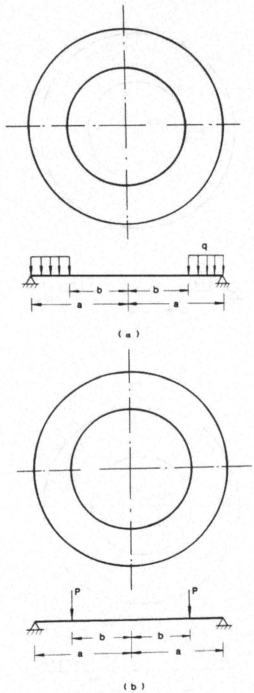

( a )

( b )

FIGURE P9.10.

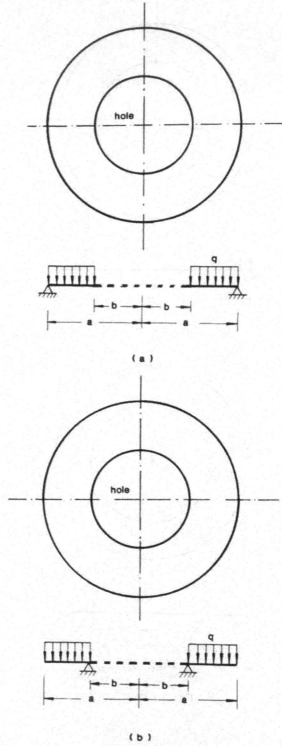

(a)

(b)

FIGURE P9.11.

(ii)
$$\begin{cases} \dfrac{M}{M_0} = \dfrac{1}{\sin \alpha} = 1.39 \\ \dfrac{V}{V_0} = \dfrac{1}{\sin \alpha} \dfrac{h}{2L} = 1.39 \dfrac{h}{2L} \end{cases}$$

(iii)
$$\begin{cases} \dfrac{M}{M_0} = \left[\left(1-\dfrac{d}{h}\right)^2 + \dfrac{2Ld}{h^2}\right] \Big/ \left(1-\dfrac{h-d}{2L}\right) \\ \dfrac{V}{V_0} = \dfrac{1}{2}\left[\left(1-\dfrac{d}{h}\right)^2 \dfrac{h}{L} + \dfrac{2d}{h}\right] \Big/ \left(1-\dfrac{h-d}{2L}\right) \end{cases}$$

$$\dfrac{d}{h} = 1 - \dfrac{L}{h}\left(2 - \sqrt{\dfrac{2h}{L}}\right)$$

9.6. (a)
$$\begin{cases} \dfrac{M}{M_0} = 1 - 0.75\left(\dfrac{A-A_f}{A+A_f}\right)\left(\dfrac{V}{V_0}\right)^2 \\ \dfrac{V}{V_0} \le \dfrac{2}{3} \end{cases}$$

where $M_0 = \left[\left(\dfrac{A_f}{2}\right)h + \left(\dfrac{A_w}{4}\right)h\right]\sigma_0$ and $V_0 = \dfrac{1}{2}A_w\sigma_0$.

(b)
$$\begin{cases} \dfrac{M}{M_0} = 1 - \left(\dfrac{A-A_f}{A+A_f}\right)\left[1 - \sqrt{1-\left(\dfrac{V}{V_0}\right)^2}\right] \\ \dfrac{V}{V_0} \le 1 \end{cases}$$

(c)
$$\begin{cases} \dfrac{M}{M_0} = \dfrac{2A_w}{A+A_f}\dfrac{L}{h} \\ \dfrac{V}{V_0} = 1 \end{cases}$$

9.8. $q^U = 59.06 M_0/L^2$ and $\beta = 0.425$.

9.9. (b) $\phi = 28°$.

9.10. (a) $q^i = 6M_0/(a^3 - 3ab^2 + 2b^3)$.

(b) $P^c = \dfrac{M_0 a}{b(a-b)}$.

9.11. (a) $q^i = 6M_0/(a^2 + ab - 2b^2)$.
     (b) $q^c = 6M_0/(2a^2 - ab - b^2)$.

9.12. $P = \dfrac{N_0}{2R}\left[1 + \left(1 + \dfrac{16M_0R}{N_0c^2}\right)^{1/2}\right]$, $b = \dfrac{c}{2}\left[1 + \left(1 + \dfrac{16M_0R}{N_0c^2}\right)^{1/2}\right]$

# Index